Archaeozoology of the Near East XII

Proceedings of the 12th International Symposium of the ICAZ Archaeozoology of Southwest Asia and Adjacent Areas Working Group, Groningen Institute of Archaeology, June 14-15 2015, University of Groningen, the Netherlands

Archaeozoology of the Near East XII

Proceedings of the 12th International Symposium of the ICAZ Archaeozoology of Southwest Asia and Adjacent Areas Working Group, Groningen Institute of Archaeology, June 14-15 2015, University of Groningen, the Netherlands

Edited by

C. Çakırlar, J. Chahoud, R. Berthon & S. Pilaar Birch

Barkhuis Publishing &
University of Groningen
Groningen, 2018

In honor of Dr. Hijlke Buitenhuis

Book design, cover design and typesetting: Siebe Boersma, Groningen Institute of Archaeology

Photo cover: Ceramic vessel from Late Neolithic Köşk Höyük depicting a scene of equid (hydruntine) hunting (photo by A.Öztan). See Chapter 4.

International Council for Archaeozoology

rijksuniversiteit groningen / groningen institute of archaeology

ISBN 9789492444745

Contents

Preface

by the editors

We grew up in the ASWA Working Group. We were in our early 20s, still doing our MAs or fresh graduates when we attended our first ASWA meeting. ASWA was first a school; a place where, for the first time, we had the opportunity to have a drink or two with the great scientists we knew from publications, but wondered what they looked like, how they acted, and whether they would mind answering couple of questions. ASWA quickly became a platform where we could present our own research, participate in debates with peers, and expand our scholarly network. At some point ASWA became a place to meet good friends, discuss research plans with close colleagues, and initiate new collaborations. When we got our first positions, it was only 'natural' that we all came up with the idea to host a meeting at our respective home institutions at about the same time in early 2013. One of us (CC) had the most urgent 'excuse': Dr. Hijlke Buitenhuis, one of the founders of ASWA Working Group, was to officially retire in 2015 –a perfect moment to celebrate this most dedicated scholar of archaeology of Southwest Asia and bring the bi-annual meeting back to its birthplace exactly 25 years after it was started by Hijlke and his mentor Prof. Dr. Anneke Clason.

This is the short story of how the 12th Meeting of the International Council of Archaeozoology (ICAZ) Archaeozoology of Southwest Asia and Adjacent Areas Working Group came to be. The months leading up to the meeting were thrilling. We received so many abstract submissions, mostly for oral presentations, that we had to introduce a session with 3-minute 'lightning' session alongside the poster session in order to maintain ASWA's (implicit) policy of inclusivity. ASWA meetings are for everyone; this includes early-stage students, independent scholars, and retired experts. In the days and weeks before the meeting, colleagues from as far as the US made last-minute calls to register just so that they could attend the meeting without presentations. Student volunteers and GIA staff came to the rescue. We now appreciate the meaning of the presumed cliché "would not have been possible without" more than ever: Yes, none of it would have been possible without the help of student volunteers, Francesca Slim and Rianne Breider; volunteer Joris Oddens; and GIA-zooarchaeology assistants Youri van den Hurk and the magnificent Esther Scheele.

We received generous funds from the Groningen Institute of Archaeology (and actually still do) and from the Archéorient, as well as BETA Analytics. With these funds and the help of participation fees, we were able to offer financial aid to a number of student and early-career scientist presenters especially from Southwest Asia.

In a large number of sessions, including a lightning and a poster session, in three days, we discussed a huge variety of topics pertaining to humans and other animals in ancient Southwest Asia and beyond. We were enlightened by a talk by Dr. Wietske Prummel, Emeritus, Groningen Institute of Archaeology, about the archaeozoology of terps and the northern Netherlands. The most outstanding and perhaps the most important in terms of ASWA's sustainability was the final round table on the future of work in the troubled, war-torn region, which was the brainchild of Jwana Chahoud. Here, the challenges ahead were openly discussed and recorded (also see Zeder, this volume).

We visited the GIA zooarchaeology collections, escaped to the town's market during lunch for some fried cod and mussels, 'borreled' in the Dutch style, and celebrated Hijlke's lifetime achievements. Some of us stayed longer and joined the excursion to megalithic tombs and a prehistoric village in Drenthe (Thanks also to our drivers on the excursion; Dr. Nimrod Marom and Christian Küchelmann).

We were honored to organize the 12th ASWA meeting, and we are honored to present the Proceedings herein, also sponsored by the Groningen Institute of Archaeology and supported by its staff. Esther Scheele, Youri van den Hurk, and Francis Koolstra helped with the copy-editing, Siebe Boersma handled the design and the layout. We thank them all for their fantastic and flawless work. All contributions have been peer-reviewed by at least two anonymous reviewers.

Canan Çakırlar, Remi Berthon, Jwana Chahoud, and Suzanne Pilaar Birch

Hijlke Buitenhuis and ASWA

Archaeozoologist and archaeologist

The palaeontologist Bert Boekschoten introduced Hijlke Buitenhuis to Anneke Clason in 1974. At that time Hijlke was a student in biology, with an interest in archaeozoology. He was immediately sent to the excavation of Tell Hadidi, a Bronze Age tell in Syria to study the archaeozoological material. While there, he collected skeletons for the archaeozoological reference collection of the then Biologisch-Archaeologisch Instituut (BAI), now the Groningen Institute of Archaeology (GIA). Hijlke published the results of his Tell Hadidi research in his master's thesis in biology. After finishing his degree in biology, he started his PhD research on the archaeozoological material from a large number of excavations along the Middle-Euphrates in Southeast Turkey and North Syria, dating to the period 10.000 BP till AD 1400. Hijlke was awarded his Dr. title in 1988.

After he became a doctor he got a job as an archaeozoologist at the Archaeological Research Centre (ARC), which was at first connected to the University of Groningen and later became an independent archaeological company. At the ARC, Hijlke studied archaeozoological material from many Dutch sites including rescue excavations conducted before the construction of the *Betuwelijn*, the rail freight connection between Rotterdam harbour and the German rail freight system. Because archaeozoological work was not always available, Hijlke developed a new skill: a specialization in coring. He developed himself to a qualified KNA-senior prospector (KNA is the quality system for people working in Dutch archaeology).

Although his job was in Dutch archaeology and archaeozoology, Hijlke continued his archaeozoological work in the Near East with a never-dying enthusiasm. He saved up all his holidays and travelled to Turkey each summer to participate in archaeological excavations as an archaeozoologist. His work in Turkey is focused on the Neolithisation of two areas; the Sea of Marmara, based on his work at several sites excavated by the Dutch Institute in Turkey (NIT) and Central Anatolia, at well-known sites such as Aşıklı and Güvercinkayası excavated by the Istanbul University. Hijlke's involvement with the archaeozoology of Anatolian Neolithic is ever growing, with new collaborations with University of Arizona and Munich among several others, with special emphasis on sheep and goat domestication and the westward dispersal of domestic animals through Central and western Anatolia. His work on Neolithisation and domestication in Turkey is recognized and celebrated worldwide.

Hijlke was among the founders of the Archaeozoology of Southwest Asia and Adjacent Areas working group of the International Council of Archaeozoology in 1990. He hosted, together with Anneke Clason, the first ASWA meeting in Groningen in 1992. He co-edited seven ASWA proceedings since then.

Although he never held an official teaching position, Hijlke has supervised several BA, MA and PhD theses in Groningen and Leiden, and trained Turkish students in archaeozoology. Hijlke became an official affiliate of the Groningen Institute of Archaeology in 2013. His contribution to the formation of the GIA zooarchaeology collections is immense, with more than 100 skeletons from Turkey and Syria. He is currently busy with digitizing part of the collection, for use at fieldwork specifically in the Near East.

Hijlke's activities in archaeology and archaeozoology are represented in this very large and diverse bibliography. Hijlke's research activities will not seize now that he reached his 65th birthday. He will continue and his list of publication will increase!

June 2015

List of presentations

Session I: Late Pleistocene and hunting strategies

Chair: H. Buitenhuis

- Martin L.: Wildlife hunting in the Levant: The longue durée
- Yeshurun R.: Contextual taphonomy as an effective zooarchaeological tool: The case of Natufian Mount Carmel
- Henton E.: Returning gazelles to their habitat: Isotopes and microwear in enamel apatite as markers of the seasonal mobility of gazelle in order to reconstruct hunting strategies in the prehistory of the Azraq Basin, East Jordan
- Tornero C., Balasse M., Fiorillo D., Molist M., Sana M.: Re-thinking seasonal migrations patterns of hunted gazelles in Euphrates Valley: Abu Hureyra's model reviewed by stable isotopes

Session 2: Transition to domestication

Chair: J.-D. Vigne

- Allentuck A., Martin L., Roe J.: Domestic maintenance practices in the Early Epipalaeolithic of the Jordanian steppe
- Lemoine X.: Mesocarnivores in the human niche: Human settlement impacts on local wildlife in the Taurus-Zagros Arc
- Hongo H., van Neer W., Wouters W., Arai S., Takahashi R., Gundem C. Y., Miyake Y.: Unsuccessful transition to food production: Animal exploitation at Hasankeyf Höyük
- Pöllath N., Alibert P., Niedermayer J., Matthias R., Joris P. : New approaches to separating wild from domestic sheep
- Wistoft Nielsen P.: Change of lifestyle, change of choices? How did the early stage of domestication (PPNB) of ovicaprines affect the choices of species for bone tools and objects in the Southern Levant?

Session 3: Neolithic husbandry: Case studies

Chair: R. Berthon

- Arai S., Kadowaki S., Ohnishi K., Guliyev F., Nishiaki Y.: Early animal husbandry in Azerbaijan: Implications for the origin and development of Neolithic in the Southern Caucasus
- Vigne J.-D., Brunet F., Debue K., Khudzhanazarov M. : Early domestic ungulates in Central Asia: Archaeozoological results from Ajakagytma (Uzbekistan, Kel'teminar, 9th-7th millennia cal BP)
- Pilaar Birch S., Çakırlar C., Buckley M.: Herding sheep, flocking goats: Neolithic livestock mobility at the site of Ulucak, Turkey
- De Cupere B.: Hunting and herding at Neolithic Mursalevo (Bulgaria): Preliminary results of the archaeozoological analysis
- Uerpmann H.P., Uerpmann M.: Animal Economy at Dosariyah - A Neolithic Site on the east coast of Saudi Arabia

Session 4: Species exploitation

Chair: L. Bartosiewicz

- Ottoni C. et al.: Ancient DNA from cats - a paleogenetics perspective into past distributions and ancient human mediated translocations of Felis silvestris
- Pawłowska K.: Times of change: Cattle in social practices of Late Neolithic Çatalhöyük
- Hadjikoumis A., Vigne J.-D.: A multi-site approach to sheep and goat management in prehistoric Cyprus: Synthesis and new insights
- Arbuckle B.S..: Exploring traditions of equid exploitation in pre- and protohistoric Anatolia
- Goulder J.: Invisible donkeys (and cows) in the ancient Near East: New archaeological insights into the early systematic use of working animals, using modern studies in developing countries

- Peters J., Neuberger F., Zimmermann M., Grupe G., Buitenhuis H., Pöllath N.: Baselines generated with faunas from 11th-10th millennia cal. BC sites in Anatolia are prerequisite to documenting the transition from hunting to herding caprines
- Bar-Yosef Mayer D.E., Marder O., Barzilai O., Hershkovitz I., Tejero J.- M.: Shell beads and shellfish at Upper Palaeolithic Manot Cave, Israel
- Deshpande-Mukherjee A.: Fresh new insights into Harappan shell working from the Kachchh region of Gujarat India

Session 5: Animal products

Chair: N. Marom

- Arnold E., Greenfield H., Maeir A. : An examination of economic specialization in the Early Bronze Age city of Tell es-Safi using isotopic analysis of domestic animals
- Manaseryan N.: Archaeozoological investigation of the site of Shengavit, Armenia
- Vila E., Chahoud J.: The development of sheep breeds in northern Mesopotamia and the Levant during the third millennium BC
- Küchelmann Ch.: The fleas cling to the golden fleece… – A large scale analysis of domestic sheep related archaeozoological data in Central Europe and the Near East from the 7th-2nd millennium BC

Lightning and Poster Session

Chair: B. Arbuckle

- Maskour M., Eisenmann, V.: Late Pleistocene to mid Holocene Equids on the Iranian Plateau. What news?
- Sheikhi, S., Mashkour, M.: Evolution of pastoralism in South-West of Iran from the 6th to 3rd millennium BC
- Kamjan, S., Maskhour, M., Azadeh, M.F., Homa, F., Azizi Kharanaghi, H.: Animal remains of the Neolithic village of Qasr-e Ahmad, Fars Province, Iran
- Davoudi, S., Mashkour, M., Abedi, A.: Subsistence economy at Kul Tepe (North-Western Iran) from Early Chalcolithic to the Early Bronze Age
- Mohaseb, F.A., Mashkour, M., Mofidi, B.: Animal exploitation during the Middle Elamite period based on the faunal remains of Haft Tappeh (Khuzestan, Iran)
- Amiri, S., Mashkour, M., Hohaseb, F.A., Naseri, R.: A glance into the subsistence economy of Gunespan-e Patappeh (Hamedan, Iran) during the Middle Bronze Age and Median Periods
- Beizazee Doost, S., Mashkour, M., Mohaseb, F.A., Atayi, M.T.: Subsistence Economy of Teppe Qasrdasht (Fars-Iran) during the Iron Age
- Çakırlar, C.: Rooting in a new niche: Neolithic human-suid interactions at the Gates of Europe
- Chahoud, J.: New data from Neolithic sites in Lebanon
- Berthon, R.: A specialised pastoral system focused on Caprinae during the Chalcolithic in the Araxes Valley (South Caucasus): A view from Ovçular Tepesi (Azerbaijan)
- De Cupere, B., Marinova, E.: Food provisioning at Hacılar BH during the Early Bronze Age: preliminary results of the archaeozoological and archaeobotanical analyses
- Berger, A.: Exploring urban economy in Early Bronze Age Tel Beit Yerah
- Adcock, S.E., Arbuckle, B.S.: Animal Economies in Middle Bronze Age Central Anatolia
- Van den Hurk, Y., Çakırlar, C.: Interpreting the zooarchaeological remains from an intact kitchen context from Middle Bronze Age Alalakh: Palace, redistribution and feasts
- Daujat, J., Mashkour, M., Debue, K., Sheikhi, S., Amiri, S., Lhuillier, J., Bendezu-Sarmiento, J.: Animal exploitation during the Iron Age at Ulug Depe, a large proto-urban site in eastern Turkmenistan

- Farebrother, E.: Prehistoric human-animal interactions of the Shahrizor Plain, Iraqi Kurdistan
- Prust, A.: The sheep, the goat, the camel and the cowry – Animal exploitation in ancient Tayma (Saudi-Arabia)
- Yengibaryan, N., Manaseryan, N.: Knucklebones of ancient Armenia
- Manaseryan, N.: Armenia: the secular and sacred roles of dogs and wolves
- Harutyunova, L.: Molluscs from the archaeological monuments of ancient Armenia
- Sarıtaş, Ö., Ozbarsaran, M., Peters, J., Buitenhuis, H., Pollath, N.: Red Deer and Human: From Palaeolithic to Iron Age in Anatolia
- Spitzer, M., Zeder, M.A.: Seasonality and habitat exploitation at Hallan Çemi: Results of the analysis of the avifauna
- Chalendar, V.: What place for animals in the medical practices of Mesopotamia? An overview through cuneiform tablets
- Bartosiewicz, L., Gal, E.: Meet the East: 16th–17th c. Ottoman Period animal exploitation in Hungary
- Buckley, M.: Zooarchaeology by Mass Spectrometry – the capabilities and limitations of collagen fingerprinting for species identification
- Al Besso, M.: Age profiling of small populations of wild goats in Crete (Capra aegagrus cretica)
- Kaptijn, E.: Faunal economy during the Early Bronze Age in the southern Levant
- Maini, E., Curci, A.: New data on the use of sheep and goat knucklebones in the Near East during the Iron Age

Session 6: Fauna in sociocultural practices

Chair: L. Martin

- Marom N.: Ritual in non-monumental contexts: Thoughts on four zooarchaeological case studies
- Weber J.: More than just "good eats": Animals as administrative artifacts
- Sapir-Hen L.: Jerusalem animal economy in the Iron Age: The relationship between the central city and its hinterland
- Hourani Y.: Life and death of dogs in Persian Beirut: Evidence from the burials

Session 7: Archaeozoology of Africa

Chair: V. Linseele

- Weinstock, J.: Faunal remains from the Pharaonic site of Amara West, Sudan: preliminary results.
- Lesur J.: Fishing and herding in the Nile Delta during the Predynastic (4th millennium BCE)
- Kunst G.K., Saliari K., Forstner-Müller I.: Intra-site variation in animal bone assemblages from two urban areas at Tell el-Dab'a (Egypt) - searching for a pattern
- Chaix L.: Hunting hartebeests (Alcelaphus buselaphus) in Sinaï during Persian times (6th to 4th century BC)

Session 8: Hellenistic, Roman and Medieval studies

Chair: N. Sykes

- Crabtree, P.J., Campana D.V.: Subsistence and ritual—faunal remains from the Iron Age, Hellenistic, and Medieval site of Kinik Höyük, Southern Cappadocia, Turkey
- Monchot H., Gourichon L., Stotzel E.: The faunal material from the Roman fortress and its associated church of Khirbet es-Samra (Mafraq Province, Jordan)
- Corbino Ch.A.: The elite and the villagers in the 14th century AD: Zooarchaeological analysis at Tell Hesban (Jordan)

1 A return to Groningen – ASWA XII and the future of archaeozoology in Southwest Asia (and Adjacent Areas)

Melinda A. Zeder[1]

1 Program in Human Ecology and Archaeobiology, Department of Anthropology, National Museum of Natural History, Smithsonian Institution

Abstract

Since the first meeting in Groningen 25 years ago, ASWA was grown in size and in its contributions to the archaeozoology in Southwest Asia. Not only has the number of participants and presentations increased at the twelve ASWA conferences up to the 2015 Groningen meeting, the working group has expanded in the number of countries represented by participants and in the geographic and topical reach of the papers and posters presented and subsequent papers published in ASWA volumes. As ASWA returns to its birthplace, it seems fitting to take a look at the history of the working group and the inroads it has made in addressing important issues in the archaeology of the region. With the considerable challenges that our discipline faces in these turbulent times in Southwest Asia, it also seems fitting to discuss some of the future obstacles and opportunities for advancing our understanding of the history human/animal interaction is this keystone region.

Introduction

Twenty five years ago, Anneke Clason and Hijlke Buitenhuis hosted the first meeting of the ICAZ Archaeologists of Southwest Asia (ASWA) working group. Clason and Buitenhuis' goal in creating this group and hosting its first meeting was to promote communication among archaeozoologists working on sites in western Asia and adjacent areas. Over the past quarter century, the group has grown in the number of participants and their geographic representation. It has also grown in the scope of the work conducted by participants. With the publication of this volume, proceedings of eleven of the twelve ASWA conferences are now available in both hard copy and increasingly in digital format. Through the years ASWA has become the nexus for a community of scholars that has had a major impact in shaping the direction of research in this broad and vital region. The return of ASWA to Groningen for its twelfth meeting in 2015 is then the perfect occasion to review the accomplishments of this vibrant working group over the past 25 years and to consider both the opportunities and challenges now faced by archaeozoologists working in Southwest Asia and its adjacent areas.

ASWA at 25

A quick glance at Table 1 and Fig. 1 clearly demonstrates the remarkable growth in participation in ASWA conferences over the years. From a relatively modest 14 presentations at the first Groningen meeting in 1992, there was a total of 60 presentations at the 2015 Groningen meetings (just one shy of the record 61 presentations at the 2011 Brussels meeting). Perhaps even more impressive is the growth in the geographic diversity of participants over the years, with seven mostly European countries represented in 1992 to a high of 27 different countries represented at the ASWA XII in 2015 (Table 1, Table 2). Particularly notable is the increased representation of researchers from core countries in Southwest Asia. Geopolitical constraints have sometimes affected participation by certain nationals in conferences hosted in the Southwest Asia. Nevertheless, the number of researchers from these core countries and the countries represented by ASWA participants has grown steadily over the past quarter century. In addition to Israel and Iran, which have been represented essentially from the beginning, ASWA meetings in recent years have welcomed participants from Egypt, Jordan, Lebanon, Oman, the Palestinian Authority, Syria, Turkey, the UAE, and Yemen. Long time

Table I ASWA Conferences since 1992.

					Geographic Focus				Chronological Focus			
Conference	*Year*	*Location*	*Presentations*	*Countries*	*Fertile Crescent*	*Anatolia*	*Arabian Peninsula*	*Adjacent Areas*	*Paleolithic*	*Neolithic*	*Chalcolithic-Iron Age*	*Classical Historical*
I	1992	Groningen	14	7	5	2	3	2	0	4	5	4
II	1994	Tubingen	23	8	7	3	3	7	1	4	5	3
III	1996	Budapest	21	9	8	7	0	4	5	7	11	0
IV	1998	Paris	48	20	24	7	3	12	6	19	12	3
V	2000	Amman	27	14	22	1	1	0	6	10	3	2
VI	2002	London	29	13	12	8	3	5	3	9	8	3
VII	2004	Ankara	16	12	8	5	0	1	1	3	7	3
VIII	2006	Lyons	40	16	23	6	2	10	5	12	9	5
IX	2008	Abu Dhabi	37	16	14	5	6	7	3	8	11	2
X	2011	Brussels	61	26	21	15	2	16	1	7	16	1
XI	2013	Haifa	58	20	30	9	0	12	11	10	20	3
XII	2015	Groningen	60	27	26	8	2	20	4	20	26	6

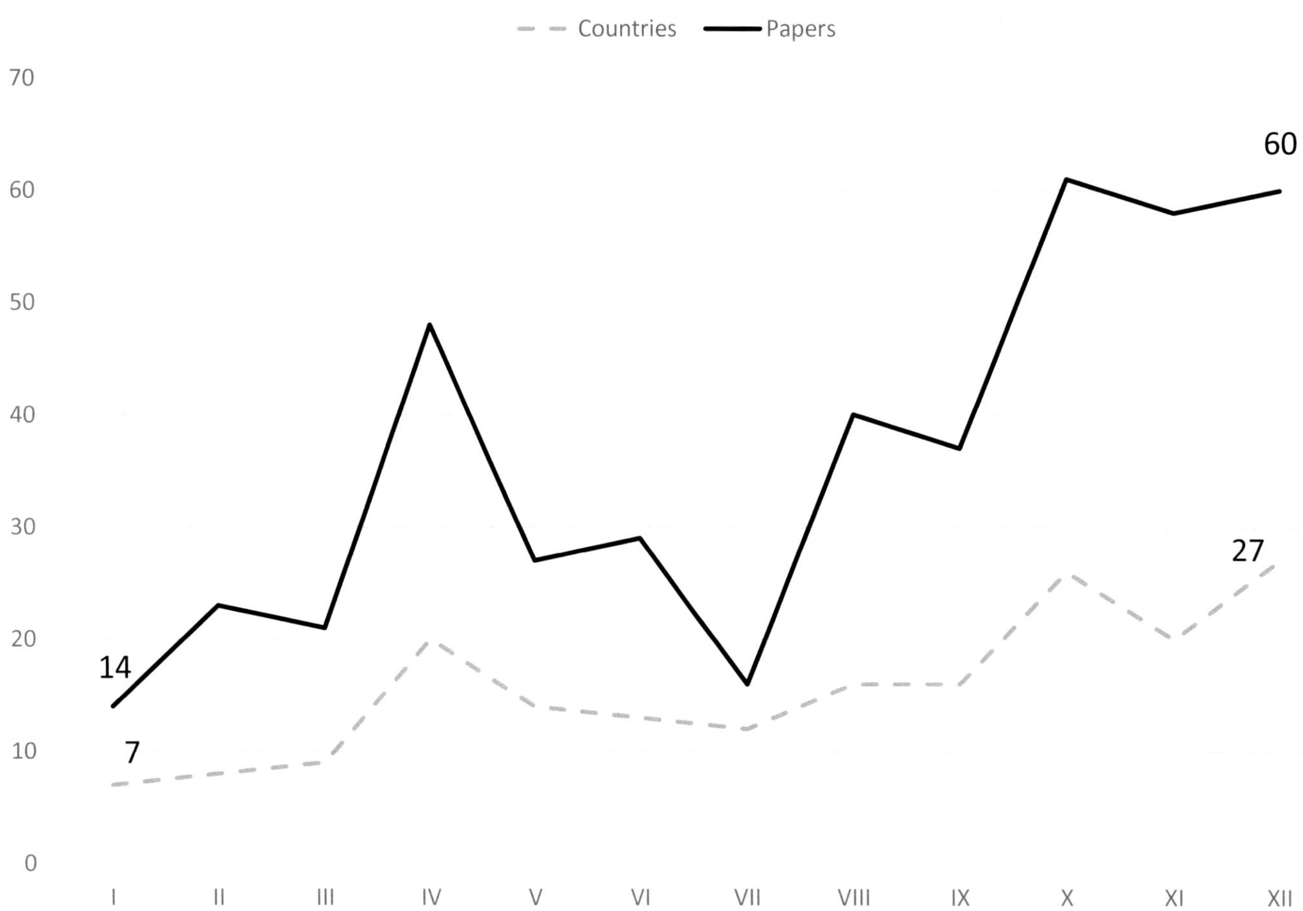

Fig. I. Numbers of Presentations and Countries Represented at ASWA Conferences.

Table 2 ASWA Conferences by Nationality of Participants.

Conference	*Year*	*Location*	*Europeans*	*North Americans*	*Near East & North Africa*	*Other Asian*
I	1992	Groningen	Belgium, France, Germany, Hungary, Netherlands		Israel	Japan
II	1994	Tubingen	Belgium, France, Germany, Hungary, Netherlands	USA	Israel	Japan
III	1996	Budapest	Belgium, France, Germany, Hungary, Netherlands, UK		Israel, Iran,	Japan
IV	1998	Paris	Belgium, France, Finland, Germany, Greece, Hungary, Italy, Netherlands, Serbia, Spain, Switzerland, UK	USA	Israel, Iran, Palestinian Authority, Turkey, UAE	India, Japan
V	2000	Amman	France, Germany, Netherlands, Spain, Switzerland, UK	Canada, USA	Israel, Iran, Jordan, Lebanon, Turkey	Japan
VI	2002	London	France, Hungary, Greece, Netherlands, Switzerland, UK	Canada, USA	Egypt, Israel, Iran, UAE	Japan
VII	2004	Ankara	Belgium, France, Greece, Netherlands, Russia, Switzerland, UK	USA	Israel, Turkey, UAE	Japan
VII	2006	Lyons	Armenia, Belgium, France, Germany, Netherlands, Portugal, Spain, Switzerland, UK	USA	Israel, Iran, Palestinian Authority, Turkey, Syria	Japan
IX	2008	Abu Dhabi	Armenia, Belgium, France, Germany, Italy, Netherlands, Romania, UK	USA	Iran, Turkey, Lebanon, Palestinian Authority, UAE	Japan
X	2011	Brussels	Armenia, Austria, Azerbaijan, Belgium, France, Germany, Greece, Ireland, Italy, Hungary, Netherlands, Poland, Spain, Switzerland, UK	Canada, USA	Israel, Iran, Jordan, Oman, Lebanon, Turkey, Syria, Yemen	Japan
XI	2013	Haifa	Armenia, Austria, Belgium, France, Germany, Greece, Hungary, Netherlands, Poland, Portugal, Spain, , UK	Canada, USA	Israel, Iran, Turkey,	India, Japan, Uzbekistan
XII	2015	Groningen	Armenia, Austria, Azerbaijan, Belgium, Cyprus, Denmark, France, Germany, Hungary, Italy, Netherlands, Poland, Spain, Switzerland, UK	Canada, USA	Egypt, Israel, Iran, Lebanon, Sudan, Turkey	Afghanistan, India, Japan, Uzbekistan

ASWA participants from Japan and India have been joined by archaeozoologists from Afghanistan and Uzbekistan, indicative of the increase of involvement by researchers (and research) from 'adjacent areas' in the mix. The increased representation of these adjacent areas is also reflected in growing numbers of ASWA participants from European countries that might be considered to fall into the 'adjacent' category, which include researchers from Armenia, Azerbaijan, Greece, Romania, and Russia.

It is also interesting to look at the geographic focus on ASWA presentations over the years. Table 1 and Fig. 2 break down ASWA presentations by geographical focus on the research featured in ASWA presentations. The Fertile Crescent represents the broad region from the Sinai, through the southern and northern Levant, Mesopotamia and Southeastern Anatolia, and the Zagros - essentially the core of Southwest Asia. Other regions represented are Anatolia, the Arabian Peninsula, and the so-called 'Adjacent Areas'. This latter category includes southeastern Europe, Cyprus, the Caucuses, North Africa and the Nile Valley, the Iranian Plateau and Afghanistan, Central Asia, and South Asia, with a single presentation on archaeozoological remains from Java at ASWA VIII in 2008 being the farthest afield 'adjacent area'.

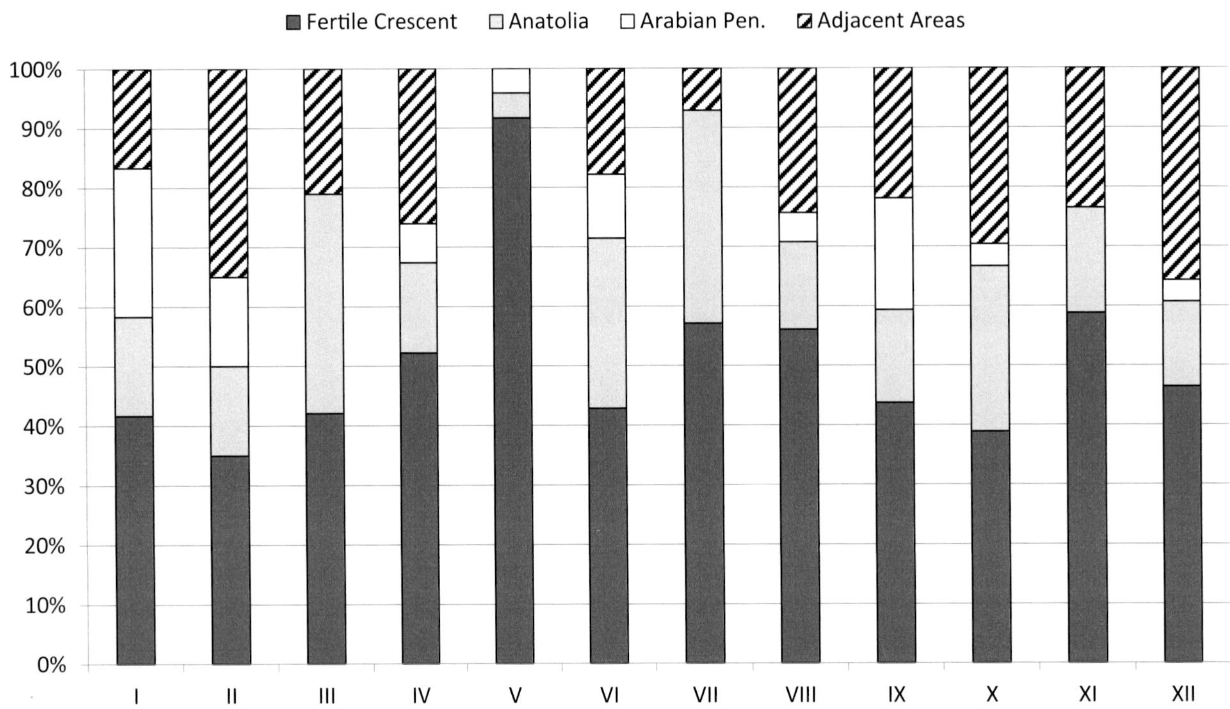

Fig. 2. Regions Represented in ASWA Presentations.

Some of the fluctuations in regional representation over the years can be attributed to the location of the conference host. The core Fertile Crescent region (especially the Levant) was particularly well represented in conferences held in Amman (ASWA V 2000) and Haifa (ASWA XI 2015); Anatolia featured prominently in the conference held in Ankara (ASWA VII 2004); and nearly 20% of the papers presented at the Abu Dhabi meeting (ASWA IX 2008) focused on the Arabian Peninsula. But perhaps the most impressive directional trend in these data is the steady increase in papers focusing on assemblages from adjacent areas in the last six conferences – from only 7% of the 14 presentations at ASWA VII (Ankara 2004) to a high of more than a third of the papers delivered at the 2015 meeting in Groningen (ASWA XII). Nearly half of these papers focused on assemblages from southern and southeastern Europe, and six of the 20 'adjacent area' papers focusing on assemblages from North Africa and the Nile Valley.

We can also break down past ASWA presentations chronologically, separating the time-line into four segments: Paleolithic, from before 20,000 to 15,000 years ago; Neolithic, which here encompasses the Late Epipaleolithic up through the Late Neolithic (c. 15,0000 to 7,000 years ago); the Chalcolithic through the Iron age (c. 5000 BC to 500 BC); and Classical/Historical period, from 500 BC to the present (Fig. 3). When we look at the distribution of papers focusing on these chronological periods, there are no obvious directional trends. Paleolithic periods were especially well represented at meetings held in the southern Levant (ASWA V in Amman and ASWA XI in Haifa), but otherwise it seems that every meeting features work in all four of these chronological segments with the Neolithic and, especially, the Chalcolithic through Iron Age particularly well represented at most meetings.

Research themes

Over the years, ASWA presentations, and subsequent publications, have explored some of the key questions for each of these large time slices putting the work of ASWA participants at the forefront on research in this pivotal region. Some of these questions are discussed here as a way of demonstrating both the past and on-going contributions of ASWA participants to addressing them. This discussion features presentations (and when possible resulting publications) from ASWA XII, linking them to work presented at earlier ASWA conferences.

Hominin dispersal

The direction, timing, and context of hominin dispersal out of Africa and across Southwest Asia and its adjacent areas remains one of the major areas of investigation in Paleolithic studies. At ASWA XII Mohamed Hamdeen Hamad used aquatic faunal remains in a paleo-climatic reconstruction of Late Paleolithic/Early Holocene Sudan (Hamad 2015). In doing so, he picked up on the work of Reuven Yeshurun at ASWA XI who used assemblages recovered during the Aswan dam salvage excavations of the 1960s for the same purpose (Yeshurun 2013). ASWA involvement in this important topic can be

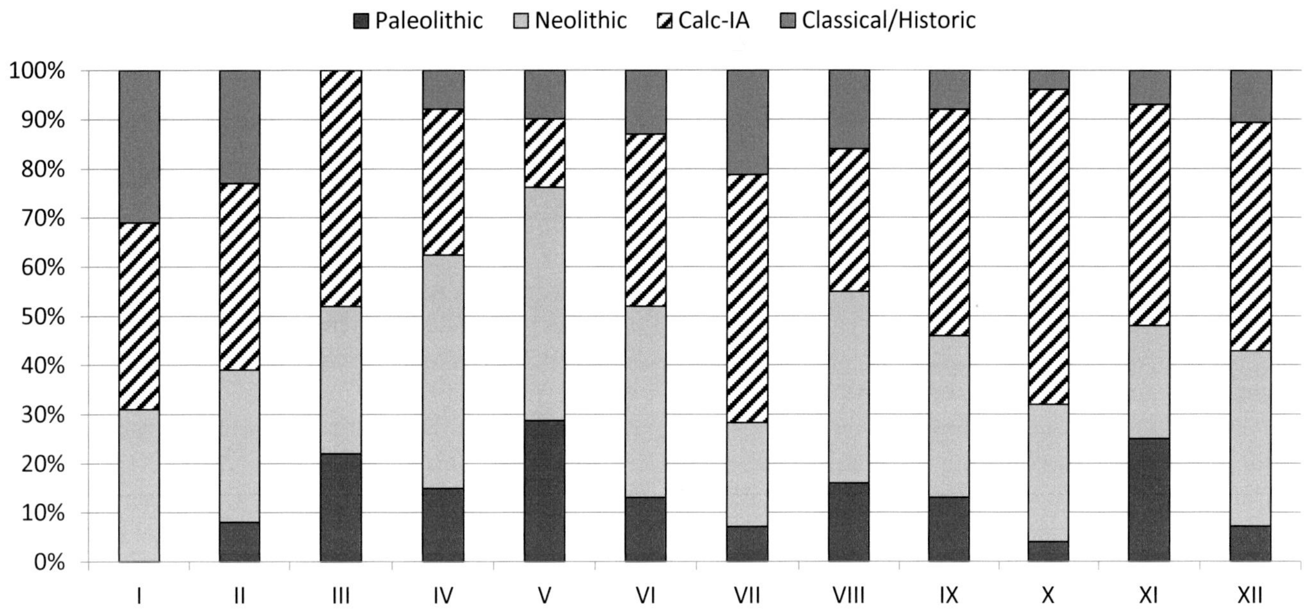

Fig. 3. Chronological Periods Represented in ASWA Presentations.

traced back to even earlier AWSA presentations focusing on hominin dispersal and adaptation in the Levant (Rabinovich and Tchernov 1994), Zagros (Bakken 2000), and the Iranian Plateau (Mashkour *et al.* 2008).

Documenting animal domestication

Domestication of animals and Neolithic emergence have been, for obvious reasons, major topics at ASWA conferences. ASWA XII continued this emphasis with the presentation by Pöllath (this volume) that uses linear metrics to capture differences in the shapes of wild and domestic sheep astragali detected through Geometric Morphometric analysis. This paper continues in the tradition of past ASWA presentations seeking to develop new markers for distinguishing wild from domestic animals in the archaeological record, either through the use of more traditional methods for reconstructing demographic profiles capable of discriminating between hunting and herding culling strategies (Zeder 1998; Lemoine *et al.* 2013), through ancient DNA (Ottoni *et al.* 2015), or isotopic studies (Pearson and Hedges 2002; Peters and Grupe 2008), with Peter's *et al.*'s (2015) presentation at ASWA XII discussing baselines for monitoring domestication in caprines.

In a review of the 2006 ASWA VII meeting in Lyon, I highlighted the need for linking proposed new markers of animal domestication clearly and explicitly to the actual process of domestication (Zeder 2006). This remains especially important today as increasingly powerful analytical techniques are applied to archaeozoological materials. No matter how sophisticated the analytical method, any marker that cannot be shown to be related solely to domestication and not to some other process (e.g., reproductive isolation, changing climate and vegetative regimes), cannot be used as a definitive marker of domestication. We need to keep in the forefront of any of this work a clear concept of how the domestic relationship between humans and animals unfolds and how this unfolding relationship may manifest itself in ways that we can detect in the archaeological record, understanding that different markers will reflect different stages of the process. And regardless of where we happen to think an animal crosses the threshold from wild to domestic, we need to remember that domestication is an extended process that may take many unexpected twists and turns along the way which we are now increasingly well equipped to trace, if only we take the time and effort to think deeply about the process.

Neolithic dispersal

Another major theme running through ASWA meetings is the dispersal of domestic animals and herding technology out of the heartland of domestication both westward into first to Cyprus (Vigne *et al.* 2000) and then across Anatolia into southeastern Europe (Benecke 1994; Buitenhuis 2002, 2004; De Cupere2002; Çakirlar 2011) and eastward across the Iranian Plateau (Mashkour *et al.* 2006); into South Asia (Meadow and Patel 1998). ASWA XII featured a number of presentations on this topic with contributions considering the case for independent domestication of pigs in western Anatolia and the movement of livestock into Eastern Europe (Aria *et al.* 2015; Çakirlar *et al.* 2015; De Cupere 2015; Pilaar Birch *et al.* 2015), papers that continue the long history of ASWA conferences featuring exciting work on the dispersal of domesticates and pastoral practices on Cyprus and other islands in the Aegean (Albesso 2015; Hadjikoumis and Vigne 2015), as well as research that tracks the movement of domesticates south and east into the Arabian Peninsula (Uerpmann 2015), and north and east across the Central Asian steppe (Vigne *et al.* 2015). Together, these and other ASWA presentations over the years paint a remarkably comprehensive and detailed picture of this pivotal process.

Hunting in Neolithic and later societies

Not only have ASWA contributors focused on domestic animals, the role of hunting in Neolithic and later societies have been frequent topics explored at ASWA conferences. The role of traps and kites in hunting wild animals was first discussed at ASWA VI in London when Carl Phillips first proposed that stone structures found in southeastern Arabia were used to capture foxes as a means of predator control and perhaps as an alternative food source (Phillips 2005). ASWA XII saw a return to the topic of animal traps in several presentations directed at understanding the social and ecological dimensions of gazelle hunting in Neolithic and later societies. Louise Martin (2015) provided an overview of the long history of gazelle hunting in the Levant and two other papers applied isotopic and micro-wear analysis to reconstruct gazelle migratory patterns and hunting strategies (Henton 2015; Tornero *et al.* 2015). Ben Arbuckle's paper on equid hunting strategies in Anatolia (Arbuckle 2015) continued the focus on the identification of the many different equid species that once lived in the region and how they contributed to Neolithic and later economies in the region going back to an early paper presented by Hijlke Buitenhuis at ASWA IV in Paris (Buitenhuis 1998), and other contributions since then (Gourichon and Helmer 2006; Orlando *et al.* 2006; Weber 2006; Bennett *et al.* 2011). Other ASWA XII contributions considered the continued role of red deer in subsistence economies from the Paleolithic to the Iron Age by amassing a comprehensive database of assemblages containing the remains of these animals (Sarıtaş *et al.* 2015). The role of wild resources in later complex societies in the region was explored by Lesur in her study of the contribution of fishing to the subsistence economies of Predynastic Egypt (Lesur this volume), while Chaix discussed hartebeest hunting in Persian occupied Sinai (Chaix this volume).

Feeding cities

Contributions to ASWA XII make it clear that long gone are the days when animal remains from urban contexts were either ignored or disposed of on the grounds they had nothing to contribute to understanding urban economies. The rational for this position had been that textual information from urban contexts gave a much more detailed picture of urban provisioning than animal bones could ever hope to provide. Earlier ASWA presentations that refute this notion include Redding's (2002) contribution to ASWA VI (London 2002) that examined study of provisioning in the worker's settlement at Giza, followed by his expanded look at Old Kingdom subsistence as it related to pyramid construction at ASWA X (Brussels) (Redding 2013). The ASWA X conference also included Tina and Haskel Greenfield's (2013) examination the effect of expansion of the Assyrian empire on subsistence in a town in northern Mesopotamia. ASWA XII continue to demonstrate the value of faunal remains in the study of complex specialized economies of early urban societies in the Near East. Two ASWA XII presentations examined urban-rural relations in the southern Levant and Anatolia (Adcock and Arbuckle 2015; Sapir-Hen 2015); two others traced the evolution of provisioning strategies over time (Berger this volume; De Cupere and Marinova 2015); while five presentations used faunal remains from urban contexts to explore the relationships between different segments of the urban population to one another and the broader economy (Arnold *et al.* 2015; van den Hurk and Çakirlar 2015; Manaseryan this volume; Mohaseb *et al.* 2015; Prust 2015). Weber's contribution to ASWA XII (Weber 2015) pressed the case that animal remains can be used to track the activities of urban administrators who served as the nexus between rural producers and urban consumers. Together the strong and varied contributions on urban provisioning from multiple different urban contexts at ASWA XII shows just how far we have come in making the case for animal bones being a primary data set for tracing the emergence of specialized urban economies.

Beyond diet: the role of animals in identity and ritual in Southwest Asia

Not all archaeozoological studies focus on the economic or ecological aspects of animal exploitation. Over the years, ASWA contributions have underscored the utility of animal remains in tackling some of the more difficult questions about the formation of ethnic and cultural identifies and the reification of social and political structures through ritual. As a matter of fact, The Haifa ASWA was subtitled *Bones and Identity* (Marom *et al.* 2016). Particular stand-outs in this vein are Weber's study of the equid sacrifices at Umm el-Marra (Weber 1998) and Bartoziewicz's (2011) examination of cattle offerings in Eleventh Dynasty Egypt. Megan Spitzer and my poster contribution to ASWA XII pushes the examination of the symbolic role of animals much further back in time in a study of the use of birds in ritual feasting and as ritual paraphernalia in Late Epipaleolithic societies in southeastern Anatolia on the brink of animal domestication (Spitzer and Zeder 2015). Crabtree and Campana's contribution to the 2015 meetings fast forwards to Persian era Cappadocia to examine the role of animal in quotidian ritual practices (Crabtree *et al.* this volume). Two contributions considered the social bonds between dogs and humans in Persian occupied Beirut (Hourani this volume) and in ritual practice among 1st century BC Kura-Axes cultures in Armenia (Manaseryan this volume). Chalendar's contribution underscored the complementary nature of texts and bones in exploring the use of animals in medicine and magic in Mesopotamia (Chalendar 2015).

Future challenges and opportunities

This overview of ASWA XII and its links to earlier ASWA meetings underscores the breadth and the vibrancy of this working group and its many varied topics and analytical approaches it explores. But there are some challenges that ASWA participants, especially the younger members of the group, must contend with in the future.

The greatest challenge is the current geo-politics of the region that is making it harder and harder for archaeozoologists to work in Southwest Asia – especially researchers from western countries. While the region has always had challenges in this arena, these seem especially heightened today. Parts of the core region of Southwest Asia are essentially no longer accessible for research - Syria, Iraq, and more recently Yemen. Countries where such work is difficult, especially for European and North American researchers, include Lebanon, Palestine, southeastern Turkey on the border with Syria and Iraq, and Iran. And even some adjacent areas which have been vibrant centers of archaeozoological research, such as Egypt and parts of Sudan, are difficult or entirely closed to this kind of work. This leaves a wide and largely accessible region of bordering adjacent areas where archaeological and archaeozoological work continues unabated, which may be the reason why there has been a steady increase of presentations focusing on these areas since 2002 - after 9/11 and the follow-on American invasions of Afghanistan and Iraq, the Arab Spring, and the civil-war in Syria.

But as is usually the case, new challenges often present new opportunities and there is no exception here. One of these is the opportunity for ASWA to continue to broaden its geographical focus beyond Southwest Asia and to focus in a more concentrated way on human/animal relationships in what it has labeled 'adjacent areas'. Rather than treating these regions as marginal recipients from a core region, a more productive approach is to view them as active centers of innovation in their own right that offer a powerful comparative perspective on parallel developments in Southwest Asia.

A second opportunity opened-up by the present geo-political challenges and the limited access of western researchers to the region is for ASWA members to expand efforts at fostering and training archaeozoologists who are native to countries in the region. As we saw earlier, ASWA has a well-established record of participation of archaeozoologists from Israel, and to a somewhat lesser extent Turkey and Iran. Though the range of researchers from core countries in Southwest Asia has expanded over the years, the participation of researchers from these countries in ASWA conferences has been relatively spotty, usually with these individuals only represented at one or two ASWA meetings over the years. At the same time, we also see a growing trend for ASWA participation by nationals from adjacent areas, with the Groningen meeting having the strongest representation of such individuals of any meeting. There is clearly a strong and increasing interest in archaeozoology in these countries and it is in the interest of all of us to support and further the involvement of researchers from countries in Southwest Asia and adjacent areas in ASWA conferences in any way that we can.

And finally, the current challenges to continued excavation in the region make it ever more important that we look to curated collections from previous excavations for our work. There are large repositories of archaeozoological collections in museums around the world – Field Museum in Chicago, the Peabody Museums at Harvard and Yale, the Museum of Anthropology at the University of Michigan, the University of Pennsylvania Museum of Archaeology and Anthropology, the Natural History Museums in London and Paris to name a few. The Smithsonian Institution, my own home, is just completing a several decades long archaeobiology collections improvement project directed at enhancing the care, documentation and the accessibility of our large holdings of archaeological animal and plant remains from around the world. These collections are surprisingly strong in assemblages from Southwest Asia. Of direct relevance to questions about initial animal domestication, for example, Smithsonian archaeofaunal assemblages include those generated by the early research in the Zagros region of Frank Hole and Kent Flannery (Yafteh Cave, Ali Kosh, Tepe Sabz, Chogha Sefid), of Ralph and Rose Solecki (Shanidar, Zawi Chemi Shanidar), and Phillip Smith (Ganj Dareh), as well as the huge assemblage of faunal remains from the more recent excavations at Hallan Çemi in southeastern Anatolia. All these legendary Near Eastern faunal assemblages are readily accessible to researchers wishing to study them. Faunal assemblages from the Braidwood excavations in Iran and Iraq are curated, and accessible for research, by the Field Museum. And while these assemblages may not have been recovered using modern excavation practices, they nevertheless represent major, largely untapped resources for future work. While some of these collections are well known and readily accessible, most are not. There is, then a pressing need for a concerted effort at locating archaezoological assemblages from this region, publicizing their whereabouts, and facilitating in increasing their accessibility.

Despite the challenges, then, I believe there are also substantial opportunities for archaeozoological studies in Southwest Asa – opportunities that ASWA is optimally positioned to help the community realize. First, ASWA can capitalize on the increasing interest in the study of human/animal interactions outside the core areas of Southwest Asia. Hosting a meeting in southeastern Europe or in Central Asia, for example, would help foster the growing understanding of the seminal role these regions played in retooling Near Eastern traditions and innovations to suit their own needs, as well as helping drive developments within Southwest Asia.

ASWA can also expand on its community's building role, especially as it involves reaching out to practicing and future potential archaeozoologists from core countries in Southwest Asia and its adjacent areas. Much the way the highly successful Zooarch list serve does for the profession as a whole, a more regionally focused ASWA list serve would be an invaluable resource for promoting communication about on-going research and helping identify and foster training opportunities for researchers in core, adjacent, and other countries.

This network could also provide the platform for a concerted effort at locating archaeofaunal assemblages from past excavations, building a directory of assemblages, where they are housed and how to access them. It would also help preserve and improve these invaluable resources by serving as a source of advice and support to under-resourced institutions as they attempt to build the capacity needed for proper curation and enhanced accessibility to these archaeofaunal assemblages. There is a critical time-sensitive element to this important effort, as archaeozoologists involved in recovering and curating these assemblages are moving on to retirement....and beyond.

Conclusions

ASWA has clearly come a long way since its founding meeting in Groningen 25 years ago. It has grown in membership, in the regional representation of participants, and in the scope and the complexity of the questions asked of archaeofaunal assemblages from this key region. It has brought together a dedicated community of researchers with global reach dedicated to the study of faunal remains from Southwest Asia and its adjacent areas. It is perfectly positioned to take on an even more potent role in leading a new generation of researchers interested in human/animal relationships in this heartland region of so much of human cultural development. I know Anneke Clason would have been proud to have seen ASWA comes so far and do so well. We all owe her a debt of gratitude for helping create this group, as we do to Lazslo Bartoseiwicz, Alice Chokye, and Marjan Mashkour who have served as editors of many of the ASWA conference volumes. But most of all, Hijlke Buitenhuis has been the heart and soul of ASWA since its inception and has been instrumental in shepherding meetings and its important conference publications over all these years. I can't think of a better tribute to Hijlke as he begins his retirement than to celebrate his many efforts on behalf of ASWA and his contributions to archaeozoology with the wonderful meeting held in Groningen in 2015 and this new conference volume.

References

Adcock, S. and Arbuckle, B.S. 2015. Animal Economies in Middle Bronze Age Central Anatolia. *Twelfth Meeting of the ASWA working group of ICAZ. Groningen, The Netherlands, 10-12 June, 2015.*

Albesso, M. 2015. Age profiling of small populations of wild goats in Crete (*Capra aegagrus cretica*). *Twelfth Meeting of the ASWA working group of ICAZ. Groningen, The Netherlands, 10-12 June, 2015.*

Arbuckle, B.S. 2015. Exploring traditions of equid exploitation in pre- and protohistoric Anatolia. *Twelfth Meeting of the ASWA working group of ICAZ. Groningen, The Netherlands, 10-12 June, 2015.*

Aria, S., Kadowaki, S., Ohniskhi, K., Guliyev, F. and Nichiaki, Y. 2015. Early Animal Husbandry in Azerbaijan: Implications for the Origin and Development of Neolithic Culture in the Southern Caucasus. *Twelfth Meeting of the ASWA working group of ICAZ. Groningen, The Netherlands, 10-12 June, 2015.*

Arnold, E., Greenfield, H. and Meir, A. 2015. An examination of economic specialization in the Early Bronze Age city of Tell es-Safi using isotopic analysis of domestic animals. *Twelfth Meeting of the ASWA working group of ICAZ. Groningen, The Netherlands, 10-12 June, 2015.*

Bakken, D. 2000. Hunting Strategies of Late Pleistocene Zarzian Populations from Palegawra Cave, Iraq, and Warwasi Rock Shelter, Iran. In,

Mashkour, M., Buitenhuis, H., Choyke, A.M. and Poplin, F. (eds) *Archaeozoology of the Near East IV : Proceedings of the 4th International Symposium on the Archaeozoology of Southwestern Asia and Adjacent Areas (Paris 1998)*: 11-17. Groningen, The Netherlands: Centre for Archaeological Research and Consultancy, Rijksuniversiteit Groningen.

Bartoziewicz, L. 2011. Eleventh dynasty cattle offering from the Temple of Montuhatep Sankhkara. *Tenth Meeting of the ASWA working group of ICAZ. Brussels, Belgium, 28 June-1 July, 2011.*

Benecke, N. 1994. The Neolithisation of southern Ukraine. *Second Meeting of the ASWA working group of ICAZ. Tübingen, Germany, 23-24 September, 1994.*

Bennett, A.E., Champlot, S., Gautier, M., Arbuckle, B.S., Balasescu, A., Davis, S., Eisenmann, V., Germonpre, M., Mashkour, M., Morales Muniz, A., Peters, J., Tournepiche, J-F., Uerpmann, H-P., Vila, Emmanuelle, Weber, J. Grange, T., Geigl, E-M. 2011. Wild asses during the Holocene and the first cross-bred hybrids. *Tenth Meeting of the ASWA working group of ICAZ. Brussels, Belgium, 28-June-1 July, 2011.*

Buitenhuis, H. 1998. New information on wild Equids in Anatolia. *Fourth Meeting of the ASWA working group of ICAZ. Paris, France. 29 June – 3 July 1998.*

Buitenhuis, H. 2002. Ilipinar, a late Neolithic/early Chalcolithic site in northwest Anatolia at the interface between the Near East and Southeastern Europe. *Sixth Meeting of the ASWA working group of ICAZ. London, UK, 30 August – 1 September 2008.*

Buitenhuis, H. 2004. The 7th millennium BC material from Yumuktepe: Diffusion or migration of the Neolithic accomplishment. *Seventh Meeting of the ASWA working group of ICAZ. Ankara, Turkey 1-4 June 2004.*

Çakirlar, C. 2011. Emergence and development of animal husbandry technologies in central-western Anatolia based on the archaeozoological case study of Ulucak Höyük. *Tenth Meeting of the ASWA working group of ICAZ. Brussels, Belgium, 28-June-1 July, 2011.*

Chalendar, V. 2015. What place for animals in the medical practices of Mesopotamia? An overview through cuneiform tablets. *Twelfth Meeting of the ASWA working group of ICAZ. Groningen, The Netherlands, 10-12 June, 2015.*

De Cupere, B. 2002. Faunal remains from the Neolithic-Early Bronze Age site of Bademağaci (Anatalya Province, SW Turkey). *Sixth Meeting of the ASWA working group of ICAZ. London, UK, 30 August – 1 September 2008.*

De Cupere, B. 2015. Hunting and herding at Neolithic Mursalevo (Bulgaria): preliminary results of the archaeozoological analysis. *Twelfth Meeting of the ASWA working group of ICAZ. Groningen, The Netherlands, 10-12 June, 2015.*

De Cupere, B. and Marinova, E. 2015. Food provisioning at Hacılar during the Early Bronze Age: preliminary results of the archaeozoological and archaeobotanical analyses. *Twelfth Meeting of the ASWA working group of ICAZ. Groningen, The Netherlands, 10-12 June, 2015.*

Greenfield, T. and Greenfield, H. 2013. Bronze and Iron Age subsistence changes in the Upper Tigris: zooarchaeology of Operation E at Ziyaret Tepe, south-eastern Turkey. In, De Cupere, B, Linseele, VL., and Hamilton-Dyer, S. (eds) *Archaeozoology of the Near East X: Proceedings of the 10th International Symposium on the Archaeozoology of Southwestern Asia and Adjacent Areas (Brussels 2011)*: 121-144. Leuven: Peeters.

Gourichon, L. and Helmer, D. 2006. The presence of *Equus hydruntinus* in the early Neolithic of northern Syria? *Eighth Meeting of the ASWA working group of ICAZ. Lyon, France, 28 June – 1 July, 2006.*

Hadjikoumis, A. and Vigne, J.D. 2015. A multi-site approach to sheep and goat management in prehistoric Cyprus: synthesis and new insights. *Twelfth Meeting of the ASWA working group of ICAZ. Groningen, The Netherlands, 10-12 June, 2015.*

Hamad, M.H. 2015. Aquatic Faunal remains form Holocene Palaeolake in El Ga'ab Depression west Dongola (Sudan). *Twelfth Meeting of the ASWA working group of ICAZ. Groningen, The Netherlands, 10-12 June, 2015.*

Henton, E. 2015. *Returning gazelle to their habitat: isotopes and microwear in enamel apatite as markers of the seasonal mobility of gazelle in order to reconstruct hunting strategies in the prehistory of the Azraq Basin, east Jordan. Twelfth Meeting of the ASWA working group of ICAZ. Groningen, The Netherlands, 10-12 June, 2015.*

Hourani, Y. 2015. Life and death of dogs in Persian Beirut: evidences from the burials. *Twelfth Meeting of the ASWA working group of ICAZ. Groningen, The Netherlands, 10-12 June, 2015.*

Lemoine, X., Bishop, K. and Rufalo, S. 2013. A new synthesis for aging prehistoric pigs (Sus scrofa): consolidating tooth eruption and attrition studies. *Eleventh Meeting of the ASWA working group of ICAZ. Haifa, Israel, 23-28 June, 2011.*

Manaseryan, N. 2015. Armenia: The secular and sacred roles of dogs and wolves. *Twelfth Meeting of the ASWA working group of ICAZ. 10-12 June, 2015, Groningen, The Netherlands.*

Marom, N., Yeshuran, R., Weissbrod, L. and Bar-Oz, G. 2016. Bones and Identity: Zooarchaeological approaches to reconstructing social and cultural landscapes in Southwest Asia. Oxford and

Philadelphia: Oxbow Books.
Martin, L. 2015. Wildlife hunting in the Levant: the longue durée. *Twelfth Meeting of the ASWA working group of ICAZ. Groningen, The Netherlands, 10-12 June, 2015.*
Mashkour, M., Vigne, J.-D. and Mohaseb, A. 2006. Neolithic Settlement in the Marvdash Plain (Fars-South west Iran). *Eighth Meeting of the ASWA working group of ICAZ. Lyon, France, 28 June – 1 July, 2006.*
Mashkour, M., Rendu, W. and Biglari, F. 2008. Qaleh Bozi (Isfahan) faunal remains: A Mousterian site of the Iranian central Plateau. *Ninth Meeting of the ASWA working group of ICAZ. Abu Dhabi, UAE, 16-20 November 2008.*
Meadow, R.H. and Patel, A. 1998. The exploitation of wild and domestic water buffalo in South Asia. In, Buitenhuis, H., Bartosiewicz, L., and Choyke A.M. (eds) *Archaeozoology of the Near East III. Proceedings of the 3rd International Symposium on the Archaeozoology of Southwestern Asia and Adjacent Areas (Budapest 1996)*: 180-199. Groningen: ARC-publication 18.
Mohaseb, F.A., Mashkour, M. and Mofidi, B. 2015. Animal exploitation during the Middle Elamite period based on the faunal remains of Haft Tappeh, (Khuzestan, IRAN). *Twelfth Meeting of the ASWA working group of ICAZ. Groningen, The Netherlands, 10-12 June, 2015.*
Orlando, L., Mashkour, M., Burke, A., Douady, C., Eisenmann, V. and Hänni, C. 2006. Geographic distribution of an extinct equid (*E. hydruntinus*, Mammalia, Equidae) revealed by morphological and genetic analysis. *Eighth Meeting of the ASWA working group of ICAZ. Lyon, France, 28 June – 1 July, 2006.*
Ottoni, C., Van Neer W., De Cupere B., Balasescu A., Becker C., Benecke N., Boivin N., Buitenhuis H., Chahoud J., Manaseryan N., Monchot H., Morales A., Onar V., Osypińska M., Peters J., Prendergast M., Putelat O., Spassov N., Studer J., Wierer U., Decorte R., Grange T. and Geigl E.-M. 2015. Ancient DNA from cats - a paleogenetics perspective into past distributions and ancient human mediated translocations of *Felis silvestris*. *Twelfth Meeting of the ASWA working group of ICAZ. Groningen, The Netherlands, 10-12 June, 2015.*
Pearson, J. and Hedges, R. 2002. A comparison of sheep and goat diet and herding strategy at Asikli Höyük and Çatalhöyük: Evidence from carbon and nitrogen analysis. *Sixth Meeting of the ASWA working group of ICAZ. London, UK, 30 August – 1 September 2008.*
Peters, J., and Grupe, G. 2008. Reconstructing hunting and farming practices in the Pre-Pottery Neolithic Upper Euphrates using stable isotopes in human and animal bone structural carbonate. *Ninth Meeting of the ASWA working group of ICAZ. Abu Dhabi, UAE, 16-20 November 2008.*
Phillips, C. 2005. Fox-traps in Southeast Arabia. In, Buitenhuis, H., Choyke, A.M., Martin, L., Bartosiewicz, L. and Mashkour, M. (eds) *Archaeozoology of the Near East VI : Proceedings of the 6th International Symposium on the Archaeozoology of Southwestern Asia and Adjacent Areas (London 2002)*: 199-207. Groningen: ARC Publication 123.
Pilaar-Birch, S. and Çakirlar, C. 2015. Herding sheep, flocking goats: Neolithic livestock mobility at the site of Ulucak, Turkey. *Twelfth Meeting of the ASWA working group of ICAZ. Groningen, The Netherlands, 10-12 June, 2015.*
Prust, A. 2015. The sheep, the goat, the camel and the cowry – Animal exploitation in ancient Tayma (Saudi-Arabia). *Twelfth Meeting of the ASWA working group of ICAZ. Groningen, The Netherlands, 10-12 June, 2015.*
Rabinovich, R. and Tchernov, E. 1994. Chronological, paleoecological, and taphonomical aspects of the Middle Paleolithic site of Qafzeh, Israel. *Second Meeting of the ASWA working group of ICAZ. Tübingen, Germany, 23-24 September, 1994.*
Redding, R.W. 2002. The fauna from the Worker's Settlement at Giza. *Sixth Meeting of the ASWA working group of ICAZ. London, UK, 30 August – 1 September 2008.*
Redding, R.W. 2013. A tale of two sites: Old Kingdom subsistence economy and the infrastructure of pyramid construction. In, De Cupere, B, Linseele, VL., and Hamilton-Dyer, S. (eds) *Archaeozoology of the Near East X: Proceedings of the 10th International Symposium on the Archaeozoology of Southwestern Asia and Adjacent Areas (Brussels 2011)*: 307-322. Leuven: Peeters.
Sapir-Hen, L. 2015. Jerusalem animal economy in the Iron Age: the relationship between the central city and its hinterland. *Twelfth Meeting of the ASWA working group of ICAZ. Groningen, The Netherlands, 10-12 June, 2015.*
Sarıtaş, Ö., Peters, J., Buitenhuis, H. and Pöllath, N. 2015. Red Deer and Human, From Palaeolithic to Iron Age in Anatolia. *Twelfth Meeting of the ASWA working group of ICAZ. Groningen, The Netherlands, 10-12 June, 2015.*
Spitzer, M. and Zeder, M.A. 2015. The Birds of Hallan Çemi: Habitat, Seasonality, and Ritual. *Twelfth Meeting of the ASWA working group of ICAZ. Groningen, The Netherlands, 10-12 June, 2015.*
Tornero, C., Balasse, M., Fiorillo, D., Molist, M. and Saña, M. 2015. Re-think seasonal migrations patterns of hunted gazelles in Euphrates Valley:

Abu Hureyras' model reviewed by stable isotopes. *Twelfth Meeting of the ASWA working group of ICAZ. Groningen, The Netherlands, 10-12 June, 2015.*

Uerpmann, H.-P. 2015. Animal Economy at Dosariyah - A Neolithic Site on the East-coast of Saudi Arabia. *Twelfth Meeting of the ASWA working group of ICAZ. Groningen, The Netherlands, 10-12 June, 2015.*

Vigne, J.-D., Carrére, I., Saliége, J.-F., Person, A., Bocherens H., Guilaine J. and Briois F. 2000. Domestic Cattle, Sheep, Goat and Pig during the 8th Millennium cal. BC on Cyprus Preliminary Results of Shillokambos (Parekklisha, Limasol). In, Mashkour, M., Buitenhuis, H., Choyke, A.M. and Poplin, F. (eds) *Archaeozoology of the Near East IV : Proceedings of the 4th International Symposium on the Archaeozoology of Southwestern Asia and Adjacent Areas (Paris 1998)*: 83-106. Groningen: ARC publications

Vigne, J.-D., Brunet, F., Debue, K. and Khudzhanazarov, M. 2015. Early domestic ungulates in Central Asia: archaeozoological results from Ajakagytma (Uzbekistan, Kel'teminar, 9th-7th millennia cal BP). *Twelfth Meeting of the ASWA working group of ICAZ. Groningen, The Netherlands, 10-12 June, 2015.*

Yeshurun, R. 2013. Late Paleolithic subsistence and ecology in the Kom Ombo Plain, Upper Egypt. *Eleventh Meeting of the ASWA working group of ICAZ. Haifa, Israel, 23-28 June, 2011.*

Weber, J. 1998. Sealing the deal : Sacrifice and burial in a Middle Bronze (ca. 1900-2000) well at Umm el-Marra, Syria. *Fourth Meeting of the ASWA working group of ICAZ. Paris, France. 29 June – 3 July 1998.*

Weber, J. 2006. Elite equids: redefining equid burials in the late 3rd millennium BC from Umm el-Marra (Syria). *Eighth Meeting of the ASWA working group of ICAZ. Lyon, France, 28 June – 1 July, 2006.*

Weber, J. 2015. More than just 'good eats': Animals as administrative artifacts. *Twelfth Meeting of the ASWA working group of ICAZ. Groningen, The Netherlands, 10-12 June, 2015.*

Zeder, M.A. 1998. Metrical Study of modern wild goats (*Capra aegagrus*) from Iran and Iraq: Implications for the study of caprine domestication in the Near East. *Fourth Meeting of the ASWA working group of ICAZ. Paris, France. 29 June – 3 July 1998.*

Zeder, M.A. 2006. Archaeozoology in Southwest Asia. *Paléorient* 32(1): 137-147.

2 'Feeding cities'? – Preliminary notes on the provisioning of animal products at Tel Bet Yerah, Israel

Alice Berger

Alice Berger (1986-2015)

Author Biography

Alice Berger gained a BA Archaeology from the Department of Archaeology and Ancient Near Eastern Cultures at Tel Aviv University (TAU) in 2010, where her academic performance put her on the *Dean's and Rector's Honor list*. She continued at Tel Aviv University for her MA Archaeology (2013). Her Master's thesis on *Plant Economy and Ecology in Early Bronze Age Tel Bet Yerah* was co-supervised by Prof Rafi Greenberg (TAU) and Prof Dorian Fuller (UCL) and won the prestigious John Evans Dissertation prize, awarded by the Association for Environmental Archaeology. Alice went on to an internship in zooarchaeology in the Institute of Prehistoric and Protohistoric Archaeology at the Christian-Albrechts-Universität in Kiel, under the mentorship of Prof Cheryl Makarewicz. Alice's academic achievements were further recognized by the double award of a UCL Overseas Research Scholarship and Graduate Research Scholarship to support her PhD research at the UCL Institute of Archaeology, which she started in 2013. Her PhD project was entitled *Social Arrangements of Subsistence Economy in Early Bronze Age Tel Bet Yerah*, and aimed to use zooarchaeology to explore Early Bronze Age urbanization and migration at the site of Tel Bet Yerah in Israel, as she had done with the archaeobotanical remains at MA level. Alice was a brilliant, original and energetic young researcher; her work was beginning to produce exciting results which she presented at ICAZ 2014 in Argentina and ASWA 2015 in Groningen, to wide acclaim. Alice was passionate about her research, was a wonderful teacher both in the classroom, laboratory and on fieldwork, and she had a stellar career ahead of her.

Alice Berger on fieldwork at Tel Bet Yerah, summer 2015.

Abstract

This paper presents preliminary zooarchaeological results from the Early Bronze Age site of Tel Bet Yerah (TBY) in the Jordan Valley, Israel. The site is multi-period, spanning the development of the settlement from a dispersed village in EBI A, to urban centre with a centralized economy in EBII, with a breakdown in centralization witnessed in EBIII deposits. Two areas are reported on here: SA-M and SA-S. A sample of over 1,300 identifiable mammalian bones (NISP) traces variation in the animal-livestock provisioning of the settlement through time, with an expected diverse representation of sheep, goat, cattle and pig in the EBI village, to a caprine dominant herding economy in EBII, likely focused on secondary products, suiting the needs of an urban centre. The presence of two distinct communities is evident in EBIII, where local indigenous inhabitants live alongside incoming 'Khirbet Kerak Ware' (KKW)-using groups. The zooarchaeological analysis, even though preliminary, shows two provisioning systems coexisting at TBY, with KKW contexts revealing unusually high cattle and equid proportions (interpreted as working animals), and also hints at distinct consumption practices.

Keywords

EBA, KKW, Southern Levant, Animal provisioning, Herd management, Centralized economy

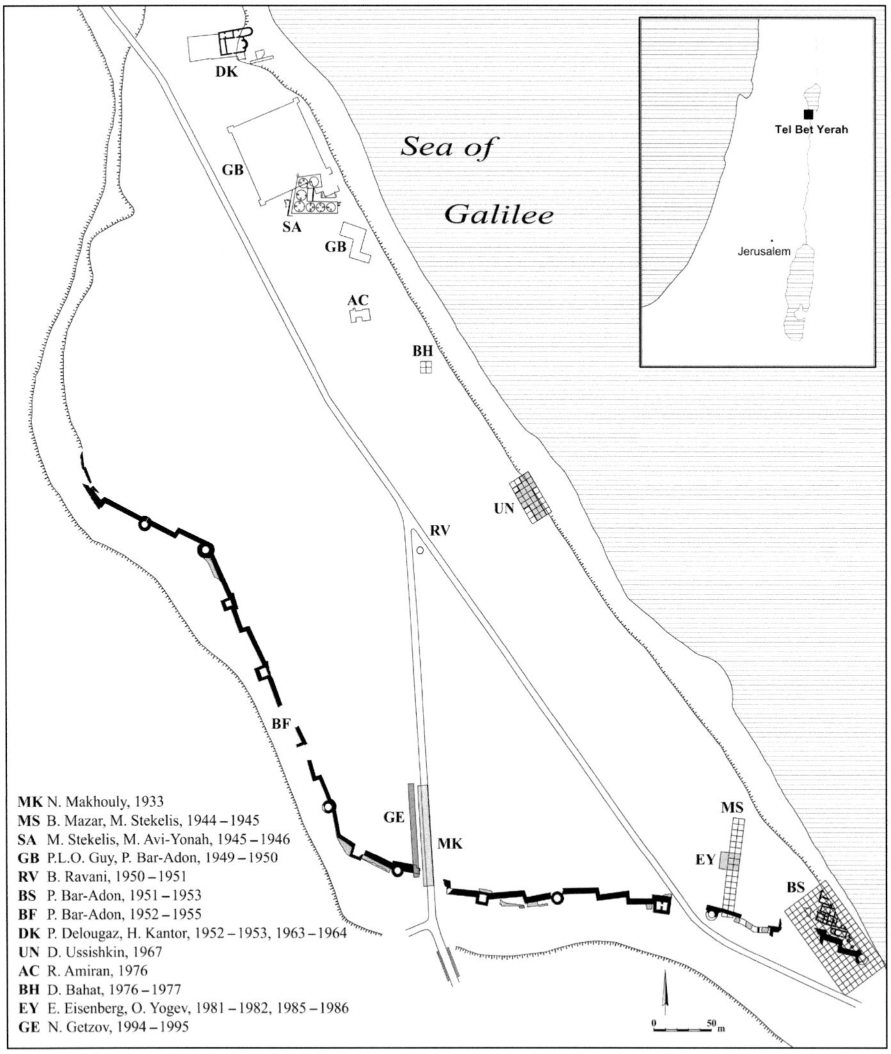

Fig. 1. Location map of Tel Bet Yerah.

Tel Bet Yerah (TBY) is a large Bronze Age mound in the Jordan Valley, located at the exit of the Jordan River from Lake Kinneret (the Sea of Galilee; Fig. 1). Extensive excavations conducted at the site since 1933 by more than 15 different excavators (Greenberg 2014a: vi-vii) have uncovered the stratified remains of a large Early Bronze I village that gives way to an Early Bronze II-III fortified town (see Table 1). The most recent series of excavations has been conducted by a joint Tel Aviv University – University College London team since 2003 (Wengrow 2009; Greenberg *et al.* 2012); the results of these excavations form the basis for the present report.

Table I. The chrono-cultural sequence at TBY.

S. Levant period	*Chronology*	*TBY Sequence*	*Contemporary Cultures*
EB IA	3700-3300 B.C.	Formation, dispersed village	Middle Uruk
EB IB	3300-3100 B.C.	Large village	Late Uruk; Egyptian predynastic
EB II	3100-2850 B.C.	Urbanism – planned streets, defensive walls, Egyptian contact	Egypt Dynasty I; Amuq G
EB III	2850-2500 B.C.	Monumental architecture, ETC newcomers	Egypt Dynasties II–IV; Amuq H; Kura-Araxes II

Excavation context and retrieval methods

The faunal remains described in this study come from the northern part of TBY, mainly from the Early Bronze Age (EBA) strata adjacent to the large, enigmatic structure known as the Granary, or Circles Building, first excavated in 1945-6 (Greenberg *et al.* 2006: 53-103). This excavation area comprises two subdivisions: Area SA-M to the north of the Circles Building, and Area SA-S to its west (Fig. 2). In area SA-S a paved street separates the Circles Building from a densely built residential quarter (Fig. 3). The street was paved in the EB II, when the site was first fortified and took on some urbanizing characteristics, and was evidently cleaned regularly as only a very thin layer accumulated during that period, while most deposits above it date to the EB III. The houses to its west are all typical to the local building tradition and, with the exception of few pits dug into them from a later level, yielded exclusively local and South Levantine Metallic Ware pottery (Greenberg and Porat 1996) ascribed to a sequence that begins in early EB II (ca. 3050-2850 BC) and ends in early EB III (ca. 2850-2700 BC). Some of the EB II layers were preserved well enough to recognize different activity areas, such as a food preparation room, an enclosed yard and a midden pit. Earlier levels were exposed only in limited soundings.

Area SA-M comprises, for the most part, an open mud-paved plaza attached to the north flank of the Circles building (Fig. 4). It was seemingly constructed simultaneously with the monumental building, in late EB II or early EB III, and was approached by a stairway from the paved street that borders Area SA-S. But, along with the Circles Building itself, it was soon occupied by bearers of an intrusive cultural assemblage associated with Khirbet Kerak Ware (KKW)—a type of highly burnished, red-black pottery with affinities in southeast Anatolia and the southern Caucasus, which arrives on the south Levantine scene at the outset of EB III and continues to be locally produced for several generations (Miroschedji 2000; Philip 1999). At least five levels of deposition were recognized in the plaza, during all of which the area maintained its open character. Discrete features in the plaza included midden accumulations, fragmented clay installations, and a cooking area marked by the presence of a hearth and andiron fragments. All of these features differ markedly from those encountered in the domestic

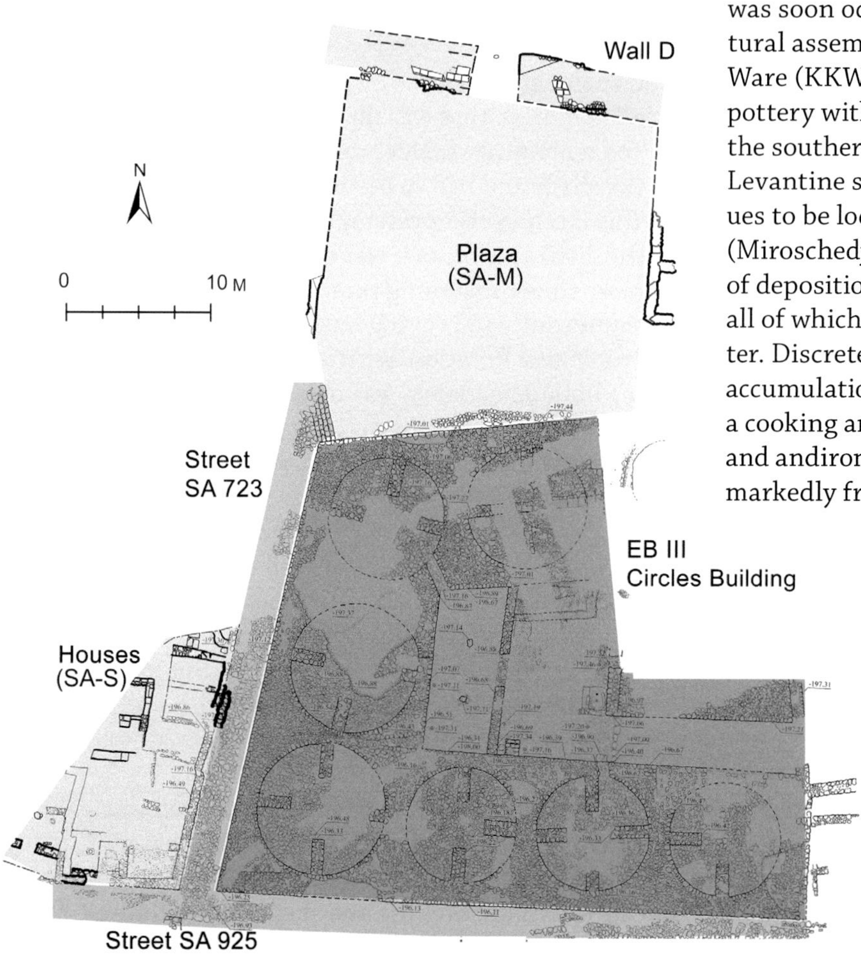

Fig. 2. Plan of the excavation area, showing location of Areas SA-S (houses) and SA-M (plaza).

Fig. 3. Area SA-S street and houses.

structures in Area SA-S and are currently interpreted as evidence of the presence of a new, culturally distinct, community that apparently coexisted with the former inhabitants (Greenberg *et al.* 2014). A study of the archaeobotanical remains confirmed that the cultural differences extended into the realm of plant processing and diet (Berger 2013).

In both excavation fields, faunal retrieval was carried out according to a standard protocol. During the excavation seasons of 2003, 2007 and 2009 bones were hand-picked during excavation, with only selected contexts sieved. Students were instructed to collect all bone fragments they encounter, but naturally a bias against smaller bones is expected (Payne 1975). In the contexts that did undergo sieving, a mesh of 4 mm was used, and volumes and percentage of the total soil from each of them are available. In 2010, sieving became routine practice at TBY for all undisturbed EBA contexts, using three different methods:

- Dry sieving consisted of processing sediments through a tier of three meshes, of 1 cm, 4 mm and 1 mm, laid on top of each other. The material in the lowermost tier, measuring 1-4 mm, was collected and sorted in the lab.
- Wet sieving was done by pouring sediments into a 1mm mesh that was dipped in a water tank. The remaining residues were then left to dry in the shade and had to be sorted later in the lab. This method of processing was abandoned after the 2010 season, as it was observed to be far more time consuming than dry sieving while not significantly improving retrieval.
- Systematic flotation, with a main aim of retrieving botanic remains, was implemented as well. The heavy fraction from those samples was collected into a 1mm mesh, dried in the shade and later sorted in the lab. This method produces similar results to wet sieving. Starting in 2012 at least 2 buckets or 10% of each defined EBA context were subjected to this treatment. While wet treatment does hold the potential to damage fragile bones, it did not cause significant losses, as attested to by the preservation of fish and rodent bones.

Since TBY is located on marl soil, and the tell was formed by the decay of mud bricks of this material, preservation in many parts of the site is rather poor. Agrilliturbation (swelling and shrinking of clays as a function of moisture) crushed some of the bones,

Fig. 4. Area SA-M plaza.

and a calcareous layer covers the surface of many. The bones are all reduced to their mineral components only, as attested by the lack of collagen even in the most robust items (E. Boaretto, pers. comm.). Due to their fragile nature the TBY bones were not soaked in water for cleaning, but were manually brushed with a dry cloth or gently scraped with dental tools when coating was persistent. During the 2013 and 2015 excavation seasons, exceptionally brittle specimens were identified and recorded prior to their removal and inevitable fragmentation. These are included in the database, although some may no longer be available for study.

In addition to the two main fields of excavation, faunal remains were hand-picked in the 2007 excavation of a deep EB IA midden pit located a short distance to the north (Greenberg *et al.* 2013).

Results

A total of 1332 bone specimens from TBY could be identified to taxonomic level, representing about 20% of the assemblage available to date. Since they originate from five different temporal or cultural contexts, the sample size for each is too small to be statistically significant (Gamble 1978) and the following presentation of the data should therefore be considered preliminary. It aims to calibrate expectations for further study and to offer projected models for analysis and interpretation of the zooarchaeological assemblage at TBY. The patterns evident in the smaller samples are presented in relation to trends emerging from other data from the site, particularly the archaeobotanical finds analyzed by this author and the material culture assemblages studied by colleagues in the Tel Bet Yerah Project (e.g., Greenberg 2014a; Greenberg *et al.* 2014; Paz 2012).

At this stage, only limited data concerning species abundance can be presented for all phases (Table 2). For the EB II sheep and goats, a sufficient sample is available to present a kill-off pattern (Table 3, Fig. 7) and attempt a prediction of the herd structure for that phase.

Early Bronze I (Fig. 5). Six species have been identified, corresponding to the common domesticates in the region (Horwitz and Tchernov 1989). Sheep and goats dominate the small assemblage at 45% (when EB IA and B are combined), with both sheep and goats identified. Next, of roughly equal representation, are cattle and pig (around 25% each), which is not an uncommon pattern in the early EB I; for example, the frequency of pigs is similarly high, or

Table 2. The number of identified animal bones from Tel Bet Yerah, by chronological or cultural phase, expressed as NISP and NISP%.

	EB IA		EB IB		EB II		EB III local		EB III KKW	
	NISP	%	NISP	%	NISP	%	NISP	%	NISP	%
Bos taurus	17	15.3	54	30.3	127	19.6	22	28.6	114	37.3
Ovis/Capra	42	37.8	103	57.9	390	59.7	50	64.9	137	44.8
Ovis aries	7		1		29	4.4	8		14	4.6
Capra hircus	4		0		64	9.8	8		20	6.5
Sus scrofa	52	46.8	9	5.1	13	2.0	4	5.2	14	4.6
Equus cf. asinus	0		9	5.1	3	0.5	0		22	7.2
Gazella gazella	0		3	1.7	14	2.1	0		2	0.7
Alcelaphus buselaphus	0		0		3	0.5	0		2	0.7
Capreolus capreolus	0		0		2	0.3	0		1	0.3
Dama mesopotamica	0		0		5	0.8	0		3	1
Cervus elaphus	0		0		0		0		4	1.3
Canis familiaris	0		0		2	0.3	0		7	2.3
Felis silvestris	0		0		0		1	1.3	0	
Total	**111**		**178**		**653**		**77**		**306**	

Table 3. Counts of sheep/goat elements from TBY EB II, grouped by Zeder's (2006) epiphyseal fusion classes.

Element	*Age of fusion (months)*		*Fused*	*Unfused*
Radius Prox.	0-6	A	4	0
Humerus Dist. Pelvis, Scapula	6-12	B	11	4
2nd phalanx 3rd phalanx	12-18	C	16	3
Tibia Dist. Metatarsal Dist., Metacarpal Dist.	18-30	D	23	7
Calcaneus, Femur Dist.+Prox. Ulna Prox., Radius Dist., Tibia Prox.	30-48	E	6	23
Humerus Prox.	48+	F	0	2

higher, at sites such as Tell es-Shuna and Azor (Croft 1994; Horwitz 1999).

There is a very low representation of equids, likely to be donkeys, important pack animals that were evidently used as beasts of burden. Hunted gazelles are also present in very small quantities.

Early Bronze II (Fig. 6). The percentage of pigs decreases dramatically from 26% to 2%, while sheep and goats increase to over 70% in this phase and entirely dominate the assemblage. Cattle maintain their importance however, at about 20%, as is the norm for the Levant throughout the entire Bronze and Iron Ages (Sasson 2010). Again, there is evidence for donkeys and some hunted wild animals.

The frequency of cattle, at about 20%, stands in contrast to a previous study on material from Tel Bet Yerah, which found sheep and goats accounting for as much as 86% of the EB II assemblage while cattle only constituted 6.6% (Cope 2006). Unfortunately, contextual information is not available for Cope's samples, so the discrepancy cannot be explained at this point. The present study includes evidence from contexts identified as roasting pits, ascribed to a very early stage of the EB II that preceded the pavement of the public street; this extraordinary context, which may not represent the site as a whole in its presumably urban phase, contributed almost a third of the cattle bones.

The survivorship curve for sheep and goats in EB II (plotted together in Fig. 7, from fusion data) is noticeably focused on the culling of young and prime adults by 30-48 months (i.e. before four years of age), with only slight indication for the culling of either youngsters or old individuals. The majority appear to be slaughtered between two and a half years and four years of age, the stage when these species achieve their maximal size.

For EB II, a comparison of the skeletal part representation shows that all body parts are present for cattle (*Bos*) and sheep/goat (*Ovis/Capra*) in the assemblage. There is an indication, however, that cattle show more skull and foot elements than sheep/goat, and fewer meaty long bone and vertebrae elements. It may be that these meat-bearing elements were deposited elsewhere on site, or removed/transported off-site for some reason, or perhaps more likely that higher levels of processing and fragmentation has left them less identifiable to element than the bones of smaller sheep and goats.

EB III local (Fig. 8). The ratio between sheep/goats and cattle is similar to the previous period. Despite the extremely small sample size of this assemblage, pigs seem to be slightly more abundant. In addition, a cat's tibia was identified.

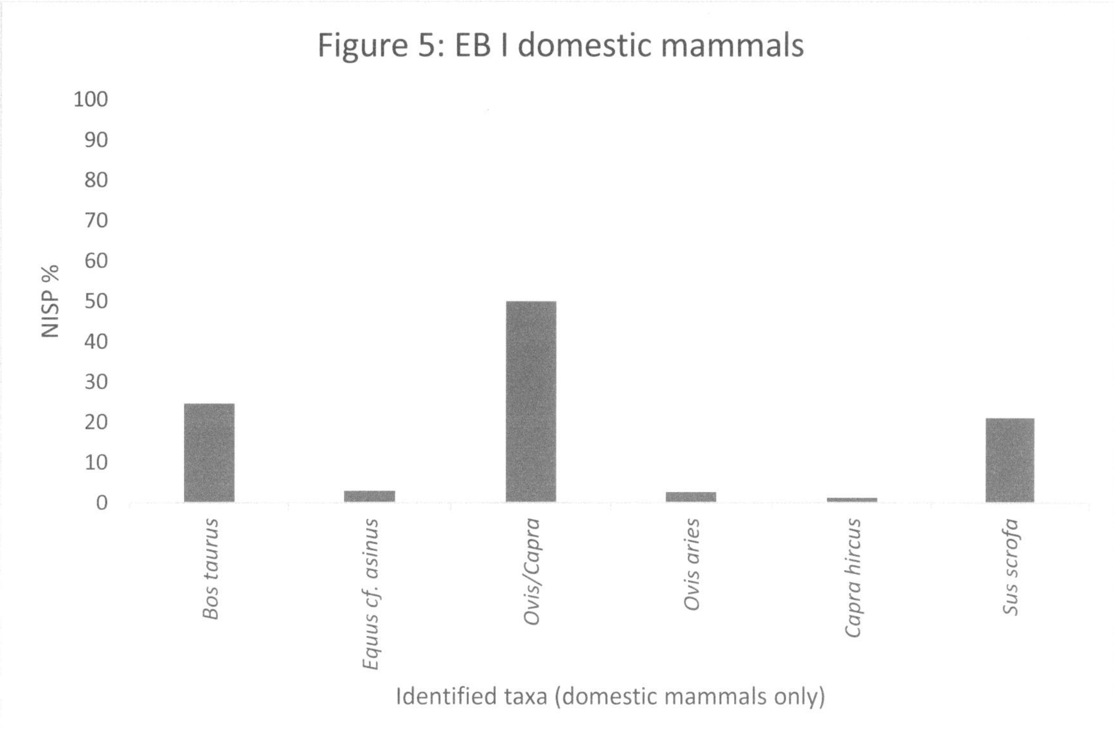

Fig. 5. Histogram showing the relative proportions of domestic mammals (NISP %) for EB I levels.

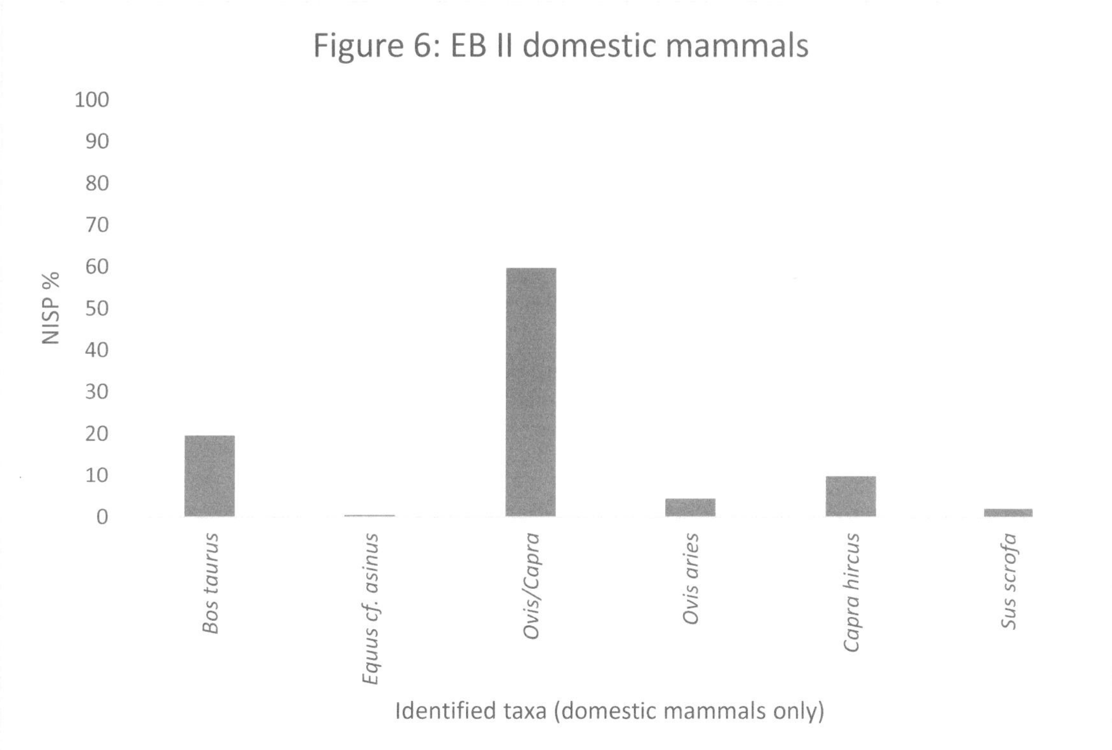

Fig. 6. Histogram showing the relative proportions of domestic mammals (NISP %) for EB II levels.

EB III KKW (Fig. 9). Untypically for the region, the number of identified cattle bones is very close to that of sheep and goats, increasing to 35%. This assemblage also contains the highest proportion of equids of the samples analysed so far. Wild animals have a minor contribution, among which cervids are dominant.

Patterns in the Tel Bet Yerah faunal assemblage

The preliminary observations described above can be discussed in light of a set of predictive approximations of the animal economy, based on our prior knowledge of the social and economic characteristics of the relevant archaeological assemblages.

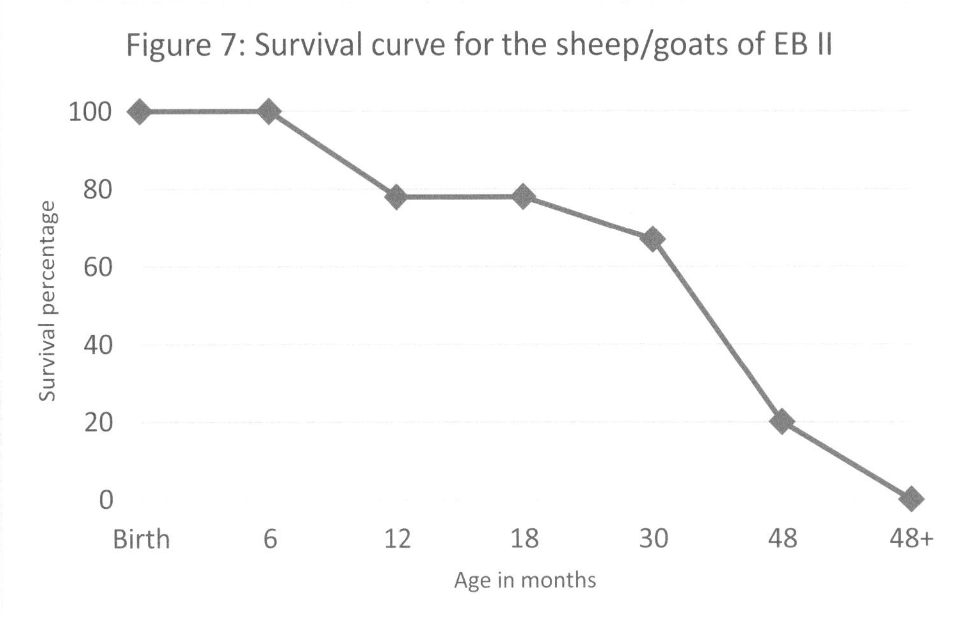

Fig. 7. Survival curve for the sheep/goats of EB II, based on the data in Table 3.

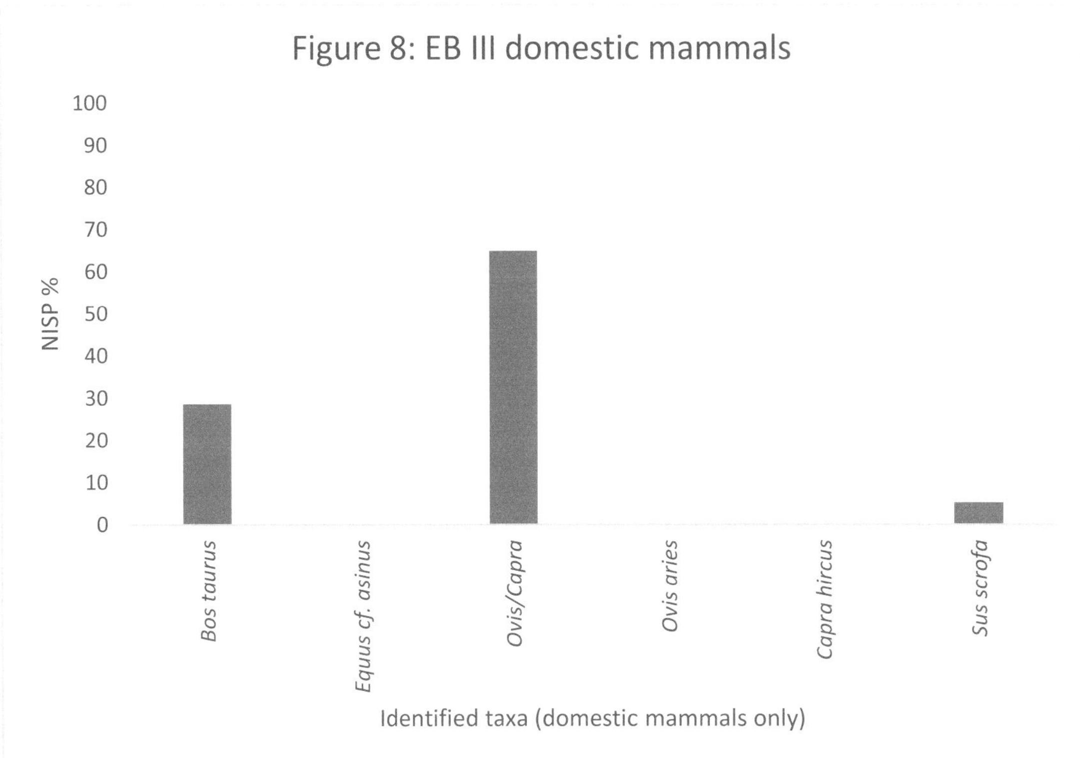

Fig. 8. Histogram showing the relative proportions of domestic mammals (NISP %) in EB III levels.

Early Bronze Age I – A dispersed village economy

Based on the dispersed nature of the site and of its agricultural production, as indicated by the archaeobotanical profile, centralization or organization of pastoral production is not expected in EB I. If animals were raised and utilized by individual households, with little motivation for surplus production, the primary consideration would have been herd security. In such a system species diversity should be high, since different animals are vulnerable to variable conditions, and if one species is subject to disease or other dangers, others may survive.

The demographic structure of the sheep and goat herds should be focused on mature females, which are expected to be kept alive so long they are able to reproduce, ensuring the continuity of the herd. The majority of males would be slaughtered by 2 years of age, as beyond that point they no longer gain weight and keeping them entails providing pasture without gaining additional return. Some young offspring

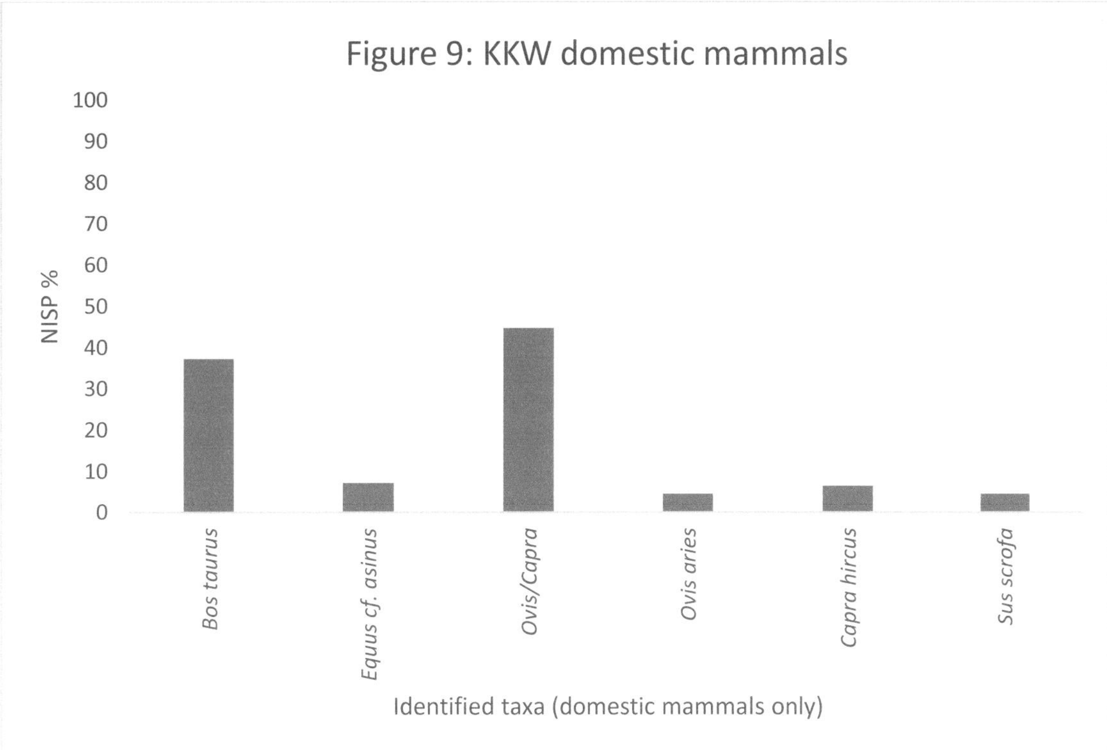

Fig. 9. Histogram showing the relative proportions of domestic mammals (NISP %) in EB III KKW levels.

might be slaughtered, in order to exploit their mothers' milk, but in a self-sufficient, non-specialized economy this is not expected to be the main focus of herders, and a sufficient number of disposable males should be kept as a 'walking larder' to be slaughtered in times of need.

It might, however, be reasonable to expect a more frequent slaughter of kids and lambs in the EB IB, when the ceramic repertoire expands significantly, suggesting perhaps diversification of food sources or cuisines (Rotem 2015). It would not be surprising if this variety was achieved by greater utilization of milk and milk products, as studies have shown that in the EBA, milking of sheep and goat became a practice common enough to deplete females of calcium on a level sufficient for bone modification (Smith and Horwitz 1984).

In a plough-using community, cattle are predicted (Sasson 2005) to make up at least 20% of the animals at the site (as, in fact, they do): the use of cattle for ploughing in the EB I is attested elsewhere both by foot pathologies (Grigson 1995: 267-268) and pictographic evidence in the form of a seal depicting a pair of oxen pulling a plough from Tel Kitan (Keel-Leu 1989). If the small EB I assemblage is broken down into EB I A and B, there is a lower frequency of cattle observed in EB IA (15%, compared to 30% in EB IB), which, if proven characteristic of the site as a whole, would indicate that intensification of agriculture only occurred in the EB IB, corresponding with the site's population growth.

A gradual decline in the frequency of pig in late EB I is anticipated for two reasons. First, if indeed the TBY economy was more reliant on secondary products, pigs would be less valued, as they produce none. Second, greater crowding within the settlement would leave less space where they could be kept. As pigs are destructive to agricultural fields, they cannot be kept off-site, unless they are penned and provided with food by their human keepers. Such a strategy would defeat the greatest advantage in keeping pigs, namely the little effort required for their raising, and is not likely to be practiced unless pigs are particularly valued for some other, culture dependant, reason.

Early Bronze Age II – Centralization

If, similarly to agricultural practice (where grain appears to have been distributed from central storage, with few remains of processing stages), pastoral production was centrally organized in EB II, very different patterns are expected to emerge in the EB II faunal assemblage. Primarily, species diversity would decrease. Most notably, pigs are predicted to remain a very small component of the meat consumed at TBY (as seen in the histogram of both predicted and actual representation). This species is ill-suited for governed economies, and normally increases in abundance when no authority regulates animal products distribution, as they are easy to maintain in very small groups (Zeder 1990: 25). In regional systems they are particularly disadvantageous as they provide no secondary products that could be stored and traded, and they are difficult to herd and move between sites (Wapnish and Hesse 1988). At TBY pigs would pose a particular problem, given the dense construction within the settlement. A near absence of pigs would also fall well within the regional narrative, as they only very rarely exceed 4% of the assemblage in any EB II south Levantine site (e.g. Davis 1976; Hesse and Wapnish 2001; Horwitz 2003; Horwitz *et al.* 1996).

The percentage of cattle in the assemblage is also likely to be reduced. When agricultural work is carried out in a centralized manner, rather than in numerous fields tended by different households, fewer oxen would be required and little incentive would exist for their private ownership. If agricultural products arrived from a supporting hinterland, this pattern would be even more pronounced, as the working beasts would be kept by farmers in other settlements. While cattle do provide more meat than other domestic animals, they also require more resources, reproduce at a lower rate, and take almost four years to achieve an optimal size for culling, so for strictly economic considerations, cattle is not profitable to raise for meat consumption alone (Bogucki 1993: 497). That said, 56% of the cattle long bones at TBY were unfused in this period, showing a relatively high cull of young and sub-adult animals, supporting the idea that apart from work oxen, animals were generally not kept until older age.

The demographic patterns observed for sheep and goat so far imply the herds might indeed have been managed by pastoralists outside TBY. The absence of old and young individuals is not characteristic for a sustainable herd, but matches the expectations of external provisioning, with TBY receiving exclusively animals of prime age and the suppliers sustaining the older reproducing females (Zeder 1988). Given the multiple lines of evidence for craft specialization and economic organization in EB II, it is reasonable to assume that TBY would be on the receiving end in a broader system based on surplus production.

As the EB II sample in this study includes only material derived from domestic contexts, the low frequency of equids should come as no surprise. These animals must have held special meaning to their owners, as attested by their appearance in contemporary art (Milevski 2011), and – perhaps more importantly – in deliberate donkey burials (Greenfield *et al.* 2012). It is plausible that their use and disposal was controlled either by the elites or by dedicated traders. Alternatively, their symbolic status might have deterred people from consuming them, or at least of disposing of their remains along with domestic refuse.

Early Bronze Age III – Breakdown

With architectural changes indicating partial abandonment of domestic areas and the archaeobotanical data pointing to a breakdown of the centralized plant food economy in EB III, a central organization of pastoral provisioning is not expected to be maintained either. In such a case, the dispersal of agricultural production would allow cattle to regain popularity, as they would be required as beasts of burden by a higher portion of the site's population.

In the bone assemblage from the Getzov excavations of 1995, pigs constitute 13% of the bones from early EB III strata, which represent local-tradition contexts (Cope 2006). In our small sample, pigs increase from 2% to 5.2%. The return of pigs to the site is indeed the pattern expected from a segmented economic system. With the demise of the central authority and lesser settlement density, pigs regain their status as a convenient, easily maintained private source of meat.

According to this expected pattern of decentralization, the inhabitants of TBY would have to tend to their own herds, with the intention of self-provisioning and with no motivation to produce surplus for trade. Since little trade is assumed, they would need to provide themselves with both meat and secondary products, therefore a more balanced slaughter curve is expected, with all age classes represented, permitting exploitation of all resources, rather than focusing on one. Overall the EB III local assemblage is expected to produce patterns closer to those of the EB I than of the prime urban period. In fact, however, the profile of local EB III species distribution falls between EB I and EB II, suggesting a more nuanced change than that predicted.

Khirbet Kerak Ware contexts

As demonstrated in the plant remains assemblage, the KKW-using community did not seem to depend on others for their provisioning. The very distinct species ratios in the preliminary study suggest this might also have been the case regarding animal-based subsistence.

Zooarchaeological assemblages from ETC sites in Anatolia and the southern Caucasus have provided no evidence for mobile pastoralism being a major component of ETC economy (Piro 2009). Given the indications for the KKW-using community's involvement in agriculture, it is not predicted to be the case at TBY either. A full, sustainable herd is expected to be represented in the demographic profiles, with variable culling ages, as appropriate for a self-sufficient economy.

Pigs are not expected to be more abundant in KKW contexts at TBY than represented in the current sample. Although they constantly appear in ETC faunal assemblages, they are never abundant

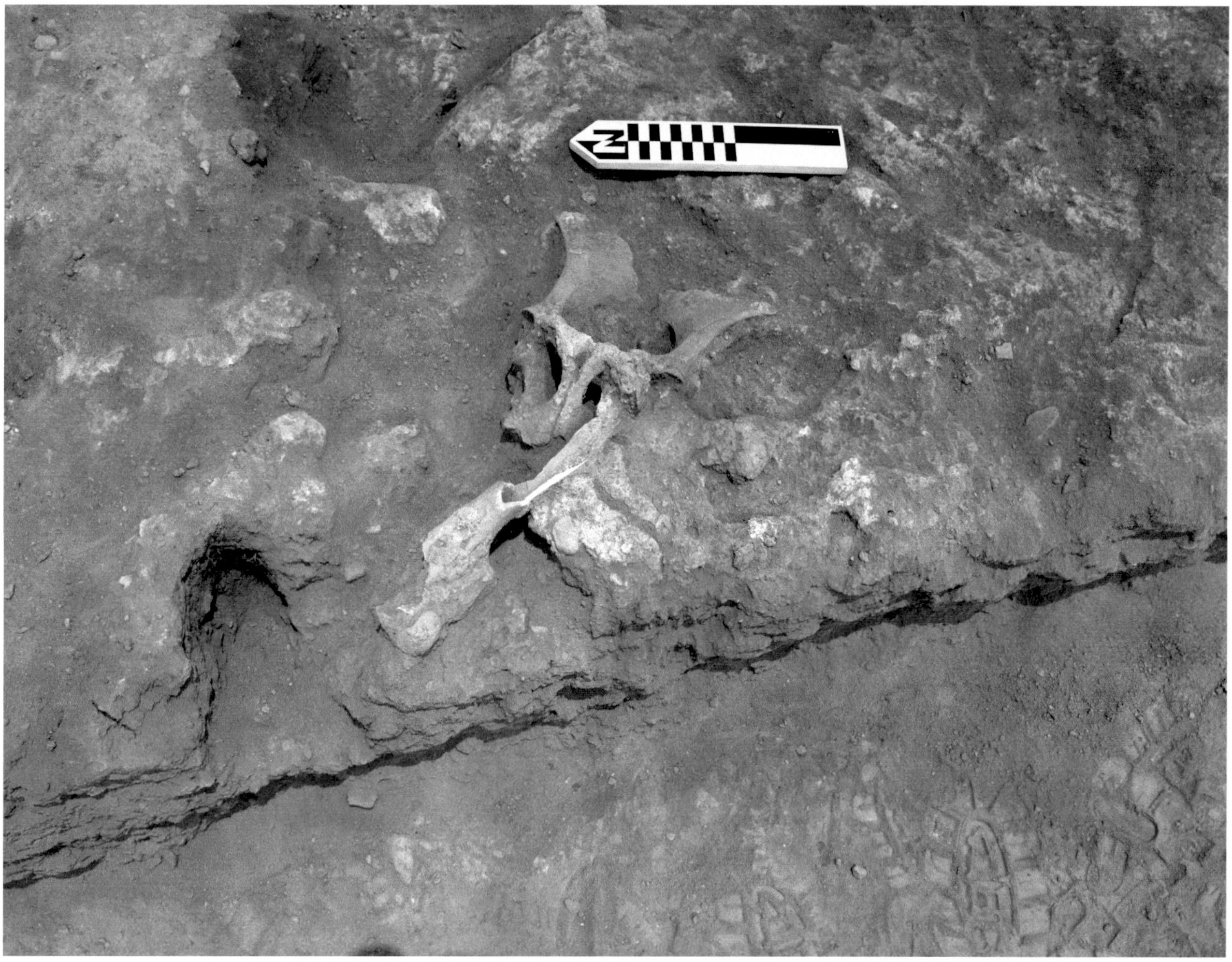

Fig. 10. Carcass parts in plaza deposits associated with the KKW-using community.

(Bartosiewicz 1998; Gadzhiev *et al.* 1997; Monahan 2007; Piro 2009), and this seems to be a persistent cultural preference, likely to be preserved in the Levant as well.

Hunted animals contribute only a small percentage to the assemblage in each phase, and this is the case for the KKW contexts as well, but the presence of cervids is not coincidental. The ETC has a tradition of antler tool crafting (Samei *et al.* 2013), and the only deer antler recovered at TBY to date, in older excavations, was indeed located in a house with KKW pottery (Paz 2009: 200).

The remarkably high abundance of cattle (at 36%) is outstanding for this period in the Levant, matched only by the finds from Tel Yaqush (Hesse and Wapnish 2001), a site with a strong KKW presence. This pattern also finds a parallel in the distant ETC site of Mokhra Blur in Armenia (Piro 2009: 280). Remote as it may be, it is situated, like TBY, in a valley bottom near a lake shore. The similarity in the sites' settings and the taxonomic distribution resonates with the seemingly flexible, adaptive strategy of the ETC communities, whose zooarchaeological assemblages vary in different environments they occupy (ibid: 300-301).

Some TBY deposits show large elements of carcasses, less fragmented than elsewhere and perhaps deriving from the same limbs and individuals (Fig. 10); these could be interpreted as evidence for communal consumption of these large animals, and beyond-household sharing of resources for binding the larger social group – an idea that needs further exploration.

Considering the recurring artistic representations of what seem to be draught cattle and two-wheeled carts in ETC communities, it is plausible to assume this mode of transportation constituted the incentive for rearing numerous oxen (and donkeys) as working animals (Greenberg 2014b). A higher proportion of equids is seen in these deposits – approximately 9% - than any other at the site, which might confirm the presence of donkeys as transport and work animals. This hypothesis may be tested in the future by examining foot pathologies (Bartosiewicz *et al.* 1997).

Conclusion

In summation, preliminary analyses of the TBY assemblages trace changes through the EBA occupations which reflect different animal provisioning

and management practices, from the dispersed village system, through to centralized production and management of the agricultural-livestock system in EB II, and into the EB III occupation where both local indigenous and incoming ETC-related communities were settled in the same location. It is intriguing that the zooarchaeological study displays not only two different livestock management systems co-existing, but also hints at distinct consumption practices.
So far, the research presented has focused primarily on shifts in the relative proportions of animals, with some ageing data allowing discussion of herd management practices. A far more detailed database underlies the results presented in this brief report. That database records evidence for butchery and carcass treatment with the aim of drawing out patterns beyond animal provisioning and management, which would allow detailed exploration of food preparation and consumption practices for TBY communities, for comparison with the archaeobotanical results already produced, and alongside a detailed contextual integration with other aspects of the material culture.

Acknowledgements

The research at Tel Bet Yerah was supported by the Israel Science Foundation, Grant no. 310/12. Alice Berger's PhD studies on the animal remains from Tel Bet Yerah were funded by a UCL Overseas Research Scholarship and a UCL Graduate Research Scholarship. The text of this article was finalized by Rafi Greenberg (Tel Aviv University) and Louise Martin (UCL) based on Alice's research documents and conference presentations. We thank Simon Longe and Lilya Berger for providing Alice's computer records for this purpose. We also thank the team at Tel Bet Yerah for helping Alice in the field laboratory, in particular Gwendolin Maurer, Mariana Ribas Albuquerque and Katie Barry.

References

Bartosiewicz, L. 1998. Interim report on the Bronze Age animal bones from Arslantepe (Malatya, Anatolia). In, Buitenhuis, H., Bartosiewicz, L. and Choyke A.M. (eds) *Archaeozoology of the Near East, Proceedings of the Third International Symposium*: 221-32. Groningen: ARC Publications.

Bartosiewicz, L., Neer, W. van, Lentacker, A. and Fabis, M. 1997. *Draught Cattle: Their Osteological Identification and History*. Tervuren.

Berger, A. 2013. *Plant Economy and Ecology in Early Bronze Age Tel Bet Yerah*. M.A. Tel Aviv University.

Bogucki, P. 1993. Animal traction and household economies in Neolithic Europe. *Antiquity* 67: 492-503.

Cope, C. P. 2006. The Fauna: Preliminary Results. In, Getzov, N. (ed.) *The Tel Bet Yerah Excavations 1994-1995*: 169-74. IAA Reports 28. Jerusalem: Israel Antiquities Authority.

Croft, P. 1994. Some preliminary comments on animal remains from the first three seasons at Shuna. *Levant* 26: 130-31.

Davis, S. 1976. Mammal Bones from the Early Bronze Age City of Arad, Northern Negev, Israel: Some Implications Concerning Human Exploitation. *Journal of Archaeological Science* 3: 153-64.

Gadzhiev, M. G., Kohl, P. L., Magomedov, R. G. and Stronach, D. 1997. The 1995 Daghestan-American Velikent Expedition: Excavations in Daghestan, Russia. *Eurasia Antiqua* 3: 181-222.

Gamble, C. 1978. Optimising Information from Studies of Faunal Remains. In, Cherry, J., Gamble, C. and Shennan S. (eds) *Sampling in Contemporary British Archaeology*, 321-353. B.A.R. Br. Ser. 50. Oxford: British Archaeological Reports.

Greenberg, R. 2014a. *Bet Yeraḥ, the Early Bronze Age Mound*, Vol. II: *Urban Structure and Material Culture, 1933-1986 Excavations*. IAA Reports 54. Jerusalem; Israel Antiquities Authority.

Greenberg, R. 2014b. Ox-Carts and the Kura-Araxes Migration. In, Narimanishvili, G. (ed.) *Problems of Early Metal Age Archaeology of Caucasus and Armenia*. Conference Proceedings, Tbilisi, 19-23 November 2014: 94-101. Tbilisi: Rostaveli National Science Foundation.

Greenberg, R., Eisenberg, E., Paz, S. and Paz Y. 2006. *Bet Yerah, the Early Bronze Age Mound*, Vol. I: *Excavation Reports, 1933-1986*. IAA Reports 30. Jerusalem: Israel Antiquities Authority.

Greenberg, R., Paz, S., Wengrow, D. and Iserlis, M. 2012. Tel Bet Yerah: Hub of the Early Bronze Age Levant. *Near Eastern Archaeology* 75: 88-107.

Greenberg, R. and Porat, N. 1996. A third millennium Levantine pottery production center: typology, petrography, and provenance of the Metallic Ware of northern Israel and adjacent regions. *Bulletin of the American Schools of Oriental Research* 301: 5-24.

Greenberg, R., Rotem, Y. and Paz, S. 2013. The earliest occupation at Tel Bet Yerah. *Tel Aviv* 40: 197-225.

Greenberg, R., Shimelmitz, R. and Iserlis, M. 2014. New evidence for the Anatolian origins of "Khirbet Kerak Ware people" at Tel Bet Yerah (Israel), ca 2800 BC. *Paléorient* 40: 183-201.

Greenfield, H. J., Shai, I. and Maeir, A. 2012. Being an "ass": An Early Bronze Age burial of a donkey from Tell es-Safi/Gath, Israel. *Bioarchaeology of the Near East* 6: 21-52.

Grigson, C. 1995. Plough and Pasture in the Early Economy of the Southern Levant. In Levy, T.E. (ed.) *The Archaeology of Society in the Holy Land*: 245-268. London: Leicester University Press.

Hesse, B. and Wapnish, P. 2001. Commodities and Cuisine: Animals in the Early Bronze Age of

Northern Palestine. In, Wolff, S. (ed.) *Studies in the Archaeology of Israel and Neighboring Lands in Memory of Douglas L. Esse*: 251-282. Chicago and Atlanta: The Oriental Institute and the American Schools of Oriental Research.

Horwitz, L. K. 1999. The fauna. In, Golani, A. and van den Brink, E.C.M. (eds) Salvage excavations at the Early Bronze Age IA settlement of Azor. *'Atiqot* 38: 33-39.

Horwitz, L. K. 2003. Fauna from Tel Qashish. In Ben-Tor, A., Bonfil, R. and Zuckerman S. (eds) *Tel Qashish. A Village in the Jezreel Valley. Final Report of the Archaeological Excavations (1978-1987)*: 427-436. Qedem Reports 5. Jerusalem: Hebrew University of Jerusalem.

Horwitz, L. K. and Tchernov, E. 1989. Animal exploitation in the Early Bronze Age of the Southern Levant: an overview. In Miroschedji, P. de (ed.) *L'urbanisation de la Palestine à l'âge du Bronze Ancien: Bilan et Perspectives des Recherches Actuelles*: 279-296. B.A.R. Int. Ser. 527. Oxford: British Archaeological Reports.

Horwitz, L. K., Hellwing, S. and Tchernov, E. 1996. Patterns of Animal Exploitation at Early Bronze Age Tel Dalit. In, Gophna, R. (ed.) *Excavation at Tel Dalit*: 193-216. Tel Aviv: Ramot.

Keel-Leu, H. 1989. Die frühesten Stempelsiegel Palästinas. Von den Anfängen bis zum Ende des 3. Jahrtausends. *Studien Zu Den Stempelsiegeln Aus Palästina/Israel*, II: 1-39. Orbis Biblicus et Orientalis 88. Freiburg: Universitätsverlag.

Milevski, I. 2011. *Early Bronze Age Goods Exchange in the Southern Levant - A Marxist Perspective.* London: Equinox.

Miroschedji, P. de 2000. La céramique de Khirbet Kerak en Syro-Palestine: état de la question. In, Marro C. and Hauptman H. (eds) *Chronol. des pays du Caucase et de l'Euphrate au IV e-III millénaires:* 255-278. *Varia Anatolica* XI. Istanbul: IFEA.

Monahan, B. H. 2007. Nomadism in the Early Bronze Age southern Transcaucasia: the faunal perspective. In Popova, L.M, Hartley C.W. and Smith A.T. (eds) *Social Orders and Social Landscapes*: 379-392. Newcastle: Cambridge Scholars.

Payne, S. 1975. Partial recovery and sample bias. In, Clason, A. (ed.) *Archaeozoological Studies Papers of the Archaeozoological Conference, Groningen 1974*: 7-17. Amsterdam and New York: North Holland Publishing Company.

Paz, S. 2009. A Home Away from Home? The settlement of Early Transcaucasian migrants at Tel Bet Yerah. *Tel Aviv* 36: 196-216.

Paz, S. 2012. Changing Households at the Rise of Urbanism: The EB I-II Transition at Tel Bet Yerah. In, Parker B. J. and Foster C. P. (eds) *New Perspectives in Household Archaeology*: 407-434. Winnona Lake IN: Eisenbrauns.

Philip, G. 1999. Complexity and Diversity in the Southern Levant during the Third Millennium BC: The Evidence of Khirbet Kerak Ware. *Journal of Mediterranean Archaeology* 12: 26-57.

Piro, J. J. 2009. *Pastoralism in the Early Transcaucasian Culture: The Faunal Remains from Sos Höyük*. PhD. New York University.

Rotem, Y. 2015 The Central Jordan Valley in the Early Bronze Age I and the Transition to Early Bronze Age II: Patterns and Processes in a Complex Village Society. PhD. Tel Aviv University.

Samei, S., Alizadeh, K. and Eqbal, H. 2013. Zooarchaeological Analyses from the KuraAraxes Site of Köhne Shahar (Ravaz) in Northwestern Iran: A Preliminary Assessment of Social Complexity, *ASOR annual meeting*. Poster presentation. Baltimore, MD, USA.

Sasson, A. 2005. Economic Strategies and the Role of Cattle in the Southern Levant in the Bronze and Iron Age. In, Maskhour, M., Choyke, A.M., Buitenhuis, H. and Poplin, F. (eds) *Archaeozoology of the Near East IV: Proceedings of the Sixth International Symposium on the Archaeozoology of Southwestern Asia and Adjacent Areas*: 208-21. Groningen: ARC Publicatie.

Sasson, A. 2010. *Animal Husbandry in Ancient Israel: A Zooarchaeological Perspective on Livestock Exploitation, Herd Management and Economic Strategies*. London: Equinox.

Smith, P. and Horwitz, L. K. 1984. Radiographic evidence for changing patterns of animal exploitation in the Southern Levant. *Journal of Archaeological Science* 11: 467-75.

Wapnish, P. and Hesse, B. 1988. Urbanization and the organization of animal production at Tell Jemmeh in the Middle Bronze Age Levant. *Journal of Near Eastern Studies* 47: 81-94.

Wengrow, D. 2009. Exploring connections: a new fieldwork collaboration at Tel Bet Yerah (Khirbet el-Kerak). *Archaeology International* 12: 31-35.

Zeder, M. A. 1988. Understanding urban process through the study of specialized subsistence economy in the Near East. *Journal of Anthropological Archaeology* 7: 1-55.

Zeder, M. A. 1990. Animal exploitation at Tell Halif. *Bulletin of the American Schools of Oriental Research Supplementary Studies* 26: 24-31.

Zeder, M. A. 2006. Reconciling Rates of Long Bone Fusion and Tooth Eruption and Wear in Sheep (Ovis) and Goat (Capra). In, Ruscillo, Deborah (ed.) *Recent Advances in Ageing and Sexing Animal Bones, Proceedings of the 9th Conference of the International Council of Archaeozoology.* Durham, August 2002, 87-118, Oxford: Oxbow.

3 Zooarchaeology of the Pre-Pottery and Pottery Neolithic site of Qasr-e Ahmad (Iran)

Kamjan Safoora[1], Mashkour Marjan[2], Mohaseb F. Azadeh[2], Fathi Homa[3], Azizi Kharanaghi Hossein[4]

1 University of Groningen, the Netherlands; S.kamjan@rug.nl
2 Archéozoologie, Archéobotanique (UMR 7209), CNRS, MNHN, UPMC, Sorbonne Universités, CP55, 55rue Buffon 75005 Paris, France; mashkour@mnhn.fr & azadeh.mohaseb@mnhn.fr
3 Payam-e Nur University, Tehran, Iran; homafathi88@gmail.com
4 ICHHTO Iranian Cultural Heritage, Handicraft and Tourism Organization, Iran; hossinazizi@yahoo.com

Abstract
The zooarchaeological study of the Neolithic site of Qasr-e Ahmad in Fars Province (southern Zagros, Iran) provides valuable evidence for a better understanding of the process of caprine domestication and dispersal in Southwest Asia.

Goat was the most commonly exploited animal in Qasr-e Ahmad during both phases of prehistoric settlement (Pre-Pottery Neolithic and Pottery Neolithic). Sheep were also present but represent a limited contribution to the subsistence economy at the site. Limited osteometric data from sheep suggest that they were already domesticated at the site, which is earlier than hitherto suggested for the region. Kill-off and osteometric data indicate that the majority of goats and sheep were managed using strategies known at other later sites in the region. Cattle and pig were not domesticated and were not regularly exploited. The location of the site next to the Qara Aqhaj permanent river, the presence of architectural remains in the PPN phases of the site, as well as the abundance of lithic tools indicate pastoral and agricultural components in the Qasr-e Ahmad subsistence economy.

Keywords
Pre-Pottery Neolithic, Pottery Neolithic, Zagros, domestication, goat, sheep, secondary products, milk

Introduction
Several studies since the mid-60s have suggested that the Zagros highlands were the natural habitat and domestication center for goats (Hole *et al.* 1969; Flannery 1969; Bökönyi 1977; Hesse 1978). Recent re-examinations of previously excavated faunal assemblages and recent investigations in the Zagros mountains highlighted the importance of this region for the initial domestication of cereals and caprines (Zeder and Hesse 2000; Zeder 2008; Matthews *et al.* 2010; Riehl *et al.* 2013; Moradi *et al.* 2016). From 8000 BC onwards, Neolithic communities dispersed from this region towards the eastern periphery of the Fertile Crescent (Broushaki *et al.* 2016). Work at various sites in southern Zagros, such as the recently excavated site of Kelek Asad Morad (Luristan) (Moradi *et al.* 2016) or Ali Kosh (Hole 1962; Zeder 2008), and further south in the Fars region, such as Mushki (Mashkour *et al.* 2006) and Rahmat Abad (Davoudi *et al.* in prep), together with the first Neolithic communities, shed light on the introduction of managed goats to southern Zagros. It seems, however, that the first domesticated sheep arrived in Zagros some time later, around 7000 BC, as shown by the Pottery Neolithic assemblages of Tepe Guran, Sarab, and Jarmo (Zeder 2008).

In this paper we present the archaeozoological results from Qasr-e Ahmad in Fars province, recently excavated by H. Azizi Kharanaghi, (Fig. 1), where the transition between the Pre-Pottery and Pottery Neolithic – (from now on PPN and PN) can be traced through the material culture and bioarchaeological data (Azizi Kharanaghi *et al.* 2012; Tenberg and Azizi Kharanaghi 2016), and address two questions:

1) Which species were exploited in southwest Zagros during both phases of the Neolithic (PPN and PN);
2) what was the domestication status of these animals, in particular the small and large ungulates?

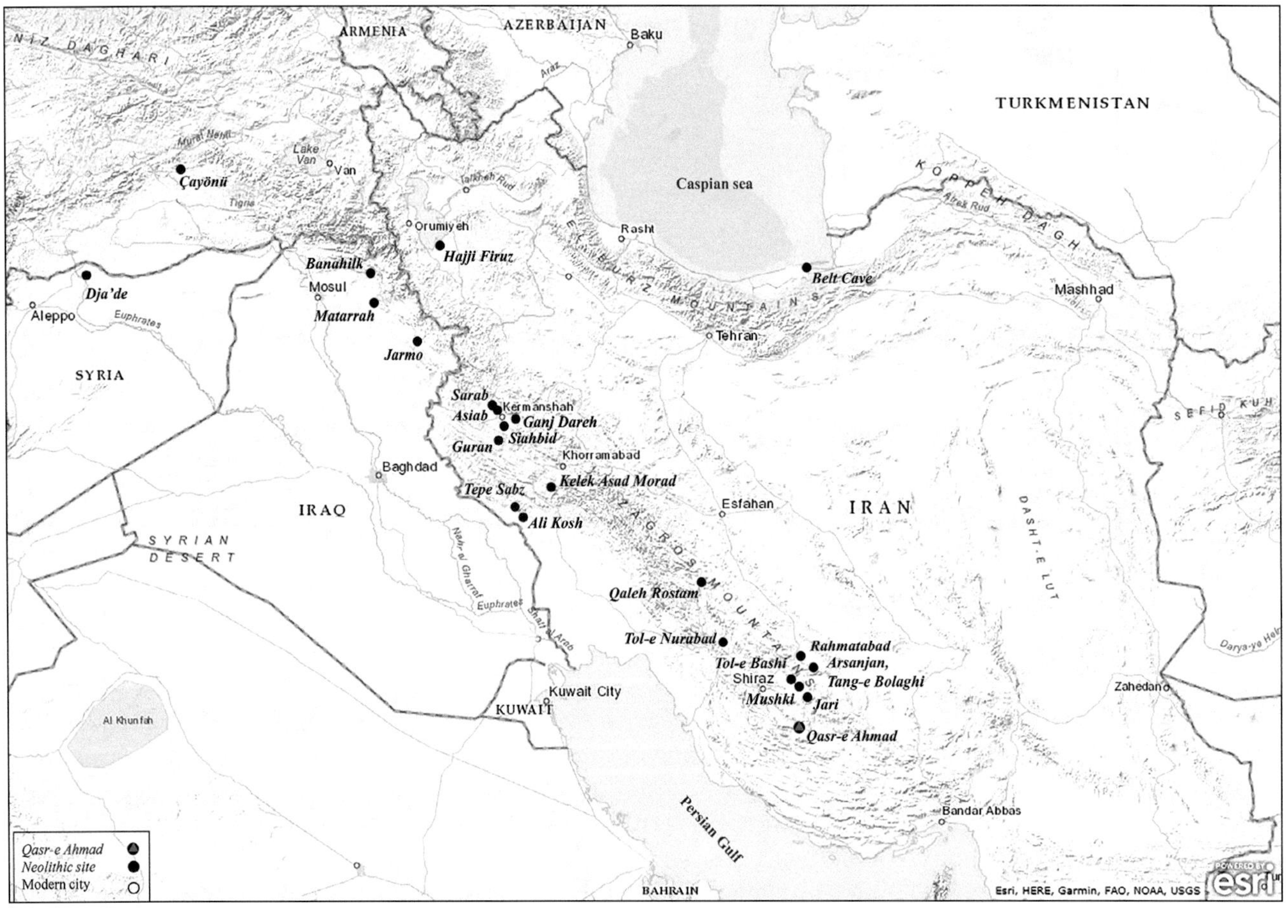

Fig. I. Location of the archaeological sites mentioned in the text.

Fig. 2. A view of Qasr-e Ahmad along the Qara Aghaj River.

The site

Qasr-e Ahmad is located along the Qara Aqhaj River near the village of Kavar in Fars province (Fig. 2). In 2003 a survey was conducted at the site by Kamyar Abdi and Reinhard Bernbeck to further archaeological investigations. For more precision on the absolute chronology of the site, a program of radiocarbon dating was initiated by M. Mashkour, H. Azizi Kharanaghi and A. Zazzo[1]. The radiocarbon dating of a sheep bone from this survey operation provided a date around 7900 BP (end of the 8th mill. BC-1st quarter of the 7th mill. cal. BC) (Mashkour unpublished data), drawing attention to the importance of this Early Neolithic settlement (Bernbeck *et al.* 2006). Unfortunately, as a result of recent bulldozing for irrigation and agricultural activities, the preserved evidence at the site only dates from the Neolithic period (Ibid.).

In winter 2011, the site was excavated during one campaign by H. Azizi Kharanaghi. The site covers an area of ca. 6.5 ha and stands at 1,515 m above sea level. Three trenches were excavated. Trench A (4x2 m) is located in the center of the mound. Trench B (2x2 m) is located in the eastern part of the site. Trench C is a one-meter-wide step trench on the northern slope of the mound (Azizi Kharanaghi *et al.* 2012).

1 Qasr-e Ahmad radiocarbon dates were performed by Antoine Zazzo at the Natural History Museum of Paris C^{14} laboratory and supported by the personal grant of Marjan Mashkour (UMR 7209 CNRS/MNHN) and will be soon published in the complete excavation report.

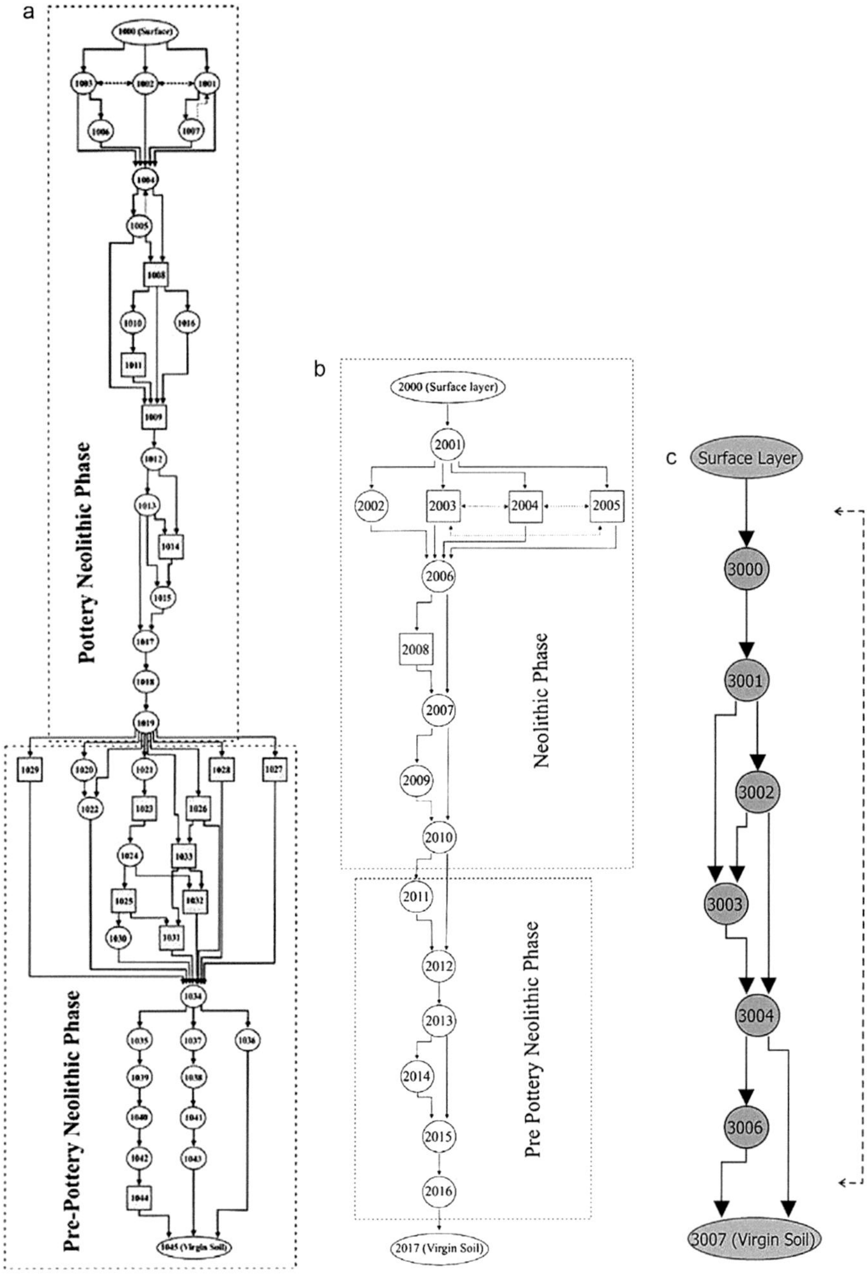

Fig. 3. Harris Matrix diagram of Trenches A (a), B (b), and C (c).

Among the 45 contexts from Trench A, 20 contexts belong to the Pottery Neolithic period (PN) and the rest to the Pre-Pottery Neolithic Period (PPN). Twelve out of 18 contexts in Trench B are attributed to the PPN. All seven contexts of Trench C belong to the PN. The PPN phase is situated immediately below the Pottery Neolithic phase, seemingly without any interruption (Figs. 3a, b, and c). The PPN and PN phases in the Fars region can respectively be dated to around the mid-8th and mid-7th millennia BC (Ibid.). The potteries from Qasr-e Ahmad are different from the other Neolithic sites in the Marvdasht Plain in terms of form and decoration. They can be divided to two phases: The earlier phase, which is represented by simple red ware bowls (formative mushki) and the later phase with potteries decorated with geometric motifs.

Neolithic subsistence economies in Southern Zagros

Evidence for Neolithic subsistence practices is limited in southern Zagros, particularly in the Fars province. A research gap is evident between the PPN sites from the mid-8th mill BC and the PN sites of the late 7th and 6th mill BC (Weeks 2013). Archaeozoological studies are very rare. Except for Tang-e Bolaghi, (Hongo and Mashkour 2008), Arsanjan A5 Cave (Mashkour and Khazaeli 2012 unpublished report) and the PPN and PN sites of Tal- i Mushki, Jari A and B (Payne 1991; Mashkour *et al.* 2006), Tol-e Nurabad (Mashkour 2009), Tol-e Bashi (Mashkour and Bailon 2010) and Rahmat Abad (Davoudi *et al.* in prep), there is no other available material for understanding the prehistoric subsistence economy in this area. However, the strategic position of the Zagros region, between the Fertile

Table 1. Faunal spectra of Qasr-e Ahmad by means of Number of Remains.

Taxa	*NR*	*Weight (g)*	*NR%*	*Weight%*
Bos cf. *Primigenius* (cattle/aurochs)	17	656	0.9	7.8
Capra hircus (goat)	292	2025	16.3	24.1
Capra cf. *aegagrus* (goat/wild goat)	28	364	1.6	4.3
Ovis cf. *aries* (sheep/wild sheep)	31	328	1.7	3.9
Caprinae (sheep/goat)	1369	4901	76.4	58.3
Gazella subgutturosa (goitered gazelle)	4	13	0.2	0.2
Sus scrofa (wild boar)	7	49	0.4	0.6
Lepus capensis (hare)	2	2	0.1	0
Rodentia	11	1	0.6	0
Alectoris chukar (chukar partridge)	4	3	0.2	0
Testudo graeca (spur-thighed tortoise)	19	51	1.1	0.6
Mollusk	8	16	0.4	0.2
Total Identified taxa	**1792**	**8409**	**27**	**45**
Large Mammal	10	54	0.2	0.3
Medium Mammal	9	86	0.0	0.5
Small Ruminants	4741	10057	71.2	53.7
Unidentified	107	126	1.6	0.7
Total Unidentified	**4867**	**10323**	**73**	**55**
Grand Total	**6659**	**18732**	**100**	**100**

Table 2. Taphonomic summary of the vertebrate remains.

Taphonomic traces	*PPN*	*PN*
Burnt fragments	19%	21%
Cut and chop marks	2%	1.6%
Tool manufacturing traces	0.3%	0.3%
Carnivore punctures	0.4%	1.7%
Rodent gnawing marks	0.2%	0.06%

Crescent and the Indus valley, which are important zones for early domestication (Hongo and Mashkour 2008), makes Qasr-e Ahmad an interesting site for studying the domestication process of caprines.

The region of Qasr-e Ahmad is located in lowlands surrounded by piedmonts where we expect to find small and medium-sized wild ungulates such as wild sheep (*Ovis orientalis*), wild goat (*Capra aegagrus*), gazelles (*Gazella subgutturosa*), and wild boar (*Sus scrofa*) in a variety of ecotones (Ziaei 1996; Etemad 1985). Among the large herbivores, hemiones (*Equus hemionus*) and red or fallow deer (*Cervus elaphus/ Dama mesopotamica*) can also live in this region. We could expect bird species to be exploited on account of the well-watered basin around Persepolis and Pasargadae (Djamali *et al.* 2009). Botanical studies (performed at the Archaeometry Laboratory of Tehran University) suggest warm and wet conditions in Qasr-e Ahmad with steppe vegetation, open tree and shrub growth. The assemblage provides evidence for both wild and domesticated wheat and barley, as well as wild/weedy-growing plants and fruits (mainly pistachio) (G, Ahadi, University of Tübingen personal communication).

Methods

The faunal assemblage for this study was collected by hand. The remains were analysed at the Archaeometry Laboratory of Tehran University in 2014. The osteological reference collection of the laboratory was used, including wild and domestic herbivores present in the region of study, as well as anatomical and osteological atlases (Barone 1986; Pales and Garcia 1981; Boessneck *et al.* 1964; Clutton-Brock *et al.* 1990; Zeder and Lapham 2010; Halstead *et al.* 2002). In this study, all the faunal remains were counted and weighed to the nearest 0.1 g. Due to the high level of fragmentation, the NR (Number of Remains) is used as the basis of quantification. The specimens were measured following the protocol suggested by Von den Driesch (1976).

In order to address the question of the wild or domestic status of caprines, size indexing and the normalizing method following Meadow (1999), improved from Uerpmann (1979) were carried out. The standard measurements are based on the measurements of Field Museum wild goats (*Capra aegagrus*) and wild sheep (*Ovis orientalis*) published in Uerpmann and Uerpmann (1994).

Another proxy for approaching the question of domestication, as demonstrated by Helmer (1992), and Zeder and Hesse (2000), is the use of kill-off patterns. Here we used the tooth wear method based on Payne (1973) with further developments introduced by Helmer (1992; 1995; 2000a; 2000b), Helmer and Vigne (2004) and Vigne and Helmer (2007). The fusion data were obtained from Zeder 2006. The survivorship of sheep and goats based on their long-bone fusion stage is assessed following Zeder (2006).

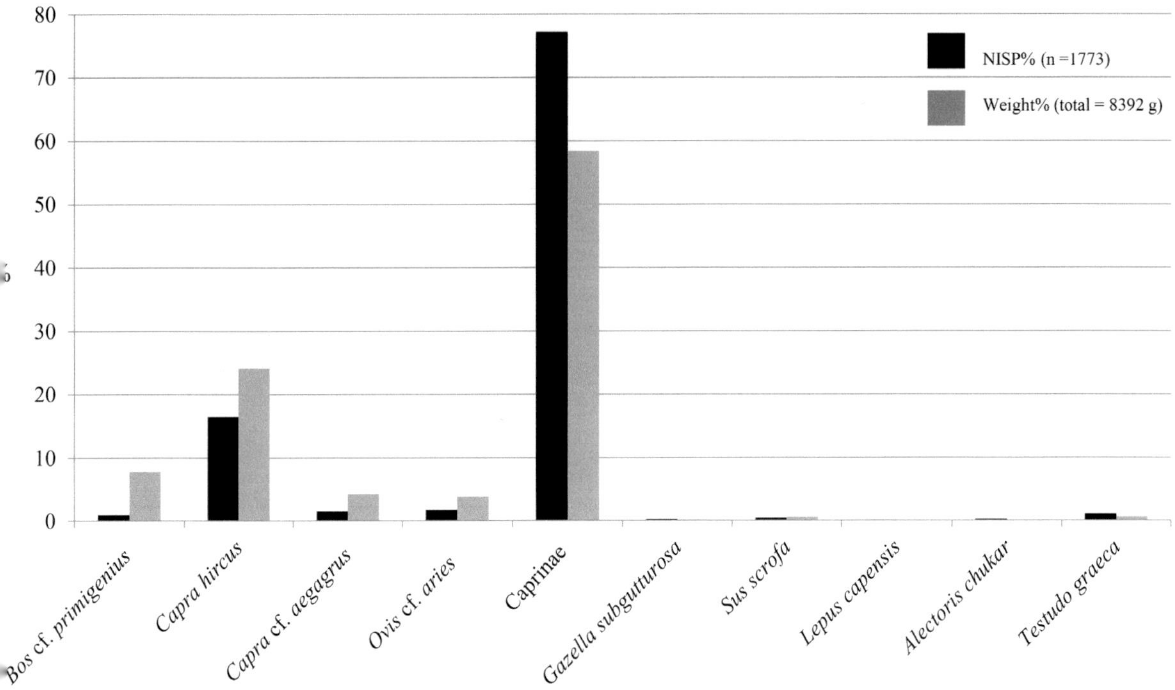

Fig. 4. Distribution of identified species by means of NISP percentage (Number of Identified Specimens).

Results

Nearly 7,000 specimens (NR) were studied (Table 1), out of which only 27% could be identified to the species level. The rest of the assemblage was divided into three size categories: large (size of cattle and equids), medium mammals (size of red deer and wild boar), and small ruminants (size of sheep, goat and gazelle). A few bird bones (n =4) were present among the remains, as well as 11 remains of intrusive rodents, such as the Indian gerbil (*Tatera indica).* Nineteen land tortoise shells (*Testudo graeca*) and eight mollusks were also found.

The considerable number of burnt bones from all three trenches is noteworthy (Table 2). These comprise about 19% of the PPN and 21% of the PN specimens. In total, 109 cases of human modification including cut marks are present, 81 of which belong to the PN. In addition, out of 49 cases of worked bones, 42 belong to the PN phases. Caprine long bones are the main raw material for manufacturing these worked bones.

Carnivore damage is common in both PPN and PN assemblages as well as rodent gnawing marks. All observations indicate that the bones were exposed before being buried. Pathological traces are observed on less than 1% of the specimens.

Faunal composition

Goat, including wild and domestic specimens (*Capra aegagrus* and *Capra hircus*)**,** is the most commonly exploited taxon in Qasr-e Ahmad (Fig. 4). It represents 17% and 19% of the identified specimens in the PPN and PN phases respectively. No major change can be detected between the PPN and PN faunal exploitation patterns.

Sheep (*Ovis* cf. *aries*) are represented in low proportions; sheep remains do not exceed 2% of the identified specimens. Cattle Bovines (cf. *Bos primigenius*) are even less common than sheep, represented by c. 1% of the identified specimens.

Only a few gazelle bones were identified in the PN assemblage, including two first phalanges, one astragalus, and one radius, accounting for less than 1% of the identified specimens. Gazelle remains could not be identified down to the species level due to the lack of diagnostic parts such as horn cores. Today, the goitered gazelle (*Gazella subgutturosa*) is present in Fars and is the most widespread species on the Iranian Plateau. This animal is a common component of the late Pleistocene and early Holocene assemblages of Fars (Mashkour *et al.* 2006; Hongo and Mashkour 2008).

Wild boar *(Sus scrofa)* is generally rare in the prehistoric faunal assemblages of the Iranian Plateau (Mashkour 2006). Only one 3rd phalanx, two isolated upper jaw teeth and one upper jaw bone from the PPN phase, together with a fused tibia and a frontal part of a cranium from the PN phase, were identified in our assemblage. It was not possible to take any measurements from these suid bones.

The inhabitants of Qasr-e Ahmad also exploited other animals, such as the hare (*Lepus capensis*), which is only present in the PN phase, and the chukar partridge (*Alectoris chukar*) during both phases of the Neolithic. The land tortoise (*Testudo graeca*) is represented by plastron and carapace fragments during both phases of the Neolithic. This could have

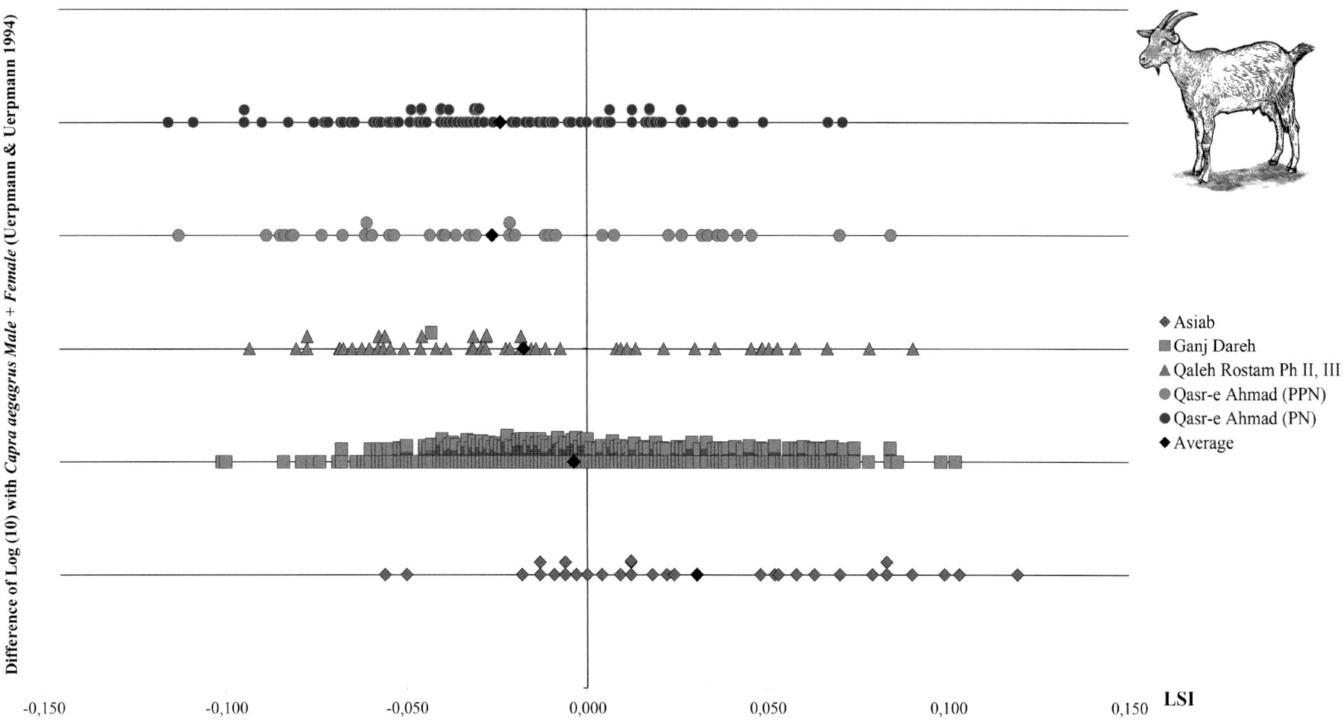

Fig. 5. Size distribution of goat population of Qasr-e Ahmad in comparison to some other Neolithic sites in Zagros.

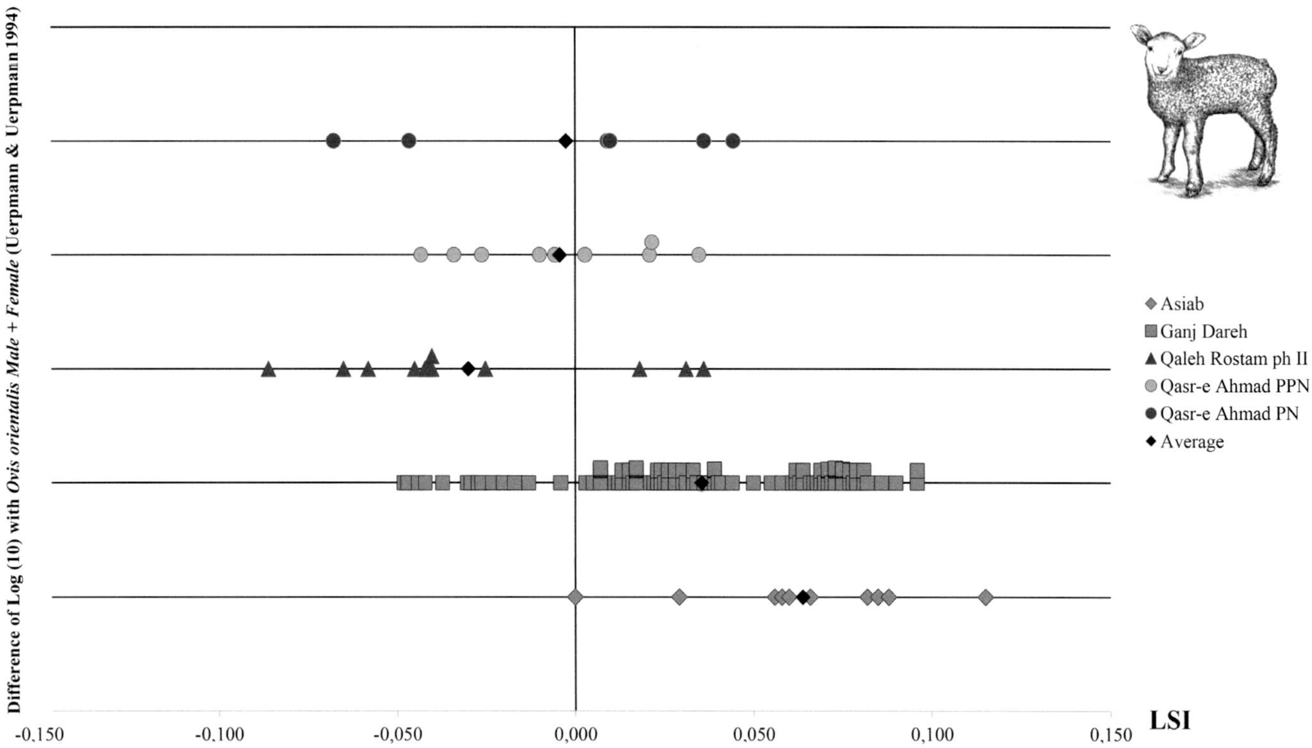

Fig. 6. Size distribution of sheep population of Qasr-e Ahmad in comparison to some other Neolithic sites in Zagros.

been consumed as a supplementary food source, used for its shell, or it could have been a commensal animal at the settlement.

Eight mollusk shells (four from the PPN and four from the PN) were recovered. In addition to the freshwater mollusks which probably originate from the vicinity of the site, *Conus* type shell fragments originating from the Persian Gulf are also present.

Finally, although no sieving was performed during the excavation campaign, some intrusive micro-vertebrate bones were present in the assemblage. These microvertebrate bones were identified as the Indian gerbil (*Tatera indica*), which is a large rodent that burrows galleries. The bones are probably not contemporaneous with the Neolithic occupation of the site.

Table 3. The *t test* and *z test* results for the comparison of sheep size from Qasr-e Ahmad with three other Neolithic sites in the region.

Site	Asiab		Ganj Dareh		Qaleh Rostam		Qasr-e Ahmad	
	t test	z test	t test	z test	t test	z test	t test	z test
Asiab	*	*	0.0105	0.0155	0.0000	0.0003	0.0000	0.0004
Ganj Dareh	*	*	*	*	0.0000	0.0002	0.0068	0.0052
Qaleh Rostam	*	*	*	*	*	*	0.0711	0.1021
Qasr-e Ahmad (PPN. PN)	*	*	*	*	*	*	*	*

Table 4. Survivorship data of sheep according to the long-bone fusion stages.

		PPN		PN	
Epiphyses (sheep)	**Age (months)**	**Unfused**	**Fused**	**Unfused**	**Fused**
Proximal radius	3-6		3		7
Distal humerus, distal scapula	6-12	1	2		1
Proximal first phalanx, proximal second phalanx	12-18				1
Distal metapodia	18-30				3
Proximal femur, proximal ulna, distal radius	30-40		1		2
Total specimens no= 21					

The morphology and age of the main ungulates

Horn core morphology

The presence of several goat horn cores in the Qasr-e Ahmad assemblage was very useful for examining the morphological changes in this species during the transition to management and possibly domestication. In the faunal assemblage of Ali Kosh and Tepe Sabz, abundant horn cores with quadrilateral cross-sections were identified as wild goats in contrast to domestic herds with lozenge-shaped horn cores (Hole *et al.* 1969). Two goat horn cores in the assemblage clearly show the presence of both wild and domestic individuals in our PN assemblage. Unfortunately, there is no horn core from the PPN phase of the site. Furthermore, there are no well-preserved sheep horn cores or hornless frontal bones from the site to assess the domestication status of the sheep.

The size of goat and sheep

The Logarithmic Size Index (LSI) graphs illustrate the change in size in sheep and goat between the PPN and PN at Qasr-e Ahmad (Figs. 5 and 6). For a better understanding of the chronological and regional evolution of goat size, the LSI profiles of Qasr-e Ahmad are compared to those from Neolithic sites in Central Zagros. These are the sites of Asiab (Bökönyi 1977), Ganj Dareh (Hesse 1978), and Qaleh Rostam (Daujat *et al.* 2016; Daujat and Mashkour 2017). The site of Qale Rostam, with three occupation phases during the Neolithic period (I-III), is the best comparison for Qasr-e Ahmad, as the settlements are almost contemporaneous, ranging in time from the start of the 7th millennium BC to the second half of the 7th millennium BC, i.e. the transition from the PPN to the PN (See Daujat and Mashkour 2017 for Qaleh Rostam,). This site is also suitable for comparison because it is situated between the central Zagros sites and Qasr-e Ahmad, in lower Zagros. Thus, any comparisons between the faunal remains from these two settlements can be informative regarding the dispersal and status of caprines, particularly sheep, towards the southern Zagros region.

The evolution of the size of goat

Extensive studies of the caprine assemblage in Ganj Dareh (Kermanshah) by Zeder and Hesse (2000) have shown that size reduction cannot be used as the only criterion for documenting domestication as apparent size change can be mimicked by the sex-selective culling of these sexually dimorphic animals. Size modification accompanied by culling data can give a better picture of the domestication status of the species. However, the Qasr-e Ahmad assemblage contains a limited number of measurable bones, and does not allow for sex ratio evaluation. Therefore, both size and culling data were used for the assessment of goat domestication status.

The goats of Qasr-e Ahmad are relatively large but compared to the average size of the goats in Asiab, they are small. The population of Qasr-e Ahmad is smaller than Ganj Dareh and even Qaleh Rostam. However, the presence of wild specimens is evidenced by high LSIs, representing large individuals (Fig. 5). All these specimens have lower LSIs than the goat average at comparable Neolithic sites.

The evolution of the size of sheep

The earliest evidence for sheep herding in Southwest Asia, with a clear shift in demographic profiles and body-size, comes from the Pottery Neolithic assemblages of Tepe Guran, Sarab, and Jarmo (Zeder 2008).

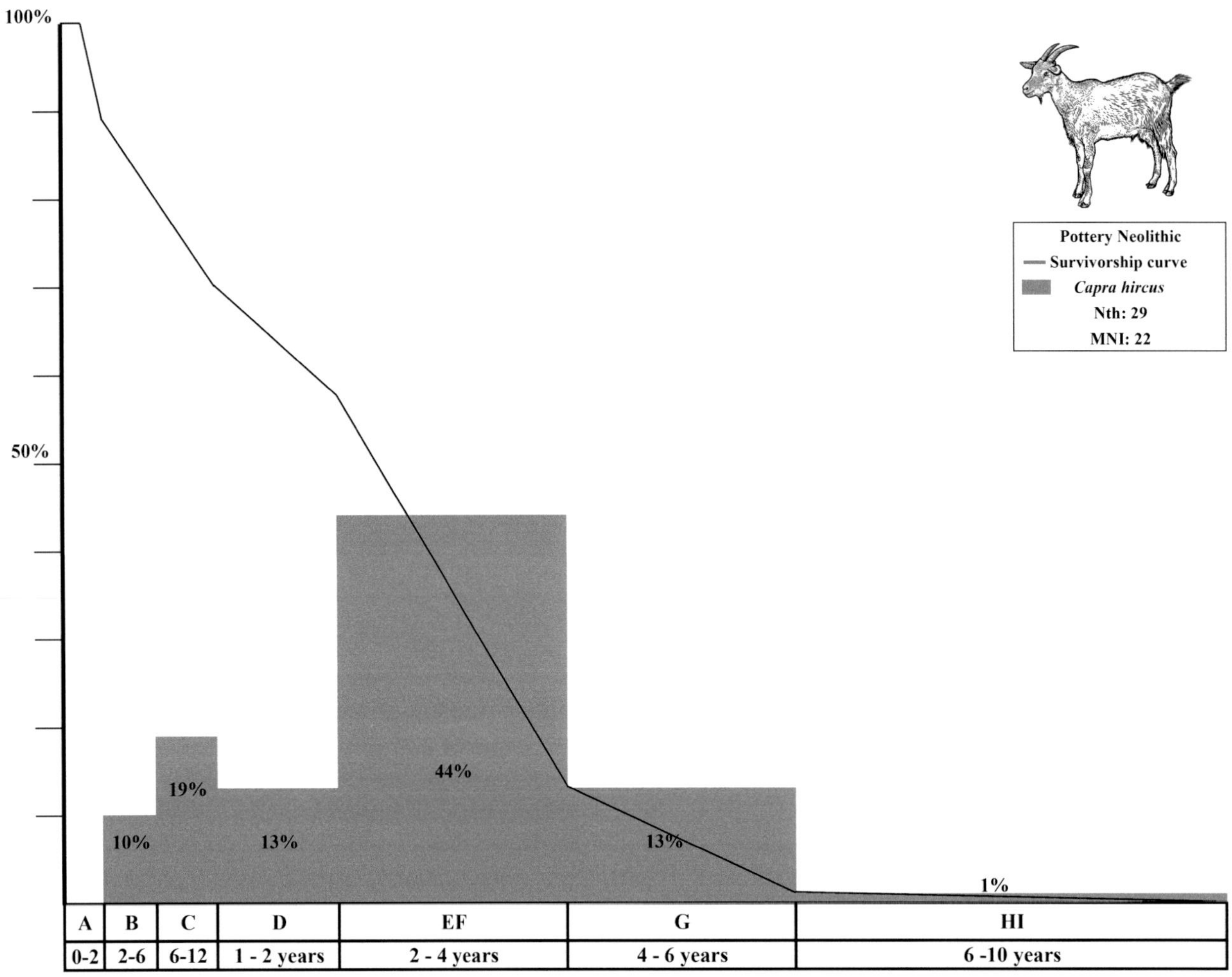

Fig. 7. Goat kill-off pattern during the pottery Neolithic.

Sheep measurements are less abundant than goat measurements in Qasr-e Ahmad and only fifteen could be measured. The LSI analyses (Fig. 6) and *t and z tests* (Table 3) for this assemblage indicate that sheep from Qasr-e Ahmad are significantly smaller than sheep in Asiab, but are similar in size to the sheep in Ganj Dareh. However, no significant difference between the average size of sheep in Qasr-e Ahmad and Qaleh Rostam has been observed.

In total, only 21 specimens provided information for epiphyseal fusion data (Table 4). However, to a great extent, this can be an artifact of hand collection methods during the excavation and presents limitations for identifying unfused long bones of caprines into separate species of sheep and goat. Except for one distal humerus, no specimen indicating the culling of the young has been identified in either occupation phase of the settlement. The present data suggest the culling of the majority of sheep in the settlement well after adulthood. However, due to the paucity of remains, any further interpretation is impossible.

Kill-off patterns for goat

Due to the limited number of tooth remains for caprine, we could only assess the kill-off pattern for the goat remains for the PN phase (Fig. 7).

A total of 29 teeth were used in this analysis, corresponding to 22 individuals (MNI) after applying the combination of tooth wear and laterality. Interestingly, no remains could be allocated to Payne's (1973) stage A (0-2 months). However, this may be due to the absence of sieving in the field or to other reasons related to traditional herding practices which will be discussed below. A small percentage (10%) of the animals were killed before 6 months (stage B). Culling peaked (19%) between 6-12 months (stage C). This means that 29% of the population were killed before the age of one year. A small percentage (13%) of the animals were killed between one and two years (stage D). The bulk of the remains (44%) belong to Stage EF corresponding to animals between their 2nd and 4th years. Animals older than four years of age are rather well represented among the goat remains (14%).

The present picture can be interpreted as a *B milk type* exploitation pattern according to the Vigne and Helmer (2007) model. In this model, a proportion of the very young animals are culled in order to control the suckling of milk for human exploitation. A high number of ewes are kept for their milk as well as

Table 5. Survivorship data of goat according to the long-bone fusion stages.

		PPN		PN	
Epiphyses (goat)	**Age (months)**	**Unfused**	**Fused**	**Unfused**	**Fused**
Proximal radius	3-6		2		9
Distal humerus, distal scapula	6-12		11		43
Proximal first phalanx, proximal second phalanx	12-18		31		30
Distal metapodia	18-30		9	9	10
Proximal femur, proximal ulna, distal radius	30-40	1	1	2	5
Total specimens no= 163					

their meat. They are then killed between the age of two and four when the peak time for lamb and milk productivity is over. The presence of animals between Stages B and D with a peak in Stage C can also be interpreted using the meat exploitation model. This age is the optimum weight time of the animals and a choice time for culling some of the young males in the herd for their meat (Ibid).

However, the high number of animals culled during Stage G is an interesting aspect of the kill-off pattern, and requires further discussion. Although there is no archaeological evidence related to textile production from goat hair in this site, such as loom weights, this practice cannot be ruled out. All of the abovementioned patterns could fit well with mixed production, including milk and hair, in addition to meat. In summary, milk production is expressed by the kill-off of young males and older ewes, and felt production by the presence of older animals.

The kill-off pattern seems to represent all seasons. As all age classes are present, the assemblage does not seem to reflect seasonal movements. A truncated harvest profile with noticeable proportions of one group and the absence of other slaughtering groups was demonstrated in seasonal settlements of south France and interpreted as an effect of seasonal herd movements (Helmer *et al.* 2005b). The presence of all ages in the culling profile of goats may point to the use of Qasr-e Ahmad as a year-round settlement.

The same exploitation pattern (meat and milk) has also been tracked in phase III-II of Qaleh Rostam for both sheep and goats (Daujat and Mashkour 2017), with however less emphasis on culling during the E-F Stage.

The fusion data of post-cranial goat bones show that about 10% of the caprines in the PPN and PN phases died prematurely. This mirrors the observations on teeth.

Unlike sheep remains, the goat yielded a higher number of long bones (no=163) for performing epiphyseal fusion data (Table 5). There is only one unfused specimen from the PPN and eleven from the PN. The fragile structure of the unfused bones decreases their chance of survival in archaeological contexts. However, the higher number of unfused specimens from the PN may be a result of the predominance of fauna from this phase in the assemblage or an indication of a different culling pattern during this phase, killing goats before the age of 18-30 months. During the PPN phase, the peak of slaughter is after 12-18 months, while a high proportion of goats were killed after the first year of their life in the PN phase. Unfortunately, the present data cannot provide further information on the killing pattern in Qasr-e Ahamd.

The size of cattle

Qasr-e Ahmad presents one of the earliest datasets for early cattle in southwest Iran, after Tal-i Mushki where cattle remains were also attested (Mashkour *et al.* 2006; Mashkour *et al.* unpublished data), but also in limited numbers. In our assemblage, the only measurable bones were two distal humeri from the PN phase. They measured 102.6 mm and 95.8 mm for the distal trochlear breadth (BT). In Helmer and Gourichon (2017), a mean of 90.98 mm for this measurement was suggested for the aurochs in Neolithic Southwest Asia. Compared to this referential measurement, the two specimens form Qasr-e Ahmad clearly fall within the size range of aurochs.

Pig in Qasr-e Ahmad

The earliest evidence for the beginning of pig management in Southwest Asia comes from Çayönü in the Upper Tigris catchment basin during the 9th millennium BC (Hongo and Meadow 1998), displayed by increasingly younger culling profiles and body size reduction throughout time. However, evidence from Zagros suggests a late adoption of pig husbandry in this region (Flannery 1983; Mashkour 2006; Price and Arbuckle 2015). The earliest evidence for domestication displayed by phenotypic changes along with changes in kill-off patterns of the pig population in Zagros comes from the Pottery Neolithic phases of Jarmo (early 7th millennium BC), or perhaps earlier during the Pre-Pottery occupation (mid to late 8th millennium BC) (Flannery 1983;

Stampfli 1983; Price and Arbuckle 2015). With a long delay of about one millennium, Matarrah during the late 7th millennium BC, Banahlik, Hajji Firuz, Siahbid during the 6th millennium BC, and Belt Cave (Pottery Neolithic phases) provide the next earliest evidence for pig domestication in the Zagros region (Coon 1951; Bökönyi 1977; Stampfli 1983; Laffer 1983; Meadow 1983). However, studies in the southern part of Zagros suggest an even later adoption of pig domestication for this area during the Chalcolithic period (Flannery 1983).

The paucity of pig remains from Qasr-e Ahmad, and their poor preservation do not allow for detailed analyses, yet their minor presence in the assemblage (six specimens) indicates their low contribution to the Neolithic subsistence economy, fitting the bigger picture of the late adoption of domesticated pigs in Iran (Mashkour 2006).

The arid climate and steppe vegetation have been suggested as the main factors making southern Zagros an unfavorable region for pig management, particularly for extensive husbandry techniques (Price and Arbuckle 2015). Although the botanical studies in Qasr-e Ahmad point to relatively warm and wet conditions with steppe vegetation for the region during the Neolithic, the unsuitable environment for wild boar populations, as well as possible cultural preferences, cannot be excluded for the late appearance of domestic pig in the subsistence economy of Zagros.

Discussion and conclusion

The faunal study of the early Neolithic site of Qasr-e Ahmad bone assemblage is an important step towards enhancing our understanding of the three major Neolithic issues in the Zagros Mountains: the main orientation of animal husbandry, the domestication status of the ungulate mammals and their spread through Southwest Asia, as well as the settlement patterns of Neolithic sites in the Zagros region. The following conclusions are possible at this stage of research:

No major changes took place in terms of the general trends of animal exploitation between the PPN and PN. In both phases, the primary animals exploited were caprines and more specifically goats, while sheep, cattle and pigs played a limited role in subsistence. The same situation is observed in the faunal remains of Tal-i Mushki in Fars where sheep and cattle are very poorly represented (Mashkour *et al.* 2006). The semi-arid climate of the region, as well as the hilly topography of the Zagros Mountains, are not propitious to aurochs and sheep populations. The delayed arrival of both domestic sheep and cattle in the region (7th millennium for sheep and 6th millennium for cattle) concurs with this picture (Zeder 2008).

Other animals were also exploited, such as gazelles, hares and possibly the land tortoise. The limited number of bird remains suggests that the wetlands were not extensively used for hunting birds. The subsistence economy was specialized in pastoralism, focusing on caprines, especially goats.

The presence of marine shells is interesting to note. This exogenous material is commonly encountered during the Neolithic resulting from trade and exchange. Another traded material during this period is obsidian (Abdi 2004), which was also found in the site and is currently undergoing analysis.

The metric analysis of goat remains and their comparison with other Neolithic sites in the Zagros Mountains show that the majority of the goats present in the assemblage were domesticated. The presence of horn cores with quadrilateral cross-sections as well as specimens in the wild goat size range suggest a wild status for a number of goat populations in Qasr-e Ahmad.

The issue of sheep domestication and spread in this area remains complicated and the assemblage from Qasr-e Ahmad is not sufficient to address this issue with any rigor. So far, a late domestication, no earlier than the Pottery Neolithic, around the 7th millennium BC, has been suggested for sheep in this region. One of the main hypotheses advocates the introduction of already domesticated sheep from the upper Euphrates and Tigris areas to Zagros (Zeder 2008). The rugged landscape of the Zagros Mountains is more favorable for goats, and probably is not very suitable for the sheep and aurochs populations. The absence of sheep and aurochs in the early Neolithic assemblages of the region backs up this assumption. In the earlier phases of Qaleh Rostam (I), sheep is absent and is only represented by a few remains during the later phases (II-III). Based on the size and the culling pattern, a domestic status has been allocated to the sheep from Qaleh Rostam (Daujat and Mashkour 2017). No kill-off pattern can be assessed for the assemblage from Qasr-e Ahmad yet the absence of any statistical difference in the size of sheep from Qaleh Rostam and Qasr-e Ahmad may constitute evidence for a domesticated status for the Qasr-e Ahmad sheep population. The presence of small sheep in Qasr-e Ahmad during both phases (PPN and PN) may imply that sheep spread between central Zagros and southern Zagros earlier than previously thought.

The presence of a very limited number of bovine remains and their large size is indicative of the limited exploitation of aurochs rather than cattle management or husbandry in Qasr-e Ahmad. On the Iranian Plateau, the history of cattle domestication has not yet been clearly documented contrary to the extensive studies which have addressed the initial domestication of cattle in the Upper Euphrates basin (Helmer and Gourichon 2017; Helmer *et al.* 2005a; Peters *et al.* 2005). The two sites with the earliest markers of cattle domestication in the Near East are Dja'de (Helmer *et al.* 2005a) and Çäyönu (Hongo *et al.* 2009; Bollongino *et al.* 2012). The paucity of cattle remains from the Early Neolithic sites in the Iranian plateau is the main obstacle for exploring the role of cattle in this area and testing present hypotheses. The low frequency of *Bos* remains prior to the 6th millennium BC in contrast to the sudden and dramatic increase of small-sized cattle during the Late Neolithic period (6th millennium BC) has been recorded in Central and Southern Zagros, Susiana plain, southeast and northern Iran. This distribution pattern has been suggested as an indicator for imported domesticated cattle from the adjacent Upper Euphrates basin to the Zagros region during the 6th millennium BC (Arbuckle *et al.* 2016).

Herding strategies are another important aspect in this assemblage. Studies of Near Eastern assemblages have revealed the prominence of mixed profiles during the early Neolithic (Helmer *et al.* 2007). Milk procurement can even be considered as one of the main motivations for the human domestication of animals as hunting does not provide this opportunity (Vigne and Helmer 2007). The goat kill-off pattern in Qasr-e Ahmad also points to milk exploitation, which is in line with the evidence from Qaleh Rostam, one of the earliest known cases for the Neolithic of southwest Iran (Daujat *et al.* 2016; Daujat and Mashkour 2017).

One of the long-lasting discussions for the prehistory of the Zagros Mountains and the Iranian plateau focuses on understanding Neolithic settlement patterns and land use. The mobility of human communities, as a response to environmental constraints, is a practice observable in variable forms in traditional communities, including nomadic and transhumant pastoralists, and has served as a reference model for several archaeologists for the interpretation of archeological material (Hole 1962; Hole *et al.* 1969; Hole and Neely 1979; Alizadeh 2004), although it is criticized by others (Potts 2014). Here we would like to examine this issue from our perspective and contribute to the debate. The location of Qasr-e Ahmad next to the permanent river of Qara Aghaj, as well as the abundance of pottery, lithics, and domesticated plants and animals are all features of the archaeological record in favor of interpreting this site as a permanently settled occupation. In Qasr-e Ahmad, two architectural structures were discovered in the 1023 and 1025 contexts of trench A (PPN phases), covered with ocher in some places. In addition, the location of the site, close to the river, could potentially facilitate farming. Wheat and barley cultivation, along with gathering wild fruit, was evidenced through botanical studies in Qasr-e Ahmad. Moreover, the kill-off pattern in Qasr-e Ahmad is not truncated, a criterion generally considered as an indication of the seasonal exploitation of animals (see Vigne and Helmer 2007). To answer this question more precisely, sequential isotopic analyses of carbon, oxygen and strontium isotopes from enamel are required for a better understanding of year-round herding practice in Neolithic Qasr-e Ahmad. The evidence for change in seasonal movements between summer and winter pasture lands can provide *in minima* evidence for seasonal herd mobility but not residency, which requires a more extensive and integral set of data, including landscape and site distribution analysis between lowlands and highlands with associated material culture and bioarchaeological material, in particular botanical and faunal remains. For the time being, based on the archaeological data and the analysis of the faunal remains, more elements point to a permanent settlement in Qasr-e Ahmad.

Acknowledgements

The authors would like to thank the members of the Bioarchaeology Laboratory of the Central Laboratory of the University of Tehran for their help and support, and in particular Dr. Haeedeh Laleh, head of the Laboratory, and Dr. Ahmad Aliyari, joint director of the Laboratory.

References

Abdi, K. 2004. Obsidian in Iran form the Epipalaeolithic Period to the Bronze Age. In, Stöllner, T., Slotta. R. and Vatandoust. A. R. (eds) *Splendor of Iran:* 148-153. Bochum.

Alizadeh, A. 2004. *The Origin of State Organizations in Prehistoric Highland Fars, Southern Iran, Excavations at Tall-e Bakun*. (Oriental Institute Publications 128). Chicago: University of Chicago, Oriental Institutes.

Arbuckle, B.S., Price, M. D., Hongo, H. and Öksüz, B. 2016. Documenting the initial appearance of domestic cattle in the Eastern Fertile Crescent (northern Iraq and western Iran). *Journal of Archaeological Science* 72: 1-9.

Azizi Kharanaghi, M.H., Khalouei, F., Khanipour, M., Khosravi, S., Razi, S., Komijani, S. and Tahmasebi, M. 2012. The First Season of Archaeological Excavation at Tell-e Qasr e Ahmad, Kavaar District, winter 2011, Fars. In, Moggadam, A. and Roustaei, K. (eds) *Abstracts Volume of the 11th Annual Symposium of Iranian Archaeology*. Tehran: Iranian Cultural Heritage Handicraft & Tourism Organisation- ICHHTO.

Barone, R. 1986. *Anatomie Comparée des Mammifères Domestiques*. Paris: Vigot.

Bernbeck, R., Abdi, K., and Gregg, M.W. 2006. A note on the Neolithic of the Qara Aghaj valley,

Fars province. *Archaeological Reports* 4: 27-36. Tehran: Iranian Center for Archaeological Research.

Boessneck, J., Müller, H. H. and Teichert, M. 1964. Osteologische Unterscheidungsmale zwischen Schaf (*Ovis aries Linne*) und Ziege (*Capra hircus Linne*). *Kühn Archiv* 78: 1-129.

Bökönyi, S. 1977. *The Animal Remains from Four Sites in Kermanshah Valley, Iran: Asiab, Sarab, Dehsavar and Siabid*. BAR, supplementary Series 34 ed. Oxford: Oxford University Press.

Bollongino, R., Burger, J., Powell, A., Mashkour, M., Vigne, J.-D. and Thomas, M.G. 2012. Modern taurine cattle descended from small number of Near-Eastern founders. *Molecular Biology and Evolution* 29(9): 2101–2104.

Broushaki, F., Thomas, M.G., Link, V., López, S., Van Dorp, L., Kirsanow, K., Hofmanová, Z., Diekmann, Y., Cassidy, L.M., Díez-del-Molino, D., Kousathanas, A., Sell, C., Robson, H., Martiniano, R., Blöcher, J., Scheu, A., Kreutzer, S., Bollongino, R., Bobo, D., Davudi, H., Munoz, O., Currat, M., Abdi, K., Biglari, F., Craig, O. E., Bradley, D. G., Shennan, S., Veeramah, K. R., Mashkour, M., Wegmann, D., Hellenthal, G. and Burger, J. 2016. Early Neolithic genomes from the eastern Fertile Crescent. *Science* 353(6298): 499-503.

Clutton-Brock, J., Dennis-Bryan, K., Armitage, P.L. and Jewell, P.A. 1990. Osteology of the Soay sheep. *Bulletin of British Museum of Natural History* 56(1): 1-56.

Coon, C.S. 1951. *Cave Explorations in Iran*. Philadelphia: University Museum Press.

Daujat, J., Mashkour, M., Emery-Barbier, A., Neef, R. and Bernbeck, R. 2016. Qale Rostam: Reconsidering the "Rise of a Highland Way of Life". An integrated bioarchaeological analysis. In, Roustaei, K., and Mashkour, M (eds) *The Neolithic of the Iranian Plateau. Recent Research and Prospects* Studies in Early Near Eastern Production, Subsistence, and Environment (SENEPSE series): 108-136. Berlin: Ex-oriente.

Daujat, J. and Mashkour, M. 2017. Faunal Remains from Middle Neolithic Site of Qaleh Rostam. In, Mashkour, M. and Beech, M. K. (eds) *Archaeozoology of the Near East* 9, *Proceedings of the 2008 Al Ain-Abu Dhabi conference*: 41-58. Oxford: Oxbow Books.

Davoudi, H., Khazaeli, R., Azizi Kharanaghi, M.H. and Mashkour, M. In prep. *Domestication Process in Southwest of Iran, The Case of Tepe Rahmat Abad*. Manuscript in preparation.

Djamali, M., Beaulieu, J.-L., Miller, N.F., Andrieu-Ponel, V., Ponel, P., Lak, R., Sadeddin, N., Akhani, H. and Fazeli, H. 2009. Vegetation history of the SE section of the Zagros Mountains during the last five millennia; a pollen record from the Maharlou Lake, Fars Province, Iran. *Vegetation History and Archaeobotany* 18: 123-136.

Etemad, A. 1985. *The Mammals of Iran*. Volume 2, National society for the conservation of natural resources and human environment, Tehran. (In Persian)

Flannery, K.V. 1969. Origins and ecological effects of early domestication in Iran and the Near East. In, Ucko, P. J. and Dimbleby, G. W. (eds) *The Domestication and Exploitation of Animals,* 73-100. London, Duckworth.

Flannery, K.V. 1983. Early pig domestication in the Fertile Crescent: A retrospective look. In, Young, T.C., Smith, P.E.L. and Mortensen, P. (eds) *The Hilly Flanks: Essays on the Pre-history of Southwestern Asia. Studies in Ancient Oriental Civilization 36*: 163–188. Oriental Institute: Chicago.

Halstead, P., Collins, P. and Isaakidou, V. 2002. Sorting the Sheep from the Goats: morphological distinction between the mandibles and mandibular teeth of adult *Ovis* and *Capra*. *Journal of Archaeological Science* 29: 545-553.

Helmer, D. 1992. *La Domestication des Animaux par les Hommes Préhistoriques*. Paris: Masson.

Helmer, D. 1995. Biometria i arqueozoologia a partir d'alguns exemples del Proxim Orient. *Cota Zero* 11: 51-60.

Helmer, D. 2000a. Discrimination des genres *Ovis* et *Capra* à l'aide des prémolaires inférieures 3 et 46. L'exemple de Dikili Tash (Macédoine - Grèce). In, *La gestion démographique des animaux à travers le temps – Animal management and demography through the ages. Colloque international de Turin, 16-18 septembre 1998. Anthropozoologica* 31: 29-38.

Helmer, D. 2000b. Étude de la faune mammalienne d'El Kown 2. In, Stordeur, D. (eds) *El Knowm 2, une île dans le Désert. La fin du Néolithique Précéramique dans la Steppe Syrienne*: 233-264. Paris: CNRS Éd.

Helmer, D. and Vigne, J.-D. 2004. La gestion des cheptels de caprinés au Néolithique dans le Midi de la France. In, Bodu, P. and Constantine, C. (eds) *Approches fonctionnelles en Préhistoire. Actes XXVe Congr. Préhist*: 397-407, *Fr., Nanterre, 24-26 nov. 2000*. Paris: Soc. Préhist. Fr. Éd.

Helmer, D., Gourichon, L., Monchot, H., Peters, J. and Sana Segui, M. 2005a. Identifying early domestic cattle from Pre-Pottery sites on the Middle Euphrates using sexual dimorphism. In, Peters, J. and Helmer, D. (eds) *The First Steps of Animal Domestication:* 86-95. Oxford: Oxbow Books.

Helmer, D., Gourichon, L., Sidi Maamar, H. and Vigne, J.-D. 2005b. L'élevage des caprinés néolithiques dans le Sud-Est de la France: saisonnalité des abattages, relations entre grottes-bergeries et sites de plein air. *Anthropozoologica* 40(1): 167-190.

Helmer, D., Gourichon, L. and Vila, M. 2007. The development of the exploitation of products from *Capra* and *Ovis* (meat, milk, and fleece) from the PPNB to the Early Bronze in the northern Near East (8700 to 2000 BC cal.). *Anthropozoologica* 42(2): 41-69.

Helmer, D. and Gourichon, L. 2017. The fauna of Tell Aswad (Damascus, Syria), early Neolithic level Comparison with northern and southern Levant sites. In, Mashkour, M. and Beech, M. K. (eds) *Archaeozoology of the Near East* 9, *Proceedings of the 2008 Al Ain-Abu Dhabi Conference*: 23-40. Oxford: Oxbow Books.

Hesse, B. 1978. *Evidence for husbandry from the early Neolithic site of Ganj Dareh in Western Iran*. PhD. Columbia University.

Hole, F. A. 1962. Archaeological survey and Excavations in Iran, 1961. *Science, New series* 137 (3529): 524-526.

Hole, F. and Neely, J. 1979. Rediscovering the Past in the Present: ethnoarchaeology in Luristan, Iran. In, Kramer, C. (ed.) *Ethnoarchaeology. Implications of Ethnography for Archaeology*: 193-216.

Hole, F.A., Flannery, K.V. and Neely, J.A., 1969. *Prehistory and Human Ecology on the Deh Luran Plain*. Ann Arbor: Museum of Anthropology, University of Michigan, Memoirs 1.

Hongo, H. and Mashkour, M. 2008. Faunal Remains from TB75. In, Tsuneki, A., and Zeidi, M. (eds) Tang-e Bolaghi: *The Iran-Japan Archaeological Project for the Sivand Dam Salvage Area*: 133-148. Tehran and Tsukuba: Iranian center for Archaeological Research and Department of Archaeology, University of Tsukuba.

Hongo, H. and Meadow, R.H. 1998. Pig exploitation at Neolithic Çayönü Tepesi (Southeastern Anatolia). *Ancestors for the Pigs*. Nelson, S. (ed.) MASCA Research Papers in Science and Archaeology 15: 77-98. Philadelphia: University of Pennsylvania Museum of Archaeology and Anthropology.

Hongo, H., Pearson, J., Öksük, B. and IIgezdi, G. 2009. The process of ungulate domestication at Cayönü, Southeastern Turkey: a multidisciplinary approach focusing on *Bos* sp. and *Cervus elaphus*. *Anthropozoologica* 44: 63-73.

Laffer, J. P. 1983. The faunal remains from Banahilk. In, L.S., Braidwood, R.J., Howe, B., Reed, C.A. and Watson, P.J. (eds) *Prehistoric Archeology along the Zagros Flanks. Braidwood*: 629-648. Chicago: Oriental Institute of the University of Chicago.

Mashkour, M. 2006. Boars and pigs: a view from the Iranian Plateau. In, Lion, B. and Michel, C. (eds) *De la domestication au tabou: Le cas des suidés dans le Proche-Orient ancient*: 155-163. Paris: Travaux de la Maison d'Archéologie et d'Ethnologie de Nanterre 1.

Mashkour, M. 2009. Faunal remains from Tol-e Nurabad and Tol-e Spid. In, Potts, D. T., Roustaei, K., Petrie, C. A., and Weeks, L. R. (eds) *The Mamasani Archaeological Project Stage One: A Report on the First Two Seasons of the ICAR-University of Sydney Expedition to the Mamasani District, Fars Province, Iran.* (BAR International Series 2044): 135-146. Oxford: Archaeopress.

Mashkour M. and Khazaeili, R. unpublished. Preliminary report of the faunal study of A5--3 Cave of Arsanjan (spring and summer 2012 campaigns). In, Tsuneki, A. and Hourshid, N. (eds) *ICHHTO*- Iran.

Mashkour, M., Mohaseb, A. and Debue, K. 2006. Towards a Specialized Subsistence Economy in the Marvdasht Plain: Preliminary Zooarchaeological Analysis of Mushki, Jari B, Jari A and Bakun A. In, Alizadeh, A. (ed.) *The origins of State Organizations in Highland Fars, Iran*: 101-105. Oriental Institute Publications 128, Chicago: University of Chicago.

Mashkour, M. and Bailon, S. 2010. Animal bones. In, Pollock, S., Bernbeck, R. and Abdi, K. (eds) *The 2003 Excavations at Tol-e Basi, Iran: Social Life in a Neolithic Village:* 215- 229. (Archäologie in Iran und Turan Band 10). Berlin: Deutsches Archäologisches Institut.

Mashkour, M., Vigne, J.-D., Bemili, C., Brehard, S., Mohaseb, A., Reynolds, W. and Nishiaki, Y. Unpublished data. The Neolithic fauna of Tal- i Mushki and Jari B (Iran).

Matthews, R., Mohammadifar, Y., Motarjem, A. and Matthews, W. 2010. Investigating the Early Neolithic of western Iran: The Central Zagros Project (CZAP). *Antiquity* 82(322).

Meadow R.H. 1983. Appendix G, The vertebrate faunal remains from Hasanlu period X at Hajji Firuz. In, Voigt, M.M. (ed.) *Hajji Firuz Tepe, Iran: The Neolithic Settlement (Hasanlu Excavation Reports I)*: 369-422. Philadelpha: The University Museum of University of Pennsylvania..

Meadow, R.H. 1999. The use of size index scaling techniques for research on archaeozoological collections from the Middle East. In, Becker, C., Manhart, H., Peters, J. and Schibler, J. (eds) *Historia Animalium Ex Ossibus: Beitrage zur Palaoanatomie, Archaologie, Agyptologie, Ethnologie, und Geschichte der Tiermedizin. Festschrift fur Angela Von den Driesch*: 285-300. Rahden/WESTF.

Moradi, B., Mashkour, M., Eghbal, H., Mohaseb, A., Ghassimi, T., Rahmati, E., Vahdati, A., Gratuze,

B. and Tengberg, M. 2016. A short account on Kelek Asad Morad, a Pre-Pottery Neolithic site in Pol-e Dokhtar, Luristan. In, Roustaei, K. and Mashkour, M. (eds) *The Neolithic of the Iranian Plateau. Recent Research and Prospects. Studies in Early Near Eastern Production, Subsistence, and Environment* (SENEPSE series): 108-136. Berlin: Ex-oriente.

Pales, L. and Garcia, M. A. 1981. *Atlas ostéologique pour servir a l'identification des mammifères du quaternaire* (Vol. 2) : Editions du centre national de la recherche scientifique.

Payne, S. 1973. Kill-off patterns in sheep and goats: The mandibles from Asvan-Kale. *Anatolian studies* 23: 281-303.

Payne, S. 1991. Early Holocene equids from Tall-e-Mushki (Iran) and Can Hassan III (Turkey). In, Meadow, R. H., and Uerpmann, H. P. (eds) *Equids in the ancient World* II: 132-177.

Peters, J., Von den Driesch, A. and Helmer, D. 2005. The upper Euphrates-Tigris basin: cradle of agro-pastoralism? In, Vigne, J.-D., Peters, J. and Helmer, D (eds) *The first steps of animal domestication: new archaeological approaches.* Proceeding of the 9th ICAZ Conference, 2005: 96-124. Oxford: Oxbow Books.

Price, M.D., and Arbuckle, B.S. 2015. Early pig management in the Zagros Flanks: Reanalysis of the fauna from Neolithic Jarmo, Norther Iraq. In, *International Journal of Osteoarchaeology* 25: 441-453.

Potts, D. 2014. *Nomadism in Iran: From Antiquity to the Modern Era.* Oxford: Oxford University Press.

Riehl, S., Zeidi, M. and Conard. N.J. 2013. Emergence of Agriculture in the Foothills of the Zagros Mountains of Iran. *Science* 341: 65–67. http://dx.doi.org/10.1126/science.1236743.

Stampfli, H. R. 1983. The fauna of Jarmo with notes on animal bones from Matarrah, the Amuq, and Karim Shahir. In, Braidwood, L.S., Braidwood, R. J., Howe, B., Reed, C. A. and Watson, P. J. (eds) *Prehistoric Archeology Along the Zagros Flanks:* 431-484. Chicago: Oriental Institute of the University of Chicago.

Uerpmann, H.P. 1979. *Probleme der Neolithisierung des mittelmeeraums* (Beihefte zum Tübinger Atlas des Vordern Orients). Wiesbaden: Ludwing Reichert.

Uerpmann, M. and Uerpmann, H.P. 1994. Animal bone finds from excavation 520 at Qala' at al-Bahrain. In, Höljund, F. and Anderson, H. (eds) *Qala'at al-Bahrain.* 1. *The Northern City Wall and the Islamic Fortress:* 417-444. Aarhus.

Vigne, J.-D. and Helmer, D. 2007. Was milk a "secondary product" in the Old World Neolithisation process? Its role in the domestication of cattle, sheep, and goats. *Anthropozoologica* 42(2): 9-40.

Von den Driesch, A. 1976. *A guide to the measurement of animal bones from archaeological sites.* Cambridge Peabody Museum Bulletin 1. Harvard University.

Weeks, L. R. 2013. The Neolithisation of Fars, Iran. In, Matthews, R., and Fazeli Nashli, H. (eds) *The Neolithisation of Iran: The Formation of New Societies*: 86-96. Oxford: Oxbow Books.

Zeder, M.A. and Hesse, B. 2000. The initial domestication of goats (*Capra hircus*) in the Zagros Mountains 10,000 years ago. *Science* 287: 2254-2257.

Zeder, M.A. 2006. Reconciling rates of long-bone fusion and tooth eruption and wear in sheep (*Ovis*) and goat (*Capra*). In, Ruscillo, D. (ed.) *Ageing and Sexing Animals from Archaeological Sites*: 87–118. Oxford.

Zeder, M.A. 2008. Animal domestication in the Zagros: An update and direction for future research. In, Vila, E., Gourichon, L., Choyke, A. M. and Buitenhuis, H. (eds) *Achaeozoology of the Near East: Proceeding of the Eighth International Symposium on the Archaeozoology of Southwestern Asia and Adjacent Areas VIII*: 243-277. Lyon.

Zeder, M.A. and Lapham, H.A. 2010. Assessing the reliability of criteria used to identify postcranial bones in sheep, *Ovis*, and goat, *Capra*. *Journal of Archaeological Science* 37: 2887- 2905.

Ziaei, H. 1996. *A field guide to the mammals of Iran.* Tehran, Iran: Department of Environment. (In Persian).

4 Horse and hemione hunting at Late Neolithic/ Chalcolithic Köşk Höyük, central Turkey

Benjamin S. Arbuckle[1*] and Aliye Öztan[2]

1 301 Alumni Building, CB#3115. Department of Anthropology, University of North Carolina at Chapel Hill, 27599, USA; bsarbu@email.unc.edu
2 Ankara Üniversitesi Dil ve Tarih-Coğrafya Fakültesi, Arkeoloji Bölümü 06100 Sıhhiye/ Ankara, Turkey
* corresponding author

Abstract
In this paper we describe the abundant equid remains from the late seventh and early sixth millennium BC site Köşk Höyük, central Turkey. We focus on identification of the equid taxa present and on contextualizing patterns of equid hunting with those at contemporary sites in Anatolia and SW Asia. We argue that both wild horses (*Equus ferus*) and hydruntines (*Equus hemionus hydruntinus*) were present in the assemblage and that the high frequencies of equid remains at Köşk are part of regional trends towards increasingly intensive wild equid exploitation characteristic of the sixth millennium BC in the arid and semi-arid grasslands of SW Asia and Anatolia.

Keywords
Horse, hemione, hydruntine, equid hunting, Anatolia, Neolithic, zooarchaeology

Introduction
The equids, including horses, hemiones and donkeys, represent one of the most important, but least well understood, groups of animals in the prehistory of SW Asia. Despite being among the largest prey animals available and their heavy exploitation in some regions, very little research has focused on human-equid relationships in the early and middle Holocene prior to the appearance of domestic horses and donkeys in the third and fourth millennia BC. Although the roles of equids have recently been addressed in more detail in Mesopotamia and the southern Levant (Grigson 2012; Vila 2006; Weber 2008; Zarins 2014), these taxa have been largely ignored in Anatolia, likely due, at least partially, to the fact that wild equids have not been part of local faunas in this region in recent centuries as well as uncertainties concerning the taxonomic identification of Anatolian equids. However, a combination of ancient texts and recent zooarchaeological work indicates that multiple species of equids were present in Anatolia for most of the Holocene and suggests that they played an important role in subsistence systems and symbolic life in this region.

In order to address the lack of information concerning this important group of animals, in this paper, we describe the equid remains from Köşk Höyük, a small village settlement in central Anatolia (Niğde province) with an occupational sequence dating from the late 7th to the early 5th millennium BC (Late Neolithic to Middle Chalcolithic). By focusing on remains of the postcranial skeleton (dental data will be described elsewhere) we provide the first assessment of which equid taxa are present at Köşk Höyük and address the nature of equid hunting at this settlement with an emphasis on contextualizing these practices within broader trends in equid exploitation in the 6th and 5th millennia BC in Anatolia and across the Near East.

Background
Recent zooarchaeological and paleogenetic work has significantly improved our understanding of the taxonomy and geographic distribution of equid populations in late Pleistocene and Holocene SW Asia (including Anatolia). Previously, it was thought that wild horses (*Equus ferus*) were widely distributed across the Fertile Crescent region of SW Asia in the Pleistocene (Davis 1980; Eisenmann *et al.* 2002; Turnbull and Reed 1974; Uerpmann 1987) but that local populations became extinct in the region by the early Holocene (with a few possible exceptions

of relict populations (e.g., Grigson 1993). In most discussions of the origins of domestic horses in SW Asia, Anatolia is not considered to have played a role in this process since it is assumed that there were no endemic wild horse populations on the Anatolian plateau (Zeuner 1963; Bökönyi 1991; Boessneck and Von den Driesch 1976a).

However, a combination of previous (largely ignored) and recent work has shown that wild horse populations were, in fact, present in Holocene Anatolia. In the 1970s, working at several sites in the Elazig region of eastern Turkey, Boessneck and Von den Driesch (1976a; 1976b) first identified the presence of Holocene wild horses in Anatolia dating back to the Chalcolithic (5th and 4th millennia BC) at Norşuntepe, Tülintepe, and Tepecik. In the same region, Bökönyi (1983; 1991) identified horse remains at Arslantepe from levels also dating to the 4th millennium BC. Bökönyi's (1991) interpretation of those remains as early domestic horses was built in part on the assumption that wild horses were not present in Anatolia and therefore horse remains must represent imported domesticates. Furthermore, three horse specimens were identified by Boessneck and Von den Driesch (1979) in NW Turkey at the Neolithic (7th millennium BC) site of Fikirtepe and Sebastian Payne recognized horse remains from the Early Chalcolithic levels (2B) of Can Hasan I located in central Turkey (French 1965).

More recent work has added significantly to our understanding of wild horses in Anatolia particularly in central Anatolia. Here, Payne (1985) identified equids at eighth millennium BC Aşıklı Höyük, while Buitenhuis (1997) more recently has confirmed that both horses and hemiones are present at that site. On the Konya plain, horses have been identified at 8th millennium BC Boncuklu, throughout the late 8th and 7th millennium BC occupation at Çatalhöyük and at 7th millennium BC Pinarbaşı (Baird *et al.* 2011; Cameron 2005; Carruthers 2005; Martin and Russell 2006). Buitenhuis (personal communication) and Arbuckle (2006) identified horses at 5th millennium BC Güvercinkayası while horses have also been identified from well-stratified contexts at Çadır Höyük, east central Anatolia, dating to the mid-5th and 4th millennia BC (Arbuckle 2009). In west central Turkey, Uerpmann (2001) identified abundant equid remains, including horses, at Orman Fidanlığı and Kes Kaya. At Orman Fidanlığı, equids are a relatively small percentage of the faunal assemblages in levels I-V (6th millennium BC) but in level VII (5th millennium BC) equids, including horse, increase dramatically in abundance. At Kes Kaya, dated by ceramic parallels to the mid-6th millennium BC, equid remains, including horse, are not quantified but they are described as 'abundant'.

Together, these results provide evidence for the presence of wild horse populations extending across the Anatolian plateau (north of the Taurus and Anti-Taurus) from Eskişehir in the west to Elazig in the east. These populations were regularly hunted from the early Neolithic (late 9th millennium BC) through the Chalcolithic (5th millennium BC) particularly in central Anatolia. Moreover, Mashkour (2002), Grigson (2012) and Kansa (Kansa *et al.* 2006) have identified horses in northern Iran and southern Israel dating back to the 5th millennium BC indicating that the early Holocene range of wild horses was much broader than previously thought.

In addition to horses, hemiones (*Equus hemionus*), a group of gracile wild asses adapted to grazing in arid environments, are also native to SW Asia and Anatolia and recent work has greatly improved our understanding of the taxonomy, geographic distribution, and hunting of these animals (Burke *et al.* 2003; Crees and Turvey 2014; Eisenmann 1986; Eisenmann and Mashkour 1999; Geigl and Grange 2012; Orlando *et al.* 2006). Hemiones have been identified at sites in arid and semi-arid steppe regions across SW Asia throughout the Holocene and they continued to be hunted into the 20th century in Syria, Israel, and Iran (Uerpmann 1987). They were heavily hunted at early Holocene sites on the Syrian middle Euphrates and in the Khabur region and were also a regular part of the animal economies at Neolithic and Chalcolithic sites in southeastern Turkey including Göbekli Tepe and Çayönü (Peters *et al.* 2005; Hongo *et al.* 2004; Gourichon and Helmer 2008).

North of the Taurus mountains, in Anatolia, hemiones are also present although their taxonomic identification has long been a source of confusion. 'Small equids' (i.e., not horse) have been identified at virtually all Neolithic and Chalcolithic sites in central Anatolia including Aşıklı Höyük (Buitenhuis 1997), Can Hasan III (Payne 1991), Neolithic and Early Chalcolithic Çatalhöyük (Martin and Russell 2006), Early and late Neolithic Pinarbaşı (Martin *et al.* 2002), Middle Chalcolithic (5th millennium BC) Can Hasan I (French 1965), and Güvercinkayası (Arbuckle 2006). However, equid remains are rare or absent in the more mountainous regions west of the Konya plain and where they have been identified, e.g., Erbaba, Höyücek and Kuruçay, only 'small equids' are reported (Arbuckle 2008; De Cupere and Duru 2003; De Cupere *et al.* 2008; Deniz and Sentuna 1988). Small wild equids (not horses or donkeys) have also been identified at Early Bronze Age Demircihöyük and Küllüoba in west central Anatolia (Efe *et al.* 2008; Von den Driesch and Boessneck 1987) and also at Middle Bronze Age Acemhöyük, central Anatolia (Arbuckle 2013).

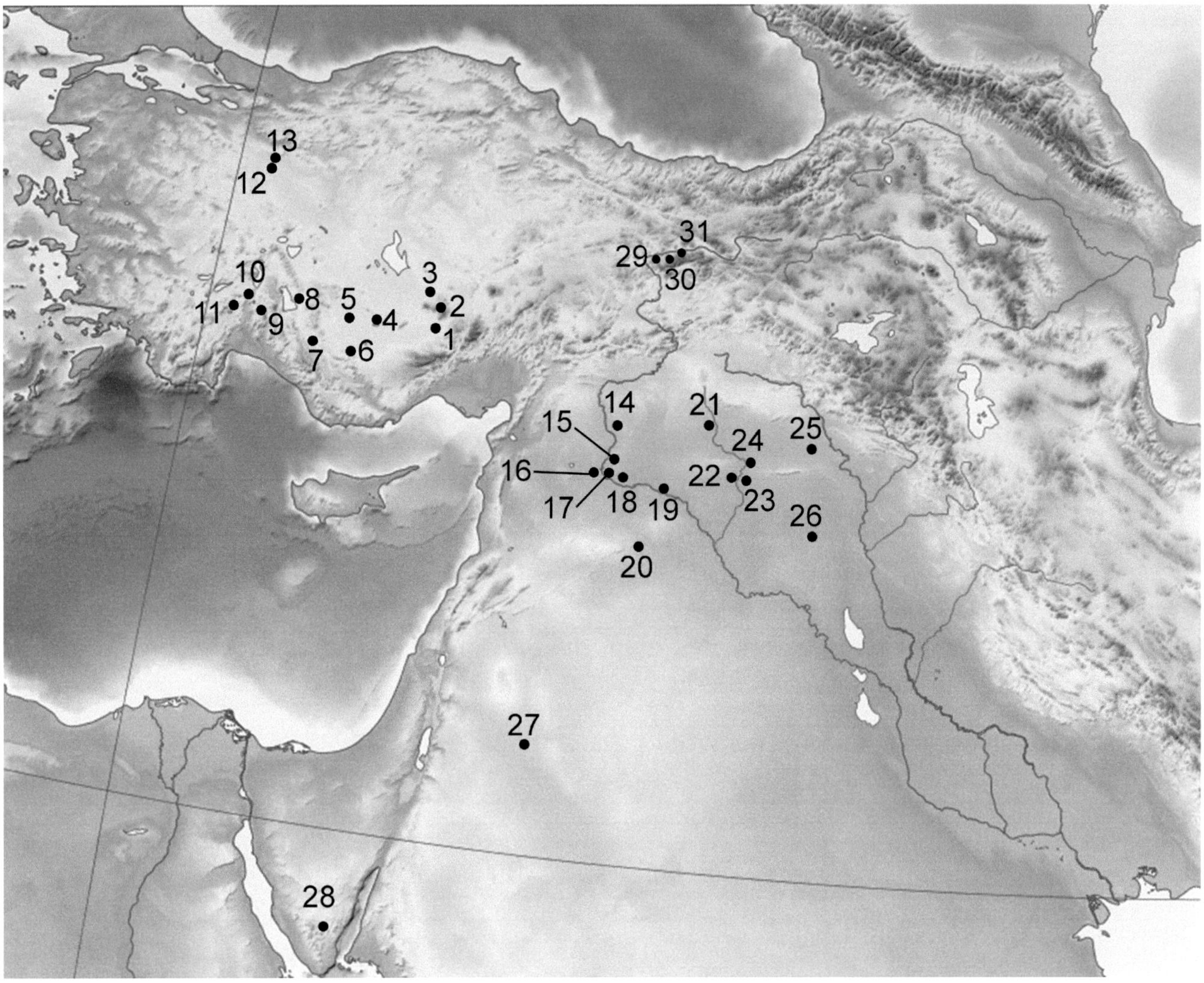

Fig. 1. Map showing the location of select sites mentioned in the text as well as early Holocene sites with high frequencies of equid remains. 1= Köşk, 2=Tepecik-Çiftlik, 3=Aşıklı, 4=Boncuklu, 5=Çatalhöyük, 6=Can Hasan, 7=Suberde, 8=Erbaba, 9=Höyücek, 10=Kuruçay, 11=Bademağacı, 12=Kes Kaya, 13=Orman Fidanlığı, 14=D'jade, 15=Jerf el Ahmar, 16=Nahr el Homr, 17=Cheikh Hasan, 18= Mureybet, 19=Zeidan, 20=El Kowm, 21=Kashkashok, 22=Qseir, 23=Mashnaqa, 24=Boueid, 25=Yarim Tepe, 26= Umm Dabaghiyah, 27=Azraq 31, 28=Wadi Tbeik, 29=Norşuntepe, 30=Tulintepe, 31=Tepecik.

Based on patterns of enamel folds in mandibular and maxillary teeth, these small Anatolian equids have been identified as representing two species including hemione (*Equus hemionus*) as well as the hydruntine (*Equus hydruntinus* Regalia 1907), or European wild ass, a poorly understood small equid known from Pleistocene Europe and the southern Levant (Bonifay 1991; Crees and Turvey 2014; Davis 1980; Regalia 1907; Stehlin and Graziosi 1935). Recent paleogenetic work (Geigl and Grange 2012; Orlando *et al.* 2006; Orlando *et al.* 2009) suggests that hydruntines, which exhibit distinctive morphological features, particularly in the mandibular molars, are characterized by a unique mitochondrial lineage but nevertheless share genetic signatures characteristic of the hemiones leading to the conclusion that hydruntines should be considered a distinctive western lineage of hemione (i.e., *Equus hemionus hydruntinus*) (also see Burke *et al.* 2003).

Although some authors have previously identified both hydruntines and non-hydruntine hemiones in Neolithic archaeofaunal assemblages in central Anatolia (Martin and Russell 2006; Carruthers 2003; although see Russell *et al.* 2013), this scenario is difficult to accept on ecological grounds and Geigl and Grange (2012) have shown that identifications of these closely related taxa based on dental traits is problematic. Instead, it is more likely that all of the small equids from early Holocene sites in central and western Anatolia represent hydruntines while non-hydruntine hemiones (e.g., *E. hemionus onager*; *E. h. hemippus; E. h. kulan*) were present south of the Taurus and Anti-taurus mountains in Syria, the southern Levant, eastern Anatolia, Iran and central Asia.

The Anatolian plateau is, therefore, the home of a poorly documented, yet apparently long-lived, tradition of both horse and hemione hunting extending from the Neolithic through the Chalcolithic. The presence of depictions of equid hunting at Çatalhöyük (Russell 2012; Mellaart 1966), Köşk Höyük (Öztan 2002; 2007), Tepecik-Çiftlik (Bıçakçı 2001) and Balkayası (and Early Chalcolithic cave

Table I. Frequencies of major mammalian taxa in the Köşk Höyük faunal assemblages.

dates (cal BC)	***6200-6000***	***6000***	***5700***	***5500***	***5200-4700***
phasing	**Pits**	**level IV**	**level III**	**level II**	**level I**
Ovis/Capra	41.7	35.6	50.6	43.1	67.5
Ovis	7.0	11.3	11.7	11.4	16.5
Capra	2.1	3.4	4.9	1.9	3.8
Bos	9.5	10.2	8.8	7.2	3.1
Deer	0.7	0.6	7.1	0.9	0.9
Sus	0.6	0.6	0.0	0.1	0.6
Equids	36.0	35.4	11.3	34.1	3.5
other mammals	2.4	2.9	5.7	1.3	4.1
Total N	**1893**	**1861**	**947**	**787**	**848**

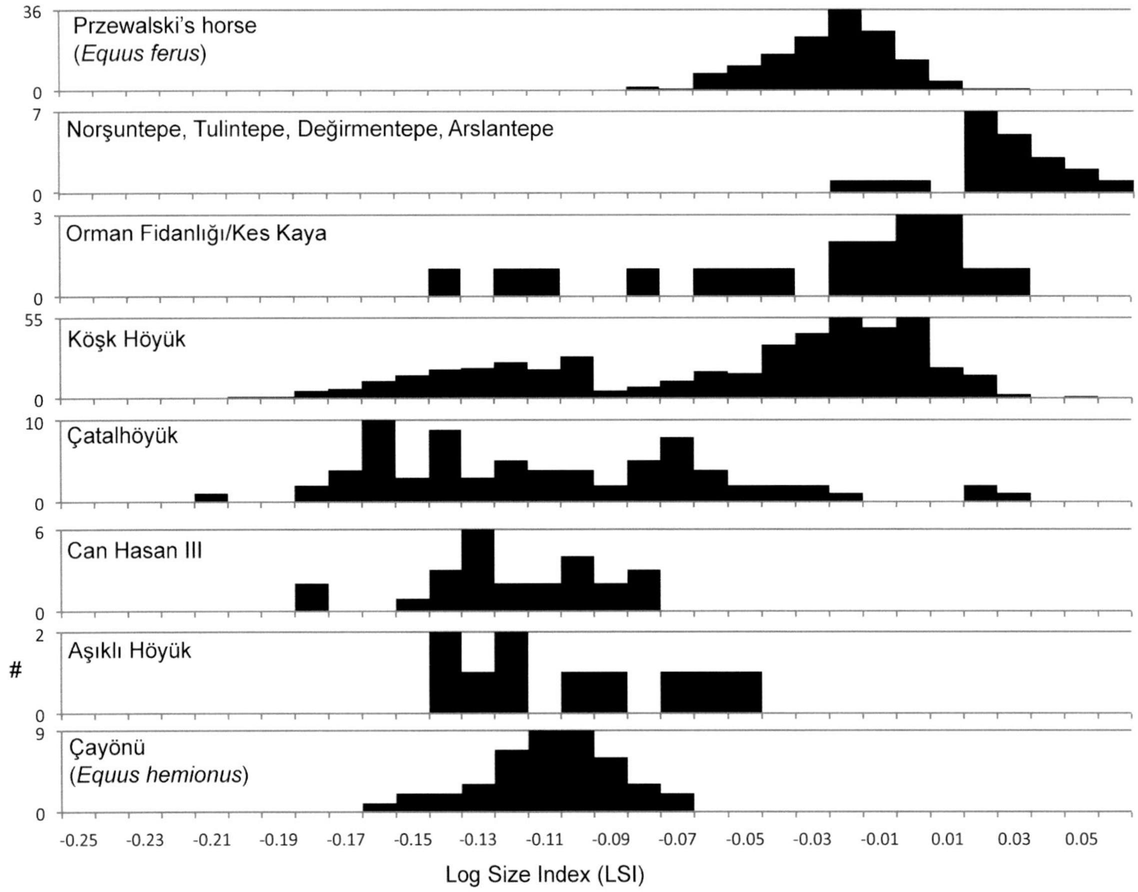

Fig. 2. Log Size Index (LSI) values for equids from Köşk Höyük compared with those from Aşıklı Höyük (Payne 1985), Can Hasan III (Payne 1991), Çatalhöyük (Cameron 2005), Orman Fidanlığı and Kes Kaya (Uerpmann 2001) as well as those from Çayönü Tepesi (*Equus hemionus*) (Meadow 1986) and recent wild horse (Zeder 1986).

painting in Eskişehir (Ali Türkcan personal communication)) further support the economic and symbolic importance of wild equids in the early Holocene cultures on the Anatolian plateau.

Materials and methods

This paper focuses on the analysis of the sizable equid assemblage recovered from Köşk Höyük. Köşk Höyük represents the remains of a small agricultural settlement located on the edge of the Bor plain (the eastern extension of the Konya plain) in close proximity to both the Melendiz mountains to the north and the Aladağ range of the Taurus mountains to the southeast (Öztan 2002; 2007). Excavations directed by Silistreli and then Öztan in partnership with the Niğde Museum since 1980 have uncovered an occupational sequence consisting of five prehistoric phases (V-I) extending from the late 6th millennium BC (late Neolithic, level V) to the early 5th millennium BC (Middle Chalcolithic, level I) (Öztan 2010; Silistreli 1989).

Despite the fact that only a fraction of the total Köşk faunal assemblage has been analyzed, the current sample of equid remains consists of 1749 specimens which represents one of the largest assemblage of prehistoric equid remains in Anatolia. This sample provides a rich resource for assessing age and biometric characteristics of the equid population and for assessing the nature of equid exploitation at the site.

Biometric data are used to confirm the identification of both large and small equids in the assemblage and to define the postcranial dimensions of these populations. In order to maximize available samples, we use the Log Size Index (LSI) to compare measurements from multiple skeletal parts (Benecke 2009; Meadow 1999). LSI values are calculated using the formula, $Log_{10}(X)$- $Log_{10}(Y)$=LSI, in which X is an archaeological measurement (e.g., humerus Bd) and Y is the corresponding measurement from a 'standard animal'. As our standard animal, we use the averaged measurements from two similarly-sized male horses from an Iron Age burial from the site of Norşuntepe (eastern Turkey, upper Euphrates) published by Boessneck and Von den Driesch (1977/1978: Table 5).

In addition, in order to assess the biometric characteristics of the Köşk equids in more detail and to compare the Köşk equids with those from other sites in the region, 108 first phalanges (including 77 from Köşk)(including measurements GL, Bp, and KD) were subject to Principal Components Analysis (PCA), using Pearson's correlation matrix, and Agglomerative Hierarchical Clustering (AHC) using XLSTAT software. PCA is widely used as an exploratory statistical tool to identify patterns in biometric data (Pinhasi 2006; Scott and Maga 2005; Van Asperen 2010). AHC is another exploratory statistical method that was used to sort first phalanx measurements into groups based on the AHC algorithm's measure of dissimilarity (defined by Euclidean distance) between samples. For the AHC aggregation method, Ward's agglomorative method was used and a dendrogram was generated to represent the relationships between specimens. We use this method in order to address the taxonomy of the Köşk equids and to identify how many morphological groups are present in the first phalanx biometric data.

Survivorship curves are used to assess the age at which both large and small equids were killed and are useful for interpreting nature of hunting practices including seasonality (Stiner 1990). In this paper survivorship is assessed based on the state of fusion of the epiphyses of long bones and the calcaneus. Estimates for the age of fusion of the skeleton were derived from Silver (1963) and Sisson and Grossman (1938) for modern horses.

Results

Species frequencies

Although sheep and goats are the most abundant taxa in the Köşk archaeofaunal assemblage in all periods, equids are, surprisingly, the second most common taxonomic group. The frequency of equid remains is remarkably high in the earliest dated deposits at the site, which include a series of pit features as well as occupational level IV where they comprise 36.0% and 35.4% of the mammalian fauna. Fewer equid remains were recovered from deposits assigned to level III (11.3%) but the frequency of equids increases again to 34.1% in level II, dating to the mid-6th millennium BC. After a hiatus in occupation, the level I occupation (c. 4800 BC) exhibits the lowest frequencies of equids at the site (3.5%) indicating a major decrease in equid exploitation and increased focus on sheep and goat pastoralism in the early fifth millennium BC (Arbuckle *et al.* 2009).

Biometrics

LSI values were generated for breadth and depth measurements of long bone epiphyses and the major tarsal bones (N=489), which were compared with similar data from other prehistoric sites in Anatolia (Fig. 2). LSI data from Köşk equids exhibit a broadly bimodal distribution with a smaller mode at -0.10 and larger mode at -0.02 on the LSI scale. The small equids at Köşk are similar in postcranial size to those from earlier Neolithic settlements in central Anatolia including Aşıklı Höyük, Can Hasan III, and Çatalhöyük and to the small equids from Chalcolithic Orman Fidanlığı. These are comparable in size to hemiones identified from the Neolithic site of Çayönü in SE Turkey.

The majority of the LSI values from Köşk (68%) represent large equids. These individuals are comparable in size to the equids identified as horse by Uerpmann (2001) from Orman Fidanlığı and Kes Kaya and to those of extant wild Przewalski's horses. They are somewhat smaller, on average, than the Standard animal (0.0 on the LSI scale), which represents the averaged measurements of two domestic Iron Age horses. A small number of equids from Çatalhöyük and Aşıklı Höyük also fall in the size range of the Köşk large equids. The sample of horse remains from Late Chalcolithic (fourth millennium BC) eastern Turkey including those from Norşuntepe, Arslantepe, Değirmentepe, and Tulintepe, exhibit considerably larger LSI values compared to the earlier large equids from central Anatolian sites, with most of the Elazig region horses overlapping only with the very largest Köşk equids.

The results of PCA based on a sample of 77 first phalanges from Köşk as well as 31 others from wild

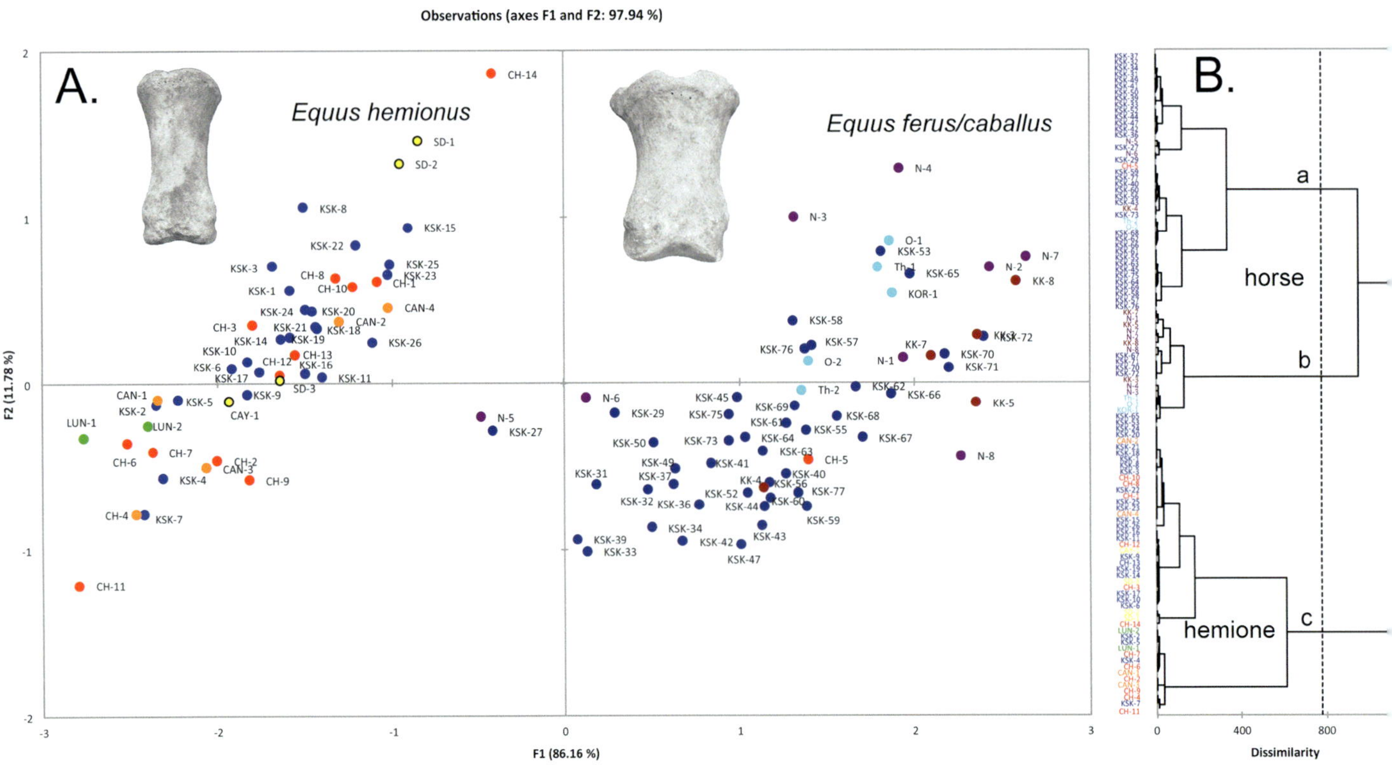

Fig. 3. A) Results of Principal Components Analysis (PCA) of equid first phalanx measurements including GL, Bp, and KD. Site codes: KSK= Köşk Höyük; N=Norşuntepe; O=Osmankayası; Th=Thebes; KOR=Körücütepe; KK=Kırklareli-Kanlıgeçit; CAN=Can Hasan III; LUN=Lunel-Viel (France); CAY=Çayönü; CH=Çatalhöyük; SD=Shams ed-Dinn (Syria). Photos show examples of Köşk large and small equids. B) Dendrogram produced from Agglomerative Hierarchical Clustering (AHC) for the same first phalanx specimens presented in the PCA graph.

Table 2. Eigenvalues associated with PCA of equid first phalanx measurements.

	F1	F2	F3
Eigenvalue	2.585	0.354	0.062
Variability (%)	86.159	11.784	2.057
Cumulative %	86.159	97.943	100.000

equids and early domestic horses in the region clearly distinguish two major groups (Fig. 3a; Eigenvalues are presented in Table 2; for measurements see Table 3). One group includes the small equids from Köşk, as well as small equids from central Anatolian sites Catalhöyük and Can Hasan III, gracile hydruntines from Pleistocene Lunel-Viel (France), as well as hemiones from Neolithic Çayönü Tepesi and Halaf Shams ed Din (Syria). These specimens group with recent hemione populations, particularly the Persian onager (*E. hemionus onager*), but are more robust than recent Syrian hemiones (*Equus hemionus hemippus*) (Dive and Eisenmann 1991). The second group consists of Köşk 'large equids' as well as wild horses from Catalhöyük, two wild(?) horses from LC/EBA Norşuntepe, and domestic horses from EBA Kırklareli-Kanlıgeçit, LBA Osmankayası, LBA Thebes (Egypt), Iron Age Norşuntepe, and Medieval Körücütepe. These specimens group with extant wild horses (*Equus przewalskii*) (Dive and Eisenmann 1991). Two specimens from Iron Age Norşuntepe and two from Köşk (KSK27, KSK29, N-5, N-6) fall in between these two main groups and exhibit intermediate biometric characteristics.

The AHC analysis identified three groups in the first phalanx biometric data (labeled A, B, and C in Fig. 3b). Groups A and B represent Köşk 'large equids' as well as wild horses from EC Çatalhöyük, LC and EBA Norşuntepe, and early domestic horses from EBA Kırklareli-Kanlıgeçit, LBA Thebes, LBA Osmankayası, Iron Age Norşuntepe and Medieval Körücütepe. Phalanges from both anterior and posterior limbs were included in this analysis and both are distributed within groupings A and B; this division, therefore, does not represent anterior versus posterior first phalanges. Group A, incorporates the most variability and includes the majority (36 out of 42) of Köşk large equid specimens, two equids identified as horse from Iron age Norşuntepe (see below), a wild horse from EC Çatalhöyük, and three 'domestic' horses from EBA Kırklareli-Kanlıgeçit, LBA Osmankayası, and LBA Thebes (Egypt). Specimens in this group are comparable in proportion to extant wild horses (Dive and Eisenmann 1991). The two 'horse' specimens from Norşuntepe (N-5 and N-6) in Group A, along with Köşk specimens KSK27 and KSK29, were identified in the PCA as outliers and are situated in the AHC dendrogram on their own branch of the dendrogram. These four specimens are

Table 3. First phalanx measurements for equids from Köşk Höyük.

Identification	ID	GL	Bp	KD	Bd
E.h.hydruntinus	KSK-1	76.8	39.8	22.3	30.9
E.h.hydruntinus	KSK-2	71.1	36	22.9	31
E.h.hydruntinus	KSK-3	77.2	37.1	23	31
E.h.hydruntinus	KSK-4	69	38.7	23.1	31
E.h.hydruntinus	KSK-5	71.6	36.6	23.2	31.1
E.h.hydruntinus	KSK-6	73.5	38.4	23.1	31.4
E.h.hydruntinus	KSK-7	67.6	38.7	23.3	32.2
E.h.hydruntinus	KSK-8	79.5	37.2	22.6	32.4
E.h.hydruntinus	KSK-9	73	39.6	23.5	32.4
E.h.hydruntinus	KSK-10	74	37.4	24.3	32.5
E.h.hydruntinus	KSK-11	74.8	41.6	24.5	32.5
E.h.hydruntinus	KSK-12	80	-	24.3	33
E.h.hydruntinus	KSK-13	67.5	-	23.6	33.4
E.h.hydruntinus	KSK-14	75.2	38.9	23.9	33.5
E.h.hydruntinus	KSK-15	80.8	39.9	25.2	33.6
E.h.hydruntinus	KSK-16	74.6	41	24.2	34.1
E.h.hydruntinus	KSK-17	73.9	38.4	24.3	34.1
E.h.hydruntinus	KSK-18	76.2	39.1	24.8	34.8
E.h.hydruntinus	KSK-19	75.4	39.9	23.5	34.9
E.h.hydruntinus	KSK-20	76.5	39.3	23.8	35
E.h.hydruntinus	KSK-21	76.2	39.3	24.7	35
E.h.hydruntinus	KSK-22	79.3	40.3	23.3	35.5
E.h.hydruntinus	KSK-23	79	41.3	24.6	35.8
E.h.hydruntinus	KSK-24	76.6	39.8	23.7	36
E.h.hydruntinus	KSK-25	79.4	40.5	25	36.2
E.h.hydruntinus(?)	KSK-26	76.9	36.6	29.2	38.4
Equus sp.	KSK-27	76.4	43.3	31.4	38.1
Equus sp.	KSK-28	73.8	-	31.2	39.6
Equus ferus	KSK-29	79.1	47.7	32.5	39.6
Equus ferus	KSK-30	78.4	-	33.6	40
Equus ferus	KSK-31	76.6	50.9	31.3	40.4
Equus ferus	KSK-32	77.4	52.1	32.5	40.7
Equus ferus	KSK-33	74.5	49.9	33.4	41.3
Equus ferus	KSK-34	76.3	53.2	32.9	41.6
Equus ferus	KSK-35	79.5	-	32.1	41.7
Equus ferus	KSK-36	77.9	51.8	35.1	42.4
Equus ferus	KSK-37	78	53.1	32.7	42.7
Equus ferus	KSK-38	80	-	35.6	42.8
Equus ferus	KSK-39	74.7	48.8	33.5	42.9
Equus ferus	KSK-40	80.3	54.9	35.4	42.9
Equus ferus	KSK-41	79.3	52.2	34.2	43.1
Equus ferus	KSK-42	76.5	51.9	35.4	43.1
Equus ferus	KSK-43	78.3	56.7	34.5	43.9
Equus ferus	KSK-44	79	53.7	36.3	44
Equus ferus	KSK-45	81.7	51.9	33.7	44.2
Equus ferus	KSK-46	84.5	-	35.2	44.3
Equus ferus	KSK-47	77.4	54.8	35.6	44.3
Equus ferus	KSK-48	86.8	-	36.6	44.3
Equus ferus	KSK-49	78.5	51.8	33.3	44.6
Equus ferus	KSK-50	78.9	50.1	33	44.7
Equus ferus	KSK-51	81	-	34.3	44.7
Equus ferus	KSK-52	79.1	52.2	36.4	45
Equus ferus	KSK-53	88.6	51.8	35.6	45.2
Equus ferus	KSK-54	81.5	-	36	45.3
Equus ferus	KSK-55	82	53.3	36.2	45.4
Equus ferus	KSK-56	79.7	56	34.2	45.5
Equus ferus	KSK-57	84.6	52	35.2	45.6
Equus ferus	KSK-58	85	50.8	34.7	45.8
Equus ferus	KSK-59	79.7	56.5	35.9	45.8
Equus ferus	KSK-60	79.3	56	34.6	46.1
Equus ferus	KSK-61	81.8	53.7	35	46.4
Equus ferus	KSK-62	84.1	56.3	34.8	46.8
Equus ferus	KSK-63	80.6	52.5	35.7	46.8
Equus ferus	KSK-64	80.7	50.5	36.1	46.9
Equus ferus	KSK-65	88.5	51.5	37.5	46.9
Equus ferus	KSK-66	84.5	56.7	36.1	47.3
Equus ferus	KSK-67	82.8	53.6	38.4	47.4
Equus ferus	KSK-68	82.9	55.8	35.2	48.3
Equus ferus	KSK-69	82.5	51.5	36.5	48.3
Equus ferus	KSK-70	86.7	55.5	38	49
Equus ferus	KSK-71	86.4	55.8	38.3	49
Equus ferus	KSK-72	87.9	55.6	39	49
Equus ferus	KSK-73	80.2	55.9	31.6	-
Equus ferus	KSK-74	77.6	-	33.6	-
Equus ferus	KSK-75	81.1	50.2	35.1	-
Equus ferus	KSK-76	84.4	50.1	36.4	-
Equus ferus	KSK-77	80	54.6	36.6	-
Equus ferus	KSK-78	73.2	50.1	-	-

unusually proportioned members of the horse group (and perhaps represent hybrid animals?).

Group B, includes 6 (out of 42) Köşk 'large equid' specimens along with one 'wild' horse from LC Norşuntepe, one wild/domestic horse from EBA Norşuntepe, four domestic horses from Iron Age Norşuntepe, three domestic horses from EBA Kırklareli-Kanliegecit, and domestic horses from LBA Osmankayası, Thebes, and Medieval Körücütepe. This group consists of robust horses, largely outside of the biometric range of extant wild horses (Dive and Eisenmann 1991), and includes most of the early domesticates represented in the biometric sample.

Group C which includes two sub-branches, includes all Köşk specimens identified as 'small equids' as well as those from Can Hasan III and Çatalhöyük; hydruntines from Lunel-Viel (Pleistocene, France); and hemiones from Çayönü Tepesi and Shams ed Din. One branch of Group C contains very gracile, small equids including four specimens from Köşk, the hydruntines from Lunel-Viel as well as many of the small equids from Can Hasan III and Çatalhöyük. These animals are similar in size to the 19th century gracile Syrian hemione, *E.h.hemippus* (Dive and Eisenmann 1991). The other branch of Group C, representing more robust small equids, contains most of the Köşk small equids (20 out of 24) some specimens from Can Hasan III and Çatalhöyük as well as hemiones from Çayönü and Shams ed-Dinn, Syria. These more robust hemiones are similar in size to extant Persian onagers (*E. h. onager*).

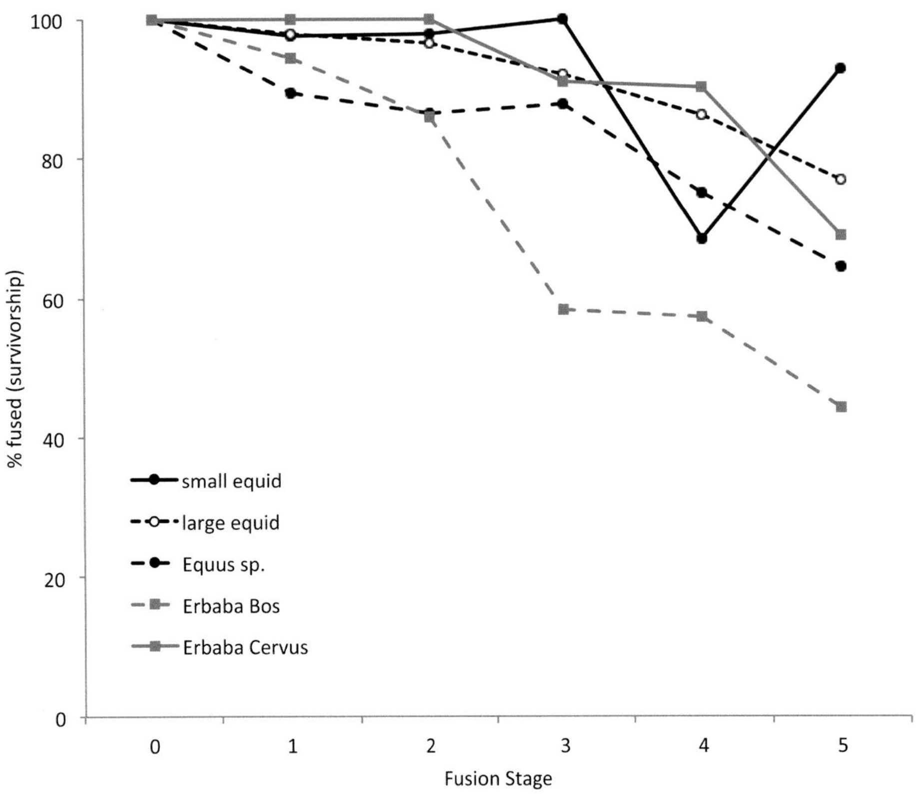

Fig. 4. Survivorship curves for small equid, large equid, and *Equus* sp. from Köşk as well as red deer and domestic cattle from Neolithic Erbaba Höyük.

Survivorship

767 specimens were available to reconstruct survivorship based on the state of fusion of the long bone epiphyses for small equids, large equids, and specimens identified more generally as 'equid' (Fig. 4; Table 4). The majority (91.5%) of equid long bones, including both early and late fusing skeletal parts exhibit fused epiphyses indicating the culled population consists primarily of skeletally mature individuals. This is comparable to the survivorship data for red deer from Neolithic Erbaba Höyük (central Turkey) where 94% of remains were fused and quite different from early domestic cattle from the same site where only 71.9% of epiphyses were fused (Arbuckle and Makarewicz 2009).

For the small equids, survivorship is very high for the earliest fusing skeletal parts (c. 97%) indicating that very few animals were killed younger than two years of age. Survivorship declines to 68% for elements in fusion stage 4 (including proximal and distal femur, proximal tibia, and calcaneus) indicating heavier culling of animals younger than three years, but then increases to 93% in fusion stage 5 (3.5 years) as a result of a low frequency of unfused proximal ulnae, distal radii, and proximal humeri.

Survivorship for large equids indicates a somewhat older cull with 86% of individuals surviving past c. three years (fusion stage 4) and 77% surviving past c. 40 months, the oldest fusion stage. Many unfused specimens could only be identified as 'equid'. As a result, the survivorship curve for *Equus* sp, is significantly lower than for the small and large equid categories with more individuals culled in fusion stages 1 and 2 (89% and 86%). However, survivorship for the later fusion stages is broadly similar to those for the small and large equid categories at 75% and 64% indicating about two thirds of the culled population survived past 40 months of age. Based on the entire equid sample, survivorship is high in the first two years (96-93%), then drops off to 78% after three and four years. This indicates that approximately a quarter of the culled population was killed before reaching four years of age. Although comparable data on wild equid survivorship are rare, Stiner (Stiner *et al.* 2014) reports that 60% of the equid tibiae from Aşıklı Höyük exhibit fusion of the distal epiphysis indicating they survived past the age of c. 20 months. In contrast, at Köşk, 89% of the equid tibiae exhibit distal fusion indicating more intensive hunting of juveniles at Aşıklı. Equid age data are also available from Çatalhöyük, where in the late 8th and early 7th millennia BC adult and subadult individuals make up 78-93% of the sample while in the mid to late seventh millennium adult and subadults decrease to 63% of the assemblage (Russell *et al.* 2013; Russell and Martin 2005). These values are broadly comparable to those from Köşk although the increase in juveniles evident in the later levels of Çatalhöyük is not evident at Köşk.

In addition, 14 specimens (out of a total equid sample of 1459) were identified by texture and size as representing the remains of perinatal individuals. A similar number of neonatal remains were identified at Neolithic Aşıklı Höyük, where 12 neonatal equid specimens were identified out of a total sample

Table 4. State of fusion of long bone epiphyses for Köşk equids. px=proximal; ds=distal. F=fused; UF=unfused. Age of fusion after Silver (1963).

Age of fusion	*Fusion*		small equid		large equid		*Equus* sp.		Total	
(months)	**Stage**	**Element**	**F**	**UF**	**F**	**UF**	**F**	**UF**	**F**	**UF**
	1	atlas	0	0	2	0	5	0	7	0
	1	axis	1	0	0	0	5	0	6	0
10	1	second phalanx	35	0	52	1	14	1	101	2
13	1	first phalanx	32	2	72	1	13	4	117	7
12	1	scapula	8	0	13	1	5	0	26	1
15	2	ds humerus	7	0	26	0	7	1	40	1
15	2	px radius	7	0	15	0	10	0	32	0
15	2	metapodial	31	1	67	4	21	5	119	10
18	3	pelvis (acetabulum)	20	0	30	0	26	3	76	3
20	3	ds tibia	16	0	38	6	10	2	64	8
36	4	px femur	4	2	8	3	1	2	13	7
36	4	calcaneus	5	3	9	1	7	2	21	6
36	4	ds femur	4	1	2	0	2	1	8	2
36	4	px tibia	0	0	6	0	11	2	17	2
40	5	px ulna	0	0	2	1	7	3	9	4
40	5	ds radius	12	1	29	8	2	2	43	11
40	5	px humerus	1	0	2	1	0	0	3	1
		TOTAL	**183**	**10**	**373**	**27**	**146**	**28**	**702**	**65**

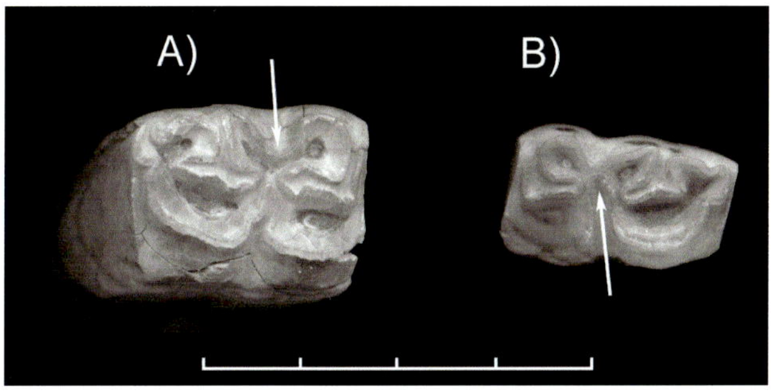

Fig. 5. Dental specimens from Köşk Höyük exhibiting typical morphological characteristics of A) horse (KSK-5, mandibular premolar 3/4; note 'U'-shaped linguaflexid) and B) hydruntine (KSK-6 mandibular M1/2; note penetration of ectoflexid)(see Geigl and Grange 2012: Fig. 3).

of 1209 equid remains (Stiner *et al.* 2014). Perinatal remains are very rare (N=1) at 7th millennium BC Çatalhöyük although specimens identified as infantile represent c. 10% of the equid remains in the early 7th millennium levels (Russell *et al.* 2013; Russell and Martin 2005). Although the birth seasonality of extinct wild Anatolian equids is not well understood, it is likely that wild equids in this region gave birth in the spring months, as is common for wild sheep in the region today (Kaya and Aksoylar 1992).

Discussion

Taxonomic identification

Based on the biometric characteristics of the Köşk equids, it is apparent that both large and small equids are present in the assemblage. Comparison with extant equids as well as with other ancient equid remains in the region strongly suggest that the Köşk large equids represent wild horse (*Equus ferus*). LSI values for the Köşk large equids compare favorably with extant wild Mongolian horses and the results of both PCA and AHC of first phalanges show a strong affiliation between Köşk large equids and ancient wild and early domestic horses. Interestingly, the 4th millennium BC horses from Norşuntepe, Arslantepe, and Değirmentepe, in eastern Turkey are much larger than those from Köşk and appear to represent a different biometric population than the 7th and 6th millennium BC horses from the Konya and Eskişehir basins. Moreover, although dental remains have not yet been subject to detailed analysis, the presence of characteristically caballine dental morphologies further supports the identification of horse at Köşk (see Fig. 5a). The early date of the Köşk settlement, the low frequencies of juvenile animals, as well as scenes of equid hunting depicted on ceramics (see Fig. 6) suggest that the horse remains at Köşk represent wild animals—there is no evidence that the Köşk horses were managed or domesticated.

We identify the small equids from Köşk as the extinct hydruntine (*Equus hemionus hydruntinus*), although this identification is more uncertain than that for horses given the presence of both hydruntines and non-hydruntine hemiones in prehistoric Turkey. Multiple lines of evidence support this

Fig. 6. Ceramic vessel from Late Neolithic Köşk Höyük depicting a scene of equid (hydruntine) hunting (photo by A.Öztan).

identification. Perhaps most importantly, in their recent study of hemione paleogenetics, Geigl and Grange (2012) found that several small equid specimens from Köşk share the distinctive mitochondrial lineage argued to be characteristic of hydruntines in Anatolia, the Balkans and western Europe. In addition, the identification of mandibular molars with characteristic hydruntine features at Köşk as well as virtually all other Neolithic and Chalcolithic sites in central Anatolia (including Aşıklı, Can Hasan III, Boncuklu, Catalhöyük, Pinarbaşı, Güvercinkayası) (see Fig. 5b), supports the identification of the Köşk small equids as hydruntines.

Although it has been suggested that both hydruntine and non-hydruntine hemiones co-existed on the central Anatolian plateau in the Neolithic, it seems unlikely on ecological grounds that two closely related hemiones (in addition to horses) would co-exist within the Konya basin of central Anatolia; there are no other examples of sympatric hemione species or sub-species (Geigl and Grange 2012; Reading *et al.* 2001; Tatin *et al.* 2003). It seems more likely that it is simply difficult to reliably distinguish between the skeletal and dental remains of hemiones and hydruntines with the exception of mandibular molars. For example, postcranial biometrics indicate that central Anatolian hydruntines are very similar in size to the non-hydruntine hemiones from Çayönü although they may be somewhat smaller than the more robust hemiones from Neolithic northern Syria (Shams ed Dinn and Mureybet (Gourichon and Helmer 2008)) (see Figs. 3 and 4).

There are only a few depictions of the extinct hydruntine in the archaeological record. It has been argued that a cave painting from the site of Lascaux represents a European hydruntine (Geigl and Grange 2012) and it is possible that the equids depicted in the cave paintings at Balkayası (western central Turkey) represent Anatolian hydruntines. A number of painted and relief decorated ceramic vessels from Köşk and also the nearby site of Tepecik-Çiftlik likely represent the most detailed images of hydruntines available (Fig. 6) (Bıçakçı 2001; Öztan 2002, 2007; Silistreli 1989). These depictions show equids (hooves consisting of a single digit are accurately depicted as opposed to cloven hooves in images of cattle and red deer) with large heads and ears, fleshy tails, a light dun colored coat, and an absence of striping on the legs (a feature typical of the African asses)—all features typical of extant Asiatic wild asses. In at least one case, a foal with dark fore- and hindquarters and a light midsection is shown following behind an adult jenny (Fig. 6). A dorsal stripe, common in most extant hemiones, is not visible in these profile-view depictions.

Equid hunting

At late 7th and 6th millennium BC Köşk Höyük, the remains of horses are approximately twice as abundant as those of hydruntines, indicating that horses were the favored, or at least, most frequently encountered equid in the vicinity of the settlement. This is also the case at the contemporary and slightly later sites of Orman Fidanlığı and Kes Kaya, where horses are clearly more abundant in the biometric data. However, at earlier sites in central Anatolia including Aşıklı Höyük, Can Hasan III, and Çatalhöyük hydruntines dominate the equid assemblages and horses are poorly represented, especially in the earliest assemblages. For example, at Can Hasan III (early 8th millennium BC), the biometric data provide no clear evidence for the presence of any horses and the dental remains described by Payne (1991) can all be attributed to hydruntines. Horses are also poorly represented in the small sample of equids described by Payne (1985) for Aşıklı Höyük (middle 8th millennium BC). At Çatalhöyük, horses are poorly represented, or absent, in the earliest deposits dating to the late 8th millennium BC but increase somewhat in relation to hydruntines through the occupational sequence (although the overall abundance of equids declines)(Martin and Russell 2006; Russell *et al.* 2013; Russell 2013).

The significance of this pattern of change over time in the frequencies of large versus small equids is difficult to assess. It may indicate a decline in hydruntines and an increase in horse populations from the eighth through the 5th millennia BC as a result of environmental changes on the central Anatolian plateau. However, most environmental reconstructions indicate increasing aridification in the region in the late Neolithic (e.g., 8.2 kbp climatic event) (Cheng *et al.* 2015; Kuzucuoğlu and Roberts 1998; Weninger *et al.* 2006) which would seem to favor hemiones which are characterized as arid adapted grazers, over horses which require wetter conditions and more productive grazing resources (Bahloul *et al.* 2001; Kaczensky *et al.* 2008). Based on observations of the behavior of sympatric equids in Mongolia and Uzbekistan (Bahloul *et al.* 2001; Kaczensky *et al.* 2008), the Konya basin may have plausibly been partitioned between horse and hemione populations with horses being more abundant around the productive margins of the plains, near permanent water sources (rivers and lakes), and within volcanic uplands (e.g., Melendiz mountains and Karadağ), while hydruntines were more common in the drier interior grasslands. Given the small number of sites from which reliable equid frequencies are available, it is possible that diachronic trends showing an increase in horses over time are not representative and that horses were simply more abundant in the immediate vicinity of the later prehistoric sites located in more productive environments whereas earlier sites (Çatalhöyük, Can Hasan) were located within the preferred habitat of hydruntines.

Given the abundance of equid remains at Köşk Höyük, equid hunting was clearly a central feature of social life in the late Neolithic settlement and played an important role in the subsistence economy. Although equid remains were recovered from all types of contexts, they frequently derive from large pit features which also contain large amounts of ash, broken ceramic vessels and the bones of wild and domestic cattle as well as smaller numbers of domestic sheep and goats (Arbuckle 2006). These deposits, in which equids represent 16-94% ($\text{mean}_{\text{level V/IVpits}}$= 40%) of the mammal remains, have been interpreted as representing large-scale feasting debris and suggest that communal hunting of and feasting on equids were important and regular community events central to the functioning of social life at Köşk (Arbuckle 2012). This is supported by the presence of equids and equid hunting scenes on decorated ceramic vessels suggesting that these events, and the animals themselves, were memorialized and likely incorporated into the social and religious fabric of this distinctive central Anatolian community.

Given the large size of equids and the large quantities of remains recovered from pit features, it seems likely that a large proportion of the Köşk community participated in feasting events. Equid hunting may have also involved community participation. Texts from Bronze Age Mesopotamia refer to hunting hemiones by digging pit traps, using snares, and stalking them at water sources (Zarins 2014). In addition, 'desert kites', large stone-lined landscape features designed to drive herd animals into enclosures, were likely used to hunt hemiones in the steppic regions from the Levant to the Caucasus perhaps extending back into the Neolithic (Bar-Oz *et al.* 2011a; Bar-Oz, *et al.* 2011b; Chahoud *et al.* 2016; Kempe and Al-Malabeh 2013). Although features similar to desert kites have not been identified in central Anatolia, it is plausible that groups of hunters were required to cooperatively track, trap and kill herds of equids and then transport carcasses back to settlements for consumption.

Previous work has shown that all parts of the equid skeleton are present in pit features at Köşk including heads, feet, fore and hind limbs, and axial skeleton (Arbuckle 2012). The lack of evidence for a

Table 5. Frequencies of equids in archaeofaunal assemblages in early Holocene central Anatolia as well as contemporary sites in SW Asia exhibiting high frequencies (>15%) of equids. Region 1= central Anatolia; 2=west Anatolia; 3= Syria, Balikh valley; 4= Syria, Khabur drainage; 5= Syria, Euphrates valley; 6=northern Iraq; 7=southern Levant. C=caballine/horse; H=hemione; A=ass/wild donkey.

Site	*region*	*date (BC)*	*% Equus*	*Taxa*	*reference*
Can Hasan I	1	4500	present	C, Hy	(French 1965)
Güvercinkayası	1	5000	2.8	C, Hy	(Arbuckle 2006)
Köşk Höyük I	1	5000	0.8	C, Hy	(Arbuckle)
Köşk Höyük II	1	5500	21.9	C, Hy	(Arbuckle)
Köşk Höyük III	1	5700	10.6	C, Hy	(Arbuckle)
Çatalhöyük South TPC	1	6000	0.7	?	(Marciniak et al. 2013)
Çatalhöyük West	1	6000	2.6	C, Hy	(Orton 2010)
Köşk Höyük IV	1	6000	35.8	C, Hy	(Arbuckle)
Köşk Höyük Pits	1	6200	35	C, Hy	(Arbuckle)
Pinarbaşı B phase E	1	6500	20.9	C, Hy	(Carruthers 2005; Baird et al. 2011)
Pinarbaşı B phase F	1	6500	29	C, Hy	(Carruthers 2005; Baird et al. 2011)
Çatalhöyük South P	1	6500	3.2	C, Hy	(Russell et al. 2013)
Erbaba	1	6500	0.2	H	(Arbuckle 2008)
Çatalhöyük South VIII	1	6800	8.3	C, H	(Russell and Martin 2005)
Çatalhöyük KOPAL	1	7000	3.9	C, Hy	(Russell and Martin 2005)
Suberde	1	7200	0	-	(Perkins and Daly 1968)
Aşıklı 2A	1	7500	0.1	C, Hy	(Stiner et al. 2014)
Aşıklı 2E	1	7900	1.2	C, Hy	(Stiner et al. 2014)
Aşıklı 2G	1	8000	2.5	C, Hy	(Stiner et al. 2014)
Can Hasan III	1	8000	present	Hy	(French 1972)
Pinarbaşı A	1	9000	5.7	Hy	(Carruthers 2003)
Pinarbaşı Epi	1	11,000	13.2	Hy	(Baird et al. 2013)
Kes Kaya	2	4500	abundant	C, Hy	(Uerpmann 2001)
Orman Fidanlığı	2	5000	2.8	C, Hy	(Uerpmann 2001)
Kuruçay (Early Chalc)	2	5500	present	?	(Deniz and Sentuna 1988)
Höyücek	2	6200	0.3	H?	(De Cupere and Duru 2003)
Bademağacı	2	6500	0	-	(De Cupere et al. 2008)
Tell Zeidan (Halaf)	3	5500	39.6	H	(Grossman and Hinman 2013)
Mashnaqa	4	4500	25	H	(Zeder 1998)
Boueid II	4	5500	45.5	H	(Sana 2002)
Umm Qseir (Halaf)	4	5500	19.2	H	(Zeder 1998)
Kashkashok III	4	5500	17	H	(Zeder 1998)
El Kowm 2	4	6500	25.8	H	(Helmer 2000)
Djade	5	8500	34.3	H	(Gourichon and Helmer 2008)
Cheikh Hasan	5	8500	30.6	H	(Helmer 1994)
Mureybet IVA	5	8500	46.4	H, A	(Gourichon and Helmer 2008)
Mureybet III	5	9000	47.7	H, A	(Gourichon and Helmer 2008)
Mureybet II	5	9000	23.3	H, A	(Gourichon and Helmer 2008)
Nahr el Homr	5	9000	85	H	(Clason and Buitenhuis 1978)
Jerf el Ahmar	5	9000	36.2	H	(Helmer 1994)
Yarim Tepe II	6	5500	17.5	H	(Bibikova 1981)
Umm Dabaghiyah	6	5500	71.7	H	(Bökönyi 1973)
Azraq 31	7	6500	25.8	H	(Martin 1992)
Wadi Tbeik	7	6500	18.4	C, H/A?	(Tchernov and Bar-Yosef 1982)

'schlepp effect' (Perkins and Daly 1968), or differential transport of portions of equid carcasses, which may weigh more than 200 kg (Nowak 1999), suggests that equid hunting took place close enough to the settlement that entire carcasses were brought back to the site rather than being subject to extensive field butchery at distant kill sites. This is supported by preliminary analysis of strontium isotopes from Köşk equid teeth (Meiggs *et al.* 2018), which indicates that hunted equids grazed in the uplands of the Melendiz mountains immediately north of the Köşk settlement. Although the seasonality data for equid hunting at Köşk is not yet well understood, the presence of perinatal remains suggests that at least some of these hunting and feasting activities took place in the spring.

Although the frequencies of equid remains at Köşk Höyük are unusually high for Anatolian faunal assemblages, equid hunting was also an important part of contemporary animal economies in other regions of SW Asia including northern Syria, eastern Jordan, and the Iraqi Jazirah. In this regards, the emphasis on equid hunting at Köşk can perhaps be seen as a part of a larger regional trend characterized by an increasingly intensive relationship with wild equids in the 7th and 6th millennia BC. The intensive hunting of equids in the Fertile Crescent region is perhaps best documented at Pottery Neolithic Umm Dabaghiyah, northern Iraq, where hemiones represent a staggering 72% of the faunal assemblage (Bökönyi 1986). However, a long tradition of intensive hemione hunting is also evident at sites on the Syrian middle Euphrates dating back to the terminal Pleistocene and early Holocene at sites including Mureybet, Jerf el Ahmar, Cheikh Hasan and D'jade (Table 5). Moreover, high frequencies of equid remains have also been identified at sites in central Syria (El Kowm), eastern Jordan (Azraq 31) and the Negev desert (Wadi Tbeik) indicating the presence of a broad arc of interior arid grasslands in which hemiones were heavily exploited through much of the early Holocene.

In addition, many sites associated with the Halaf tradition, particularly those in the Balikh and Khabur drainages of northern Syria, exhibit frequencies of hemiones similar to those evident at Köşk (see Fig. 1; Table 5). This increasingly intensive exploitation of wild equids in the 6th millennium BC has rarely been addressed but may be associated with an increasingly intensive use of steppe environments by sheep and goat pastoralists suggested for the 6th millennium BC (e.g., Miller 1996). It may have also led to experiments with capturing, taming and managing hemiones; practices which are attested in both texts and zooarchaeological remains from later periods (Weber 2008; Zarins 2014).

Although the 6th millennium BC seems to represent a peak in wild equid exploitation at Köşk and more generally across the arid regions of the Fertile Crescent, equid exploitation decreases dramatically in the following 5th and 4th millennia BC as evident at Köşk I and post-Halaf sites in Greater Mesopotamia (e.g., Grossman and Hinman 2013). However, as a result of the dearth of faunal assemblages from this crucial period, combined with the widespread appearance of domestic equids in the 3rd millennium BC, it is difficult to assess the timing of the extirpation of local wild Anatolian horses and hydruntines. It is possible that Anatolian wild horses disappeared by the end of the 4th millennium BC and were replaced by imported domesticates in the 3rd millennium. It is currently unclear if those horse remains dating to the mid and late 4th millennium BC from sites in eastern Turkey including Çadır Höyük (Yozgat), Arslantepe (Malatya) and Norşuntepe (Elazig) represent the last wild horses in the region or an early wave of domesticates (Arbuckle 2009; Boessneck and Von den Driesch 1976b; Bökönyi 1991; Shev 2016).

For hydruntines, the situation is similarly unclear. Small numbers of hydruntines have been identified at EBA Demirci Höyük and Küllüoba in west central Turkey dating to the late 4th and early 3rd millennium BC and several mandibular molar teeth with hydruntine features have been identified from late 3rd and early 2nd millennia BC deposits at Acemhöyük (central Turkey) (Arbuckle 2013). Scholars have also claimed that hydruntines were present in the Balkans and the Iberian peninsula well into the 1st millennium BC (although see Geigl and Grange 2012). Furthermore, Classical authors including Pliny and Varro indicate that wild hemiones were well known on the central Anatolian plateau into the first centuries AD (Devereux 1965) suggesting that hydruntines may have outlasted wild horses in the region by a considerable extent.

Conclusions

In this paper we describe the abundant equid remains from Köşk Höyük, central Turkey. These remains, which represent up to 35% of the fauna from late 7th and 6th millennia BC deposits, indicate that equids were an abundant part of the endemic fauna of this region of the Anatolian plateau and that they were a central focus of the economic and social systems within this community. We argue that both Anatolian wild horses (*Equus ferus*) and hydruntines (*Equus hemionus hydruntinus*) were exploited at Köşk although the remains of horses dominate the assemblage. The dominance of horses over hydruntines at Köşk may represent a regional diachronic pattern in equid hunting practices characterized by a shift from hunting hydruntines, which were the dominant equid in 8th and early 7th millennia BC faunal assemblages, to a focus on the exploitation of wild horses in the late 7th and 6th millennia BC.

Although equid remains have rarely been addressed in prehistoric Anatolia, the Köşk data, when combined with data from other sites in the region, indicate the presence of a significant tradition of intensive horse and hemione hunting on the Anatolian plateau spanning the Neolithic and Chalcolithic periods (c. 8000-4000 BC). Documenting this long tradition of equid hunting, which has parallels in other regions of SW Asia including northern Syrian and the Iraqi Jazirah, is important for understanding the ancient distribution of these taxa as well as the human relationship with two of the largest prey taxa in the SW Asian biome. Further exploration of traditions of equid hunting in Anatolia are also relevant for understanding the origins of traditions of taming and breeding hemiones in captivity—practices attested in texts in the 3rd millennium BC in northern and southern Mesopotamia—and also for reconstructing the processes involved in the eventual appearance of domestic horses in the Near East.

Acknowledgements

Excavations at Köşk Höyük were carried out with the permission of and funded by the Turkish Ministry of Culture and Tourism and TÜBITAK. Funding for zooarchaeological work was provided to BA by NSF grants BCS-0530699 and BCS-1311551, grants from the American Research Institute in Turkey (2004 and 2011) and the National Geographic Society (2011), as well as support from Baylor University. Special thanks to the director and staff of the Niğde Museum for facilitating this work.

References

Arbuckle, B. S. 2006. *The Evolution of Sheep and Goat Pastoralism and Social Complexity in Central Anatolia*. PhD. Harvard University.

Arbuckle, B. S. 2008. Caprine exploitation at Erbaba Höyük: a Pottery Neolithic village in Central Anatolia. In, Gourichon, L. and Vila, E. (eds) *Archaeozoology of Southwest Asia and Adjacent Areas VIII*: 345-365. Paris: Travaux de la Maison de l'Orient.

Arbuckle, B. S. 2009. Chalcolithic caprines, Dark Age dairy and Byzantine beef. *Anatolica* 35: 179-224.

Arbuckle, B. S. 2012. Animals and inequality in Chalcolithic Anatolia. *Journal of Anthropological Archaeology* 31: 302-313.

Arbuckle, B. S. 2013. Zooarchaeology at Acemhöyük. *Anadolu/Anatolia* 39: 55-68.

Arbuckle, B. S. and Makarewicz, C. A. 2009. The early management of cattle (*Bos taurus*) in Neolithic central Anatolia. *Antiquity* 83: 669-686.

Arbuckle, B. S., Öztan, A. and Gulçur, S. 2009. The evolution of sheep and goat husbandry in central Anatolia. *Anthropozoologica* 44: 129-157.

Bahloul, K., Pereladova, O. B., Soldatova, N., Fisenko, G., Sidorenko, E. and Sempéré, A. J. 2001. Social organization and dispersion of introduced kulans (*Equus hemionus kulan*) and Przewalski horses (*Equus przewalski*) in the Bukhara Reserve, Uzbekistan. *Journal of Arid Environments* 47: 309-323.

Baird, D., Asouti, E., Astruc, L., Baysal, A., Baysal, E., Carruthers, D., Fairbairn, A., Kabukcu, C., Jenkins, E. and Lorentz, K. 2013. Juniper smoke, skulls and wolves' tails. The Epipalaeolithic of the Anatolian plateau in its South-west Asian context; insights from Pınarbaşı. *Levant* 45: 175-209.

Baird, D., Carruthers, D., Fairbairn, A. and Pearson, J. 2011. Ritual in the landscape: evidence from Pınarbası in the seventh-millennium cal BC Konya Plain. *Antiquity* 85: 380-394.

Bar-Oz, G., Nadel, D., Avner, U. and Malkinson, D. 2011a. Mass hunting game traps in the southern Levant: the Negev and Arabah "Desert Kites". *Near Eastern Archaeology* 74: 208-215.

Bar-Oz, G., Zeder, M. and Hole, F. 2011b. Role of mass-kill hunting strategies in the extirpation of Persian gazelle (*Gazella subgutturosa*) in the northern Levant. *Proceedings of the National Academy of Sciences* 108: 7345-7350.

Benecke, N. 2009. On the beginning of horse husbandry in the southern Balkan peninsula: the horses from Kırkareli-Kanlıgeçit (Turkish Thrace). *TÜBA-AR* 12: 13-24.

Bibikova, V. I. 1981. Animal husbandry in northern Mesopotamia in the fifth millennium BC [in Russian]. In, Munchaev, R. M. and Merpert, N. Y. (eds) *Rannezemledel'cheskie Poseleniya Severnoj Mesopotamii*: 299-307. Moscow: Nauka.

Bıçakçı, E. 2001. Tepecik-Çiftlik Höyüğü (Niğde) kazısı ışıgında orta Anadolu tarihöncesi külturleri ile ilgili yeni bir değerlendirme. *TÜBA-AR* 4: 25-41.

Boessneck, J. and Von den Driesch, A. 1976a. Die Wildfauna der Altinova in vorgeschichtlichler Zeit, wie sie die Tierknochenfunde vom Norsuntepe und anderen Siedlungshülgeln erschliessen. *Keban Projesi 1972 Çalısmaları'ndan Ayrıbaşım*: 91-100.

Boessneck, J. and Von den Driesch, A. 1976b. Pferde im 4/3. Jahrtausend v. Chr. in Ostanatolien. *Saugetierkundliche Mitteilungen* 24: 81-87.

Boessneck, J. and Von den Driesch, A. 1977/1978. Die zoologische Dokumetation von drei Pferdskeletten und anderen Tierknochenfunden aus einem Kammergrab auf dem Norsun-Tepe (Ostanatolien). *Istambuler Mitteilungen* 27/28: 73-91.

Boessneck, J. and Von den Driesch, A. 1979. *Die*

Tierknochenfunde aus der Neolithischen Siedlung auf dem Fikirtepe bei Kadiköy am Marmara Meer. München: Institut für Palaeoanatomie, Domestikationsforschung und Geschichte der Tiermedizin der Universität München.

Bökönyi, S. 1973. The fauna of Umm Dabaghiyah: a Preliminary report. *Iraq* 35: 9-11.

Bökönyi, S. 1983. Late Chalcolithic and Early Bronze I animal remains from Arslantepe (Malatya): a Preliminary report. *Origini* 12: 581-598.

Bökönyi, S. 1986. The equids of Umm-Dabaghiyah, Iraq. In, Meadow, R. H. and Uerpmann, H.-P. (eds) *Equids in the Ancient World*: 302-318. Wiesbaden: Dr. Ludwig Reichert Verlag.

Bökönyi, S. 1991. Late Chalcolithic horses in Anatolia. In, Meadow, R. H. and Uerpmann, H.-P. (eds), *Equids in the Ancient World Volume II*: 123-131. Wiesbaden: Dr. Ludwig Reichert Verlag.

Bonifay, M.-F. 1991. *Equus hydruntinus* Regalia *minor* n. ssp. from the caves of Lunel-Viel (Herault, France). In, Meadow, R. H. and Uerpmann, H.-P. (eds) *Equids in the Ancient World Volume II*: 178-276. Wiesbaden: Dr. Ludwig Reichert Verlag.

Buitenhuis, H. 1997. Aşıklı Höyük: A 'protodomestication' site. *Anthropozoologica* 25-26: 655-662.

Burke, A., Eisenmann, V. and Ambler, G. K. 2003. The systematic position of Equus hydruntinus, an extinct species of Pleistocene equid. *Quaternary Science* 59: 459-469.

Cameron, I. 2005. Equid of Çatalhoyuk: An insight into the theory and methodology in the identification of equidae. MSc. University College London.

Carruthers, D. 2003. Hunting and herding in Anatolian prehistory: the 9th and 7th millennium site at Pınarbaşı. PhD. University of Edinburgh.

Carruthers, D. 2005. Hunting and Herding in Central Anatolian Prehistory: the 9th and 7th millennium site at Pınarbaşı. In, Buitenhuis, H., Choyke, A. M., Martin, L., Bartosiewicz, L. and Mashkour, M. (eds) *Archaeozoology of the Near East VI*: 85-95. Groningen: ARC Publication 123.

Chahoud, J., Vila, E., Bălăşescu, A. and Crassard, R. 2016. The diversity of Late Pleistocene and Holocene wild ungulates and kites structures in Armenia. *Quaternary International* 395: 133-153.

Cheng, H., Sinha, A., Verheyden, S., Nader, F., Li, X., Zhang, P., Yin, J., Yi, L., Peng, Y. and Rao, Z. 2015. The climate variability in northern Levant over the past 20,000 years. *Geophysical Research Letters* 42: 8641-8650.

Clason, A. T. and Buitenhuis, H. 1978. A preliminary report on the faunal remains of Nahr el Homr, Hadidi and Ta'as in the Tabqa Dam region of Syria. *Journal of Archaeological Science* 5: 75-83.

Crees, J. and Turvey, S. 2014. Holocene extinction dynamics of *Equus hydruntinus*, a late-surviving European megafaunal mammal. *Quaternary Science Reviews* 91: 16-29.

Davis, S. J. 1980. Late Pleistocene and Holocene equid remains from Israel. *Zoological Journal of the Linnean Society* 70: 289-312.

De Cupere, B. and Duru, R. 2003. Faunal remains from Neolithic Höyücek (SW-Turkey) and the presence of early domestic cattle in Anatolia. *Paléorient* 29: 107-120.

De Cupere, B., Duru, R. and Umurtak, G. 2008. Animal husbandry at the Early Neolithic to Early Bronze Age site of Bademağacı (Antalya province, SW Turkey): evidence from the faunal remains. In, Vila, E., Gourichon, L., Choyke, A. and Buitenhuis, H. (eds) *Archaeozoology of the Near East VIII*: 367-406. Lyon: Maison de l'Orient et de la Mediterranee.

Deniz, E. and Sentuna, C. 1988. Kuruçay Höyük Kazısı Arkeobiyolojik Materyalının Tum Değerlendirilmiş. *Arkeometri Sonuçlari Toplantısı* 4: 169-185.

Devereux, G. 1965. Homer's wild she-mules. *The Journal of Hellenic Studies* 85: 29-32.

Dive, J. and Eisenmann, V. 1991. Identification and discrimination of first phalanges from Pleistocene and modern *Equus*, wild and domestic. In, Meadow, R. H. and Uerpmann, H.-P. (eds) *Equids in the Ancient World Volume II*: 278-333. Wiesbaden: Dr. Ludwig Reichert Verlag.

Efe, T., Fidan, E. and Gundem, C. Y. 2008. Complex two in the Early Bronze II Upper Town of Külüoba near Eskişehir. *Anatolica* 34: 67-102.

Eisenmann, V. 1986. Comparative osteology of modern and fossil horses, half-asses, and asses. In, Meadow, R. H. and Uerpmann, H.-P. (eds) *Equids in the Ancient World*: 67-116. Wiesbaden: Dr. Ludwig Reichert Verlag.

Eisenmann, V. and Mashkour, M. 1999. The small equids of Binagady (Azerbaidjan) and Qazvin (Iran): *E. hemionus* Binagadensis nov. subsp. and *E. hydruntinus*. *GEOBIOS* 32: 105-122.

Eisenmann, V., Helmer, D. and Sequi, M.S. 2002. The big Equus from the Geometric Kebaran of Umm El Tlel, Syra: *Equus valeriani, Equus capensis* or *Equus caballus*? In, Buitenhuis, H. Choyke, A.M., Maskour, M. and Al-Shiyab, A.H. (eds) *Archaeozoology of the Near East V, Proceedings of the Fifth International Symposium on the Archaeozoology of Southwestern Asia and Adjacent Areas*: 62-73. Groningen.

French, D. 1965. Excavations at Can Hasan: Fourth preliminary report, 1964. *Anatolian Studies* 15: 87-94.

French, D. 1972. Settlement distribution in the Konya Plain, south central Turkey. In, Ucko, P. J., Tringham, R. and Dimbleby, G. W. (eds)

Man, Settlement and Urbanism: 231-238. London: Duckworth.

Geigl, E.-M. and Grange, T. 2012. Eurasian wild asses in time and space: morphological versus genetic diversity. *Annals of Anatomy* 194: 88-102.

Gourichon, L. and Helmer, D. 2008. Étude archéozoologique de Mureybet. In, Ibánez, J. J. (ed.) *Le site néolithique de Tell Mureybet* (Syrie du Nord): 115-228. Oxford: BAR International Series 1843.

Grigson, C. 1993. The Earliest Domestic Horses in the Levant?—New Finds from the Fourth Millennium of the Negev. *Journal of Archaeological Science* 20: 645-655.

Grigson, C. 2012. Size matters: donkeys and horses in the prehistory of the southernmost Levant. *Paléorient* 38: 194-233.

Grossman, K. and Hinman, M. 2013. Rethinking Halaf and Ubaid animal economies: Hunting and herding at Tell Zeidan (Syria). *Paléorient* 39: 201-219.

Helmer, D. 1994. La domestication des animaux d'embouche dans le Levant nord (Syrie du nord et Sinjar) du milieu du IXe millenaire BP à la fin du VIIIe millenaire BP. Nouvelles donnees d'apres les fouilles recents. *Anthropozoologica* 20: 41-54.

Helmer, D. 2000. Étude de la faune mammalienne d'El Kowm 2. In, Stordeur, D. (ed.) *Une île dans le désert : El Kowm 2 (Néolithique précéramique, 8000-7500 BP Syrie)*: 233-264. Paris: CNRS Editions.

Hongo, H., Meadow, R. H., Öksüz, B. and Gülçin, I. 2004. Animal exploitation at Çayönü Tepesi, southeastern Anatolia. *TÜBA-AR* 7: 107-119.

Kaczensky, P., Ganbaatar, O., Von Wehrden, H. and Walzer, C. 2008. Resource selection by sympatric wild equids in the Mongolian Gobi. *Journal of Applied Ecology* 45: 1762-1769.

Kansa, E. C., Kansa, S. W. and Levy, T. E. 2006. Eat like an Egyptian? A contextual approach to an Early Bronze I "Egyptian colony" in the southern Levant. In, Maltby, M. (eds) *Integrating Zooarchaeology*: 76-97. Oxford: Oxbow.

Kaya, M. A. and Aksoylar, M. Y. 1992. Bozdağ Konya'da Yasayan Yaban Koyunun, *Ovis orientalis* anatolica Valenciennes 1856'nin Davranışları. *Turkish Journal of Zoology* 16: 229-241.

Kempe, S. and Al-Malabeh, A. 2013. Desert kites in Jordan and Saudi Arabia: Structure, statistics and function, a Google Earth study. *Quaternary International* 297: 126-146.

Kuzucuoğlu, C. and Roberts, N. 1998. Évolution de l'environnement en Anatolie de 20,000 a 6000 BP. *Paléorient* 23: 7-24.

Marciniak A., Pawlowska, K., Russell, N., Twiss, K., Martin, L. et al. (2013) Çatalhöyük Area TP Main Zooarchaeological Dataset. [online] Open Context: Open Context. Available at: <http://opencontext.org/tables/84332d5b790ecfbbf9a341273ea8d142>.

Martin, L. 1992. The faunal remains. *Levant* 24: 25-27.

Martin, L. and Russell, N. 2006. The equid remains from Neolithic Çatalhöyük, central Anatolia: A preliminary report. In, Olsen, S. J. (ed.) *Horses and Humans: the Evolution of Human-Equine Relationships*: 115-126. Oxford: British Archaeological Reports International Series 1560.

Martin, L., Russell, N. and Carruthers, D. 2002. Animal remains from the Central Anatolian Neolithic. In, Gérard, F. and Thissen, L. (eds) *The Neolithic of Central Anatolia: Internal Developments and External Relations During the 9th-6th millennia cal BC*: 193-216. Istanbul: Ege Yayınları.

Mashkour, M. 2002. Chasse et elevage au nord du plateau central Iranien entre le Neolithique et l'age du fer. *Paléorient* 28: 27-42.

Meadow, R. H. 1986. Some equid remains from Çayonu, Southeastern Turkey. In, Meadow, R. H. and Uerpmann, H.-P. (eds) *Equids in the Ancient World*: 266-301. Wiesbaden: Dr. Ludwig Reichert Verlag.

Meadow, R. H. 1999. The use of size index scaling techniques for research on archaeozoological collections from the Middle East. In, Becker, C., Manhart, H., Peters, J. and Schibler, J. (eds) *Historia animalium ex ossibus: Beitrage zür Palaoanatomie, Archaologie, Agyptologie, Ethnologie, and Geschichte der Tiermedizin*: 285-300. Rahden: Verlag Marie Leidorf.

Meiggs, D., B. S. Arbuckle, A. Öztan (2018) The Pixelated Shepherd: Identifying detailed local land use practices at Chalcolithic Köşk Höyük, central Turkey, using a strontium isotope (87Sr/86Sr) isoscape. In, Ventresca-Miller, A. and Makarewicz, C. (eds) *Isotopic Investigations of Pastoralism*: 77-96. Routledge: Oxford.

Mellaart, J. 1966. Excavations at Çatal Hüyük, 1965: Fourth Preliminary Report. *Anatolian Studies* 16: 165-191.

Miller, N. F. 1996. Seed eaters of the ancient Near East: human or herbivore? *Current Anthropology* 37: 521-528.

Nowak, R. M. 1999. *Walker's mammals of the world*. Baltimore: JHU Press.

Orlando, L., Mashkour, M., Burke, A., Douady, C. J., Eisenmann, V. and Hanni, C. 2006. Geographic distribution of an extinct equid (*Equus hydruntinus*: Mammalia, Equidae) revealed by morphological and genetical analyses of fossils. *Molecular Ecology* 15: 2083-2093.

Orlando, L., Metcalf, J. L., Alberdi, M. T., Telles-Antunes, M., Bonjean, D., Otte, M., Martin, F., Eisenmann, V., Mashkour, M. and Morello, F.

2009. Revising the recent evolutionary history of equids using ancient DNA. *Proceedings of the National Academy of Sciences* 106: 21754-21759.

Orton, D. 2010. Animal Bones Report Trench 5. Çatalhöyük Archive Report 2010. [online] Available at <http://www.catalhoyuk.com/archive_reports/>. [Accessed 10 January 2012].

Öztan, A. 2002. Köşk Höyük: Anadolu arkeolojisine yeni katkılar. *TÜBA-AR* 5: 57-72.

Öztan, A. 2007. Köşk Höyük: Niğde--Bor Ovası'nda bir Neolitik yerleşim. In, Ozdoğan, M. and Başgelen, N. (eds) *Türkiye'de Neolitik Donem: Yeni kazılar, yeni bulgular*: 223-236. Istanbul: Arkeoloji ve Sanat Yayınları.

Öztan, A. 2010. Archaeological investigations at Köşk Höyük, Niğde. In, d'Alfonso, L., Balza, M. E. and Mora, C. (eds) *Geo-archaeological Activities in Southern Cappadocia, Turkey*: 83-96. Pavia: Italian University Press.

Payne, S. 1985. Animal bones from Aşıklı Hüyük. *Anatolian Studies* 35: 109-122.

Payne, S. 1991. Early Holocene equids from Tall-i-Mushki (Iran) and Can Hasan III (Turkey). In, Meadow, R. H. and Uerpmann, H.-P. (eds) *Equids in the Ancient World. Volume 2*: 132-177. Wiesbaden: Dr. Ludwig Reichert Verlag.

Perkins, D. P. and Daly, P. 1968. A hunters' village in Neolithic turkey. *Scientific American* 219: 96-106.

Peters, J., Von den Driesch, A. and Helmer, D. 2005. The upper Ephrates-Tigris basin: Cradle of agro-pastoralism? In, Vigne, J.-D., Peters, J. and Helmer, D. (eds) *The First Steps of Animal Domestication: New archaeological approaches*: 96-124. Oxford: Oxbow.

Pinhasi, R. 2006. Neolithic skull shapes and demic diffusion: a bioarchaeological investigation into the nature of the Neolithic transition. *Documenta Praehistorica* 33: 61-70.

Reading, R. P., Mix, H. M., Lhagvasuren, B., Feh, C., Kane, D. P., Dulamtseren, S. and Enkhbold, S. 2001. Status and distribution of khulan (*Equus hemionus*) in Mongolia. *Journal of Zoology* 254: 381-389.

Regalia, E. 1907. Sull *Equus (Asinus) hydruntinus* Regalia della grotta di Romanelli (Castro, Lecce). *Archivo per l'Antropologia e l'Etnologia* 37: 375-390.

Russell, N. 2012. *Social Zooarchaeology: Humans and Animals in Prehistory. Cambridge*: Cambridge University Press.

Russell, N. 2013. Mammals from the BACH Area. In, Hodder, I. (eds) *Humans and Landscapes of Çatalhöyük. Reports from the 2000-2008 seasons*: 225-242. London: British Institute at Ankara.

Russell, N. and Martin, L. 2005. The Çatalhöyük mammal remains. In, Hodder, I. (ed.) *Inhabiting Çatalhöyük: Reports from the 1995-1999 seasons*: 33-98. Cambridge: McDonald Institute for Archaeological Research.

Russell, N., Twiss, K., Orton, D. C. and Demirergi, A. 2013. More on the Çatalhöyük Mammal Remains. In, Hodder, I. (ed.) *Humans and Landscapes of Çatalhöyük. Reports from the 2000–2008 Seasons*: 213-258. Los Angeles: Cotsen Institute of Archaeology Press.

Sana, M. 2002. The faunal remains. *Subartu* 11: 125-140.

Scott, R. S. and Maga, M. 2005. Paleoecology of the Akkasdagı hipparions (Mammalia, Equidae), late Miocene of Turkey. *Geodiversitas* 27: 809-830.

Shev, E. 2016. The Introduction of the domesticated horse in Southwest Asia. *Archaeology, Ethnology & Anthropology of Eurasia* 44: 123-136.

Silistreli, U. 1989. Les fouilles de Köşk Höyük. In, Emre, K., Mellink, M., Hrouda, B. and Özgüç, N. (eds) *Anatolia and the Ancient Near East: Studies in Honor of Tahsin Özgüç*: 461-463. Ankara.

Silver, I. A. 1963. The aging of domestic animals. In, Brothwell, D., Higgs, E. and Clark, G. (eds) *Science in Archaeology*: 250-268. London: Thames and Hudson.

Sisson, S. and Grossman, J. D. 1938. The Anatomy of the Domestic Animals. Philadelphia: W. B. Saunders.

Stehlin, H. G. and Graziosi, P. 1935. *Richerche sugli Asinidi fossili d'Europa*. Bale: Memoires de la Sociéte Paléontologique Suisse Vol 56.

Stiner, M. C. 1990. The use of mortality patterns in archaeological studies of hominid predatory adaptations. *Journal of Anthropological Archaeology* 9: 305-351.

Stiner, M. C., Buitenhuis, H., Duru, G., Kuhn, S. L., Mentzer, S. M., Munro, N. D., Pöllath, N., Quade, J., Tsartsidou, G. and Özbaşaran, M. 2014. A forager–herder trade-off, from broad-spectrum hunting to sheep management at Aşıklı Höyük, Turkey. *Proceedings of the National Academy of Sciences* 111: 8404-8409.

Tatin, L., Darreh-Shoori, B. F., Tourenq, C., Tatin, D. and Azmayesh, B. 2003. The last populations of the Critically Endangered onager *Equus hemionus onager* in Iran: urgent requirements for protection and study. *Oryx* 37: 488-491.

Tchernov, E. and Bar-Yosef, O. 1982. Animal Exploitation in the Pre-Pottery Neolithic B Period at Wadi Tbeik, Southern Sinai. *Paléorient* 8: 17-37.

Turnbull, P. F. and Reed, C. A. 1974. The fauna from the terminal Pleistocene of Palegawra cave, a Zarzian occupation site in northeastern Iraq. *Fieldiana*: Anthropology 63: 81-146.

Uerpmann, H.-P. 1987. *The ancient distribution of ungulate mammals in the Middle East*. Wiesbaden: Beihefte zum Tübinger Atlas des Vorderen Orients Reihe A (Naturwissenschaften) 27.

Uerpmann, H.-P. 2001. Remarks on faunal remains from the Chalcolithic sites "Orman Fidanlığı" and "Kes Kaya" near Eskişehir in North-Western Anatolia. In, Efe, T. (ed.) *The Salvage Excavations at Orman Fidanlığı: A Chalcolithic Site in Inland Northwestern Anatolia*: 187-210. Istanbul: TASK Vakfı Yayınları.

Van Asperen, E. N. 2010. Implications of age variation and sexual dimorphism in modern equids for middle Pleistocene equid taxonomy. *International Journal of Osteoarchaeology* 23:1-12.

Vila, E. 2006. Data on equids from late fourth and third millennium sites in northern Syria. In, Mashkour, M. (ed.) *Equids in Time and Space: Papers in honour of Vera Eisenmann*: 102-123. Oxford: Oxbow Books.

Von den Driesch, A. and Boessneck, J. 1987. Gesamtergebnis der Untersuchungen an den Tierknochenfunden vom Demircihüyük. In, Korfmann, M. (ed.) *Demircihüyük. Die Ergebnisse der Ausgrabungen 1975-1978. Band II Naturwissenschaftliche Untersuchungen*: 52-66. Mainz: Verlag.

Weber, J. A. 2008. Elite equids: redefining equid burials of the mid to late 3rd millennium BC from Umm el-Marra, Syria. In, Vila, E., Gourichon, L., Choyke, A. and Buitenhuis, H. (eds) *Archaeozoology of the Near East VIII*: 499-520. Lyon: Maison de l'Orient et de la Mediterranee.

Weninger, B., Alram-Stern, E., Bauer, E., Clare, L., Danzeglocke, U., Joris, O., Kubatzki, C., Rollefson, G. O., Todorova, H. and Van Andel, T. 2006. Climate forcing due to the 8200 cal yr BP event observed at Early Neolithic sites in the eastern Mediterranean. *Quaternary Research* 66: 401-420.

Zarins, J. 2014. *The Domestication of Equidae in Third-Millennium BCE Mesopotamia*. Cornell University Studies in Assyriology and Sumerology, vol. 24.

Zeder, M. 1998. Environment, economy and subsistence in northern Mesopotamia. In, Fortin, M. and Aurenche, O. (eds) *Espace Naturel, Espace Habité en Syrie de Nord (10e-2e millénaires Av. J-C.)*: 55-66. Quebec: Canadian Society for Mesopotamian Studies Bulletin 33.

Zeder, M. A. 1986. The equid remains from Tal-e Malyan, southern Iran. In, Meadow, R. H. and Uerpmann, H.-P. (eds) *Equids in the Ancient World*: 366-412. Wiesbaden: Dr. Ludwig Reichert Verlag.

Zeuner, F. E. 1963. *A History of Domesticated Animals*. London: Hutchinson and Company.

5 Fishing and herding in the Nile Delta during the Predynastic (4th millennium BC)

Joséphine Lesur[1]

1 Archéozoologie, Archéobotanique: Sociétés, Pratiques et Environnements (UMR 7209), Sorbonne Universités, Muséum national d'Histoire naturelle, CNRS. C.P. 55, 55, rue Buffon 75005 Paris, France. Tel.: + 33 1 40 79 56 17. Email: jolesur@mnhn.fr

Abstract
In the Nile Delta, as in Upper Egypt, the 4th millennium BC was a period of subtle evolutions. Recent archaeological data has revealed that the societies of the Delta evolved independently from those of Upper Egypt, due to their privileged relationship with the Near East. In terms of food production, the Delta provided favourable conditions for agriculture and livestock herding (especially pigs and cattle). Nevertheless, wild resources, especially fish from the Nile, continued to play an important part in the human diet. Using archaeozoological data, mainly from the ongoing excavations of Tell el-Iswid, this article aims to illustrate the characteristics of animal exploitation, its evolution and the region's relationship with Upper Egypt throughout the 4th millennium BC.

Keywords
Predynastic, Egypt, Nile Delta, herding, fishing

Introduction
In the Nile Delta, Holocene human occupation began at the end of the 6th millennium BC (Tristant and Midant-Reynes 2011). Unfortunately, alluvial deposits have buried the archaeological data under thick layers of silt, especially in the western part, therefore the main sites available to study are located in the eastern part where they occupy *geziras:* Natural mounds that remain dry during flooding (Fig. 1; Tristant 2006). During the 5th millennium, human settlements developed in the area along with advances in farming and herding (Lesur 2013). Culturally, the area presents various influences, for example, the different occupation phases of the large site of Merimde Beni Salame are related to both the Saharo-Sudanese and Near East traditions (Eiwanger 1999; Tristant and Midant-Reynes 2011). At the beginning of the 4th millennium, the cultures of Lower Egypt began to develop (a phase formerly called "Maadi-Buto") which maintained strong connections with the contemporary Levantine cultures (Guyot 2010; Tristant and Midant-Reynes 2011). During the second half of this millennium, many sites show an increased influence of Naqadian cultural traits. And progressively, through subtle social and cultural dynamics, the native culture of northern Egypt blended with the main traits of the southern Naqada culture (Buchez and Midant-Reynes 2007).

As with other large sites such as Buto, Tell el-Iswid offers a unique archaeological opportunity to refresh our vision of these successive cultures which, during the formative period of the Egyptian state, occupied the key region of the Delta, at the crossroads of Saharan Africa and the Levant. The excavations began in 2007 (Midant-Reynes and Buchez 2014; Midant-Reynes *et al.* 2015) and are still ongoing, but we have already obtained interesting results from the fauna due to the large assemblage of bones. These have allowed us to assess the evolution of animal exploitation in the rich natural environment of the Delta during this period of major cultural and social change. Food production, developed mainly during the 5th millennium in this area, deeply modified the landscape and its relationship with faunal diversity. However, and contrary to other regions of Northeastern Africa, the varied aquatic resources still provided a large part of the human diet during the 4th millennium (Lesur 2013).

After a brief description of the origin and diffusion of herding in Egypt, which had a strong influence on the social and economic dynamics of

Predynastic cultures, we will focus on the faunal results from Tell el-Iswid. These results are imperative to the understanding of animal exploitation in the Delta, and the whole of Egypt, during this crucial period.

Development of herding in Egypt

The origin of domesticates in Africa is still controversial. The first reliable evidence for cattle occurs at the end of the 7th millennium, in the region of Nabta Playa and Bir Kiseiba (Gautier 1984, 2001). From the 6th millennium onwards, they spread to other parts of Northern Africa (e.g. Hassan 2002; Jousse 2004; Riemer 2007). During the same period, evidence of sheep (*Ovis aries*) and goats (*Capra hircus*) emerge from different sites in Egypt and Libya (Close 2002; Kindermann *et al.* 2006; Linseele *et al.* 2010).

Since there are no wild ancestors for sheep and goat in Africa, it is highly probable that they were introduced from the Near East, either through northern Sinai and the Nile Delta or via southern Sinai across the Gulf of Suez (Close 2002). The origin of cattle is still strongly debated since its ancestor, the aurochs (*Bos primigenius*), is present in North Africa and the Egyptian Nile Valley until the Middle Holocene (Linseele 2004), suggesting that it could have been domesticated locally. For the moment however, neither archaeological nor genetic data can securely confirm its local or Near-Eastern origin (Gifford-Gonzalez and Hanotte 2011; Smith 1986, 2013; Riemer 2007; Di Lernia 2013).

Though scarce, this evidence suggests that livestock appeared during the time of hunter-gatherer societies, a period that is characterized by the exploitation of wild bovids, fish, wild plants and considerable mobility, but nevertheless still poorly understood.

During the 6th and 5th millennium BC, livestock spread throughout major parts of the Sahara and the Sahel (Gautier 1987, 2001; Hassan 2002; Marshall and Hildebrand 2002; Garcea 2004; Jousse 2004; Honegger 2005; Chaix 2011). Varying climatic regimes were certainly influential in driving the first herders across the Sahara, especially with the beginning of aridification at the end of the 6th millennium (Kuper and Kröpelin 2006).

The first phase of herding development in Egypt, at the beginning of the 6th millennium, is characterised by the introduction of animals (goat, sheep and possibly cattle and dog - *Canis familiaris*) adapted to mobile systems. During the 5th millennium, there was a second stage of introduction from the Near East that seems to have been directed towards more sedentary farming with the arrival of pigs (*Sus domesticus*) and cereals. Recent genetic studies on human populations and their movements into Africa from the Near East seem to corroborate this bimodal pattern. According to A.C. Smith (2013), the first wave seems to have taken place during the 8th or 7th millennium BC, and could correspond to the arrival of cattle herders who already had the lactase persistence haplotype. The second wave, strongly associated with the adoption of agriculture in Egypt, occurred during the 6th and 5th millennium BC (Smith A.C. 2013: 141), and could be linked to movements initiated by drought (Hassan 2000).

From that period onwards we see the development, first in Lower Egypt (Boessneck and Von den Driesch 1990; Von den Driesch and Boessneck 1985), then a few hundred years later in Upper Egypt (Van Neer 2001; Vermeersch *et al.* 2004), of sedentary communities that grew crops, herded pigs and cattle, but still took a large part of their diet from aquatic resources (Lesur 2013). Although we have almost no data for the first half of the Holocene in the Nile Valley (probably because of alluvial sedimentation; Kuper and Kröpelin 2006), it seems that both here and in the Nile Delta, food production emerged directly from sedentary farming. And, probably because of its particular geographic features, the Nile Valley seems to be more similar to the Fertile Crescent of the Near East than to other parts of Africa (Linseele 2010).

In the Egyptian western desert, during the 5th millennium, we also notice mobile communities, such as at Nabta and Bir Kiseiba (Gautier 2001), who appear to have relied mainly on wild species and others, such as at the sites KS43 and KS52 on the south edge of the Kharga Oasis in Egypt (Briois *et al.* 2012, in prep.), who relied primarily on caprine herding. At these sites, evidence from the faunal spectrum is dominated by livestock and herding practices directed towards meat and milk production (Lesur *et al.* 2011). Much of the data suggests seasonal occupation during winter favoured by the presence of an artesian source in a semi-arid environment. Links with the Nile Valley are clearly established by artefacts such as pottery, emmer wheat (*Triticum dicoccum*) seeds, and remains of fish and shells from the Nile (Briois *et al.* 2012; Lesur *et al.* 2011). These two sites, though isolated in the current regional context, show the presence of pastoral groups circulating with their herds between the Nile Valley and the western desert during the 5th millennium BC.

From that latter period onwards, various food production strategies were maintained alongside major regional differences (relating to biogeographic conditions and cultural diversity) throughout Egypt. In Lower Egypt, animal husbandry can be strongly associated with important changes in agricultural production and the impact of human activities on the natural landscapes.

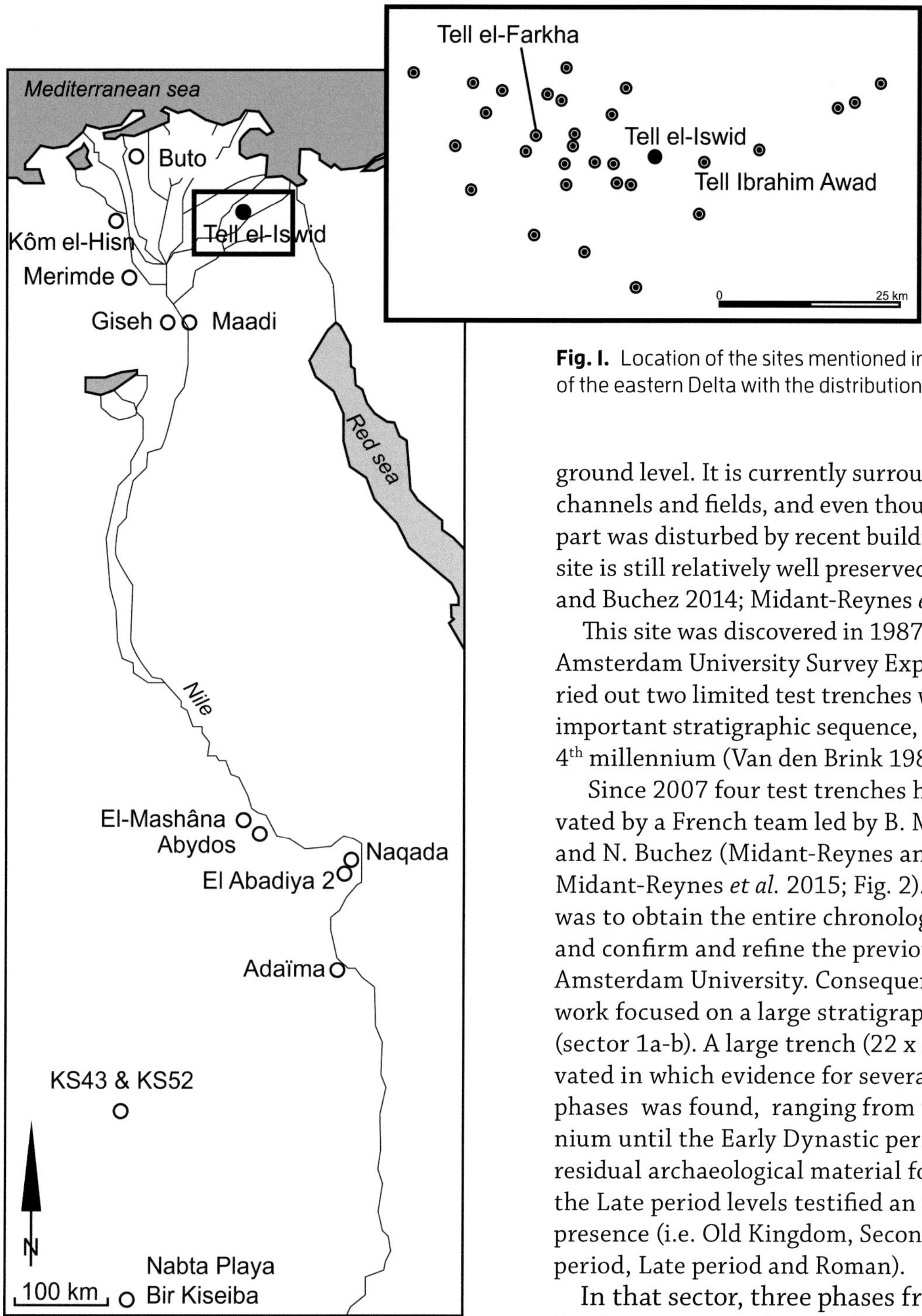

Fig. 1. Location of the sites mentioned in the text and details of the eastern Delta with the distribution of Predynastic sites.

Tell el-Iswid during the 4th millennium in the Nile Delta

Within this context, the site of Tell el-Iswid, situated in a zone of dense human occupation, provides new archaeological data with which to tackle the question of cultural and economic patterns among Predynastic cultures.

The site is located in the Sharqiya Province, 14 km north-west of Faqus (Fig. 1). Within a radius of 5 km, there are four different mounds: Tell el-Iswid (South), Tell el-Iswid (North), Tell el-Abassieh and Tell Umm ʿAgram. In this paper, we present the results of the ongoing excavations of Tell el-Iswid (South): A 400 x 300 m mound, rising 3-4 m above ground level. It is currently surrounded by irrigation channels and fields, and even though the eastern part was disturbed by recent building activity, the site is still relatively well preserved (Midant-Reynes and Buchez 2014; Midant-Reynes *et al.* 2015).

This site was discovered in 1987 by the Amsterdam University Survey Expedition, who carried out two limited test trenches which exposed an important stratigraphic sequence, especially for the 4th millennium (Van den Brink 1989).

Since 2007 four test trenches have been excavated by a French team led by B. Midant-Reynes and N. Buchez (Midant-Reynes and Buchez 2014; Midant-Reynes *et al.* 2015; Fig. 2). Our main goal was to obtain the entire chronological sequence, and confirm and refine the previous findings of the Amsterdam University. Consequently, the initial work focused on a large stratigraphic sequence (sector 1a-b). A large trench (22 x 2.5 m) was excavated in which evidence for several occupational phases was found, ranging from the 4th millennium until the Early Dynastic period. However, residual archaeological material found within the Late period levels testified an intermediary presence (i.e. Old Kingdom, Second Intermediate period, Late period and Roman).

In that sector, three phases from the Predynastic period were identified; the earliest was related to the Lower Egyptian Cultures (LEC), then Naqada IIIA-B and Naqada IIIC-D. This last phase was characterized by a succession of mud-brick structures that were preserved up to 1.5 m. The end of the Predynastic occupation was marked by the conversion of the space from domestic to funerary: Ten pit burials have been found, as well as a tomb constructed of mud- bricks (Midant-Reynes and Buchez 2014; Midant-Reynes *et al.* 2015).

In 2009, magnetic prospection helped locate high concentrations of Predynastic remains. A large building was subsequently excavated dating to the end of the Predynastic period (Naqada IIIA-B), which greatly advanced our

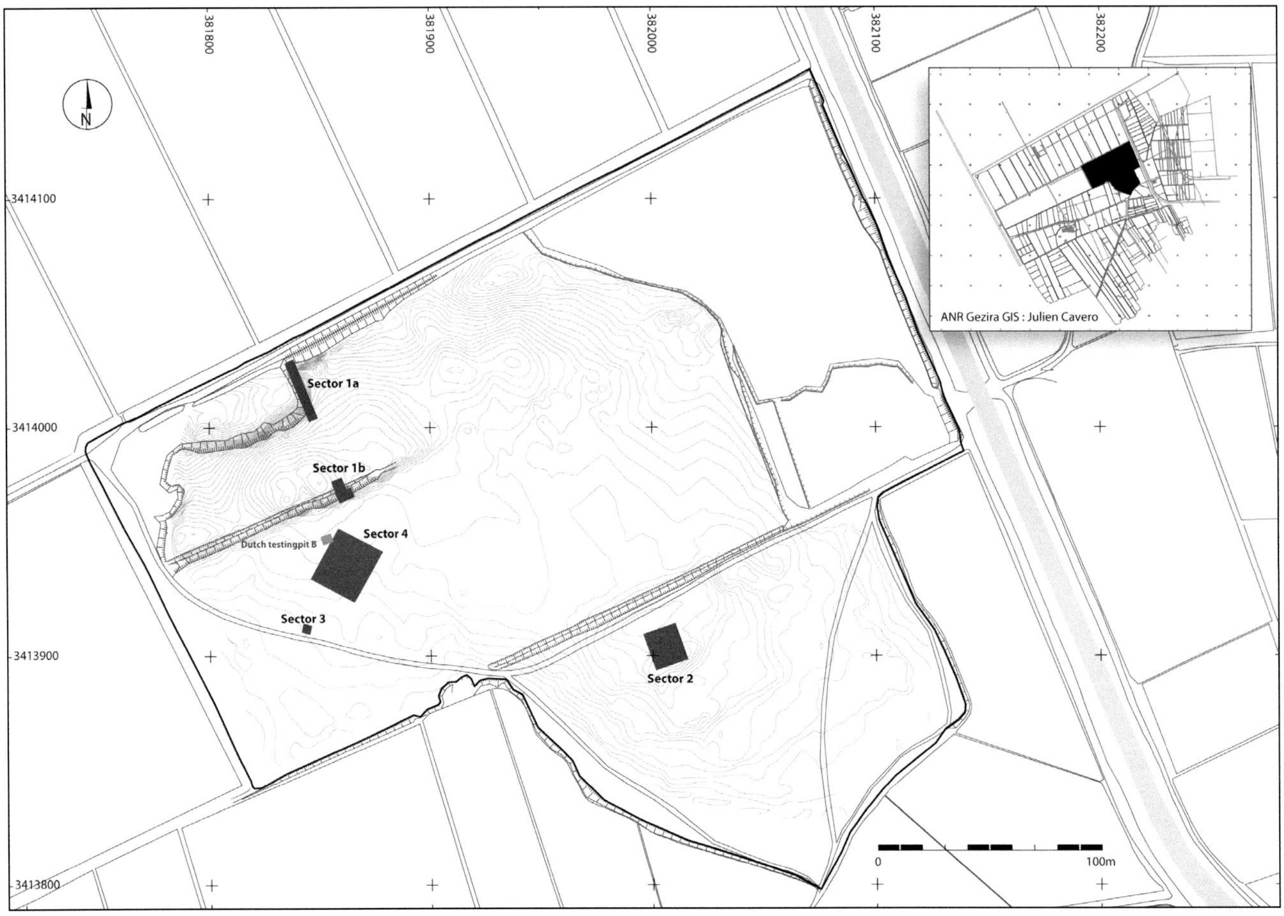

Fig. 2. Topographical map showing the excavated areas (Map: B. Fabry, S. Aussel, F. Vinolas, J. Cavero).

knowledge of housing techniques for this period. The clearing of this building (sector 4) began in 2009 and is still ongoing. It was set upon older levels, and is currently being excavated, confirming the presence of extended LEC occupation levels to a considerable depth, just above the water level (Midant-Reynes *et al.* in prep.).

Faunal assemblages and chronological phases

Animal bone fragments came from the two main areas excavated: Sector 1 and 4. They were handpicked during the excavations and collected by dry sieving through a 2 mm mesh. Bone identifications were completed during the course of the fieldwork in 2008, 2010, 2012 and 2015, with the aid of bone atlases. To increase the samples for analysis and because both sectors delivered material/remains of the same chrono-cultural phases, we merged them for the following study. The results from sector 1 presented here have already been published (Lesur 2014), but the data from sector 4 is only partially available as it is still being excavated Even though three phases were identified for the Predynastic period (i.e. LEC, Naqada IIIA-B and Naqada IIIC-D), we combined the assemblages from the two Naqada III phases. The first reason was to increase the sample size and therefore their statistical validity. The second reason was linked to the ongoing work on the chrono-cultural attributes from the stratigraphic units, which has not provided us with any definitive results yet.

The site produced an abundance of faunal remains, more than 17,000 fragments (Table 1), most of which came from sector 4. Unfortunately, their preservation was poor due to various post-depositional processes that had caused severe damage: The acidity of the sediment and the presence of salty concretions had caused heavy corrosion of the bone surface, and acute fragmentation had resulted in a high percentage of splinters. These post-depositional processes limited both bone identification (only 36% of identified remains), and the possibility of any anthropic traces (such as cut and burn marks) being preserved.

Faunal spectrums showed a clear predominance of fish accounting for 63% of the assemblage (Fig. 3). From these at least seven species were identified: Catfish dominated with the presence of *Synodontis* sp., *Bagrus* sp., *Clarias* sp. and *Heterobranchus* sp. but several remains of Nile perch (*Lates niloticus*), African tetras (*Alestes/Brycinus*) and tilapias (Tilapinii) were also identified (Fig. 4).

Aside from fish, domesticates represented an important part of the assemblage, especially pigs (*Sus domesticus*), followed, in terms of percentage, by

Table I. Faunal spectrum from sectors I and 4 of Tell el-Iswid, according to the NISP (Number of Identified Specimens).

Taxa	*Sector 1*	*Sector 4*
Donkey (*Equus cf asinus*)	10	27
Hippopotamus (*Hippopotamus amphibius*)	1	1
Cattle (*Bos taurus*)	35	262
Large ungulate		4
Hartebeest (*Alcelaphus buselaphus*)		2
Caprine	30	166
Sheep (*Ovis aries*)	1	7
Goat(*Capra hircus*)	3	9
Pig (*Sus domesticus*)	222	852
Dog (*Canis familiaris*)	1	18
Middle-sized mammal		47
Unidentified rodent	3	31
Unidentified bird	5	19
Unidentified reptile		2
Softshell turtle(*Trionyx triunguis*)	14	46
Crocodile (*Crocodylus niloticus*)		3
Tetras (*Alestes / Brycinus*)	7	10
Catfish 1 (*Clarias sp.*)	20	154
Catfish 2 (*Heterobranchus sp.*)		21
Clariid catfish (Clariidae)	269	1113
Catfish 3 (*Bagrus sp.*)	63	99
Catfish 4 (*Synodontis sp.*)	8	783
Nile perch (*Lates niloticus*)	34	60
Tilapia (Tilapinii)	39	145
Unidentified fish	71	971
Total identified vertebrate	**836**	**4852**
Bulinus truncatus		3
Unidentified gastropod		12
Cardiidae		1
Muleta sp.		1
Spathopsis sp.	18	116
Unio sp.	9	51
Unidentified bivalve	49	259
Total identified invertebrate	**76**	**443**
Unidentified	*489*	*10373*
TOTAL	**1401**	**15668**

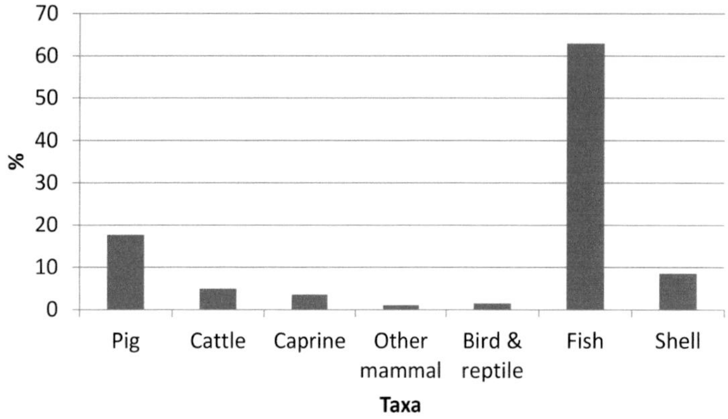

Fig. 3. Representation of the faunal spectrum from Tell el-Iswid, according to the %NISP.

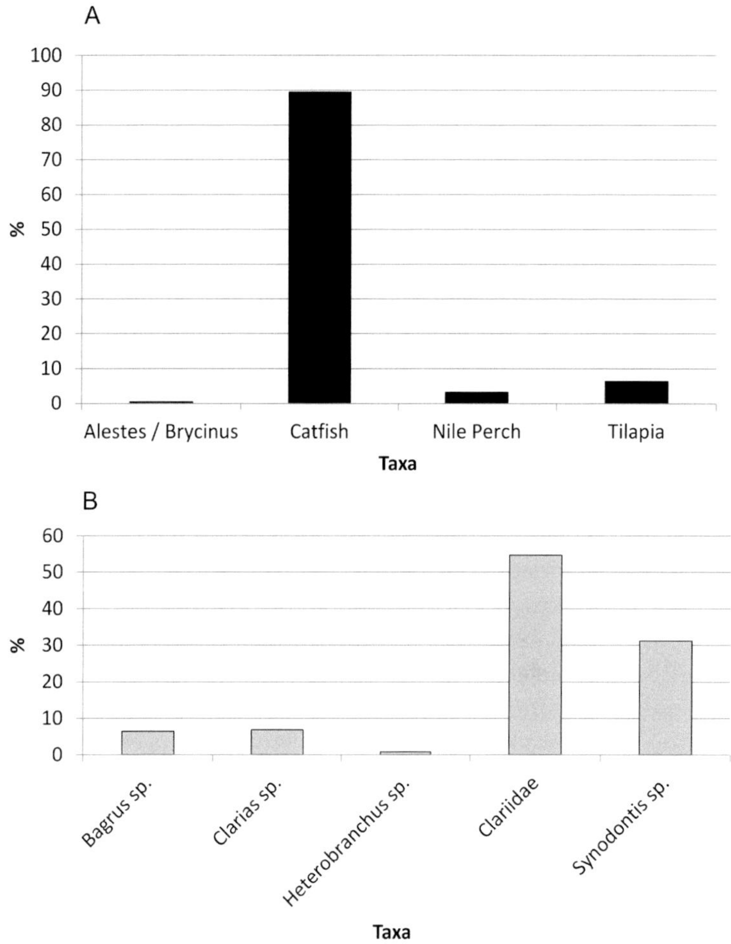

Fig. 4. A- Fish taxa identified from Tell el-Iswid, based on the %NISP. B- Catfish taxa identified from Tell el-Iswid, based on the %NISP.

cattle (*Bos taurus*) and caprines (Table 1 and Fig. 3). For the latter, both goats (*Capra hircus*) and sheep (*Ovis aries*) were present, although poor preservation of the assemblage limited the specific identification of most of the caprine remains. Other domestic animals included dogs (*Canis familiaris*) and probably donkey (*Equus cf asinus*).

Other mammals included hartebeest (*Alcelaphus buselaphus*) and hippopotamus (*Hippopotamus amphibius*). The latter was the most important hunted species during the Predynastic in Lower Egypt since they found perfect living conditions with large waterbodies and ample grassland (Linseele and Van Neer 2009). Two long bone fragments found at the site suggest hunting, perhaps to protect fields and/or cultivated crops, as was suggested for the site of El Omari (Boessneck and Von den Driesch 1990). One object made from a hippopotamus tusk was also discovered on site, which could have belonged to a knife or a throwing weapon (Minotti 2014).

The presence of the hartebeest is tenuous as in the other Predynastic sites of both Lower and Upper Egypt. Although it was hunted prominently during the Late Palaeolithic, climatic changes and

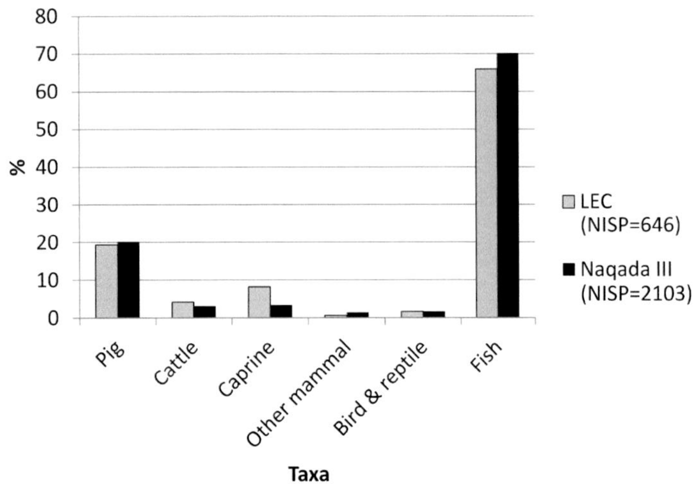

Fig. 5. Proportion of the main vertebrates' taxa for the two cultural occupation phases of Tell el-Iswid, based on the % NISP. LEC: Lower Egyptian Cultures.

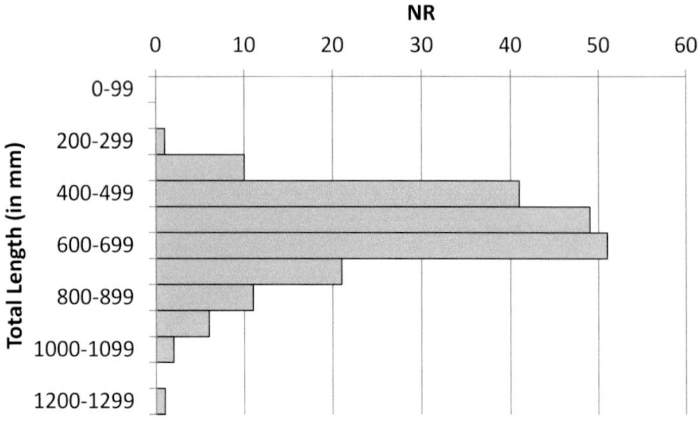

Fig. 6. Size reconstructions for the Clariidae of Tell el-Iswid based on 193 observations of measurement I of the articular.

increasing human pressure drastically impacted its suitable habitat, consequently its distribution across Egypt became scarce (Linseele and Van Neer 2009).

Among the rest of the vertebrates, we noted the presence of some bird bones, although too fragmented to be fully identified. Remains of the Nile softshell turtle (*Trionyx triunguis*) and crocodile (*Crocodylus niloticus*) were also identified.

Invertebrates represented almost 10% of the identified assemblage. However, these comprised mainly of large Nilotic bivalves, such as *Spathopsis rubens* and *Unio abyssinicus* (Table 1 and Fig. 3), in spite of the large diversity of shells present in the Delta. We also noted the presence of one marine shell (Cardiidae).

Although excavations are still ongoing, parts of the assemblage can be attributed to one of the two main occupations phases of the site: Naqada III and the LEC. If we look at the main taxa across the two phases, few changes seem to have occurred (Fig. 5). These differences were statistically tested with the mean of a global Chi^2 test (df= 6). The result was highly significant ($Chi^2 = 26.6$; $p < 0.001$), with a contribution of 93.4% (24.8) of caprines to the Chi^2 value. This indicates a change in herding strategies with a decrease of caprine consumption during the Naqada phase. However, as excavations are not yet complete, especially for the LEC phases, we await further data to confirm this tendency.

Fishing at Tell el-Iswid

Due to the predominance of remains, the faunal spectrum suggests that fish were an important dietary requirement for the inhabitants of Tell el-Iswid (Fig. 3). Catfish, and in particular Clariidae, are highly dominant, accounting for more than 90% of the fish remains (Fig. 4).

The variety of species suggests fishing in different types of aquatic habitats. Some species such as the bagrid catfish, the Synodontis catfish and the Nile perch live in well-oxygenated water, but most fish, especially clariid catfish and tilapia, are typical of shallow waters (Van Neer 2004).

When reconstructing the size of the clariid catfish (Van Neer and Lesur 2004), we noticed the presence of large sexually mature specimens, with total lengths averaging about 62 cm (Fig. 6). This suggests that they were captured when spawning in shallow waters at the beginning of the floods and the weeks following, probably between July and September. Clariids reproduce in shallow waters. When breeding is over, the adult fish migrate back into deeper waters and at the end of the floods, when the waters recede, they resort to the main channel (Van Neer 2004). The spawning season is therefore the best period to capture clariid catfish, as they come very close to the river bank and can be easily caught by hand, with sticks or other capturing techniques.

The Nile perch and the other catfish live exclusively in well-oxygenated water. Therefore, their exploitation may have occurred mainly during low-water levels in spring, when the main channel was more easily accessible.

If the main season for fishing was probably at the beginning of the floods, as suggested by the high number of adult clariid catfish, the proximity of water in the Delta facilitated all-year fishing practices.

Indeed, several artifacts, probably used for fishing, were found at the site: refitted potsherds, probably used as floats; one copper harpoon and the pectoral spines of a catfish, which may have been used as equipment (Minotti 2014). These objects are not enough to reconstruct the variety of techniques used to capture fish, but their presence illustrates the inhabitants' capacity to exploit the various aquatic environments that surrounded them.

If we look at the evolution of fish remains throughout the two main occupation phases, we see that catfish are always highly dominant. However,

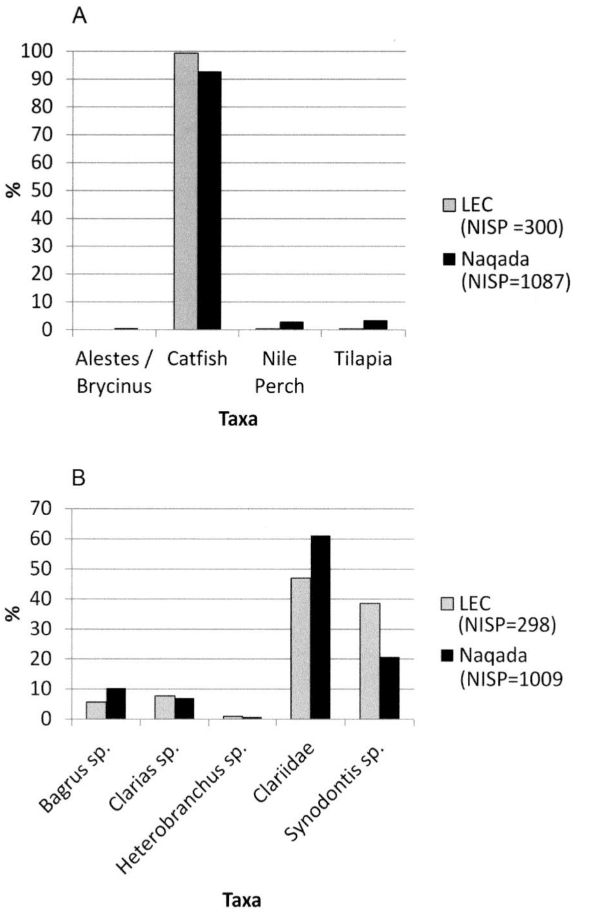

Fig. 7. A- Proportion of fish taxa for the two cultural occupation phases of Tell el-Iswid, based on the % of NISP. LEC: Lower Egyptian Cultures. B- Proportion of catfish taxa for the two cultural occupation phases of Tell el-Iswid, based on the % of NISP. LEC: Lower Egyptian Cultures.

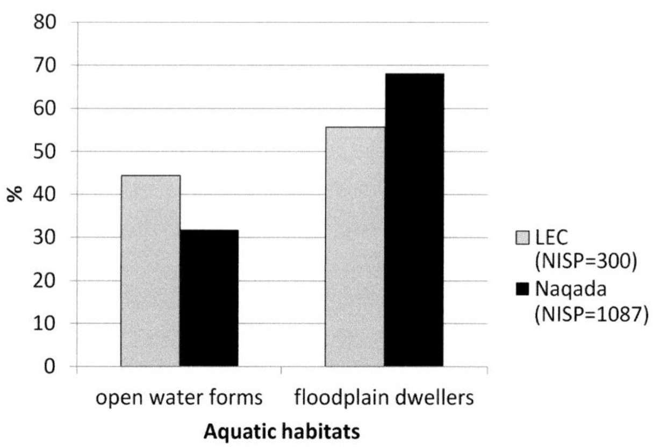

Fig. 8. Exploitation of different aquatic habitats during the two cultural occupation phases of Tell el-Iswid, based on the % of NISP. LEC: Lower Egyptian Cultures.

the diversity of species changes during the Naqada phase, with an increase of clariids and a decrease of Synodontis catfish (Fig. 7). The evidence for these two taxa, requiring different habitats, could mean a change in fishing seasons or practices. To test this hypothesis, we grouped the different species according to their ecological behaviour: open water forms or

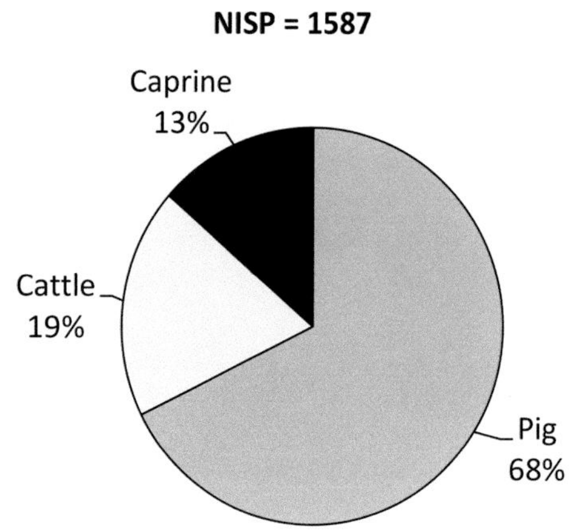

Fig. 9. Proportion of livestock of Tell el-Iswid according to the % of NISP.

floodplain dwellers (Fig. 8). It seems that during the Naqada phase, the floodplain dwellers, such as clariids, increased due to a reduction of open water forms. These differences were statistically tested with the mean of a global Chi^2 test. The result was significant ($Chi^2 = 16.7$; $p < 0.001$) with a contribution of 60% (10.2) of the floodplain dwellers to the Chi^2 value.

It appears, therefore, that during the Naqada period the inhabitants of the site focused their fishing season during the beginning of the floods when large clariids were easier to catch in shallow waters. Open water forms appear to have been captured more sporadically than during earlier periods. But, as previously mentioned, the excavations are still ongoing, therefore it is impossible to confirm this hypothesis.

Herding at Tell el-Iswid

Aside from fishing, herding was another important source of animal products. Pig was the most dominant, representing 68% of livestock remains (Fig. 9).

The skeletal parts distribution shows numerous long bone fragments, probably from meat consumption, but also significant numbers of pelvis, skull and tooth parts (Fig. 10). The loss of the more porous bones (e.g. from the rachis) could be due to poor preservation conditions caused by the acidic and salty sediments. However, this distribution could also suggest the transportation of meat from its place of production to other sites, and would explain the butchery refuse, such as skulls, that was discovered at Tell el-Iswid. Analysis of the assemblage for age at death indicates a dominance of young adults (6 to 24 months), which suggests a typical consumption pattern (Fig. 11).

As for cattle, sheep and goats the current data is not robust enough to predict their exploitation patterns. However, the presence of all skeletal parts from mainly young adult individuals for these three species, suggests meat production strategies, probably alongside the consumption of milk.

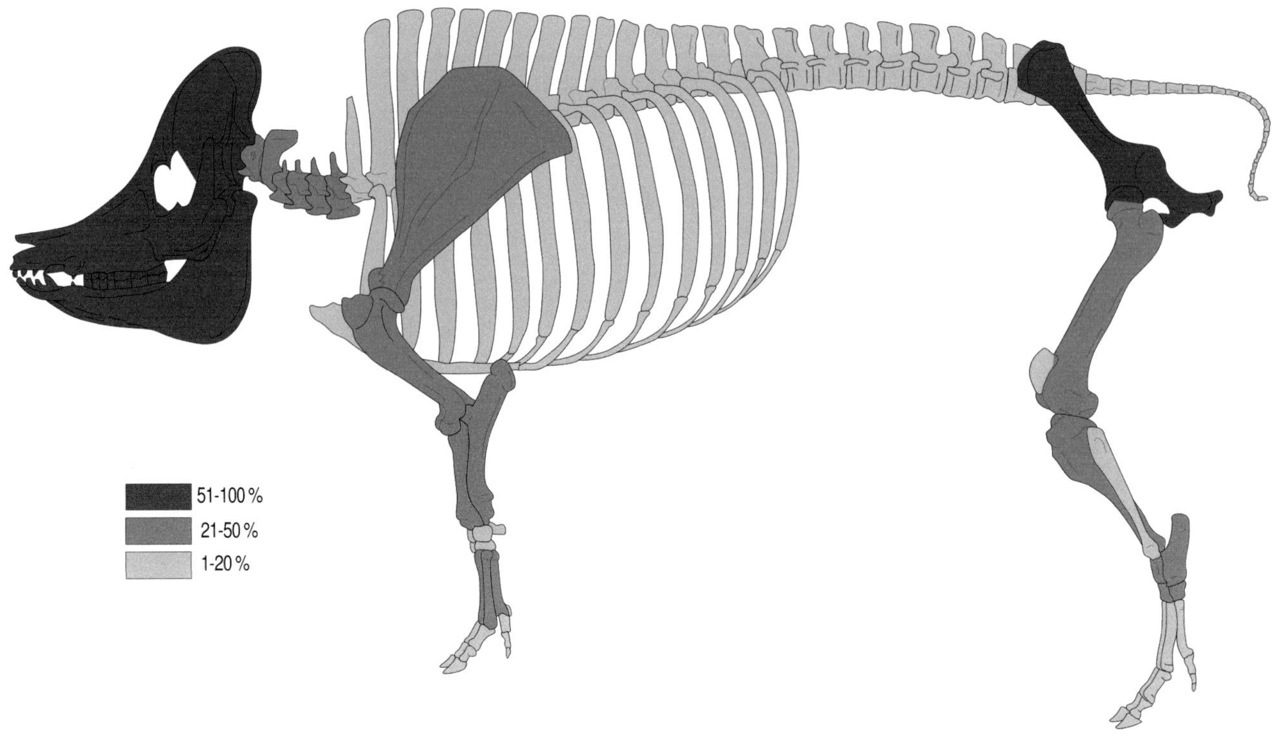

Fig. 10. Representation of the percentage of skeleton parts of pigs (calculated according to Dodson and Wexlar 1979).

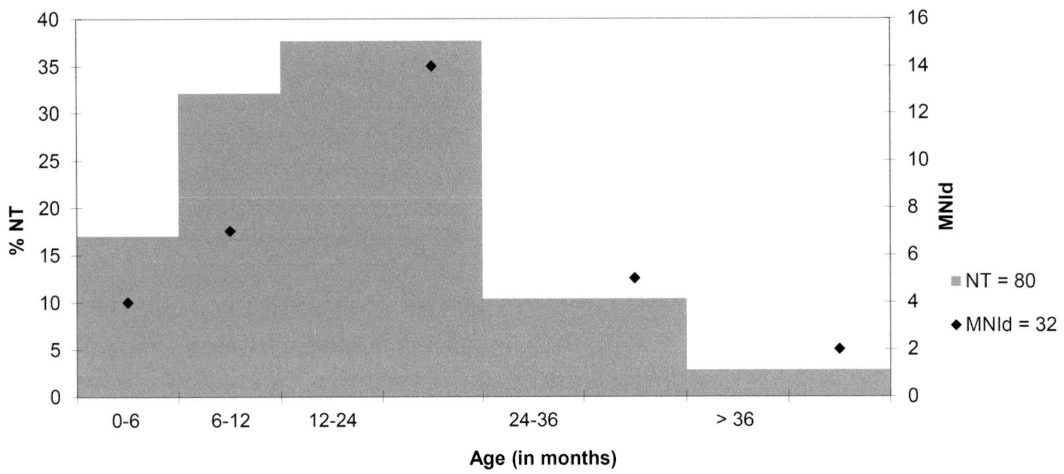

Fig. 11. Kill-off pattern for the pigs from Tell el-Iswid (NT, Number of teeth; dMNI, dental Minimal Number of Individuals).

The presence of almost 40 remains of donkey is very interesting as it provides new data regarding the question of donkey domestication in Africa and their use during the Predynastic. It is a difficult subject to approach because like most of the species that are rarely eaten but used for transportation or work, they are generally poorly represented in faunal assemblages (Rossel *et al.* 2008).

However, recent genetic studies have begun to clarify the origin of its domestication with sub specific forms of its ancestor, *Equus africanus*, having been found across arid zones of Africa and the Arabian Peninsula. Indeed, both of these regions could have been possible domestication centres. Analysis of the wild and domestic mitochondrial DNA (mtDNA) sequences of this species has identified two maternal lineages within the domestic donkey (Beja-Pereira *et al.* 2004; Vilà *et al.* 2006; Kimura *et al.* 2010). The 'Nubian lineage' was undoubtedly domesticated in Africa, as indicated by its close relationship with the Nubian African wild ass sequences (*E. africanus africanus*). The origin of the second lineage, closer to the Somali wild ass (*E. africanus somaliensis*), is less clear as this particular wild ass was also present in the Arabian Peninsula (Gifford-Gonzalez and Hanotte 2011).

Some Egyptian sites from the 5th and the 4th millennium have yielded the earliest potential domestic donkeys, with some bones smaller than the wild ass ones (Boessneck and Von den Driesch 1990; Boessneck *et al.* 1989; Rossel *et al.* 2008). However, widespread morphological change was slow to

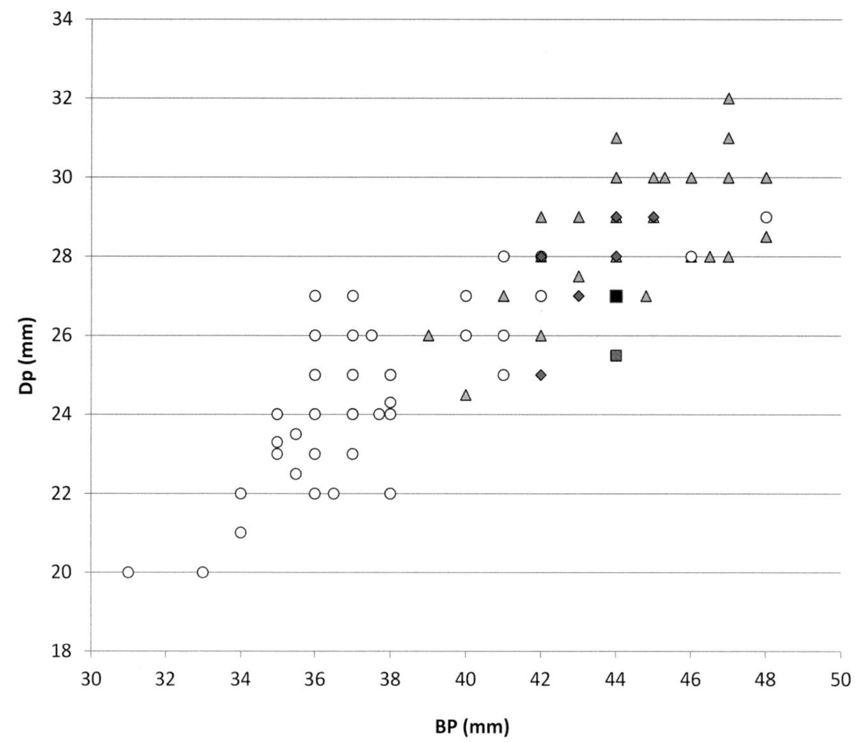

Fig. 12. Comparison of the width and breath measurements (BP and DP, after Von den Driesch 1976) of the proximal metacarpal from a possible donkey from Tell el-Iswid (TEI), compared with archaeological data from Maadi (Boessneck *et al.* 1989), Abydos (Rossel et al. 2008) and with modern specimens of wild ass and donkey (Rossel *et al.* 2008; Eisenman 2007).

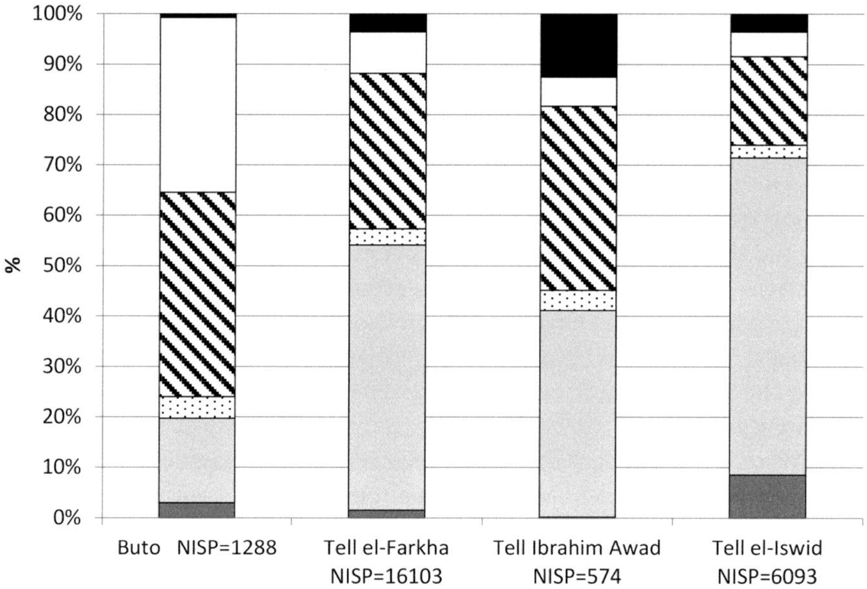

Fig. 13. Percentage of the main taxa identified from the sites of Buto (von den Driesch 1997), Tell el-Farkha (Ablamowicz 2002; 2004; 2006), Tell Ibrahim Awad (Boessneck and von den Driesch 1992) and Tell el-Iswid (this study).

develop in ancient Egypt and on some sites at the end of the 4th millennium, these animals were not yet morphologically distinguishable from the African wild ass. This is the case at the Early Dynastic site of Abydos, where evidence of bone pathologies suggests the use of donkey to carry heavy loads (Rossel *et al.* 2008; Marshall and Weissbrod 2011).

We compared the measurement of one proximal metacarpal from Tell el-Iswid with data from the 4th millennium sites of Maadi and Abydos and some modern data (Fig. 12). We found that our (large) specimen fit in the same range as the Maadi and Abydos specimens, and the modern wild ass specimen. We did not notice any bone pathology suggesting the use of the animal as beast of burden or working animal and, contrary to Abydos, we could not confirm its domestic status. However, its presence in this moist area is very different from the natural arid habitat, which could suggest that it was indeed a domestic animal showing wild ass morphology. More data from sector 4 will hopefully clarify this.

Food strategies in the Delta during the 4th and the 3rd millennium BC

Though many Predynastic sites are known from the eastern Delta (Fig. 1), few have yielded large faunal assemblages. In addition, poor preservation due to the acidity of the sediment and the presence of salty concretions has limited archaeozoological analysis. However, if we compare our data with that of the three other important sites (Fig. 13) the predominance of livestock and fish is still apparent,

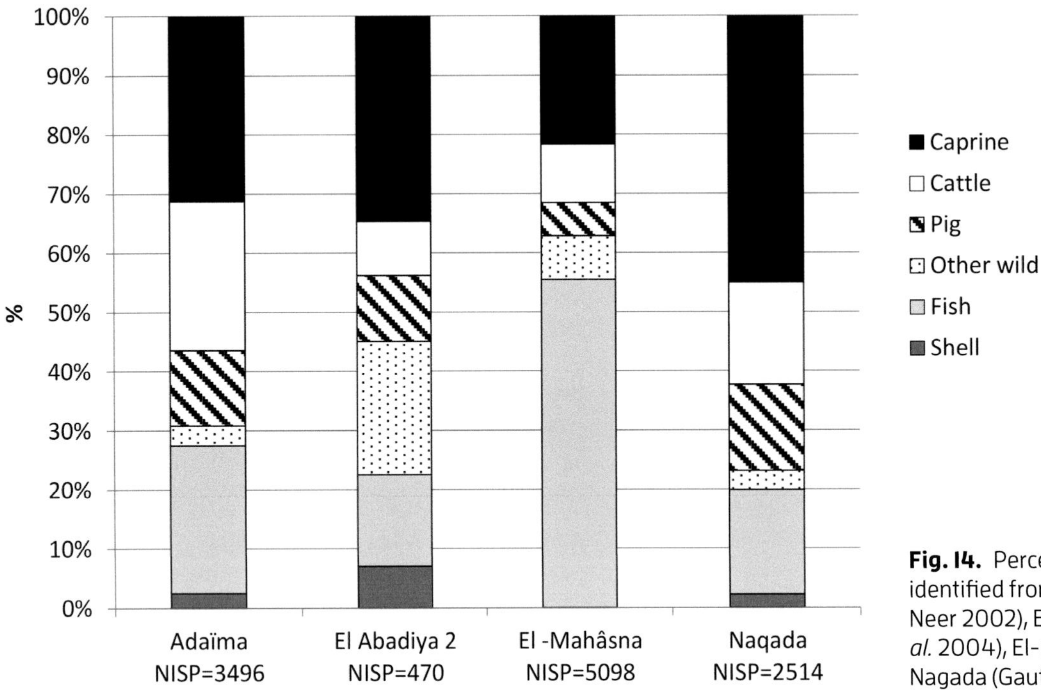

Fig. 14. Percentage of the main taxa identified from the sites of Adaïma (Van Neer 2002), El Abadiya 2 (Vermeersch *et al.* 2004), El-Mahâsna (Rossel 2006) and Nagada (Gautier and Van Neer 2009).

though with some variance between them. Among herding resources, pig remains dominate the assemblage (40%) of Buto (Von den Driesch 1997) and Tell Ibrahim Awad (Boessneck and Von den Driesch 1992). Cattle are well represented, especially at Buto, whereas caprines appear not to have been overly exploited. This all points to management of the landscape and water resources that favoured the settlement, which provided good pasture for livestock, especially pigs and cattle. It could also explain the scarcity of wild mammals, such as hartebeest and aurochs that were more abundant during the 5th millennium (Linseele and Van Neer 2009). However, contrary to Tell el-Iswid, other sites have delivered higher amounts of small wild mammal remains, such as small carnivores and rodents, which could have been hunted or trapped in order to protect the fields and crops. On most sites, as at Tell el-Iswid, the exploitation of aquatic resources, such as shells and fish, was a highly developed activity that was practiced all year round.

Few archaeozoological studies for the 3rd millennium (Early Dynastic and Old Kingdom) have been undertaken across Lower Egypt. Nevertheless, the available evidence suggests that animal production underwent major changes due to an increased population and a move to a more state-controlled economy (Yokell 2004). For example, at the site of Kôm el-Hisn in the western Delta, the animal remains are mainly comprised of fish, caprines and pig, but few bovines (Wenke *et al.* 1988; Redding 1992). This data is surprising given that this is a known cattle production centre, and that other sites in the region have provided numerous cattle remains (Yokell 2004). According to the model suggested by Redding (1992:104-106), data from Kôm el-Hisn reflects both a local and regional production system. Pigs and fish provided meat for local inhabitants. Caprines, especially goats, provided meat but were mainly kept for milk. As for cattle and sheep, also raised in the area, the distribution of their meat was controlled by the state with animals being transported to other regions, such as Giza, to feed the workers of the pyramids (Redding 2010; Yeomans 2011).

In conclusion, from the 5th to the 3rd millennium the sites of Lower Egypt and the Delta show that animal consumption was based primarily on fishing and pig herding, due to the availability of a favourable aquatic environment and good pastures. Even amidst the development of the state-controlled economy during the Old Kingdom, these taxa remained important for local consumption whereas some animals, such as cattle, were integrated into a more regional resource management.

Predynastic animal exploitation in Upper Egypt

The Badarian and Naqadian cultures spread across Upper Egypt with a food economy close to that of the Delta but more adapted to this warm and arid region. The main animal resources were still fishing and herding. However, the data from four of the main regional sites (Fig. 14) indicate that livestock percentages increased significantly from Lower Egypt, and reached almost 80% in the Naqada area (Gautier and Van Neer 2009). Another significant difference with the Delta concerns the proportion of domesticates. Caprines, well adapted to arid environments, are highly predominant whereas pigs appear to have barely contributed to food production, probably because of their inadequacy for an arid and mobile life.

Herding was mainly oriented towards meat production, but caprines and cattle were also exploited for milk. The use of the latter for ploughing and transportation is sometimes noticed on foot bones that exhibit pathologies and deformation, for example at Adaïma (Van Neer 2002: Fig. 6).

As for fishing, the presence of many different species such as Nile perch, bagrid catfish, tilapia and clariids suggest year-round fishing in various aquatic environments. Most of the sites also contained evidence of different fishing techniques such as harpoons, hooks and floats.

Finally, on most sites, hunting appears to have been sporadic with evidence for aquatic reptiles and mammals (i.e. crocodile, Nile softshell turtle and hippopotamus), and also terrestrial mammals such as gazelle, hartebeest and Barbary sheep. It has been suggested that these species were killed when they came to drink and feed on the alluvial plain during the dry season (Linseele and Van Neer 2009), where they were an easy target for opportunistic hunters.

Conclusion

The 4th millennium was a crucial period for cultural and social changes in Lower Egypt with the development of local cultures influenced by both the Saharan and Levantine traditions, but also the cultures of Upper Egypt. It was also significant in terms of the evolution of food production strategies and, more generally, in the relationship between man and the rich environment of the Delta.

The introduction of domesticates - pig, caprines, maybe cattle, cereals and leguminous plants - from the Near East since the 6th and the 5th millennium BC, deeply changed food consumption patterns. The landscape was also strongly impacted through the management of water resources and the development of large areas of cultivated fields and pastures. However, people also continued to exploit wild resources. Though most of the large mammals, such as hartebeest and aurochs progressively disappeared because of human activities, aquatic resources, especially fish and shells, remained available as a major component of the food supply. The evidence of fishing equipment such as harpoons, hooks, nets and boats suggests the exploitation of aquatic environments throughout changing seasons.

Tell el-Iswid, as with other contemporary sites such as Buto, Tell el-Farkha and Tell Ibrahim Awad, offers copious data which illustrates the different aspects within the subsistence economy and shows that this was mainly based on pig herding and the exploitation of catfish. Until the excavations are finished, the possible evolution of animal exploitation between the LEC and Naqada phases, the reduction of caprine husbandry and the increase of clariid fishing is still difficult to confirm, but this tendency could be a consequence of the large global cultural changes that occurred during the second half of the 4th millennium.

Nevertheless, during the following period and the development of the Egyptian state, herding practices became a significant part of regional exchange trades. Although human activities continued to impact the landscapes of the Delta, the inhabitants continued with the consumption of local products such as fish and shells.

There is still much to be discovered at Tell el-Iswid. However, current archaeozoological data has provided us with a reliable image of a transitional period, based on shared food production strategies and local resources management.

Acknowledgements

This work is part of the archaeological project instigated in 2006 by the Institut Français d'Archéologie Orientale (IFAO) and led by B. Midant-Reynes and N. Buchez, with the financial support of the French foreign ministry (MAE) and the cooperation of the Inrap (Institut national de recherche en archéologie préventive). From 2008 to 2013, the excavation at Tell el-Iswid was part of an ANR programme (ANR_8_BLAN_0312_01) devoted to Human Occupation in the Nile Delta during the 4th millennium. (see http://gezira.tge-adonis.fr). The English was corrected by Jill Cucchi. I'm extremely grateful to the two anonymous reviewers whose criticisms and comments have helped improve the quality of the manuscript.

References

Ablamowicz, R. 2002. Animal remains. In, Chlodnicki, M and Cialowicz, K.M. (eds) *Polish Excavations at Tell el-Farkha (Ghazala) in the Nile Delta. Preliminary report 1998-2001*: 109-114. Poznan: Institute of Archaeology and Ethnology, Polish Academy of Sciences.

Ablamowicz, R. 2004. Animal remains. In, Chlodnicki, M and Cialowicz, K.M. (eds) *Polish Excavations at Tell el-Farkha (Ghazala) in the Nile Delta. Preliminary report 2002-2003*: 70-74. Poznan: Institute of Archaeology and Ethnology, Polish Academy of Sciences.

Ablamowicz, R. 2006. Archaeozoological research. In, Chlodnicki, M and Cialowicz, K.M. (eds) *Polish Excavations at Tell el-Farkha (Ghazala) in the Nile Delta. Preliminary report 2004-2005*: 109-115. Poznan: Institute of Archaeology and Ethnology, Polish Academy of Sciences.

Beja-Pereira, A., England, P., Ferrand, N., Jordan, S., Bakhiet, A., Abdalla, M., Mashkour, M., Jordana, J., Taberlet, P. and Luikart, G. 2004. African

Origins of the Domestic Donkey. *Science* 304: 1781.

Boessneck, J. and Von den Driesch, A. 1990. Tierreste aus der vorgeschichtlichen siedlung von El-Omari bei Heluan/Unterägypten. In, Debono, F. and Mortensen, B. (eds) *El Omari. A Neolithic Settlement and Other Sites in the Vicinity of Ouadi Hof, Helwan*: 99-107. Mainz am Rhein: Philipp von Zabern.

Boessneck, J. and Von den Driesch, A. 1992. Weitere Tierknochenfunde vom Tell Ibrahuim Awad im östlichen Niledelta. In, Van den Brink, E.C.M. (ed) *The Nile Delta in Transition: 4th-3rd Millennium BC, Proceedings Seminar held in Cairo, 21-24 October 1990 at the Netherlands Institute of Archaeology and Arabic Studies*: 97-110. Tel Aviv: Ramot.

Boessneck, J., Von den Driesch, A. and Ziegler, R. 1989. Die Tierreste von Maadi und Ouadi Digla. In, Rizkana, I and Seeher, J. (eds) *Maadi III. The Non-Lithic Small Finds and the Structural Remains of the Predynastic Settlement*: 87-128. Mainz am Rhein: Philipp von Zabern.

Briois, F., Midant-Reynes, B., Marchand, S., Tristant, Y., Wuttmann, M., De Dapper, M., Lesur, J. and Newton, C. 2012. Neolithic occupation of an artesian spring: KS043 in the Kharga Oasis, Egypt. *Journal of Field Archaeology* 37(3): 178-191.

Buchez, N. and Midant-Reynes, B. 2007. Le site prédynastique de Kom el-Khigan (Delta Oriental): Données nouvelles sur les processus d'unification culturelle au IVe millénaire. *Bulletin de l'Institut Français d'Archéologie Orientale* 107: 43-70.

Chaix, L. 2011. A Review of the History of Cattle in the Sudan throughout the Holocene. In, Jousse, H. and Lesur, J. (eds) *People and Animals in Holocene Africa. Recent Advances in Archaeozoology*: 13-26. Frankfurt: Africa Magna Verlag.

Close, A. 2002. Sinai, Sahara, Sahel: The Introduction of Domestic Caprines to Africa. In, Jennerstrasse 8 (ed.) *Tides of the Desert. Contributions to the Archaeology and Environmental History of Africa in Honour of Rudolph Kuper*: 459-470: Köln: Africa Praehistorica 14.

Di Lernia, S. 2013. The emergence and spread of herding in northern Africa. A critical reappraisal. In, Mitchell, P. and Lane, P. (eds) *The Oxford Handbook of African Archaeology*: 527-540. Oxford: Oxford University Press.

Dodson, P. and Wexlar, D. 1979. Taphonomic investigations of owl pellets. *Paleobiology* 5: 275-84.

Eisenmann, V. 2007. *Equidés actuels et récemment éteints.* [online] (updated March 2016). Available at: < http://www.vera-eisenmann.com/-Equus-actuels-et-recemment-eteints> [accessed 20 April 2016].

Eiwanger, J. 1999. Merimde Beni-salame. In, Bard K.A. (ed.) *Encyclopedia of the Archaeology of Ancient Egypt*: 501-505. London, New York: Routledge.

Garcea, E. 2004. An Alternative Way Towards Food Production: The Perspective from the Libyan Sahara. *Journal of World Prehistory* 18 (2): 107-154.

Gautier, A. 1984. Archaeozoology of the Bir Kiseiba region, eastern Sahara. In, Wendorf, F., Schild, R. and Close, A. (eds) *Cattle-keepers of the Eastern Sahara: the Neolithic of Bir Kiseiba*: 49-76. Dallas: Southern Methodist University Press.

Gautier, A. 1987. Prehistoric men and cattle in North Africa: a dearth of data and a surfeit of models. In, Close, A. (ed.) *Arid North Africa. Essays in Honor of Fred Wendorf*: 163-187. Dallas: Southern Methodist University Press.

Gautier, A. 2001. The Early to Late Neolithic from nabta and Bir Kiseiba. In, Wendorf, F. and Schild, R. (eds) *Holocene Settlement of the Egyptian Sahara. Volume 1. The Archaeology of Nabta Playa*: 609-635. New York: Kluwer Academic / Plenum Publishers.

Gautier, A. and Van Neer, W. 2009. Animal remains from predynastic sites in the Nagada region, Middle Egypt. *Archaeofauna* 18: 27-50.

Gifford-Gonzalez, D. and Hanotte, O. 2011. Domesticating Animals in Africa: Implications of Genetic and Archaeological Findings. *Journal of World Prehistory* 24: 1-23.

Guyot, F. 2010. Les dynamiques d'échanges entre l'Egypte prédynastique er le Levant sud au 4e millénaire. *Archéo-Nil* 20: 87-96.

Hassan, F.A. 2000. Climate and Cattle in North Africa. In, Blench, M. and MacDonald, K.C. (eds) *The Origins and Development of African Livestock: Archaeology, Genetics, Linguistics and Ethnography*: 61-86. London: UCL Press.

Hassan, F.A. 2002. *Droughts, Food and Culture. Ecological Change and Food Security in Africa's Later Prehistory.* New York: Kluwer Academic.

Honegger, M. 2005. Kerma et les débuts du Néolithique africain. *Geneva* NS 53:239-249.

Jousse, H. 2004. A New Contribution to the History of Pastoralism in West Africa. *Journal of African Archaeology* 2 (2): 187-202.

Kimura, B., Marshall, F., Chen, S., Rosenbom, S., Moehlman, P.D., Tuross, N., Sabin, R.C., Peters, J., Barich, B., Yohannes, H., Kebede, F., Teclai, R., Beja-Pereira, A. and Mulligan, C.J. 2010. Ancient DNA from Nubian and Somali wild ass provides insights into donkey ancestry and domestication. *Proceedings of the Royal Society B* 278: 50-57.

Kindermann, K., Bubenzer, O., Nussbaum, S., Riemer, H., Darius, F., Pöllath, N. and Smettan, U. 2006. Palaoenvironment and Holocene land-use of Djara, Western Desert of Egypt. *Quaternary Science Reviews* 25: 1619-1637.

Kuper, R. and Kröpelin, S. 2006. Climate-Controlled Holocene Occupation in the Sahara: Motor's of Africa Evolution. *Science* 313: 803-807.

Lesur, J. 2013. Des animaux et des hommes en Égypte au Néolithique et Prédynastique: les apports de l'archéozoologie. *Archéo-Nil* 23: 33-54.

Lesur, J. 2014. La faune de Tell el-Iswid sud, secteur 1a et 1b. In, Midant-Reynes, B. and Buchez, N. (eds) *Tell el-Iswid, 2006-2009*: 275-280. Le Caire: FIFAO 73, Editions de l'IFAO.

Lesur, J., Briois, F., Midant-Reynes, B. and Wuttmann, M. 2011. Domesticates and wild game in the Egyptian Western Desert at the end of the 5th millennium BC: the fauna from KS 43, Kharga Oasis. In, Jousse, H. and Lesur, J. (eds) *People and Animals in Holocene Africa. Recent Advances in Archaeozoology*: 59-74. Frankfurt: Africa Magna Verlag.

Linseele, V. 2004. Size and Size Change of the African Aurochs during the Pleistocene and Holocene. *Journal of African Archaeology* 2(2): 165-186.

Linseele, V. 2010. Did Specialized Pastoralism Develop Differently in Africa than in the Near East? An Example from the West African Sahel. *Journal of World Prehistory* 23: 43-77.

Linseele, V. and Van Neer, W. 2009. Exploitation of desert and other wild game in ancient Egypt: The archaeozoological evidence from the Nile Valley. In, Reimer, H., Förster, F., Herb, M. and Pöllath, N. (eds) *Desert Animals in the Eastern Sahara*: 47-78. Köln: Heinrich-Barth-Institut.

Linseele, V., Marinova, E., Van Neer, W. and Vermeersch, P.M. 2010. Sites with Holocene dung deposits in the Eastern Desert of Egypt: Visited by herders? *Journal of Arid Environments* 74: 818-828.

Marshall, F.B. and Hildebrand, E. 2002. Cattle Before Crops: the Beginnings of Food Production in Africa. *Journal of World Prehistory*, 16 (2): 99-141.

Marshall, F.B. and Weissbrod, L. 2011. Domestication Processes and Morphological Change Through the Lens of the Donkey and African Pastoralism. *Current Anthropology* 52 (4): 397-413.

Midant-Reynes, B. and Buchez, N. 2014. *Tell el-Iswid 2006-2010*. Le Caire: FIFAO 73, Editions de l'IFAO.

Midant-Reynes, B., Buchez, N., Bréand, G., Briois, F., Cavero, J., Coupey, A.-S., De Dapper, M., Delhopital, N., Emery-Barbier, A., el-Hajaoui, R., Guérin, S., Guyot, F., Hochstrasser-Petit, C., Lesur, J., Minotti, M., Regulski, I., Robitaille, J., Torchy, L. and Tristant, Y. 2015. TELL EL-ISWID 2006-2010. Archaeology of the Eastern Delta in the 4th millennium BC. In, Jucha, M.A., Debowska-Ludwin, J. and Kolodziejczyk, P. (eds) *Aegyptus est Imago Caeli. Studies Presented to Krzysztof M. Cialowicz on his 60th Birthday*: 37-56. Cracovie: Institut of Archaeology, Jagiellonian University of Krakow.

Minotti, M. 2014. Le petit mobilier. In, Midant-Reynes, B. and Buchez, N. (eds) *Tell el-Iswid, 2006-2009*: 243-265. Le Caire: FIFAO 73, Editions de l'IFAO.

Redding, R.W. 1992. Egyptian Old Kingdom patterns of animal use and the value of faunal data in modeling socioeconomic systems. *Paléorient* 18 (2): 99-107.

Redding, R.W. 2010. Status and Diet at the Workers Town, Giza, Egypt. In, Campana, D., Crabtree, P.J., deFrance, S.D., Lev-Tov, J. and Choyke, A.M. (eds) *Anthropological Approaches to Zooarchaeology*: 65-75. Oxford: Oxbow Books.

Riemer, H. 2007. When hunters started herding: Pastro-foragers and the complexity of holocene economic change in the Western Desert of Egypt. In, Bollig, M., Bubenzer, O., Vogelsang, R. and Wotzka, H.P. (eds) *Aridity, Change and Conflict in Africa. Proceedings of an International ACACIA Conference in 2003*: 105-144. Köln: Heinrich-Barth-Institut.

Rossel, S. 2006. *The Development of Productive Subsistence Economies in the Nile Valley: Zooarchaeological Analysis at El-Mahâsna and South Abydos, Upper Egypt*. PhD. Harvard University.

Rossel, S., Marshall, F., Peters, J., Pilgram, T., Adams, M.D. and O'Connor, D. 2008. Domestication of the donkey: Timing, processes and indicators. *PNAS* 105 (10): 3715-3720.

Smith, A.B. 1986. Cattle Domestication in North Africa. *African Archaeological Review*, 4: 197-203.

Smith, A. B. 2013. An appraisal of the terms 'Neolithic' and 'Neolithisation' for use in North Africa in the 21st century. In, Shirai, N. (ed.) *Neolithisation of Northeastern Africa*: 5-22. Berlin: Studies in Early Near Eastern Production, Subsistence, and Environment 16.

Smith, A. C. 2013. Unraveling the prehistoric ancestry of the present-day inhabitants of Northeast Africa: An archaeogenetic approch to neolithisation. In, Shirai, N. (ed.) *Neolithisation of Northeastern Africa:* 121-147. Berlin: Studies in Early Near Eastern Production, Subsistence, and Environment 16.

Tristant, Y. 2006. *L'occupation Humaine dans le Delta du Nil aux 5e et 4e Millénaires. Approche Géo-archéologique à Partir de la Région de Samara (Delta Oriental)*. PhD. Ecole des Hautes Etudes en Sciences Sociales.

Tristant, Y. and Midant-Reynes, B. 2011. The predynastic cultures of the Nile Delta. In, Teeter, E. (ed.) *Before the Pyramids. The Origins of Egyptian*

Civilization: 45-54. Chicago: Oriental Institute Museum Publication 33.

Van den Brink, E.C.M. 1989. A Transitional Late Predynastic – Early Dynastic Settlement Site in the Northeastern Nile Delta, Egypt. *MDAIK* 45: 55–108.

Van Neer, W. 2001. Restes fauniques du site Badarien de Mahgar Dendera 2 (Haute-Egypte). In, Hendrickx, S., Midant-Reynes, B. and Van Neer, W. (eds) *Le site Prédynastique de Mahgar Dendera 2*: 91-102. Leuven: Egyptian Prehistory Monographs.

Van Neer, W. 2002. Matériel faunique. In, Midant-Reynes, B. and Buchez, N. (eds) *Adaïma 1. Économie et Habitat*: 521-566. Le Caire: Institut français d'archéologie orientale.

Van Neer, W. 2004. Evolution of prehistoric fishing in the Egyptian Nile Valley. *Journal of African Archaeology* 2: 251–269.

Van Neer, W. and Lesur, J. 2004. The ancient fish fauna from Asa Koma (Djibouti) and modern osteometric data on 3 Tilapiini and 2 *Clarias* catfish species. *Documenta Archaeobiologiae* 2: 141-160.

Vermeersch, P.M., Van Neer, W. and Hendrickx, S. 2004. El Abadiya 2, a Naqada site near Danfiq, Upper Egypt. In, Hendrickx, S., Friedman, R., Cialowicz, K.M. and Chlodnicki, M. (eds) *Egypt at its Origins. Studies in Memory of Barbara Adams:* 213-276. Leuven: Peeters.

Vilà, E., Leonard, J.A. and Beja-Pereira, A. 2006. Genetic Documentation of Horse and Donkey Domestication. In, Zeder, M., Bradley, D., Emshwiller, E. and Smith, B.D. (eds) *Documenting Domestication. New Genetic and Archaeological Paradigms*: 342-353. Berkeley: University of California Press.

Von den Driesch, A. 1976. *A Guide to the Measurement of Animal Bones from Archaeological Sites.* Cambridge: Harvard University.

Von den Driesch, A. 1997. Tierreste aus Buto im Nildelta. *Archaeofauna* 6: 23-39.

Von den Driesch, A. and Boessneck, J. 1985. *Die Tierknochenfunde aus der Neolithischen Siedlung von Merimde-Benisalâme am Westlichen Nildelta.* Munich: Uni-Druck.

Wenke, R.J., Long, J.E. and Buck, P.E. 1988. Epipaleolithic and Neolithic subsistence and settlement in Fayum Oasis of Egypt. *Journal of Field Archaeology*, 15: 29-51.

Yokell, C. 2004. *Modelling Socioeconomic Evolution and Continuity in Ancient Egypt. The Value and Limitations of Zooarchaeological Analyses.* Oxford: BAR International Series.

Yeomans, L. 2011. Stews, Meat and Marrow: Extracting protein and fat for the Lost City. *Aeragram* 12 (2): 13-15.

6 Archaeozoological investigation of the site of Shengavit, Armenia (Based on the archaeological excavations of 1965-1980; 2003-2007)

Nina Manaseryan[1]

1 Institute of Zoology, Research Center for Zoology and Hydroecology of Armenian National Academy of Science, 7 Sevak str. Yerevan, Armenia; ninna_man@yahoo.com

Abstract

This paper presents the results of the study of osteological material from the site of Shengavit (Armenia). The faunal spectrum includes wild and domestic animal species that are usually found in the ancient settlements located in the territory of modern Armenia.

The following wild species have been identified: *Canis lupus, Vulpes vulpes, Martes foina, Sus scrofa, Cervus elaphus, Ovis orientalis, Capra aegagrus, Gazella subgutturosa, Capreolus capreolus, Meles meles, Lepus europaeus and Vormela peregusna*. Domestic animals are represented by all the main species such as cattle, horse, sheep, goat, pig and dog.

Keywords

Armenia, faunal remains, wild and domestic animals, mammals

Introduction

One of the most important Early Bronze Age settlements in Armenia is Shengavit, founded on the left bank of the river Hrazdan, now on the right shore of Yerevan Lake. The exploration of Shengavit by archaeologists began in 1936 when Eugenie Bayburtian did some excavations there and discovered a culture unknown at that time, which he rightly dated to the Early Bronze Age and named after the name of the site 'Shengavitian'.

Shengavit consists of a settlement, a fortress with a citadel and a necropolis. In archaeological terms, this represents an integral complex which allows a comprehensive study of life, customs, traditions and rites of our ancestors. The monument is dated to the beginning of the 4th millennium BC. Faunal remains were discovered during the excavations led by G. Sardaryan between 1958 and 1980, and by H. Simonyan between 2003 and 2007 (Simonyan 2013). The osteological assemblage was sent to the Institute of Zoology of the Armenian National Academy of Sciences for archaeozoological analyses (Manaseryan and Mirzoyan 2005).

The faunal material

The material presented here comprises more than 15000 remains. Among them, 4070 specimens were identified to the level of genus or species. Both wild and domestic taxa are represented. The material consists of animal remains such as bones and fragments of bones. The skulls, apart from the isolated molars which had dropped out of the dental alveoli due to destruction of the jaws, are composed of fragments of upper and lower jaws. Other parts of the skull and horn cores are unfortunately not well represented. It should be noted that the remains are abundant and include individuals of different age classes. Although the majority of the remains belong to adult individuals, unfused metacarpal and metatarsal bones were occasionally found.

Remains of wild animals

Looking at the relative representation of the different species in the assemblage, it is possible to determine the importance of hunting in the economy, and the approximate composition of herds of domestic animals. It should be noted that the study of wild mammal species is very limited (Manaseryan and Mirzoyan 2013). The presence of 18 different wild species in the assemblage (13.5%), shows that the region was rich and diverse. The wild mammalian fauna is represented by remains of hare, beaver, fox and wolf, four Mustelidae species, two Cervidae species, wild ass and wild pig, gazelle, bezoar goat and moufflon. Apart from the bones

Table I. List of wild and domestic animal species identified at site Shengavit (based on the Number of Identified Specimens - NISP).

Species	1958-1980	2003-2007	Total
Brown hare – *Lepus europaeus* Pall.	1	-	1
Beaver – *Castor fiber* L.	1	1	2
Fox – *Vulpes vulpes* L.	40	39	79
Wolf – *Canis lupus* L.	28	-	28
Canis lupus or *Canis familiaris*	170	73	243
Marten – *Martes foina* Erxleben.	2	1	3
Weasel – *Mustela nivalis* L.	1	-	1
Marbled polecat – *Vormela peregusna* Gueldenst.	1	-	1
Badger – *Meles meles* L.	17	3	20
Wild ass – *Equus hemionus* Pall.	2	-	2
Roe deer – *Capreolus capreolus* L.	4	6	10
Red deer – *Cervus elaphus* L.	53	44	97
Wild pig – *Sus scrofa* L.	2	-	2
Gazelle – *Gazella subgutturosa* Gueldenst.	2	-	2
Moufflon – *Ovis orientalis* Gmel.	16	4	20
Bezoar goat – *Capra aegagrus* Erxleben.	-	2	2
Fish – Pisces	1	1	2
Tortoise – *Testudo* sp. or *Testudo graeca*	1	1	2
Birds – Aves	-	33	33
Wild animals total	**342**	**208**	**550**
Horse – *Equus caballus* L.	128	47	175
Ass – *Equus asinus* L.	11	-	11
Pig – *Sus domesticus* L.	41	37	78
Cattle – *Bos taurus* L.	821	513	1334
Domestic Goat – *Capra hircus* L.	80	15	95
Domestic Sheep – *Ovis aries* L.	60	23	83
Ovis aries or *Capra hircus*	1056	623	1679
Dog - *Canis familiaris* L	57	8	65
Domestic animals total	**2254**	**1266**	**3520**
Grand Total	**2596**	**1474**	**4070**

of fox and badger, all other post-cranial remains of wild mammals belong to ungulates such as red deer, roe deer and, moufflon. These species were hunted mainly for meat but perhaps also for sports hunting. Bone remains of some wild animals, previously widely distributed, are now relicts of the past.

Lepus europaeus Pallas, 1778 – Brown hare
Fragments of mandibles. In Armenia the contemporary brown hare is encountered in almost all landscapes, except central areas of the forest.
Vulpes vulpes L. - Fox
A skull and bones of extremities. A common animal in Armenia nowadays. It inhabits various landscape zones.
Martes foina Erxl. - Marten
A skull. An animal which is valued for its fur.
Mustela nivalis L. - Weasel
A skull. The weasel is common in Armenia everywhere except for dry subtropical areas.
Vormela peregusna G. – Marbled polecat
A skull. The marbled polecat occurs in mountainous steppes, semi desert areas, and in the mountainous regions of Armenia.
Meles meles L. - Badger
A skull and mandible fragments. It was hunted, perhaps, for the fur, fat (used by folk medicine see Amasiaci 1478-82) and partially for meat.
Capreolus capreolus L. - Roe deer
A horn core and bones of extremities.
Bos primigenius Boj. – Aurochs
It has not been indubitably identified at Shengavit, but we cannot exclude the possibility that some remains attributed to domestic bull, might belong to a female aurochs.
Capra aegagrus Erxl. - Bezoar goat
Horn core fragments. The remains are not numerous.
Ovis orientalis Gmelin - Armenian moufflon
Horn core fragments and bones of extremities.
Cervus elaphus maral Gray – Caspian red deer
Antler of a young deer with a well defined crown. During the post-Paleolithic era, red deer was a beloved subject for artists and sculptors to decorate weapons, dresses, and pottery with. The analysis of wild animals shows that the most important game species was Red deer (71.8% of the wild ungulates NISP).

Table 2. Measurements of sheep horn cores (in mm following Tsalkin 1960; 1961).

	Male		*Female*	
Site	**Circumference of the basis**	**Length on the greatest curve**	**Circumference of the basis**	**Length on the greatest curve**
Shengavit n=9	153.8 (140-172)	225	Shengavit n=7	77.0 (53-95)
Mokhrablur n=7	158.4 (137-181)	222 , 225	-	-

Castor fiber L. – Eurasian beaver

Lower jaw with damaged teeth. It is a rare and very important find, confirming the presence of this animal in Armenia

Gazella subgutturosa Gul. – Goitered gazelle

A few horn core fragments. These remains show that *Gazella subgutturosa* was also present in Armenia.

Equus hemionus Pall. - Asian wild ass

A mandible fragment, and separate teeth. This species fully disappeared during the Middle Holocene (Mejlumian 1988).

Remains of domestic animals

Looking at the relative importance of various domestic species, we can estimate the role of each species in the subsistence and economic strategies of the ancient population at Shengavit (Table 1).

Sheep

Sheep and goats represent more than 50% of the domestic animal NISP. In addition, sheep/goat is the most abundant taxa in the domestic assemblage. All skeletal elements are present, but fragments of skulls and horn cores are less frequent. Because no complete skulls were found it is difficult to define which specific sheep breeds existed during that time. It is difficult to make any judgement based on the size of the horn core as they show a great variation, especially because the horn core of the wild genus member – the mouflon – could be classified as domestic.

As for the post-cranial skeleton, the most interesting results come from the analyses of the metapodials (metacarpal and metatarsal bones).

Metapodials, metatarsal bones in particular, suggest a substantial variability in size. Judging from metapodials, the sheep height at withers ranged within 61-75 cm. Sheep metapodials of similar sizes and proportions (withers height of 67-70 cm) were previously identified in the faunal remains from Khatunarkh (70cm) and Mokhrablur (66-70cm) (Manaseryan 1972).

Evidently, a large-size and fairly similar breed of sheep were raised on the plains of Armenia during the Bronze Age (Manaseryan 1984).

As far as sheep are concerned, it can be concluded from the measurements that Early Bronze Age populations attained withers height between 64 and 70 cm. Keeping in mind that the withers height of

Table 3. Measurements of sheep metacarpal and metatarsal bones (in mm following Tsalkin 1960; 1961).

Maximum length (mm)	*Shengavit 1965-1980*	*Withers height (cm)*	*Shengavit 2003-2005*	*Withers height (cm)*
Metacarpal	-	-	118-120	72-73
Metatarsal	142-149	67-70	130-145	61-75

Table 4. Withers height of sheep (in cm, following Tsalkin 1961) based on the length of metapodia.

Site	*N*	*min-max*	*M*
Shengavit	4	67-70	68.2
Mokhrablur	7	66-70	68.1
Sepasar	8	62-70	65.7

Table 5. Measurements of horn core of goat (in mm, following Tsalkin 1960; 1961).

Horn core	*Shengavit*	*Bezoar goat*
Circumference of the basis	71 - 85	65-96
Length measured on the greatest curve	138-150	67-140

domesticated sheep of aboriginal Armenian species (named Mazekh) falls within the limits of 61-74 cm, it can be assumed that the sheep from the old settlements are similar to the modern ones.

Goat

Bones of goats are extremely rare in the assemblage, however it should be noted that several fragments of horn cores were identified thanks to the barely noticeable positive twist and were classified as horn core belonging to female domestic goats (Gromova 1953; Manaseryan 1996).

There is no doubt that the horn cores belong to large species of domesticated goats, as their horn core basal circumference does not differ from those of present day wild goats.

The length of the goat metatarsal from Shengavit and Mokhrablur varies within a narrow limit: 104-107 mm. The length of metatarsals for Bezoar goats varies in males from 130 to 140 mm, and in females from 118 to 142 mm. As noted, the goat bones from the assemblage are much shorter in length than those of Bezoar goats, hence there are grounds for considering them as domesticated species.

Table 6. Horn core circumference size comparison between Bronze Age cattle from different sites (in mm, following Tsalkin 1960; 1961).

Site	N	Horn core circumference (mm)
Shengavit	3	146-155
Mokhrablur	7	118-165
Shirakavan	1	140-165
Lchashen	1	132-165

Table 8. Withers height size comparison between Bronze Age cattle from different sites (following Tsalkin 1960).

Site/Age	Withers Height Size
Early Bronze Age	
Shengavit	113-138 cm
Middle Bronze Age	
Aruch	136-143 cm
Lori Berd	130-149 cm
Aghavnatun	136-143 cm

Table 7. Bone measurements of cattle (in mm, following Tsalkin 1960; 1961).

Maximum length, mm	Shengavit 1965-1980	Shengavit 2003-2004	Aratashen Neolithic	Aruch III-II millennia BC	Lori Berd XIX
Radius /Ulna	-	184	-	-	-
Astragalus	56-82	62-76	65-70	58-71	60-68
Calcaneus	134-137	132-149	-	146-148	132-139
Metacarpale	179-188	211-221	-	218	208-234
Metatarsale	182-238	235- 245	-	235-267	243-257
I Phalanx	54-75	57-66	50-66	66-75	59-72
Withers height (cm)	120-135	135-140	-	135-140	130-145

Metapodials from Shengavit and Mokhrablur allows us to calculate the withers height for domesticated goats around 50-52cm (Manaseryan 2004). This corresponds to the withers height of modern Bezoar goats. It can be assumed that the domesticated goats of the settlements originate from the population of Bezoar goats in the Armenian mountains. The curling of the horns is a domestic change in the form of the horn, for the wild *Capra aegagrus* this can be the of a result of mutation

Cattle

Cattle bones are common among the excavated material, domestic bulls and cows represent 46% of the domestic animal NISP. Despite a relatively large number of bones (1364 specimens), none of the skulls among them were found intact, which greatly limits the possibility of osteological (primarily cranial) description of livestock. Only a few fragments of horn cores suggest that the ancient domestic cattle had horns.

Post-cranial bones are relatively abundant. They are represented by complete metapodials, calcaneus, and phalanges.

Summarizing the data on the length of post-cranial bones from Shengavit, we found a great variability. In all cases we see bones of relatively small specimens showing dimensions that are usually characteristic of domestic livestock. At the same time there are fairly large-size bones reaching the limits of the upper range on the variables established in aurochs. Therefore, it can be argued that both wild and domestic animal remains are found at Shengavit.

This assumption was confirmed by comparing the size of cattle post-cranial bones from archaeological sites that are the closest in time to Shengavit. At the same time, it is quite likely that a number of specimens showing bones of same length as those of aurochs in fact belong to domestic animals, just like the bones in some cases belong to aurochs which have been exclusively related to domestic livestock.

A series of surviving metapodials provide an opportunity to evaluate the withers height of the ancient cattle. Using established parameters based on the length of metapodia (Tsalkin 1960), we establish that the withers height of the cattle varied from 130 to 145 cm and was higher or lower only in relatively rare cases.

Previous investigations have provided a large amount of osteometric data which permits some conclusions concerning the size of the domestic animals kept here by Early Bronze Age and Middle Bronze Age farmers (Manaseryan 2003). As can be seen from Table 8, the Early Bronze Age cattle were of similar size as cattle populations from the Middle Bronze Age.

Horse

Apart from the remains of cattle and small ruminants found in the faunal assemblage, a limited number of horse bones have also been identified (5% of the domestic mammal NISP). The proximal phalanges of horse are highly variable both in size and shape. This is confirmed by the figures which indicate the length and the overall proportions of bones. To illustrate this, the total length of the long pastern bones ranges from 70 mm to 85 mm (Von den Driesch 1976).

Proximal phalanges of similar dimensions were observed in horses from the archaeological

Table 9. Bone measurements of horse (in mm, following Von den Driesch 1976).

Name	Aratashen - Neolithic	Shengavit	Sepasar	Mokhrablur	Lori Berd XIX
GH - Astragalus	60-67	53-63	54-55	63-66	59-60
GL - Calcaneus	-	101-105	-	-	103-104
GL - Metacarpal	-	-	223	-	-
GL - I Phalanx	-	63-84	72-82	79-80	85-95
GL - II Phalanx	-	41-45	44	-	41-50
GL - III Phalanx	-	74-85	-	-	-

Table 10. Measurements of the maxilla of Pig (mm).

Site	Length of the cheek tooth row	Length of the premolar row	Length of the molar row	Length of M^3	Breadth of M^3	Height of M^3
Shengavit 23	120	72	47	33	17	-
Shengavit 8	119	71	47	36	19	-
Shengavit 20	117	70	39	33	18	12

sites of different years at Shengavit (63-84 mm), Mokhra Blur (79-80 mm), and Sepasar (72-82mm) (Manaseryan 2011).

These figures suggest that long pastern bones are distinguished by relatively small size. Some of the specimens found among them are relatively thin and slim, while very thick and bulky specimens are relatively rare. Variability limits in long pastern bones are relatively stable, for example, the ratio of the width of the diaphysis to the total length of the bone ranges from 32 to 38 percent (Manaseryan 2006).

Pig

Pigs remains are rare. The discovered remains are extremely decayed and represented by cranial fragments, mandible and maxilla. The preservation of these remains is poor and bones that would allow any information about osteological peculiarities are missing. Available measurements are represented in the (Table 10).

Dog

Dog is represented at Shengavit by 65 bones, making up less than 1.8% of the domestic animals NISP. Dog bones were differentiated from wolf remains based on their smaller size. 243 finds of intermediate size were identified as "dog or wolf". Numerically dogs do not play a significant role among the bone remains from Shengavit. All parts of the skeleton are present, but there are no indications that dogs were eaten.

Conclusion

These are the main results from the study of the faunal remains recovered from the archaeological investigations at Shengavit. This assemblage is not comprehensive, further excavations will undoubtedly produce new and interesting material in order to make a more detailed account of the economic life of the ancient population. Based on our results it is evident that cattle was an important component of the subsistence economy at Shengavit . All the important species of agricultural animals were discovered among the excavated materials: cattle, sheep, goat, and horse. Existence of such a substantial complex of domesticated animals is less in harmony with the existing image of cattle breeding. The mentioned situation lets us consider cattle breeding as one with a high level of development at these populations.

After existing for more than 2000 years, the settlement at Shengavit was deserted for unknown reasons. But life in the territory of nowadays Yerevan continued.

Acknowledgements

I have to express my thanks to Armenian archeologists for availability of the excavated material.

References

Amasiaci, A. 1478-82. *Useless for Ignorant* (in Russian). Translated from Armenian by S. A. Vardanian, 1990. Moskow: Nauka.

Von den Driesch, A. 1976. *A Guide to the Measurement of Animal Bones from Archaeological Sites.* Cambridge: Harvard University.

Gromova, V. 1953. Osteological differences of the genera *Ovis* and *Capra* (in Russian). *Proceedings for the Study of the Quaternary Period* 10(1): 3-122.

Manaseryan, N. 1972. Materials for the study of small cattle in the economy of the ancient tribes inhabiting the territory of Armenia from VI to II millennium BC (in Russian). *Biological Journal of Armenia* 10: 82-5.

Manaseryan, N. 1984. Small cattle from Bronze Age monuments in Armenia (in Russian). *Biological Journal of Armenia* 11: 966-75.

Manaseryan, N. 1996. To the question about the differential characteristics of the genera *Capra* and *Ovis. Biological Journal of Armenia* 49(3-4): 12-15.

Manaseryan, N. 2003. Cattle of the most ancient monuments of Armenia. *Materials of regional scientific conference "Research and protection of wildlife in South Caucasus"*.

Manaseryan, N. 2004. Ancient domestic animals of Armenia and neighboring regions. *Journal of History and Philology* 2(166): 282-90.

Manaseryan, N. 2006. The Stature of Horses in Armenian Bronze and Early Iron Age

Burials. In, Olsen, S. L., Grant, S., Choyke, A. M. and Bartosiewicz, L. (eds) *Horses and Humans: The Evolution of Human-Equine Relationships*: 271-74. Oxford: British Archaeological Reports.

Manaseryan, N. 2011. Horses in ancient and medieval Armenia (in Russian). *Biological Journal of Armenia* 63(1): 30-39.

Manaseryan, N. and Mirzoyan, L. 2005. Collections of subfossil mammals. *Culture of Ancient Armenia* 13: 200-205.

Manaseryan, N. and Mirzoyan, L. 2013. Armenia: Animal remains from Neolithic and Bronze Age settlements and burials. *Studii de Preistorie* 10: 131-53.

Mejlumian, S. 1988. *The Holocene mammal fauna of Armenia* (in Russian). Yerevan: Academy of Sciences of Armenia.

Simonyan, H. 2013. Shengavit – an ordinary settlement or an early city (in Armenian). *Hushardzan* 8: 395-406.

Tsalkin, V. 1960. Metapodial variation and its significance for the study of ancient

Horned Cattle (in Russian). *Bulletin of Moscow Society of Naturalists, Biological Series, V* 65(1): 109-26.

Tsalkin, V. 1961. The variability of metapodia in sheep (in Russian). *Bulletin of Moscow Society of Naturalists, Biological Series,* V, 65(5): 115-32.

7 Invisible donkeys in ancient Mesopotamia: New insights from modern studies in Sub-Saharan Africa

Jill Goulder[1]

1 Institute of Archaeology, University College London, 31-34 Gordon Square, London WC1H 0PY; jill@jgoulder.com

Abstract

To date, there has been very little focus on systematic use of working donkeys in late 4th and 3rd-millennium BC Mesopotamia. Models of working animal adoption have largely emphasised ploughing oxen, drawing in particular from urban, elite-commissioned texts. An additional barrier to the recognition of working-donkey presence has been taphonomic and zooarchaeological issues with evidence in settlement-oriented fieldwork, as in many cultures donkeys are not eaten. A currently underutilised resource in archaeology is the large body of modern development studies in sub-Saharan Africa and elsewhere, which provides invaluable accounts of the practical, social and economic implications of adoption of working donkeys and cattle.

Keywords

Donkeys, Mesopotamia, secondary products, working animals, ethology

Introduction

In Mesopotamia in the 4th and 3rd millennia BC, the systematic use of animals for traction and transport provided vital functional underpinning to the rapid development of complex societies (Algaze 2013: 72). Ploughing oxen have often held centre stage in archaeological models of this period, following Sherratt (1981)'s proposition of the Secondary Products Revolution. Acknowledgement of the key role in Mesopotamia of pack-donkey caravans is now growing, and the presence of ploughing donkeys is clear from 3rd-millennium BC texts; but there has as yet been little attention to their likely employment in the many invaluable local transportation roles that donkeys perform to this day on farms and in villages. This, exacerbated by the marked rarity of donkey remains in the archaeological record, has resulted in a profile which is at odds with donkey use in modern developing regions.

In this paper I outline the practical and conceptual factors contributing to the scarcity of donkeys in both the archaeological record and interpretative commentary on social and economic developments in 4th- and 3rd-millennium BC Mesopotamia, and suggest a new approach enabling more informed discussion of their presence and role. My research makes use of the large body of recent published work relating to modern societies in developing countries which rely significantly upon working animals; a further useful resource consists of material on the physiological and behavioural characteristics of donkeys, which assists in revisions to common assumptions about their likely role and value.

The 'Secondary Products Revolution'

An overview of the Near Eastern archaeological literature indicates that debate on donkeys often centres on their domestication and hybridisation rather than on their systematic usage for work. In Mesopotamia this stems particularly from the emphasis placed on ploughing oxen by Sherratt (1981; 1983; 2003; 2006) in his far-reaching theory concerning the advent of secondary products – traction, milk and wool – in the 4th millennium BC. He originally argued that these products evolved or were adopted as a package, including an ox-plough/cart 'traction complex, with its own technology, ideology, and attitude to domestic livestock' (Sherratt 1997: 240); his model included a distinct but roughly contemporary enlistment of equids for long-distance pack transport in the southern Levant (Sherratt 1981: 295; Fig. 1) – though he later acknowledged the spread of pack donkeys to Mesopotamia (2003: 238). Sherratt's model drew explicitly upon 4th-millennium BC pictograms of ploughs and carts and contemporary seal depictions

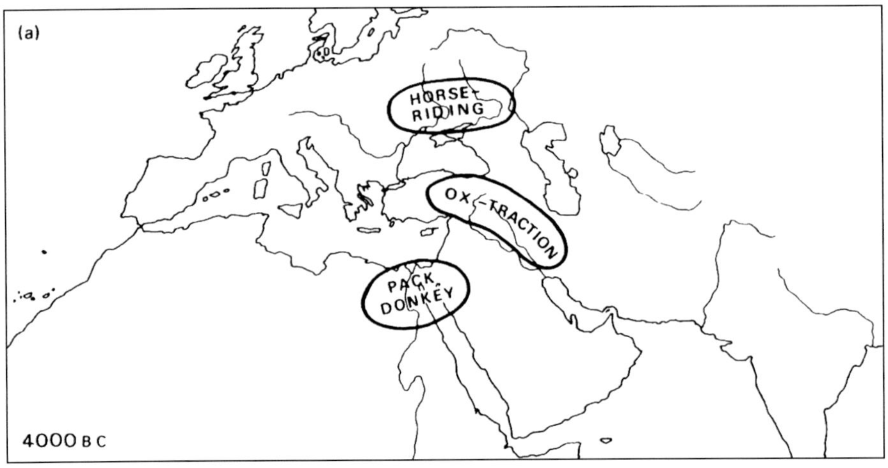

Fig. I. Sherratt 's (1981, 288) ox-centric view of early working animal use in Mesopotamia, since refuted.

of draught oxen (Sherratt 1981: 263-6), and on pack-donkey figurines from the southern Levant (Fig. 2). These sources, while providing valuable information, inevitably reflect the interests of an urban elite rather than the rural majority, notably low-status women (e.g. Asher-Greve 2013: 359). I argue in this paper that the advent of the donkey provided an important new resource at grass-roots level, both for ploughing and for the unremarked daily transport tasks around farms and villages for which donkeys are overwhelmingly used today in developing regions.

Sherratt's model focused on the transmission of specific hardware – the plough and the cart – between the Near East, Europe and the Central Asian steppe in approximately the 4th millennium BC (Sherratt 1981: 266, 288). Sherratt initially included the horse in his model (1981: 272-4), suggesting a steppe 'package' of horse and ox-cart (the former for herding, the latter for transport of households to follow the herds (Anthony 1991: 266)) and this may well also have influenced his focus on ox-carts rather than pack-donkeys for transport in Mesopotamia (Sherratt 1981: 295). Archaeological thought now (e.g. Halstead and Isaakidou 2011: 62), while acknowledging the highly significant impact of the use of working animals on human society, generally takes the view that there was neither a moment of innovation nor Sherratt's proposed package involving the close interrelation of several secondary products and traction modes diffusing as a unit.

The ploughing ox has become closely associated with the later Uruk (c.3400-3200 BC), based on Sherratt's (1981: 290-98; 2006: 343) and other depiction- and text-based models identifying it as the enabler of centrally-controlled production of an agricultural surplus upon which the new social complexity depended for its survival. Ploughing donkeys on state agricultural land are regularly recorded in 3rd millennium BC texts (Heimpel 1995: 91-6, 110-11) – and in modern developing regions all ploughing donkeys additionally or even mainly fulfil local transport functions – but they are still rarely acknowledged in broader discussion of Mesopotamian culture and activities. Recognition is meanwhile rising among commentators of the role of the pack-donkey as a valuable complementary technology to the southern Mesopotamian system of rivers and canals which has often dominated transport models. This is illustrated for example by the change of focus in Algaze's commentaries on the Uruk period, from dismissal of the donkey in 2001 (Algaze 2001: 204) to its confirmation as a key transport mode in 2008 (Algaze 2008: 55, 66).

The possible presence of domesticated donkeys in Mesopotamia from the late 4th millennium BC is hinted at by zooarchaeological evidence (e.g. Payne 1988: 99; Uerpmann 1987: 141; Vila 2006: 102), though with obstacles as outlined later; by the 3rd millennium BC, zooarchaeological and textual evidence of donkey use is common (Clutton-Brock 2012: 28; Heimpel 1995: 91-97, 134-5; Maekawa 1979; Postgate 1986: 194, 200; Potts 2011: 171; Sallaberger 2014: 350; The Electronic Text Corpus of Sumerian Literature, Enmerkar and the lord of Aratta, c.1.8.2.3; Vila 1998: 132; Way 2011: 133, 141; Zarins 2014: 67-78, 111, 189). Donkeys – for ploughing or pack – are rarely mentioned in elite contexts in texts, unlike the expensive and prized hemione-donkey hybrids identified in 3rd-millennium BC texts, representations and burials (Postgate 1986: 194-200; Weber 2008; Zarins 1986: 185-7). A noted phenomenon in the Near East at this period is the dichotomy between the despised status of the working donkey from earliest times (donkey-related insults appear in Egyptian and Mesopotamian texts from the 3rd millennium BC (e.g. Way 2011: 944)) and the widespread incidence of burial of pairs of donkeys in elite circumstances (Way 2011). Donkeys were a valuable resource to merchants benefiting economically and socially from their ability to move goods long-distance at low cost. I suggest that while perhaps lowly and despised in themselves, a symbolic pair taken from a 'fleet' of pack-donkeys could

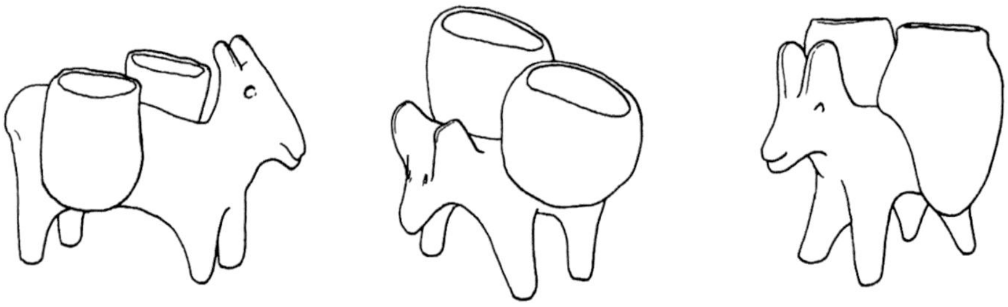

Fig. 2. Early Bronze Age (later 4th millennium BC) figurines from the southern Levant (From Figure 7.5, P207, Sherratt, A.G. (1997), Economy and Society in Prehistoric Europe: Changing Perspectives. Princeton, NJ: Princeton University Press).

have been appropriate adjuncts to the burial of one whose power derives from control of this facility (Greenfield *et al.* 2012: 25; Milevski 2011: 233; Shai *et al.* 2016: 6-7), as with slaves interred with their masters.

The invisible donkey in Ancient Near Eastern archaeology

There is still little discussion in archaeology or even zooarchaeology of the on-the-ground use and impact of donkeys for ploughing and particularly local transport, perhaps a legacy of European/ Mediterranean agricultural traditions in which ploughing and by extension all farm work is carried out with oxen. The donkey is even missing from the list of major domesticates in Russell's recent work *Social Zooarchaeology* (2012: 208), and from her consideration of the likeliest early pack species (2012: 228). In addition to the legacy of Sherratt's animal-traction scenario, with its strong emphasis on oxen, and to the comparative lack of focus on the donkey in elite-generated ancient records, there are cogent practical explanations for the striking invisibility of the working donkey in Ancient Near Eastern models including the scarcity of donkey remains in settlements and the problem of identification.

Detection of the presence of working donkeys is considerably impeded by the low incidence of donkey skeletal remains in the Near East from the 4th and 3rd millennia BC (Grigson 1995: 258; Rossel *et al.* 2008: 3715; Sasson 2010: 2; Zarins 2014: 69, 82). It is sometimes insufficiently recognised that the presence of animals used for work but not eaten can be seriously underestimated. Mesopotamian archaeology during the 20th century AD was necessarily influenced by cultural preoccupations of the time in its focus on monumental architecture and elite material culture. It also reflected the means of site identification available at the time, with its inevitable focus on tells – the occupation mounds that dot the region. This emphasis is changing with growing detection facilities such as satellite photography, and an increased focus on humans in the landscape (Wilkinson 2003: 76), but the archaeological legacy still results in most animal-bone finds coming from food-refuse middens, often in urban environments.

Such middens form a key source of evidence of working cattle, as they are almost universally used for meat at the end of their working lives. But in many regions throughout history and to the present day, there is marked evidence of cultural (and practical) reluctance to eat working equids (Blench 2004: 24; Jones 2008: 6; Payne 1988: 103; Smith and Pearson 2005: 1; Vila 2006: 103; Zarins 2014: 82, 202). Dead donkeys today are commonly dragged well out of camps and villages (MacDonald 2000: 11; Marshall and Weissbrod 2009: 72), and donkeys no longer able to work are abandoned in the bush/ desert rather than killed (e.g. du Boulay 1974: 87; Fielding 1987: 27; Waithanji 2009: 36). Their appearance in the archaeological record in Mesopotamia and elsewhere is correspondingly rare.

An additional obstacle is that *Equus asinus* skeletal remains are difficult to distinguish osteologically from those of hemiones (*E. hemionus*), which are similar in appearance to the donkey but effectively non-domesticable, and native to the arid belt from Syria to Mongolia, including Mesopotamia. Hemiones were hunted for meat and perhaps hides from early times, including in 4th-3rd millennium BC northern Mesopotamia, and the presence of their remains can impede detection of domesticated donkeys (e.g. Kirkbride 1974: 85; Vila 1998: 132-48; Zarins 1986: 179; Zarins 2014: 15-16).

A primary identifier of working donkeys would be their location outside their natural habitat (e.g. Arnold *et al.* (2016: 1-5) describing isotopic analysis suggesting the Egyptian origin of a 3rd-millennium BC donkey skeleton found at Tell es-Safi in the southern Levant). Donkeys are native to northeast Africa (Beja-Pereira *et al.* 2004; Kimura *et al.* 2010; Vilà *et al.* 2006), though there are suggestions of wild donkeys moving through Arabia to the southern Levant and even Mesopotamia (Potts 2011: 167-8; Uerpmann 1987: 30; 1989: 164; 1991: 12-30; Vila 1998: 144; 2006: 101, Zarins 2014: 33-6). In such regions, specific evidence of domestication and work use may additionally be required. Unlike most livestock species, domesticated donkeys are not easily distinguishable morphologically from their wild ancestors, as from earliest to modern times donkey-using cultures have encouraged free breeding with the wild and made little use of selective breeding; unlike species kept for meat or milk, the size and vigour of wild donkeys are a prized virtue (Groves 1974: 159; Kimura *et al.* 2010: 6; Marshall and Weissbrod 2011:

S405-7; Marshall *et al.* 2014: 6153-5; Rossel *et al.* 2008: 3715, 3719; Shackelford *et al.* 2013: 4170-71). Stronger clues as to domestication status may be obtained from bone pathologies resulting from heavy pack or traction use, and a range of osteological and particularly dental indications of constraints upon natural moving and eating (Baker 1984: 253-4; Bartosiewicz *et al.* 1997: 11-12; Clutton-Brock and Davies 1993: 214-5; Grigson 1995: 258, 267; Levine 1999: 74; Rossel *et al.* 2008: 3719; Russell 2012: 228; Shackelford *et al.* 2013: 4170-71; Shai *et al.* 2016: 11; Weber 2008: 501; Zarins 2014: 42). However, while there have been useful identifications through these means, an issue remains of the rarity of finding donkey skeletons in circumstances indicating normal death and disposal. Where these do occur, key skeletal elements may be missing, and in the general circumstances of limited comparison material for domesticated donkeys, interpretation of pathologies is also a concern.

A number of factors have therefore combined to make the donkey almost invisible from archaeological view, giving limited material for examining the impact of donkey adoption and use through assessment of cultural evidence in association with zooarchaeological remains. In a move to address this issue, I propose a complementary ethnographic approach to assist in creating new models of the presence of working donkeys in the Ancient Near East.

Use of modern working-animal studies

The use of ethnographies for study of ancient societies is an established technique, valuable in addressing potential bias and gaps by providing dimensions beyond individual present-day experience. Sensitively used, they can explore the range of responses to situations, possible behaviours and interpretations, assisting in 'eliminating error and assessing likelihood, improving credibility and delimiting uncertainty' (Wylie 2002: 145). Archaeological discussion of social and economic development in the Ancient Near East has made only very limited use of accounts from extant cultures of interaction with working animals on a day-to-day basis at farm and household level. Pioneering work has been done for other regions by Bogucki (1993) and Halstead (1995), who use interpretations respectively from modern Africa and traditional Greek farming for interpretation of scenarios in Neolithic Europe and the Aegean Bronze Age – though Halstead warns (2014: 2) against 'uncritical' analogy.

Ethnographic and other modern study findings can usefully be reinforced by ethology – the study of innate animal behaviour and psychological responses to situations – and animal physiology. A large body of published practical experimentation and specialist commentary exists on the biological capabilities and constraints of donkeys in relation to pack and traction, including work rates and capabilities, and food/water needs and preferences. Dijkman (1992: 153-5), Hagmann and Prasad (1995: 231, 235-7), Pearson and Vall (1998: 309-12, 315) and Renger (1990: 269, 271-2) report for example on physiological experiments on working-energy efficiency of donkeys and oxen, while Yousef *et al.* (1972) analyse the anatomical and mechanical factors contributing to the low energy-expenditure of working donkeys. Brodie (2008: 303) argues in the context of the Aegean Bronze Age that donkeys are physiologically and behaviourally more suited than oxen to traction and transport use by small farming households in regions of light soil. Also of value are studies of the social behaviour of donkeys in the wild, relating to their propensity for taming and handling, pack-train and traction activity, and breeding in captivity (e.g. French 1997).

A prime zone for relevant investigation of responses to the advent of working animals is sub-Saharan Africa, where in many regions working cattle and donkeys were only introduced within the last 100 or even 50 years (Goody 1976: 25-33; Münzinger 1982: 21-49). Usage of working animals is increasing significantly in many countries in this region, as a result of a shift of NGO and governmental focus away from tractors since the 1970s-80s. In some regions there has been a direct transfer from hoe agriculture and human porterage to use of working animals, with the opportunity (with careful caveats) to study uptake processes and problems and to draw tentative interpretations relating to working animal use in the Ancient Near East. This is enabled by the new focus by government bodies, development agencies and specialist charity organisations on the value of promoting use of working animals in this region, resulting in a series of major recent research initiatives, international colloquia and expert consultations. A smaller number of studies address use of working animals in Asia, Latin America and the Near East.

For my doctoral research I have accumulated several hundred such published studies in the categories outlined above. Although these are rarely ethnographic in original aim, they provide on-the-ground accounts of situations outside the experience of many Near Eastern archaeologists attempting to interpret the fragmentary clues as to the social and economic impact of the advent of working animals into a community. Through detailed qualitative analysis of this material – with careful attention to caveats about modern analogy – I aim to extract some of the physiological and ecological irreducibles which can be applied to use of working animals in Ancient Near Eastern environments, and also to develop ethnographically-based predictions as to the archaeological evidence and deductions potentially available for this phenomenon.

Emerging themes

One of the most striking themes emerging from these modern sources is the difference in social and economic trajectories arising from adoption of working donkeys versus cattle, and in particular the impact of donkey use on the empowerment of women. In modern sub-Saharan Africa there are often cultural (male) prejudices in favour of cattle where ownership and usage is prestige-related, but women have considerably more cultural access to lowly donkeys. Traditional female daily tasks such as fuel- and water-collection are made lighter and speedier, allowing many women time to make or grow items for sale and transport them to market on donkey-back (Admassu and Shiferaw 2011: 37-40; Marshall and Weissbrod 2009: 59; Sylwander 1994).

Can any parallels be drawn with possible implications for women in Mesopotamian society? While there is text and representational evidence of Mesopotamian ox-ploughing associated with elite individuals, as in the 2nd-millennium BC Sumerian poem *'Disputation between the Hoe and the Plough'* (Black *et al.* 1998), there are as yet no clear clues to the association of donkeys with women or other disenfranchised groups. The low profile of both donkeys and (low-status) women in the archaeological and textual record in 4th- and 3rd-millennia BC Mesopotamia is a significant obstacle, and future work may need to focus on linkages in re-interpretation of contemporaneous social and economic changes. For example, it could be speculated that the advent of the donkey for burden-carrying might have provided the Late Uruk southern Mesopotamian state with a newly-available labour resource that enabled the rapid development of 'centrally administered weaving establishments exploiting the labor of various categories of dependent women in both cities and the countryside' (Algaze 2007: 356).

Continuing the rural theme, discussion of early ploughing has often assumed the use of oxen related to the desirability of deeper ploughing, burying weeds and pests too deep to survive; but extensive experience in Africa demonstrates that in light, dry soils this can instead cause soil compaction, erosion and nutrient depletion. Oates and Oates (1976: 119) note that shallow ard-ploughing is still used in modern Mesopotamia to avoid over-cultivation, and in arid areas with light soil in modern Africa and elsewhere there is a clear preference in many instances for the lighter tillage that can be performed with donkeys (e.g. Blench 1997: 52; Starkey 1994: 79). Donkeys, widely acknowledged as low-investment, low-maintenance and multi-function, are commonly favoured over oxen on small farms in particular. While Mesopotamian texts clearly indicate massive centralised agricultural operations (e.g. Charles 1990: 47-9), the widespread presence of individual smallholdings outside the remit of the texts is argued increasingly by commentators from archaeobotanical and other evidence (Adams 2008: 13; Halstead 1995: 19; Stone 1999: 204-12).

The central economic importance of donkeys for short-distance transport – from field to farm, and farm to market – has also been largely ignored in models of use in antiquity. Donkeys used today for ploughing in practice spend the majority of their time on farm or household transportation tasks (e.g. Aganga and Seabo 2004: 156; Barrett *et al.* 1982: 6; Pearson *et al.* 2001: 20), suggesting that the teams of ploughing donkeys described in 3rd-millennium BC Mesopotamian texts are very likely to have been employed for short-distance transport in addition. Zarins (2014: 188) points out that the 'plough' sign in Mesopotamian texts may sometimes denote farm work generally, rather than solely ploughing, just as now a tractor image often stands generally for farming. Speedy transport of crops from the field to storage, processing or market has a marked economic effect as it hugely reduces harvest loss from decay, pests and theft. Donkeys can carry considerably more per bodyweight than cattle and negotiate narrow, uneven paths (e.g. Digard 1982: 138-9). Today they carry heavy implements and manure to distant fields, and bulky fodder to farmyard animals: a study in Ethiopia found that crop residue is transported from the fields by equids but fed to oxen (Admassu and Shiferaw 2011:33). Raw materials (ore, clay) and construction materials can also be conveyed further and in much greater quantities, altering the parameters for their sourcing and for manufacturing and construction sites.

On a related issue, modern studies demonstrate the economic importance of year-round utilisation of working animals (e.g. Blench 1997: 23; Jaeger and Matlon 1990: 35). A common modern feature of working-animal use is therefore the hiring and lending of animals, and arrangements such as communal ownership of donkeys, which may assist interpretation of the early Mesopotamian textual references to working animal supply systems and shed light on the practicalities of early investment in working animals. An additional slant on year-round utilisation concerns the use of (female) cows for ploughing: very little attention is as yet given archaeologically to these, though along with donkeys they are very widely employed on small farms in particular, in Asia and other regions, and are known to have been used in ancient Knossos and Egypt (Halstead and Isaakidou 2011: 65-6; Johannsen 2011: 14; Matthewman 1987: 216-7).

Fig. 3. Thrice-weekly donkey-market in western Ethiopia, selling donkeys from the breeding regions of Gonder and Gojam (© J. Goulder 2014).

Another feature of working animal adoption, the proliferation of related occupational activities, is also unmentioned in classic models constructed by Boserup (e.g. 1990) and Goody (1976) concerning wealth disparity, social status, labour use, community and kinship interaction emerging from adoption of ox-ploughing. In the modern industrialised world there has been rapid forgetting, for example, of the complex nature and central importance until very recent times of working-equid breeding and trading, equivalent perhaps to the vehicle industry today. In sub-Saharan Africa today there are sophisticated systems of regional breeding and supply of donkeys (e.g. Blench 2000: 340-42), and in my brief research visits to rural areas of Burkina Faso (2013) and Ethiopia (2014; see Fig. 3), I found the scale of long-distance supply to be very apparent. Donkeys are native to the rugged, dry, often hilly terrain of North-East Africa, and in modern times their large-scale breeding zones reflect this environment. In modern West Africa, donkeys are bred in large numbers by specialist groups in arid Saharan and northern Sahelian countries, and are supplied profitably, via regional markets, to more humid regions further south, where donkey fertility is lower and where humidity-related diseases, unsuitable foodstuff and cultures unused to donkey management lead to short lifespans (e.g. Blench 2000: 340-42). The animals are taken on foot hundreds of kilometers, using equipment and communication methods largely available in antiquity.

References to donkey supply in the 3rd- and 2nd-millennium BC Near East underline the similarly large-scale geographical complexity of the equid-breeding industry then. 3rd-millennium BC donkey supply centres are indicated at Dêr and Gutium in the Zagros region, Mari, and Kish in central Mesopotamia (Prentice 2010: 106; Sallaberger 2014: 350; Zarins 2014: 160, 199). In 3rd-millennium BC texts on the hybridising of donkeys with hemiones, donkeys are recorded as being imported to breeding centres in or close to regions with indigenous wild hemiones, in northern Iran and further west as in Hamazi in Iraq and Nagar (Tell Brak) in north-eastern Syria (Archi 1998: 10; Zarins 2014: 171). Zarins (2014) summarises the textual references to donkey supply and demand in 3rd-millennium BC Mesopotamia in terms that closely describe the modern African situations described above:

Fig. 4. Pack animals being hauled up a precipice in Africa (Binger 1892: 363).

> *[M]ortality rates were high and longevity was at a minimum. Therefore, the net local effect was negative, and, as a result, constant purchase from the peripheral areas of Mesopotamia was necessary to supplement local breeding. (Zarins 2014: 245)*

Finally in this brief overview of modern and ancient use of donkeys, there is the enigmatic subject of long-distance pack-donkey caravans. Until recent years there has been very little scholarly focus on the logistics of donkey transport in antiquity: how the new donkey trains were organised and equipped, and how and where the animals were bred, kept and disposed of. Some clues to their early organisation may be gathered from the invaluable texts from early 2nd-millennium BC Kaneš on the large-scale system of donkey-caravans between Aššur in Mesopotamia and Kaneš in Anatolia (Larsen 1967, Veenhof 1972 and others). However, this is one area where modern ethnography has only limited light to shed, as – now as doubtless throughout history – caravaneers have had a strong interest in conducting profitable activities under the radar of authorities. Modern caravan-drivers worldwide seek invisibility not only from tax-hungry authorities and preying bandits, but also from the unwanted attentions of NGOs who would regulate and publicise their treatment of working animals. The advent of the donkey for long-distance transport may have provided such an evasionary facility from earliest times of its use: laden donkeys can negotiate rocky, precipitous paths, and advantage was taken of this by Assyrian merchants in the early 2nd millennium BC for taking difficult back-routes into Kaneš in Anatolia to avoid tax-payment on their goods (Larsen 2015: 157-8, 173, 179; Veenhof 1972: 34, 323-38). Modern entrepreneurs in Africa and elsewhere have taken this a step further, making profitable use of donkeys' excellent memory for routes, good night vision, and ability to work unsupervised by sending them unaccompanied across national borders laden with smuggled goods, through desert, mountains or dense vegetation where capture is unlikely (e.g. Sosovele 2004: 109).

Most descriptions of donkey-caravans are therefore garnered from anecdotal accounts of travellers in recent centuries. Binger (1892; Fig. 4) describes a long-established figure-of-eight donkey-caravan route whereby the Yarse of central West Africa convey cotton goods and cereals northwards to Saharan markets such as Timbuktu and return with salt slabs; some of these are then taken to southern West Africa, and from Ghana Hausa traders use the donkeys to carry kola nuts (highly valued for chewing) back to the Saharan region. Other accounts describe the long-established salt caravans in Ethiopia travelling from the Danakil depression to highland markets (e.g. Abir 1966; Pankhurst 1968), and Förster *et al.* (2013: 201, 210-12) give information on the practices and logistics of a Koro-Timbuktu caravan.

Other themes emerging in my continuing analysis of the role of donkeys are the issue of grazing and the major industry of provision of fodder to working animals, and illustrations of supply meeting demand in the adoption of donkeys for work at zones of interaction between nomadic donkey-using groups and sedentary urbanising farmers.

What can be achieved, specifically in terms of understanding of day-to-day use of donkeys, through synergy between modern working-animal studies and archaeological interpretation? Renger (1990: 267) blazed an initial trail by studying the use of draught animals in Africa with the aim of better understanding 3rd-millennium BC Mesopotamian texts on agricultural practices. More recently, Marshall and Weissbrod (2009: 72, 75) examined the settlements of donkey-using Kenyan Maasai pastoralists in terms of potentially-detectable evidence of donkey presence, concluding that while bones are rare, there are significant accumulations of donkey dung in the enclosures where donkeys were kept at night. This line of investigation was followed for example by Fedele (2014: 190-91) in his work on a caravan waystop at 1st-millennium BC Barāqish in Yemen, where donkey and camel dung concentrations were found immediately outside the walls. Meanwhile, an example of on-the-spot ethnography is given by Palmer *et al.* (2007: 369, 373), who accompanied archaeological work in Wadi Faynan in Jordan with a survey of modern pastoralists in the region, noting animal pens, donkey-tethering locations, and a decomposing donkey in a wadi where rubbish was dumped.

While some use has been made of detection of the advent of long-distance pack-animals through study of changes in routes, settlements and location of goods, little has yet been done in considering the archaeological traces of donkeys working in local agricultural environments. With new appreciation of the practicalities of donkey adoption, site directors might be alerted for example to possible evidence of light ploughing, and of rearrangement of spatial relationships between fields, threshing-floors and storage facilities.

Conclusions

Mesopotamian archaeology in the 21st century AD is adjusting to accelerating change in fieldwork constraints, technical capabilities and interpretative challenges. New models for the developments in exploitation of secondary products in the 4th and 3rd millennia BC must now include a reassessment of the vital supporting role of the 'other' – the non-humans added to the workforce who supported social and economic transformation at the highest and lowest levels. While there has been acknowledgement of the role of cattle for ploughing, the working donkey in particular walks between the high-level anthropological models of Goody and Boserup with their focus on ploughing oxen, and site-focused zooarchaeological reports with their scarcity of non-food animals. The use of data and observations from widely-available modern agricultural, social and economic development studies, in regions of significant working-animal use such as sub-Saharan Africa, is a largely-untapped resource for Ancient Near Eastern archaeologists. With suitable caveats it provides a means of rebooting archaeological thought and placing working donkeys within a newly-assessed social and economic framework focusing on practicalities and on household-level responses to change.

References

Abir, M. 1966. Salt, trade and politics in Ethiopia in the "Zämänä Mäsafent". *Journal of Ethiopian Studies* 4/2: 1-10.

Adams, R.McC. 2008. An interdisciplinary overview of a Mesopotamian city and its hinterlands. *Cuneiform Digital Library Journal*: 1.

Admassu, B. and Shiferaw, Y. 2011. *Donkeys, Horses and Mules – Their Contribution to People's Livelihoods in Ethiopia.* Addis Ababa, Ethiopia: The Brooke.

Aganga, A. A. and Seabo, D. 2004. A survey of donkey use by small-scale farmers in south-east Botswana highlighting gender differences. In, Fielding, D. and Starkey, P. (eds) *Donkeys, People and Development: A Resource Book of the Animal Traction Network for Eastern and Southern Africa ATNESA*: 154-7. Wageningen, Netherlands: Technical Centre for Agricultural and Rural Cooperation CTA.

Algaze, G. 2001. Initial social complexity in Southwestern Asia: the Mesopotamian advantage. *Current Anthropology* 42: 199-233.

Algaze, G. 2007. The Sumerian takeoff. In, Stone, E.C. (ed.) *Settlement and Society: Essays Dedicated to Robert McCormick Adams*: 343-68. Los Angeles, CA: Cotsen Institute of Archaeology, University of California, and Chicago, IL: Oriental Institute, University of Chicago.

Algaze, G. 2008. *Ancient Mesopotamia at the Dawn of Civilization: the Evolution of an Urban Landscape.* Chicago, IL: University of Chicago Press.

Algaze, G. 2013. The end of prehistory and the Uruk period. In, Crawford, H. (ed.), *The Sumerian World*: 68-94. London: Routledge.

Anthony, D.W. 1991. The domestication of the horse. In, Meadow, R.H. and Uerpmann H.-P. (eds) *Equids in the Ancient World Vol* II: 250-77. Wiesbaden: Reichert.

Archi, A. 1998. The regional state of Nagar according to the texts of Ebla. *Subartu* 4/2: 1-15.

Arnold, E.R., Hartman, G., Greenfield, H.J., Shai, I., Babcock, L.E. and Maeir, A.M. 2016. Isotopic evidence for early trade in animals between Old Kingdom Egypt and Canaan. *PLoS ONE* 11/6, e0157650.

Asher-Greve, J.M. 2013. Women and agency: a survey from Late Uruk to the end of Ur III. In, Crawford, H. (ed.) *The Sumerian World*: 359-77 London: Routledge.

Baker, J.R. 1984. The study of animal diseases with regard to agricultural practices and man's attitude to his animals. In, Clutton-Brock, J. and Grigson, C. (eds) *Animals and Archaeology. Vol. 4 Husbandry in Europe*: 253-7. Oxford: Archaeopress. British Archaeological Reports.

Barrett, V., Lassiter, G., Wilcock, D., Baker, D. and Crawford, E. 1982. *Animal Traction in Eastern Upper Volta: A Technical, Economic, and Institutional Analysis.* East Lansing, MI: Department of Agricultural Economics, Michigan State University.

Bartosiewicz, L., Van Neer, W. and Lentacker, A. 1997. *Draught Cattle: Their Osteological Identification and History.* Tervuren, Belgium: Musée royal de l'Afrique centrale.

Beja-Pereira, A., Englund, P.R., Ferrand, N., Jordan, S., Bakhiet, A.O., Abdalla, M.A., Mashkour, M.,

Jordana, J., Taberlet, P. and Luikart, G. 2004. African origins of the domestic donkey. *Science* 304/5678: 1781.

Binger, L.G. 1892. *Du Niger au Golfe de Guinée par le pays de Kong et le Mossi.* Paris: Hachette.

Black, J.A., Cunningham, G., Robson, E. and Zólyomi, G. 1998. *The Electronic Text Corpus of Sumerian Literature.* Oxford: Faculty of Oriental Studies, University of Oxford.

Blench, R.M. 1997. *Animal Traction in West Africa: Categories, Distribution and Constraints on its Adoption and Further Spread: a Nigerian Case Study.* London: Overseas Development Institute. ODI Working Paper 106.

Blench, R.M. 2000. A history of donkeys, wild asses and mules in Africa. In, Blench, R.M. and MacDonald, K.C. (eds) *The Origins and Development of African Livestock: Archaeology, Genetics, Linguistics and Ethnography*: 339-54. London: University College London Press.

Blench, R.M. 2004. The history and spread of donkeys in Africa. In, Fielding, D. and Starkey, P. (eds) *Donkeys, People and Development: A Resource Book of the Animal Traction Network for Eastern and Southern Africa ATNESA*: 24-32. Wageningen, Netherlands: Technical Centre for Agricultural and Rural Cooperation CTA.

Bogucki, P. 1993. Animal traction and household economies in Neolithic Europe. *Antiquity* 67: 492-503.

Boserup, E. 1990. Economic change and the roles of women. In, Tinker, I. (ed.). *Persistent Inequalities: Women and World Development*: 14-24. Oxford: Oxford University Press.

Brodie, N. 2008. The donkey: an appropriate technology for Early Bronze Age land transport and traction. In, Brodie, N., Doole, J., Gavalas, G. and Renfrew, C. (eds), *Horizon: A Colloquium on the Prehistory of the Cyclades*: 299-304. Cambridge: McDonald Institute for Archaeological Research.

Charles, M.P. 1990. Traditional crop husbandry in southern Iraq, 1900-1960 AD. *Bulletin on Sumerian Agriculture* 5: 47-64.

Clutton-Brock, J. 2012. *Animals as Domesticates: A World View Through History.* East Lansing, MI: Michigan State University Press.

Clutton-Brock, J. and Davies, S. 1993. More donkeys from Tell Brak. *Iraq*: 209-21.

Digard, J.P. 1982. *Techniques des Nomades Baxtyari d'Iran.* Paris: Maison des Sciences de l'Homme.

Dijkman, J.T. 1992. A note on the influence of negative gradients on the energy expenditure of donkeys walking, carrying and pulling loads. *Animal Production* 54: 153-6.

du Boulay, J. 1974. *Portrait of a Greek Mountain Village.* Oxford: Clarendon Press.

Fielding, D. 1987. Donkey power in African rural transport. *World Animal Review* 63: 23-30.

Fedele, F.G. 2014. Camels, donkeys and caravan trade: an emerging context from Barāqish, ancient Yathill Wadi al-Jawf, Yemen. *Anthropozoologica* 492: 177-94.

Förster, F., Riemer, H. and Mahir, M. 2013. *Donkeys to El-Fasher or how the Present Informs the Past.* Köln: Heinrich Barth Institut.

French, J. 1997. Social behaviour. In, E.D. Svendsen (ed.) *The Professional Handbook of the Donkey*: 125-36. Sidmouth: The Donkey Sanctuary.

Goody, J. 1976. *Production and Reproduction: A Comparative Study of the Domestic Domain.* Cambridge: Cambridge University Press.

Greenfield, H.J., Shai, I. and Maeir, A. 2012. Being an "ass": An Early Bronze Age burial of a donkey from Tell es-Safi/ Gath, Israel. *Bioarchaeology of the Near East* 6: 21-52.

Grigson, C. 1995. Plough and pasture in the early economy of the southern Levant. In, Levy, T.E. (ed.) *The Archaeology of Society in the Holy Land*: 245-68. New York, NY: Facts on File.

Groves, C.P. 1974. *Horses, Asses, and Zebras in the Wild.* London: David and Charles.

Hagmann, J. and Prasad, V.L. 1995. Use of donkeys and their draught performance in smallholder farming in Zimbabwe. *Tropical Animal Health and Production* 27/4: 231-9.

Halstead, P. 1995. Plough and power: the economic and social significance of cultivation with the ox-drawn ard in the Mediterranean. *Bulletin on Sumerian Agriculture* 8: 11-22

Halstead, P. 2014. *Two Oxen Ahead: Pre-mechanized Farming in the Mediterranean.* Chichester: Wiley-Blackwell.

Halstead, P. and Isaakidou, V. 2011. Revolutionary secondary products: the development and significance of milking, animal-traction and wool-gathering in later prehistoric Europe and the Near East. In, Wilkinson, T.C., Sherratt, S. and Bennet, J. (eds) *Interweaving Worlds: Systemic Interactions in Eurasia, 7th to the 1st millennia BC*: 61-76. Oxford: Oxbow.

Heimpel, W. 1995. Plow animal inspection records from Ur III Girsu and Umma. *Bulletin on Sumerian Agriculture* 8: 71-171.

Jaeger, W.K. and Matlon, P.J. 1990. Utilization, profitability, and the adoption of animal draft power in West Africa. *American Journal of Agricultural Economics* 72: 35-48.

Johannsen, N.N. 2011. Past and present strategies for draught exploitation of cattle. 13-19 In, Albarella, U. and Trentacoste, A. (eds) , *Ethnozooarchaeology: the Present and Past of Human-animal Relationships.* Oxford: Oxbow.

Jones, P.A. 2008. *Major Differences Between Work Animals and Food Animals.* Paper prepared for

SASAS KZN. and DAAIG joint symposium on Understanding Communal Animal Agriculture. Rome: Food and Agriculture Organization of the United Nations.

Kimura, B., Marshall, F.B., Chen, S., Rosenbom, S., Moehlman, P.D., Tuross, N., Sabin, R.C., Peters, J., Barich, B., Yohannes, H., Kebede, F., Teclai, R., Beja-Pereira, A. and Mulligan, C.J. 2010. Ancient DNA from Nubian and Somali wild ass provides insights into donkey ancestry and domestication. *Proceedings of the Royal Society of London Series B Biological Sciences*: 1-8.

Kirkbride, D. 1974. Umm Dabaghiyah: a trading outpost? *Iraq* 36: 85-92.

Larsen, M.T. 1967. *Old Assyrian Caravan Procedures*. Istanbul: Nederlands Historisch-Archeologisch Instituut Voor Het Nabije Oosten.

Larsen, M.T. 2015. *Ancient Kanesh: A Merchant Colony in Bronze Age Anatolia*. New York, NY: Cambridge University Press.

Levine, M. 1999. Botai and the origins of horse domestication. *Journal of Anthropological Archaeology* 18: 29-78.

MacDonald, K.C. 2000. The origins of African livestock: indigenous or imported? In, Blench, R.M. and MacDonald, K.C. (eds) *The Origins and Development of African Livestock: Archaeology, Genetics, Linguistics and Ethnography*: 2-17. London: UCL Press.

Maekawa, K. 1979. The ass and the onager in Sumer in the late third millennium BC. *Acta Sumerologica Hiroshima*. 1: 35-62.

Marshall, F. and Weissbrod, L. 2009. The consequences of women's use of donkeys for pastoral flexibility: Maasai ethnoarchaeology. In, Grupe, G., McGlynn, G. and Peters, J. (eds) *Tracking down the Past: Ethnohistory meets Archaeozoology*: 59-79. Rahden, Germany: Marie Leidorf. Documenta Archaeobiologiae no. 7.

Marshall, F. and Weissbrod, L. 2011. Domestication processes and morphological change: through the lens of the donkey and African pastoralism. *Current Anthropology* 52/S4: S317-S413.

Marshall, F., Dobney, K., Denham, T. and Capriles, J.M. 2014. Evaluating the roles of directed breeding and gene flow in animal domestication. *Proceedings of the National Academy of Sciences of the United States of America* 111/17: 6153-8.

Matthewman, R.W. 1987. Role and potential of draught cows in tropical farming systems: a review. *Tropical Animal Health and Production* 19: 215-22.

Milevski, I. 2011. *Early Bronze Age Goods Exchange in the Southern Levant: A Marxist Perspective*. London: Equinox Publishing.

Münzinger, P. 1982. Development and situation of animal traction in Africa and conditions limiting further promotion. In, Münzinger, P. (ed.) *Animal Traction in Africa*: 15-61. Eschborn, Germany: Vieweg/ GATE/ GTZ.

Oates, D. and Oates, J. 1976. Early irrigation agriculture in Mesopotamia. In, Sieveking G. de, Longworth, I.H. and Wilson, K.E. (eds) *Problems in Economic and Social Archaeology*: 109-35. London: Duckworth.

Palmer, C., Gilbertson, D., el-Rishi, H., Hunt, C., Grattan, J., McLaren, S. and Pratt, B. 2007. The Wadi Faynan today: landscape, environment, people. In, Barker, G., Gilbertson, D.D. and Mattingly, D.J. (eds) *Archaeology and Desertification: the Degradation and Well-being of the Wadi Faynan Landscape, Southern Jordan*: 38-58. Oxford: Oxbow.

Pankhurst, R. 1968. *Economic History of Ethiopia, 1800-1935*. Addis Ababa: Haile Selassie I University Press.

Payne, S. 1988. Animal bones from Tell Rubeidheh. In, Killick, R.G. (ed.) *Tell Rubeidheh: An Uruk Village in the Jebel Hamrin*: 98-135. Warminster: British School of Archaeology in Iraq and the Directorate of Antiquities: Aris & Phillips.

Pearson, R.A. and Vall, E. 1998. Performance and management of draft animals in agriculture in sub-Saharan Africa: a review. *Tropical Animal Health and Production* 30/5: 309-24.

Pearson, R.A., Alemayehu, M., Tesfaye, A., Allan, E.F., Smith, D.G. and Asfaw, M. 2001. *Use and Management of Donkeys in Peri-urban Areas of Ethiopia*. Roslin: Centre for Tropical Veterinary Medicine, Draught Animal Power Technical Report 5.

Postgate, J.N. 1986. The Equids of Sumer, again. In, Meadow, R.H. and Uerpmann, H.-P. (eds) *Equids in the Ancient World Vol. I*: 194-206. Wiesbaden: Reichert.

Potts, D.T. 2011. Equus asinus in highland Iran: evidence old and new. In, Conard, N.J., Drechsler, P. and Morales, A. (eds) *Between Sand and Sea: the Archaeology and Human Ecology of Southwestern Asia. Festschrift in Honor of Hans-Peter Uerpmann*: 167-75. Tübingen: Kerns Verlag.

Prentice, R. 2010. *The Exchange of Goods and Services in Pre-Sargonic Lagash*. Münster: Ugarit-Verlag.

Renger, J. 1990. Report on the implications of employing draught animals. *Bulletin on Sumerian Agriculture* 5: 267-79.

Rossel, S., Marshall, F., Peters, J., Pilgram, T., Adams, M.D. and O'Connor, D. 2008. Domestication of the donkey: timing, processes, and indicators. *Proceedings of the National Academy of Sciences of the United States of America* 105/10: 3715-20.

Russell, N. 2012. *Social Zooarchaeology: Humans and Animals in Prehistory*. Cambridge: Cambridge

University Press.

Sallaberger, W. 2014. Urban organizations for offerings, overland traffic and the Euphrates trade at pre-Sargonic Mari. In, Butterlin, P., Margueron, J.-C., Muller, B., Al-Maqdissi, M., Beyer, D., Cavigneaux, A. (eds) *Mari, ni Est, ni Ouest. Supplement Syria* II, *Vol.* I: 341-54. Beyrouth: Presses de l'Ifpo.

Sasson, A. 2010. *Animal Husbandry in Ancient Israel: A Zooarchaeological Perspective on Livestock Exploitation, Herd Management and Economic Strategies.* London: Equinox Publishing.

Shackelford, L., Marshall, F. and Peters, J. 2013. Identifying donkey domestication through changes in cross-sectional geometry of long bones. *Journal of Archaeological Science* 40: 4170-4179.

Shai, I., Greenfield, H.J., Brown, A., Albaz, S. and Maeir, A.M. 2016. The importance of the donkey as a pack animal in the Early Bronze Age southern Levant: a view from Tell es-Safi / Gath. *Zeitschrift des Deutschen Palastina-Vereins* 132/1: 1-25.

Sherratt, A.G. 1981. Plough and pastoralism: aspects of the secondary products revolution. In, Hodder, I., Isaac, G. and Hammond, N. (eds) *Pattern of the Past: Studies in Honour of David Clarke*: 261-305. Cambridge: Cambridge University Press.

Sherratt, A.G. 1983. The secondary exploitation of animals in the Old World. *World Archaeology* 15/1: 90-104.

Sherratt, A.G. 1997. Wool, wheels and ploughmarks: local developments or outside introductions in Neolithic Europe? [1997 notes to 1987 original paper]. In, Sherratt, A.G. (ed.) *Economy and Society in Prehistoric Europe: Changing Perspectives*: 240-1. Princeton, NJ: Princeton University Press.

Sherratt, A.G. 2003. The horse and the wheel: the dialectics of change in the circum-Pontic region and adjacent areas. In, Levine, M., Renfrew, C. and Boyle K. (eds) *Prehistoric Steppe Adaptation and the Horse*: 233-52. Cambridge: McDonald Institute Monographs.

Sherratt, A.G. 2006. La traction animale et la transformation de l'Europe néolithique. In, Pétrequin, P., Arbogast, R.-M., Pétrequin, A.-M., Van Willigen, S. and Bailly, M. (eds) *Premiers Chariots, Premiers Araires: La Diffusion de la Traction Animale en Europe Pendant les IVe et IIIe Millénaires avant Notre Ère*: 329-60 Paris: Centre National de la Recherche Scientifique.

Smith, D. and Pearson, R.A. 2005. A review of the factors affecting the survival of donkeys in semi-arid regions of Sub-Saharan Africa. *Tropical Animal Health and Production* 37 Suppl. 1: 1-9.

Sosovele, H. 2004. Donkey traction in Tanzania: some critical issues. In, Fielding, D. and Starkey, P. (eds) *Donkeys, People and Development: A Resource Book of the Animal Traction Network for Eastern and Southern Africa ATNESA*: 107-12 Wageningen, Netherlands: Technical Centre for Agricultural and Rural Cooperation CTA.

Starkey, P. 1994. A world-wide view of animal traction highlighting some key issues in eastern and southern Africa. In, Starkey, P., Mwenya, E. and Stares, J. (eds) *Improving Animal Traction Technology*: 66-81. Wageningen, Netherlands: ATNESA.

Stone, E.C. 1999. The constraints on state and urban form in ancient Mesopotamia. In, Hudson, M. and Levine, B.A. (eds) *Urbanization and Land Ownership in the Ancient Near East*: 203-27. Cambridge, MA: Peabody Museum of Archaeology and Ethnology, Harvard University.

Sylwander, L. 1994. Women and animal traction technology. In, Starkey, P., Mwenya, E. and Stares, J. (eds) *Improving Animal Traction Technology*: 260-5. Wageningen, Netherlands: ATNESA.

The Electronic Text Corpus of Sumerian Literature. 2006. *Enmerkar and the Lord of Aratta.* [online] (updated November 2016) Available at: <http://etcsl.orinst.ox.ac.uk> c.1.8.2.3 [Accessed 8 August 2016]. Oxford: Faculty of Oriental Studies, University of Oxford.

Uerpmann, H.-P. 1987. *The Ancient Distribution of Ungulate Mammals in the Middle East.* Wiesbaden: L. Reichert Verlag.

Uerpmann, H.-P. 1989. Problems of archaeo-zoological research in eastern Arabia. In, Costa, P.M. and Tosi, M. (eds) *Oman studies: Papers on the Archaeology and History of Oman*: 163-8. Rome: Serie Orientale Roma 63.

Uerpmann, H.-P. 1991. *Equus africanus* in Arabia. In, Meadow, R.H. and Uerpmann, H.-P. (eds) *Equids in the Ancient World, Vol* II: 12-33. Wiesbaden: Beihefte zum Tübinger Atlas des Vorderen Orients A 19/2.

Veenhof, K.R. 1972. *Aspects of old Assyrian trade and its Terminology.* Leiden: Brill.

Vila, E. 1998. *L'exploitation des Animaux en Mésopotamie aux IVe et IIIe Millénaires avant J.-C.* Paris: Centre National de la Recherche Scientifique. Monographie du CRA 21.

Vila, E. 2006. Data on equids from late fourth and third millennium sites in Northern Syria. In, Mashkour, M. (ed.) *Equids in Time and Space: Papers in Honour of Véra Eisenmann*: 101-23. Oxford: Oxbow.

Vilà, C., Leonard, J.A. and Beja-Pereira, A. 2006. Genetic documentation of horse and donkey domestication. In, Zeder, M.A., Bradley, D.G., Emshwiller, E. and Smith, B.D. (eds) *Documenting Domestication: New Genetic and Archaeological*

Paradigms: 342-53. Berkeley, CA: University of California Press.

Waithanji, E.M. 2009. *Findings of a Preliminary Anthropological Survey on Donkey Welfare in Mwea, Lari, and Limuru, Kenya.* Nairobi: KENDAT.

Way, K.C. 2011. *Donkeys in the Biblical World: Ceremony and Symbol.* Winona Lake, IN: Eisenbrauns.

Weber, J.A. 2008. Elite equids: redefining equid burials of the mid- to late 3rd millennium BC from Umm el-Marra. In, Vila, E., Gourichon, L., Choyke, A.M. and Buitenhuis, H. (eds) *Archaeozoology of the near East* VIII: 499-520. Lyon: Maison de l'Orient et de la Méditerrannée.

Wilkinson, T.J. 2003. *Archaeological Landscapes of the Near East.* Tucson: University of Arizona.

Wylie, A. 2002. *Thinking from Things.* Berkeley, CA: University of California Press.

Yousef, M.K., Dill, D.B. and Freeland, D.V. 1972. Energetic cost of grade walking in man and burro, *Equus asinus*: desert and mountain. *Journal of Applied Physiology* 33: 337-40.

Zarins, J. 1986. Equids associated with human burials in third millennium B.C. Mesopotamia: two complementary facts. In, Meadow, R.H. and Uerpmann, H.-P. (eds) *Equids in the Ancient World. Vol* I: 164-93 Wiesbaden: Riechert.

Zarins, J. 2014. *The Domestication of* Equidae *in Third-millennium BCE Mesopotamia.* Bethesda, MD: CDL Press.

8 Kura-Araxes exploitation of animal resources in North-western Iran and Nakhchivan

Hossein Davoudi[1,2,3], Rémi Berthon[4], Azadeh Mohaseb[4,2], Shiva Sheikhi[4], Akbar Abedi[5], Marjan Mashkour[4,2]

1 Post-Doctoral Fellow, Department of Archaeology, Tarbiat Modares University, Tehran, Iran; davoudih1@gmail.com
2 Bioarchaeology Laboratory, Central Laboratory, University of Tehran, Iran
3 National Museum of Iran, Tehran, Iran
4 Archéozoologie, Archéobotanique (UMR 7209), Muséum national d'Histoire naturelle, CNRS; CP56, 55rue Buffon 75005 Paris, France; mashkour@mnhn.fr; azadeh.mohaseb@mnhn.fr; remi.berthon@mnhn.fr
5 Islamic Art University of Tabriz, Department of Archaeometry, Tabriz, Iran; akbar.abedi@tabriziau.ac.ir

Abstract
The Early Bronze Age in north-western Iran and adjacent regions is known for the Kura-Araxes or Early Trans-Caucasian Culture, spreading through the Caucasus and the Lake Urmia Basin. The subsistence economy during this important period is not well known, especially in Iran. This paper deals with the characterization of Kura-Araxes herding strategies by examining the faunal remains from six settlements (Tepe Hasanlu, Kul Tepe, Haftavan Tepe, Kohnehh Tepesi, Kohneh Pasgah Tepesi and Ovçular Tepesi). The results indicate that all these settlements shared similar herding strategies. The exploitation of animal resources was mainly based on domestic ruminants.

Keywords
Iran, Azerbaijan, Kura-Araxes Culture, Herding strategy, Animal husbandry.

Introduction
The material culture of the Early Bronze Age in Iran and Nakhchivan belongs to the Kura-Araxes (also known as Early Trans-Caucasian) culture (second half of fourth to the end of the third millennium BC), which spread through the Caucasus and the Urmia Basin. Its origins are unknown, but it has been observed in the valleys and foothills of three Caucasian republics (Azerbaijan, Armenia and Georgia), as well as north-western and western Iran, eastern Anatolia, Dagestan in the Russian Federation and the Levant (Abedi *et al.* 2014; Kohl 2007; Kushnareva 1997; Rothman 2003; Sagona 1984). New data were obtained during recent archaeological excavations of the Kura-Araxes culture sites in north-western Iran and Nakhchivan (Abedi *et al.* 2014; Abedi and Omrani 2015; Alizadeh *et al.* 2015; Khaksar *et al.* 2015; Marro *et al.* 2014; Maziar 2015; Motarjem and Niknami 2011). They allow a better understanding of the Late Chalcolithic to Early Bronze Age transition. New radiocarbon dating also shed light on the emergence, duration, and expansion of the Kura-Araxes culture (Abedi and Omrani 2015: Tab. 2; Maziar 2010: 171, 174; Marro *et al.* 2011: Fig. 6).

A lack of bioarchaeological studies, such as archaeozoology and archaeobotany, is quite clear for the Kura-Araxes culture in north-western Iran and Nakhchivan (Decaix *et al.* in press a). The archaeozoological studies carried out in adjacent regions for Kura-Araxes faunal remains reflect sedentary agro-pastoral societies (e.g. Decaix *et al.* in press a; Howell-Meurs 2001a; 2001b; Piro 2009).

This paper deals with the faunal remains from six formerly and recently excavated sites in north-western Iran and Nakhchivan: Tepe Hasanlu, Kul Tepe, Haftavan Tepe, Kohneh Tepesi, Kohneh Pasgah Tepesi and Ovçular Tepesi (Fig. 1). These sites are situated along major corridors connecting the Zagros Mountains with Mesopotamia, Anatolia, and the Caucasus. Tepe Hasanlu, Kul Tepe and Haftavan Tepe are multi-period mounds, inhabited up to the Historic and Islamic eras. Kul Tepe has the longest chronological sequence starting from the Middle Chalcolithic period. In this site, the Early Bronze Age layers were found within a continuous cultural sequence after the Late Chalcolithic and before the Middle Bronze age occupations. Ovçular Tepesi is a smaller site occupied only during the Late

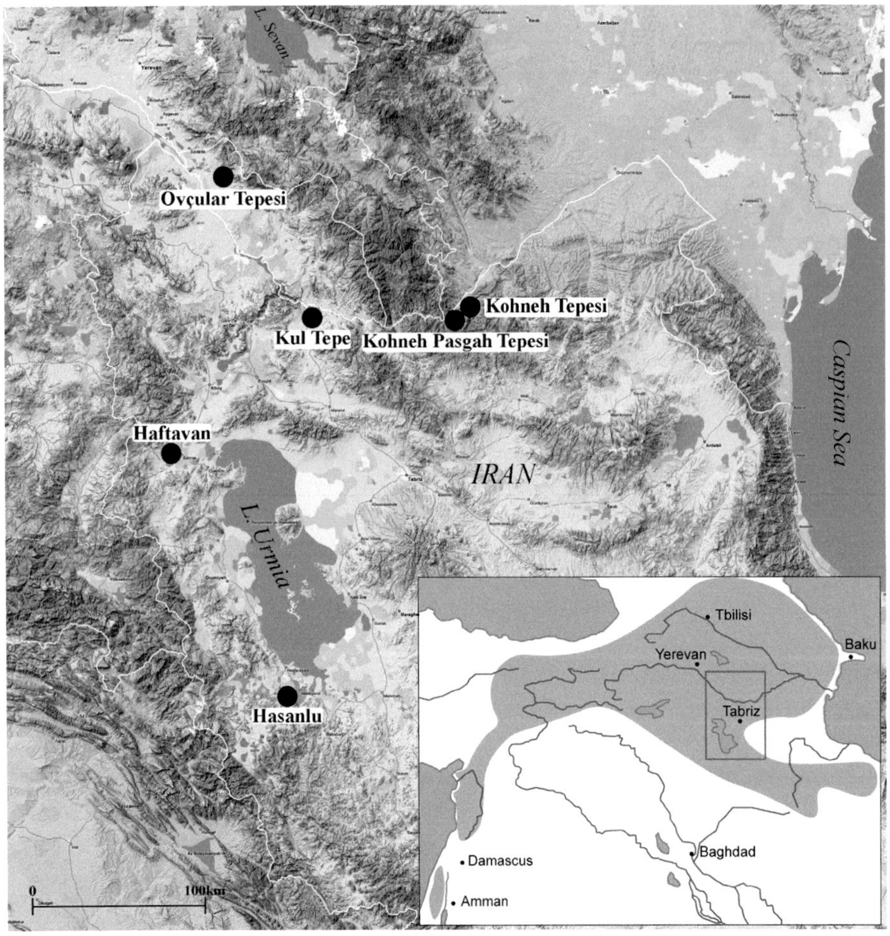

Fig. 1. Distribution of Kura-Araxes Culture sites mentioned in the text in north-western Iran and Nakhchivan. In the frame is the maximal expansion of the Kura-Araxes culture. Red rectangle is the area concerned in this study.

Chalcolithic and Early Bronze Age periods. Kohneh Pasgah Tepesi presents a continuity of occupations from the Late Chalcolithic to the end of the Kura-Araxes period, while Kohneh Tepesi was occupied only during the Kura-Araxes period (Table 1).

The periodisation used in this paper, for the beginning of the Kura-Araxes Culture (KA I), follows the cultural sequence of Kul Tepe (Abedi *et al.* 2014: Fig. 6) and Kohneh Pasgah Tepesi (Maziar 2010: Tab. 1) on the south bank of Araxes River (Iran) and Ovçular Tepesi in the Nakhchivan (Marro *et al.* 2014). For the Kura-Araxes II-III (Early Bronze Age) we used the chronological framework of north-western Iran such as Hasanlu (Danti *et al.* 2004: Tab. 1), Kul Tepe, Haftavan (Summers 2014), Kohneh Pasgah Tepesi, Kohneh Tepesi (Maziar 2015: Tabs. 1 & 2). Regarding the dearth of radiocarbon dating for the end of Early Bronze Age (coinciding with start of Middle Bronze Age), the paper relies on the new chronology of Kul Tepe (Abedi *et al.* 2014: Fig. 6) and Hasanlu and Dinkhah, comparing with Kordlar, Geoy, and Haftavan (Danti 2013a: Tab. 17.1) (see Table 1).

The examination of faunal remains from Kura-Araxes occupations at these sites is the main purpose of this research in order to characterize the subsistence patterns and herding strategies, as well as the exploitation of wild animals.

The archaeological contexts

Tepe Hasanlu

Among the key sites of north-western Iran, Tepe Hasanlu is of high importance due to its long-term occupation and well-defined stratigraphy. Hasanlu is located in the Solduz valley on the southern shore of Lake Urmia, Naqadeh city, in Western Azerbaijan province of Iran (Fig. 1). The marshy areas offer an appropriate environment for the development of mixed farming and herding economies. The site consists of two distinct topographic zones: a High Mound which is called 'Citadel Mound', surrounded by a Low Mound, also known as the 'Outer Town'. The High Mound is 200 m in diameter and rises 25 m above surrounding plain. The Low Mound measures 600 m across its widest point and it rises 6-7 meters above modern plain. Tepe Hasanlu has a long history of excavations and research (see Dyson 1989; Muscarella 2006; Danti 2013b).

According to the excavations' results, the site was intermittently occupied during 10 different cultural periods from the Late Neolithic (period X) to the Ilkhanid dynasty (period I) (Danti 2013a; Dyson and Muscarella 1989). Material culture from the Neolithic to the Middle Bronze Age has been discovered in deep soundings and stratigraphic trenches. Hasanlu Period VII can be linked to the Kura-Araxes

II-III, coinciding with the earliest strata (strata 19-42) of High Mound reached in controlled excavations in the U22 Deep Sounding in 1972 and 1974. It has to be noted that Early Bronze Age material was first encountered in 1957 on the Low Mound during excavations in Operation V. The stratigraphic excavations at Well Sounding in the centre of the High Mound produced 6.5 m of deposits assigned to period VII, which suggested a relatively long occupation with an increasing settlement size. Period VII was divided into three sub-periods (VIIC-A) related to the time span from the first half of the third millennium to the late third millennium BC (3000-2100 cal. BC) (Danti *et al.* 2004; Danti 2013b: 12-13; Dyson 1958: 27; Dyson and Pigott 1975) (Table 1). Ceramics of period VIIC indicated the strong Kura-Araxes influence at Hasanlu. Sub-periods VIIB and VIIA are characterized by a distinctive Painted Orange Ware (POW) component at Hasanlu with only a few Kura-Araxes sherds. In these sub-periods, there is a division, or "shatter zone," in the Southern Lake Urmia Basin (Danti 2013b: 12-13). The analysis of faunal remains of Hasanlu period VII will result in understanding the animal exploitation during Early Bronze Age in the southern part of Lake Urmia.

Fortunately, numerous well-preserved animal remains from periods VII to I, were housed and curated at the National Museum of Iran (Mashkour *et al.* 2012). This large collection belongs to the last seasons of Hasanlu excavations during 1970, 1972, and 1974 and was not studied before; Sandor Bököny reported 21 animal species present on the site on the basis of the 1972 campaign, however no archaeozoological report was ever published (Dyson 1973: 195). The analysis of the faunal remains from Hasanlu began officially in September 2013 at the Osteology Department of the National Museum of Iran within the PhD project of H. Davoudi in Tarbiat Modares University, Iran (Davoudi in prep; Davoudi *et al.* 2014). This paper deals with the faunal remains from Early Bronze Age layers in soundings U22 (strata Nos. 19 to 38), V22 (stratum 39), and W23 (strata 20 to 22).

Kul Tepe

The archaeological site of Kul Tepe is located near the city of Hadishahr, Jolfa County, 10 km south of the Araxes River in the Western Azerbaijan province of Iran. The site is a multi-period mound about 6 ha in size and rising 19 m above the surrounding lands. Kul Tepe is located precisely in the north-western corner of Iran which represents the gateway between the southern Caucasus and north-western Iran (Fig. 1). This strategic location at the crossroad of major routes linking the Iranian Plateau to Anatolia and the Caucasus to northern Mesopotamia suggests relations and interactions between human communities and makes it possible to compare the results with other contemporaneous sites, especially within the Kura-Araxes Culture (Abedi *et al.* 2014: 33-34).

The site was excavated by Akbar Abedi and Hamid Khatib Shahidi in 2010. Eight main occupation periods were identified: Early, Middle and Late Chalcolithic, Early, Middle and Late Bronze Age, Iron III, Urartian, and Achaemenid periods (Abedi *et al.* 2014). The phase V related to the 'Proto Kura-Araxes' and 'Kura-Araxes I' represents 3.5 m of deposits in the stratigraphic trench III. This phase is important to increase the archaeological knowledge concerning the second half of the fourth millennium BC (3350-3000 BC) in north-western Iran. Two architectural phases were obtained from this period: a mud-brick circular structure and a pottery kiln. Also, Kura-Araxes II, III (3000-2500 BC) material culture was identified in 8 m of archaeological deposits in layers of step trenches I and II at Kul Tepe, suggesting a long period of settlement by sedentary groups (Abedi *et al.* 2014: 46-47).

Animal remains are exceptionally well preserved in all layers at Kul Tepe. For this study, however, we selected the assemblages that covered a period from Kura-Araxes I (phase V) to the end of Early Bronze Age (phase IV) (3350-2500 BC) (Table 1). These periods have the highest proportion of remains and allow a better insight into the understanding of animal exploitation during the Kura-Araxes Culture. The study of the faunal remains began in September 2014 at the Archaeozoology laboratory of the University of Tehran and is still in progress. We hope to complete this study with more faunal remains from the second excavation season at Kul Tepe.

Haftavan Tepe

Haftavan Tepe is located in the Salmas plain, close to the north-western corner of Lake Urmia, in the West Azerbaijan province (Fig. 1). It is one of the four largest mounds in the Urmia basin, 550 m from east to west and about 400 m from north to south (18 hectares). The highest point on the 'citadel' of the mound is 27 m above the datum (Edwards 1983). Haftavan Tepe was excavated by Charles Burney in 1971-1978 and the principal objective of the excavations was the investigation of occupation levels dated to the second millennium B.C., to fill the gap between the excavations at Yanik Tepe, near Tabriz, where continuous occupation ended around 2000 B.C. (Burney 1970; 1972; 1973; 1975).
The earliest levels reached to date are assigned to the Early Bronze II (Kura-Araxes III / ETC III), Haftavan Tepe period VIII, in Trial Trench 9 (T.T.9). The extent of the Kura-Araxes III settlement is unknown but the contours of the Tepe suggest that

Table 1. Chronological Sequence of Tepe Hasanlu, Kul Tepe, Haftavan Tepe, Ovçular Tepesi, Kohneh Pasgah Tepesi and Kohneh Tepesi from the Late Chalcolithic to the Middle Bronze Age (chronology table after Danti *et al.* 2004: Tab. 1; Danti 2013a: Tab. 17.1; Danti 2013b: Fig. 2.2; Marro *et al.* 2014; Summers 2014: Tab. 1; Abedi *et al.* 2014: Fig.6; Abedi and Omrani 2015: Tab. 1; Maziar 2010: Tab. 1; Maziar 2015: Tabs. 1 & 2).

Cultural period	Cal. BC	Settlement phases					
		Tepe Hasanlu	Kul Tepe	Haftavan Tepe	Ovçular Tepesi	Kohneh Pasgah Tepesi	Kohneh Tepesi
Middle Bronze I	2100-1900	Vic	III	VIc	-	-	-
Early Bronze III	2200-2100	VIIa	IV	VII	-	IV-V	-
Kura-Araxes III (EB II)	2500-2200	VIIb	IV	VIII	-	III	X
Kura-Araxes II (EB I)	3000-2500	VIIc	IV	-	EBKA	II	X
Kura-Araxes I	3350-3000	-	V	-	EBKA	I	-
Late Chalcolithic	4500-3750	VIII	VII & VI	-	LC Phase I &II & LCKA	I (KAI)	-

Table 2. Identified (NISP) and unidentified animal bone fragments in percent of the Number of remains (NR) and weight of remains (WR).

Site	Hasanlu		Kul Tepe		Haftavan		Ovçular		Kohneh Tepesi		Kohneh Pasgah Tepesi		Mean	
Percent (%)	NR	WR	NR	WR	NR	WR	NR	WR	NR	WR	NR	WR	NR	WR
Identified	41	62	48	73	58	54	36	63	71	86	18	39	45	63
Unidentified	59	38	52	27	42	46	64	37	29	14	82	61	55	37

it was not smaller than the one at Yanik Tepe. The round houses excavated at Haftavan Tepe level VIII yielded a range of plain burnished pottery very similar to the ones from the later round houses at Yanik Tepe (Summers 2004). The period VII, Early Bronze III, occupation may have stretched over most of the mound, if the evidence from T.T.6 and 6B have been correctly interpreted (Summers 1982: 5-6; 2014: 159). The period VI is the longest and most important in the history of Haftavan Tepe. Probably the settlement reached its maximum size during this period. The most ancient part of this period, VIC was excavated mainly in C1, C2 and C4.

Haftavan Tepe has yielded one of the largest faunal assemblages in western Azerbaijan, especially for the Middle and Late Bronze Age. The archaeozoological study of the animal remains showed that the bone preservation in general was satisfactory (Mohaseb 2012). In this paper we are dealing with Kura-Araxes faunal remains that only represent a small part of faunal assemblage from Haftavan Tepe.

Kohneh Pasgah Tepesi and Kohneh Tepesi

Kohneh Pasgah Tepesi is a small archaeological site (*ca* 0.5 ha) localized in Kaleybar district in the East Azerbaijan province, on the south bank of the Araxes valley in the Khoda-Afarin plain. The site was partially excavated by Sepideh Maziar in 2006 during the Salvage excavations in Khoda-Afarin dam project. The Late Chalcolithic and Kura-Araxes periods were excavated at this site and circular mud-brick architecture from the Kura-Araxes phase was uncovered (Maziar 2010).

Kohneh Tepesi is another small site (*ca* 0.2 ha) located just 100 m away from Kohneh Pasgah Tepesi. The site was excavated by Ali Zalaghi, Bayram Aghlari and Sepideh Maziar as part of a Khoda-Afarin dam rescue project in 2006 and 2007. On the basis of the pottery and other material the site can be dated to the Kura-Araxes II or perhaps part of III (Maziar 2015). Based on the study of the faunal remains, the bone preservation in Kohneh Tepesi was satisfactory while the faunal remains were highly fragmented in the assemblage from Kohne Pasgah Tepesi (Decaix et al. in press b; Sheiki, unpublished data).

Ovçular Tepesi

Ovçular Tepesi is located in the valley of the Arpaçay River, 8 km north-east of the city of Şərur in the Nakhchivan Autonomous Republic of Azerbaijan. The site lies on top of a natural hill rising above the river. Extensive archaeological investigations started in 2006 within the framework of a Franco-Azerbaijani joint project (Marro *et al.* 2009; 2011). Ovçular Tepesi is a small settlement (< 2 ha) occupied only during the Late Chalcolithic and Early Bronze Age periods. The discovery of a Kura-Araxes potsherd-scatter in a Late Chalcolithic habitat evidenced the presence of Kura-Araxes groups in the region at least since ca. 4300 cal. BC (Marro *et al.* 2014). Features related to a Late Chalcolithic Kura-Araxes occupation (also called LCKA) are very

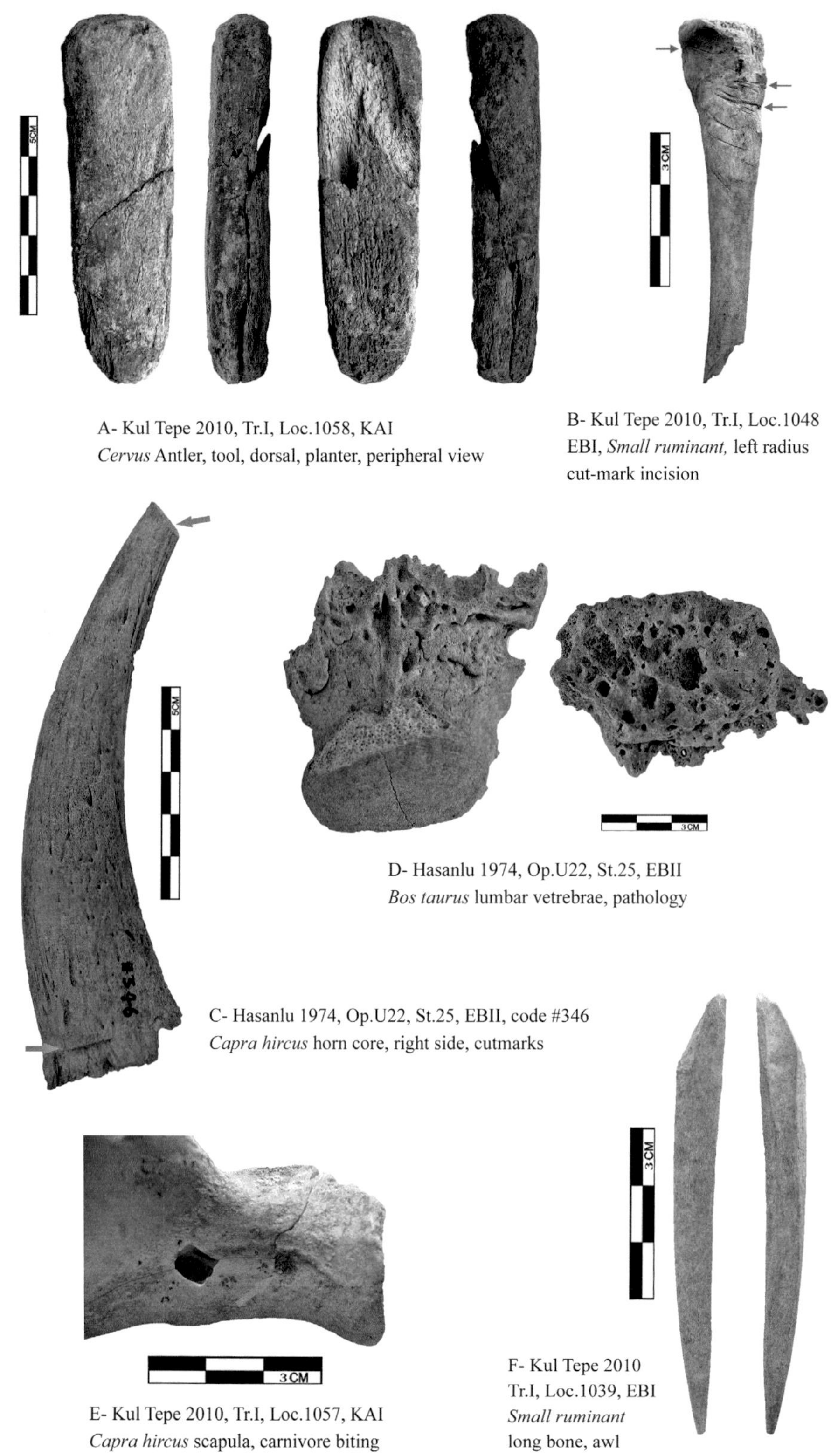

Fig. 2. Examples of anthropogenic and animal traces on the bones.

rare in Ovçular Tepesi and yielded only a handful of bones. The faunal assemblage included in this paper is coming from Early Bronze Age Kura-Araxes features (EBKA) dated from 3200 to 2400 cal. BC (Marro *et al.* 2014) (Table 1).

Faunal assemblage

In total 17750 bones (165 kg) from Kura-Araxes contexts were registered in this study. The identified remains represent from 18 to 71% of the Number of Remains (NR) and, on average, 63% of the total weight of the examined collection (Table 2). This indicates that the preservation of the faunal remains is variable between sites and that the unidentified remains are rather small. The lack of dry sieving during the excavations in Hasanlu and Haftavan could explain the lack of bird, fish, amphibian and reptilian remains. It has to be noted that at Kul Tepe

Table 3. Taxonomic distribution of animal remains in the Kura-Araxes assemblages in number of remains (NR).

Taxa	*Tepe Hasanlu*	*Kul Tepe*	*Haftavan Tepe*	*Ovçular Tepesi*	*Kohneh Pasgah Tepesi*	*Kohneh Tepesi*	*TOTAL*
	NR	**NR**	**NR**	**NR**	**NR**	**NR**	**NR**
Capra hircus	119	89	4	24	23	162	421
Ovis aries	128	85	33	45	40	196	527
Caprinae	664	316	106	262	161	4362	5871
Bos taurus	227	168	75	89	87	1272	1918
Sus cf. *domesticus*					12	255	267
Equus sp.	4	24	12			2	42
Canis familiaris	5	4	1			11	21
Total domestic mammals	**1147**	**686**	**231**	**420**	**323**	**6260**	**9067**
Capra aegagrus		13					13
Ovis orientalis		7					7
Bos primigenius		4					4
Sus scrofa	12	1	10	1		254	278
Gazella sp.	3	3	4			10	20
Cervus elaphus	1	5			5	200	211
Carnivora						7	7
Canidae	5				2	2	9
Vulpes vulpes				1		12	13
Ursus arctos					1	8	9
Felidae					2	6	8
Panthera cf. *leo*					1		1
Mustelidae						2	2
Lepus sp.				1	1	4	6
Total wild mammals	**21**	**33**	**14**	**3**	**12**	**505**	**588**
Reptilia					1		1
Testudo sp.		2		1		61	64
Aves	7	1		5	8	28	49
Ichtyo					3	38	41
Molluscs	83						83
Total other taxa	**90**	**3**	**0**	**6**	**12**	**127**	**238**
Total identified	**1258**	**722**	**245**	**429**	**347**	**6892**	**9893**
Large Mammal	497	166	79	116	179	678	1715
Small Ruminant	1069	571	69	394	544	701	3348
Unidentified	224	37	27	247	833	1426	2794
Total unidentified	**1790**	**774**	**175**	**757**	**1556**	**2805**	**7857**
Grand Total	**3048**	**1496**	**420**	**1186**	**1903**	**9697**	**17750**

the examination of the heavy fractions yielded from a large volume of flotation may result to the identification of these taxa.

Traces of anthropogenic and animal activities, as well as post-depositional processes, are represented in the faunal assemblages of all the sites. The most common traces belong to human activities such as burning, tool making, and incisions of cut-marks during chopping and butchering process (Fig. 2).

Faunal spectrum

According to preliminary results, the identified faunal remains are divided into three categories: the domestic mammals (n=9067), the wild mammals (n=588), and the other taxa that include birds, reptiles, and mollusc (n=238). Domestic mammals are dominant in the assemblages and the low numbers of taxa other than mammals is certainly due to a lack of sieving. The results of the identifications from each site are summarized in Table 3 and 4, and Fig. 3.

Animal resources during the Kura-Araxes Culture

Domestic mammals

Most of the faunal remains belong to domestic mammals with an average of 92% of the NISP. Among domesticates, two major sub-families, Caprinae and Bovinae, are dominant.

Sheep and goat: The identification between sheep (*Ovis aries*) and goat (*Capra hircus*), was based on morphological features (Boessneck 1969; Clutton-Brock *et al.* 1990), while Caprinae bones with no

Table 4. Taxonomic distribution of animal remains in the Kura-Araxes assemblages in weight of remains (WR).

Taxa	Tepe Hasanlu	Kul Tepe	Haftavan Tepe	Ovçular Tepesi	Kohneh Pasgah Tepesi	Kohneh Tepesi	TOTAL
	WR	WR	WR	WR	WR	WR	WR
Capra hircus	1328.2	1587.1	22	175.6	338	2095	5545.9
Ovis aries	1757	1618.1	201	396.1	376.4	3045	7393.6
Caprinae	4662.98	3809.9	520	1153.2	860	27070	38076.10
Bos taurus	11197.9	6345.7	1389.4	1563.2	1910	33713	56119.2
Sus cf. *domesticus*					128	2970	3098
Equus sp.	85.3	1763.7	116			122	2087
Canis familiaris	67.9	81.2	4			107	260.1
Total domestic mammals	**19099**	**15206**	**2252**	**3288**	**3612**	**69122**	**112580**
Capra aegagrus		426.9					426.9
Ovis orientalis		675.3					675.3
Bos primigenius		860.9					860.9
Sus scrofa	166.1	42	186	0.8		2969	3363.9
Gazella sp.	16.3	98.4	18			80	212.7
Cervus elaphus	83.4	220.3			40	4632	4975.7
Carnivora						39	39
Canidae	4.8				32	13	49.8
Vulpes vulpes				1.7		67	68.7
Ursus arctos					22	347	369
Felidae					6	31	37
Panthera cf. *leo*					21		21
Mustelidae						30	30
Lepus sp.				5.2	1	8	14.2
Total wild mammals	**271**	**2324**	**204**	**8**	**122**	**8216**	**11144**
Reptilia					0.2		0.2
Testudo sp.		81.4		0.7		195	277.1
Aves	5	1.3		2.3	7.2	164	179.8
Ichtyo					18	87	105
Molluscs	109.7						109.7
Total other taxa	**115**	**83**	**0**	**3**	**25**	**446**	**672**
Total identified	**19485**	**17612**	**2456**	**3299**	**3760**	**77784**	**124396**
Large Mammal	8086	3400	1467	664.1	2552	8148	24317.1
Small Ruminant	2913.38	2771.1	514	823.3	2086	2128	11235.78
Unidentified	817.4	272.9	131	486.2	1238	2278	5223.5
Total unidentified	**11817**	**6444**	**2112**	**1974**	**5876**	**12554**	**40776**
Grand Total	**31301**	**24056**	**4568**	**5272**	**9636**	**90338**	**165172**

distinguishable features were assigned to the subfamily (Fig. 5). Sheep and goats are the most common domestic mammals in these six Kura-Araxes sites with about 70% of the NISP (Fig. 3). The sheep to goat ratio is close to 1:1 in Hasanlu, Kul Tepe and Kohneh Tepe. At Kohneh Pasgah Tepesi and Ovçular Tepesi, sheep are more common than goats with ratios of 1.7:1 and 1.9:1 respectively. In Haftavan Tepe, sheep largely outnumber goats with a ratio of 8:1 (Table 3).

Cattle: The bones of Bovinae were exclusively assigned to the wild or domestic cattle (*Bos primigenius / Bos taurus*) (Fig. 5). Although the water buffalo (*Bubalus bubalis*) is at home in north-western Iran and the eastern shore of Caspian Sea (Uerpmann 1987), we could not assign any specimen to this species. Cattle (*Bos taurus*) is exploited mainly as food provider but also as work force. In all sites, cattle remains are the most numerous after the ones of Caprinae with about 20% of the NISP (Fig. 4; Table 3). Cattle bones represent 45% of the WISP. It was therefore the most important species for the procurement of animal resources such as milk or meat (Table 4; Figs. 3 and 4). Preliminary results from the analysis of Late Chalcolithic animal bones at Kul Tepe indicate an increase of cattle proportion during the KA I period.

Pig: Pig remains were only found at Kohneh Pasgah Tepesi and Kohneh Tepesi and represent respectively 3.5 and 3.7 percent of the NISP of these assemblages. Suidae remains from Kohneh Pasgah Tepesi were tentatively assigned to the domestic pig due to

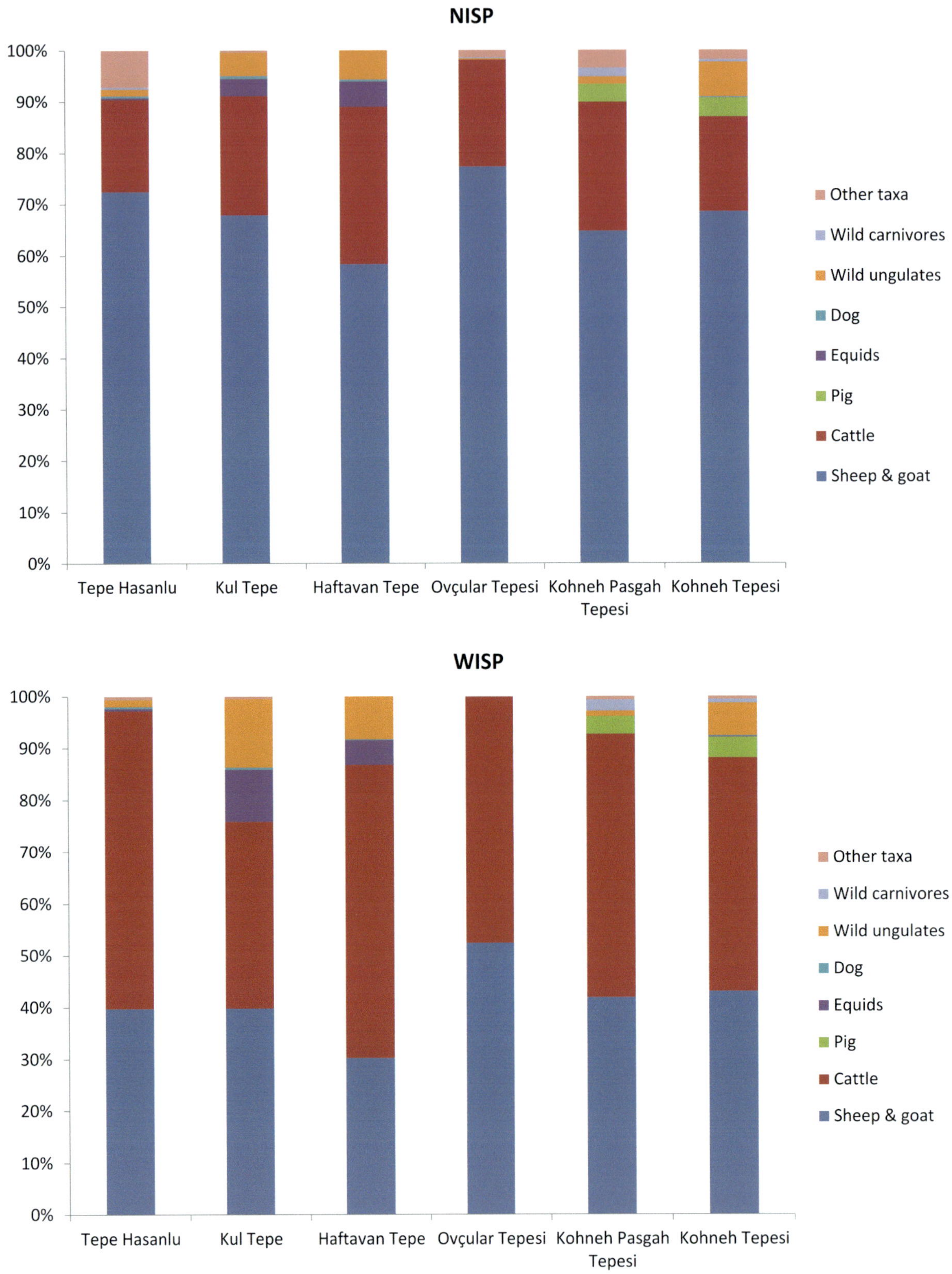

Fig. 3. Faunal spectra in percent of the Number of Identified Specimens (NISP) and of the Weight of Identified Specimens (WISP). See Table 3 and 4 for row counts.

their small size, although they were heavily fragmented and could not be analysed with osteometrical methods. The mandibles from Kohneh Tepesi were examined metrically in order to assess their wild or domestic status. The size of the second and third lower molars indicates the presence of the domestic and wild forms in similar proportions (Fig. 12). Therefore, awaiting further analyses, we tentatively attributed 50 percent of the Suidae remains from Kohneh Tepesi to the domestic pig (*Sus* cf. *domesticus*) and the 50 percent to the wild boar.

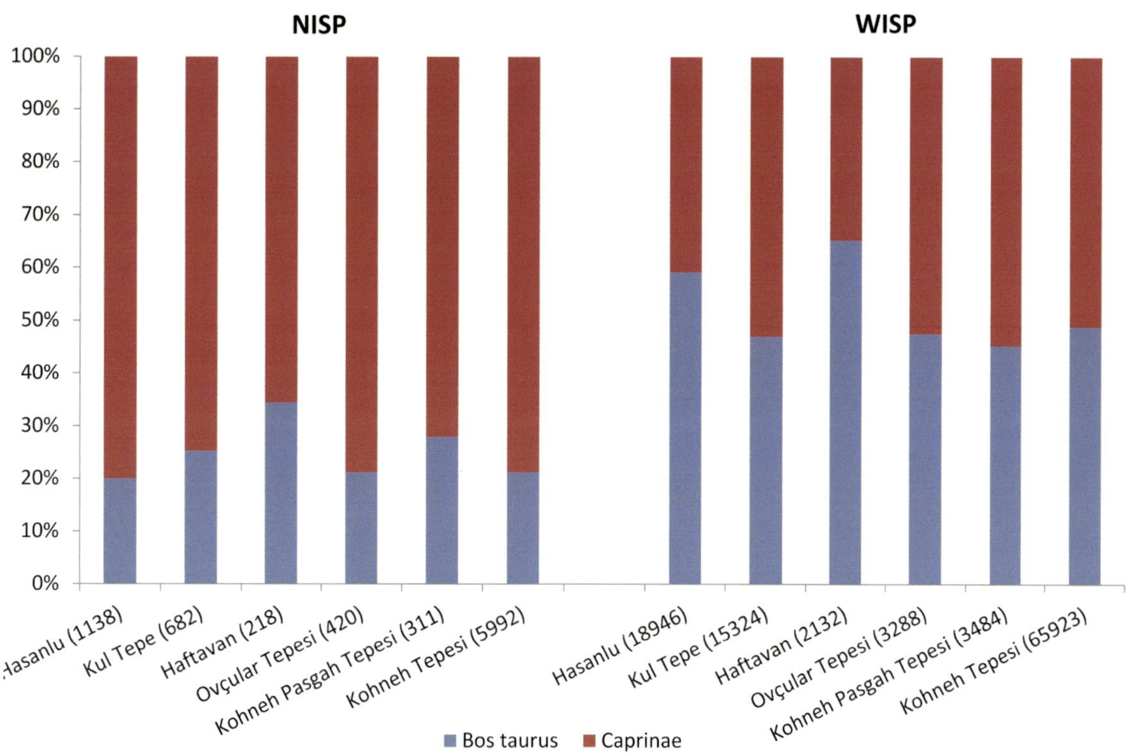

Fig. 4. Percentages of number and weight of identified specimens of Caprinae and Bos taurus.

A- Kul Tepe 2010, Tr.I, Loc.1048, EBI code #325, *Capra hircus* horn core, left side lateral, medial, transversal view

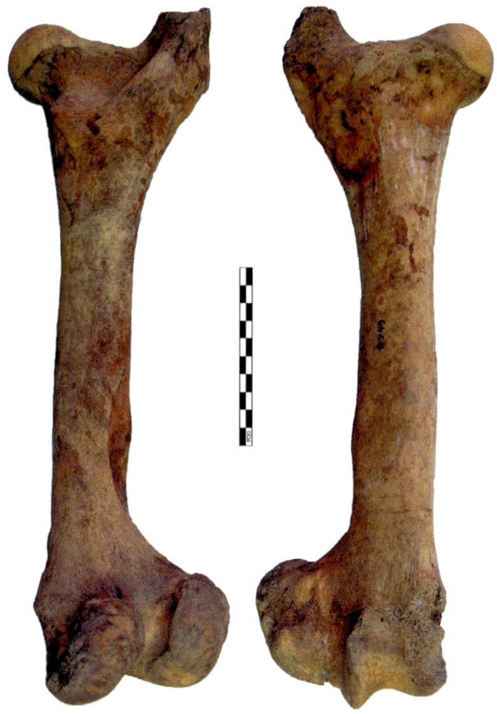

C- Hasanlu 1972, Op.U22, St.22, EBII code #949, *Bos taurus* femur, right side anterior, posterior view

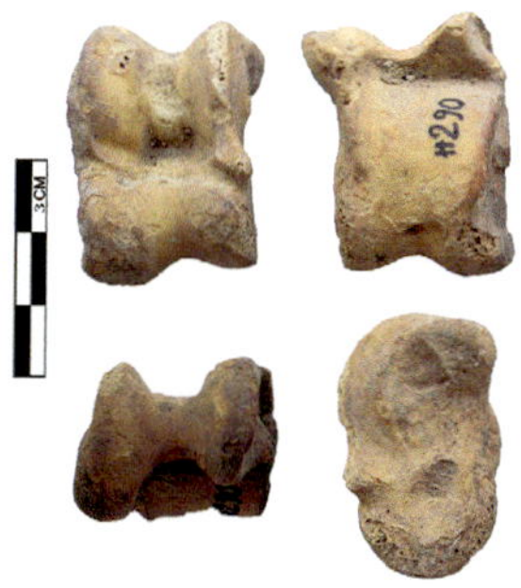

B- Kul Tepe 2010, Tr.I, Loc.1043, EBI code #290, *Ovis aries* talus, right side dorsal, plantar, superior, medial view

D- Kul Tepe 2010, Tr.I, Loc.1049, EBI code #317, *Bos taurus* metapodial, epiphysis distal right side, anterior, proximal, distal view

Fig. 5. Examples of Caprinae (sheep and goat) and cattle bones.

A- Kul Tepe 2010, Tr.II, Loc.2034, EBI
code #56, *Equus sp.* tibia, right side
dorsal, distal view

B- Kul Tepe 2010, Tr.II, Loc.2034, EBI
code #55, *Equus sp.* first phalanx
dorsal, proximal, distal, & planter view

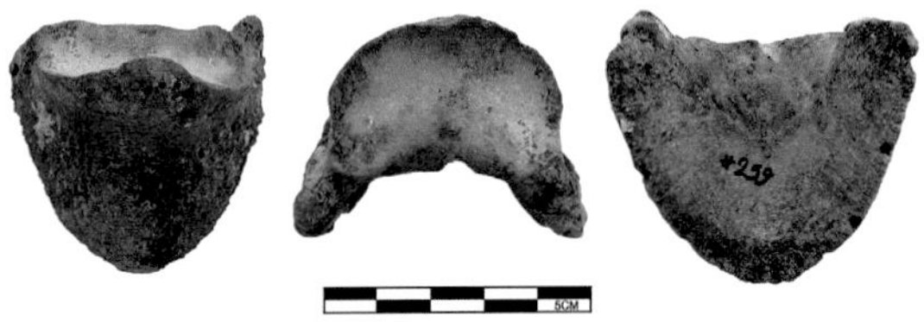

C- Kul Tepe 2010, Tr.I, Loc.1055, KAI
code #259, *Equus sp.* third phalanx
dorsal, proximal, planter view

Fig. 6. Examples of Equids bones.

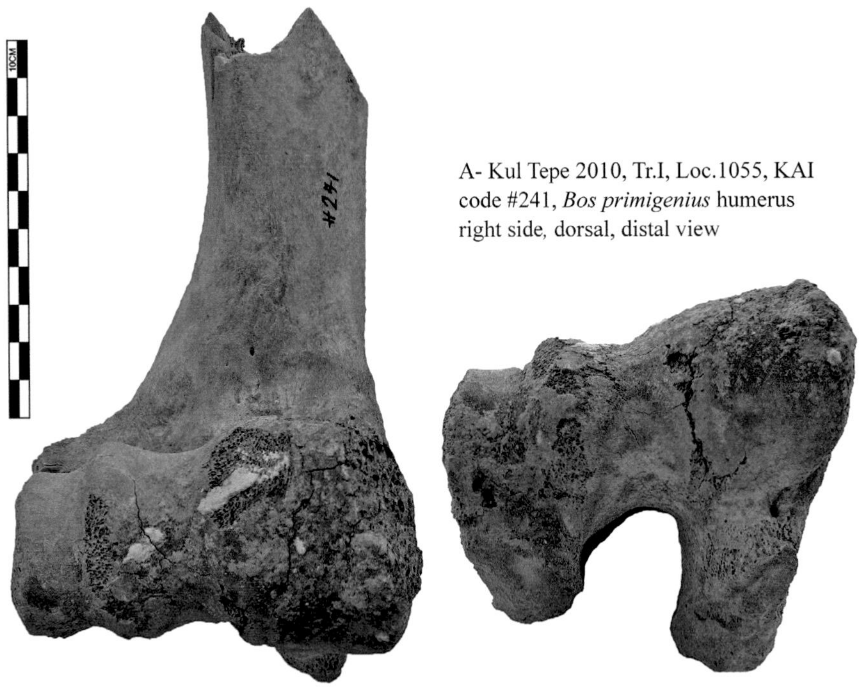

A- Kul Tepe 2010, Tr.I, Loc.1055, KAI
code #241, *Bos primigenius* humerus
right side, dorsal, distal view

B- Kul Tepe 2010, Tr.I, Loc.1055, KAI
Cervus antler with human modification

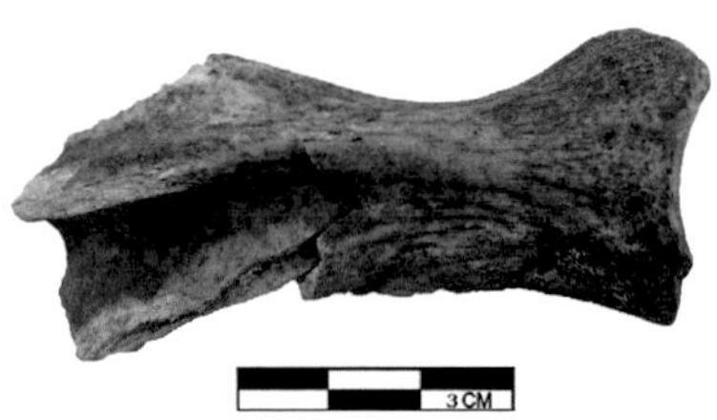

C- Hasanlu 1974, Op.U22, St.25, EBII
code #345, *Sus scrofa* scapula
right side, lateral view

Fig. 7. Examples of wild mammals bones: aurochs, red deer and boar.

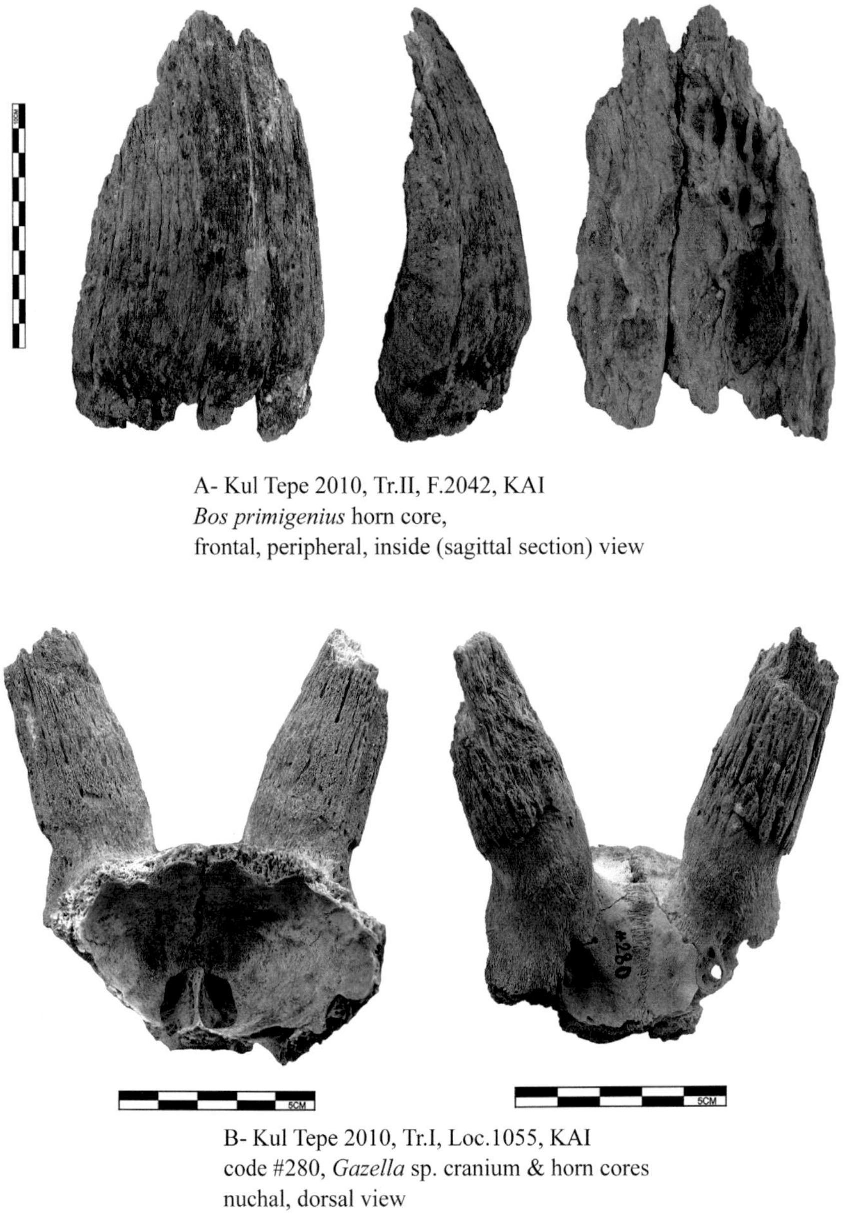

Fig. 8. Examples of wild mammals bones: aurochs and gazelle.

Equidae: In total, 42 bones (0.4% of the NISP) were identified as equids (Table 3 and 4; Fig. 6). They are absent in Ovçular Tepesi and Kohneh Pasgah Tepesi, and very rare in Tepe Hasanlu and Kohneh Tepesi. In Kul Tepe and Haftavan they represent respectively 3.3% and 4.9% of the NISP (Table 3). In this area, during this period, we can expect both wild and domestic horse (*E. ferus* and *E. caballus*), wild ass (*E. hemionus*), domestic donkey (*E. asinus*) and hybrids between these different species. The equid remains are under more detailed investigation, although most of them are generally large in the size. It has to be noticed that at Haftavan Tepe, a hybrid (donkey x horse) was identified in the Kura-Araxes period (Mohaseb 2013).

Dog: In total, 21 bones (0.2% of NISP) belong to the dog (Table 3; Fig. 3). They were found in all the sites excepting Ovçular Tepesi and Köhne Pasgah Tepesi. However, their presence is attested in all the settlements by gnawing and biting marks on the bones of other species. Dogs were not consumed, as none of their bones display any cut mark or fire trace.

Wild mammals

Game mammals are represented by 588 remains (6% of the NISP). Their frequency is very variable from one site to another (Fig. 3). However, hunting didn't represent a substantial source of animal products in any settlement. Among these remains, the majority come from herbivores (Table 3; Fig. 3). Hunted species represent different habitats: a forested/riparian landscape (*Sus scrofa*, *Cervus elaphus*), open grasslands (*Gazella*, *Bos primigenius*), and hilly/mountainous areas (*Ovis orientalis*, *Capra aegagrus*) (Figs. 7 and 8). Bones of wild sheep (*Ovis orientalis*), wild goat (*Capra aegagrus*) and aurochs (*Bos primigenius*) were identified using morphological and metrical criteria (Helmer 1992; Uerpmann 1987;

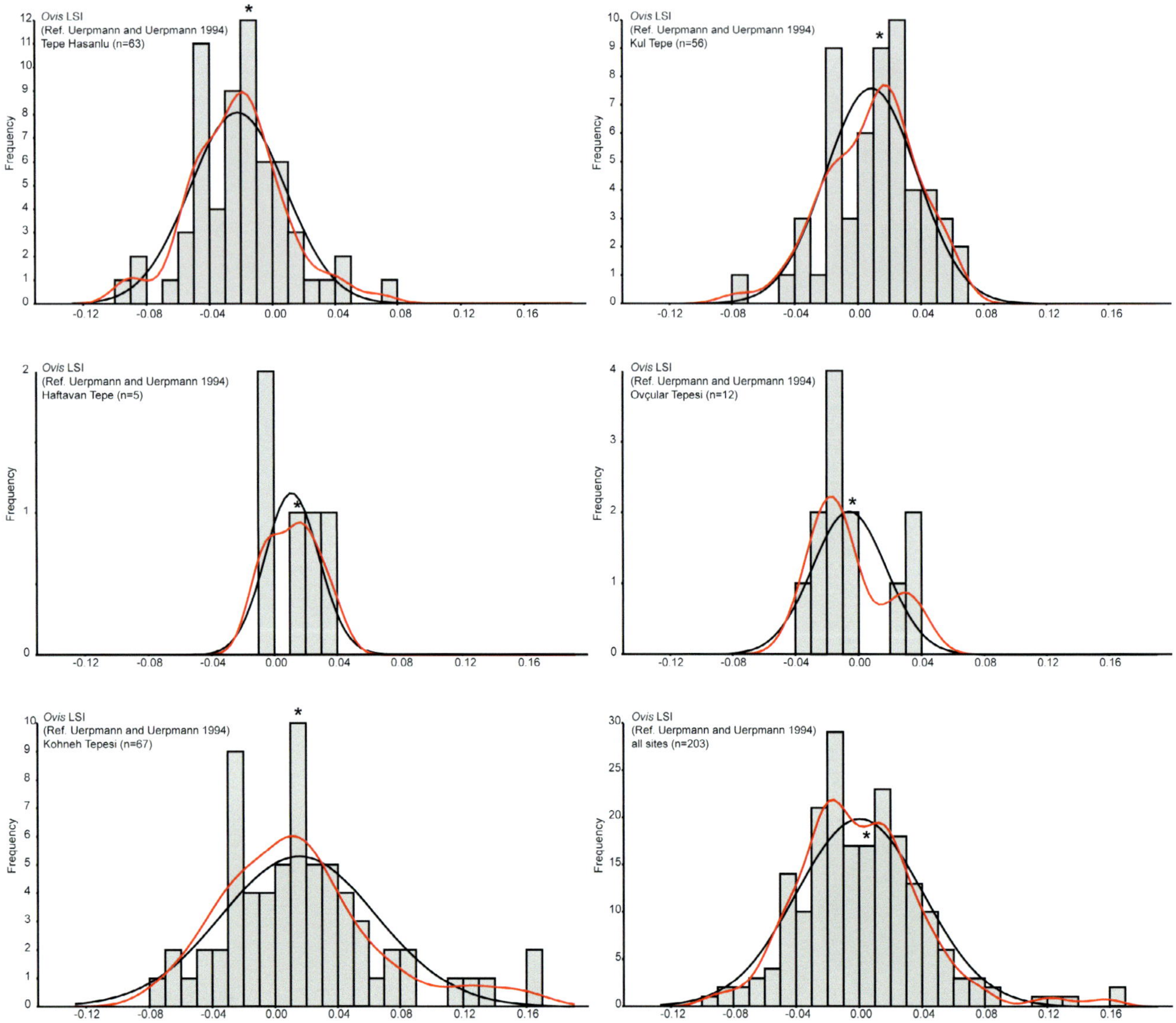

Fig. 9. Size variation in the Sheep (*Ovis* sp.) remains from the Kura-Araxes layers at Tepe Hasanlu, Kul Tepe, Haftavan Tepe, Ovçular Tepesi and Kohneh Tepesi. The black curve represents the normal distribution. The red curve is a Kernel density. The star represents the mean.

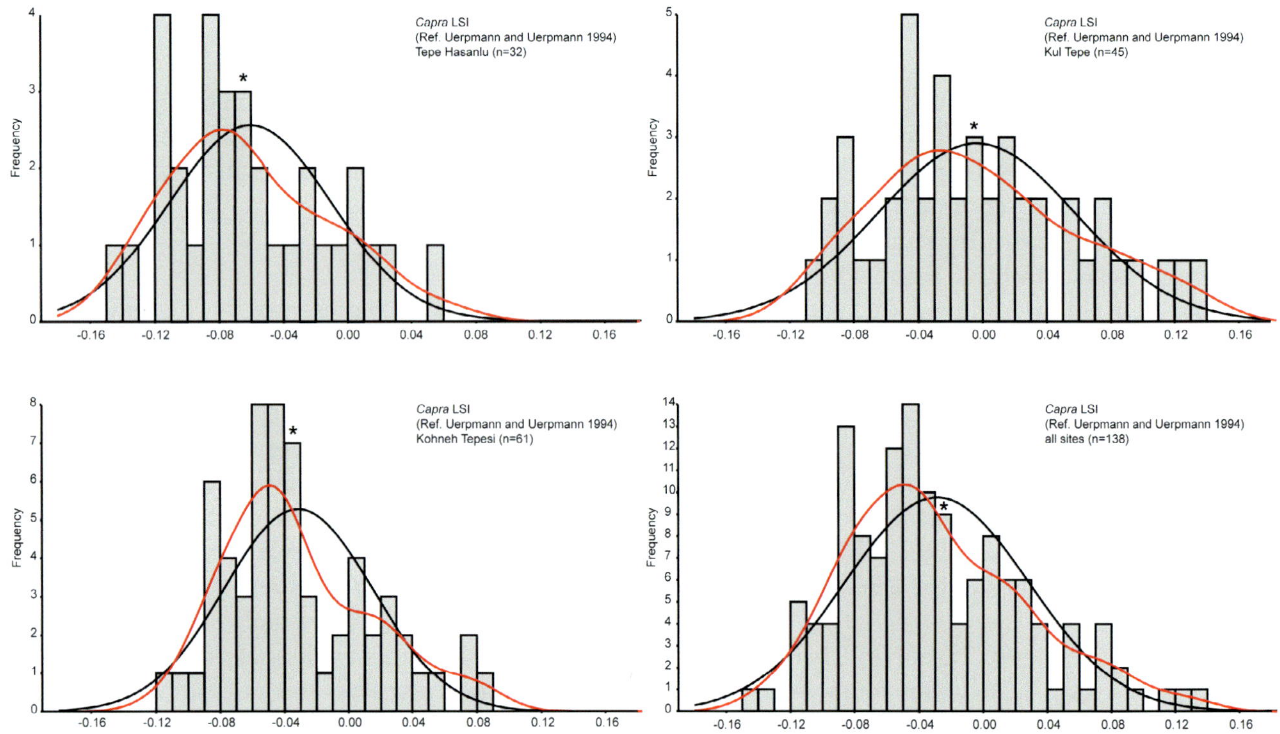

Fig. 10. Size variation in the Goat (*Capra* sp.) remains from the Kura-Araxes layers at Tepe Hasanlu, Kul Tepe, and Kohneh Tepesi. The black curve represents the normal distribution. The red curve is a Kernel density. The star represents the mean.

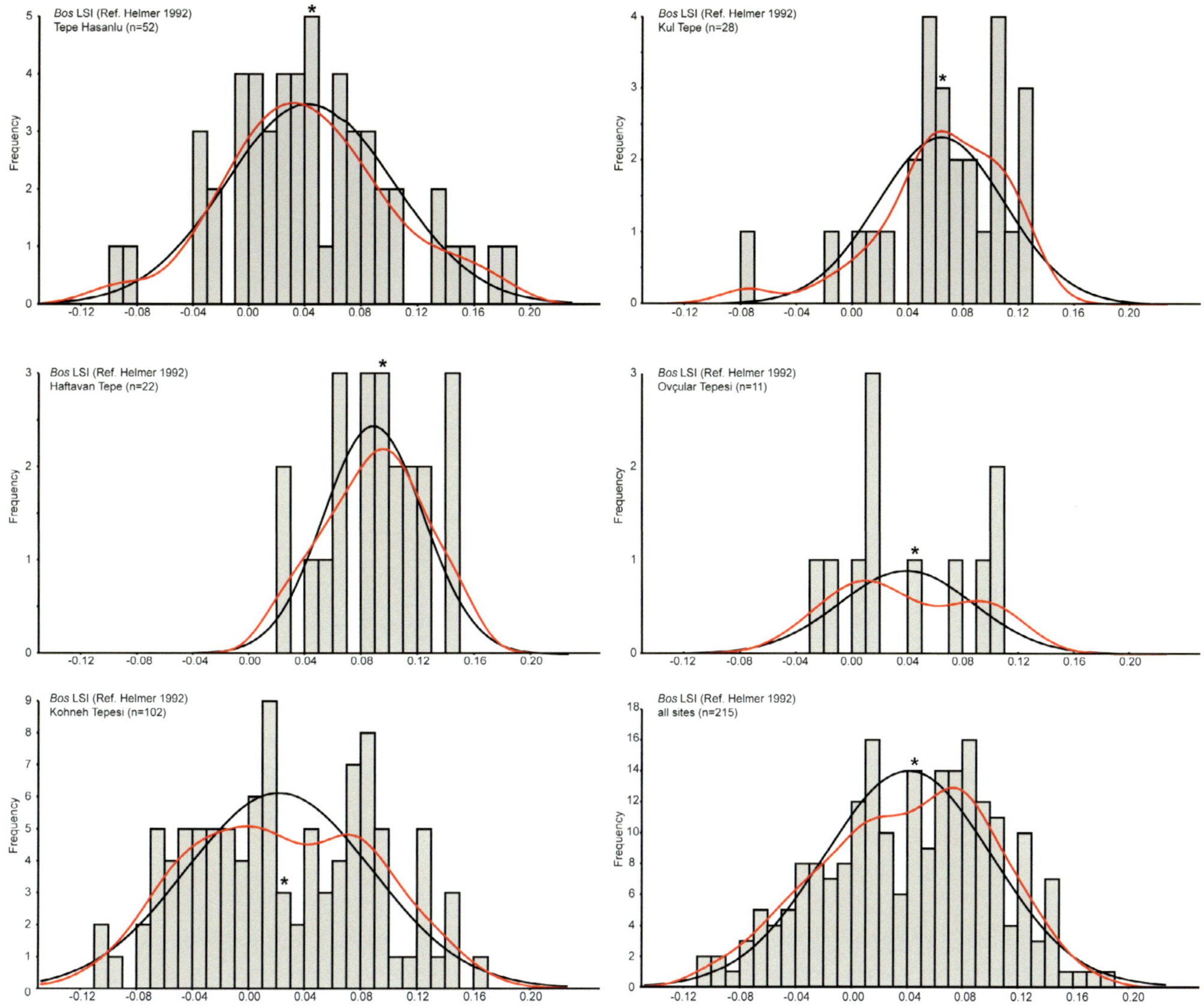

Fig. 11. Size variation in the Cattle (*Bos* sp.) remains from the Kura-Araxes layers at Tepe Hasanlu, Kul Tepe, Haftavan Tepe, Ovçular Tepesi and Kohneh Tepesi. The black curve represents the normal distribution. The red curve is a Kernel density. The star represents the mean.

Uerpmann & Uerpmann 1994). Interestingly, these three species were only found at Kul Tepe. On the contrary, red deer (*Cervus elaphus*), gazelle (*Gazella* sp.) and wild boar (*Sus scrofa*) were identified in most of the assemblages (Table 3).

Other taxa

Remains of fish, birds, reptiles, and molluscs were occasionally found in the assemblage with 2.4% of the NISP (Table 3). The frequencies of these taxa are very variable from one assemblage to another. This is certainly due to the different sampling and sieving methods applied in the field. For this reason, these taxa are not good indicators of inter-site differences in their environments or in the socio-economic strategies of their inhabitants.

Size of sheep, goat, cattle and pig.

Measurements of post-cranial bones of sheep, goat, cattle and pig are used for the taxonomic distinction and morphological analysis within a species (Von den Driesch 1976; Payne and Bull 1988). In this section, the measurements of the mentioned taxa were compared by using 'difference of logs' or 'log size index' (LSI) method within five of the six sites (Uerpmann 1979; Meadow 1999). The measurements of a wild sheep and a wild goat (Uerpmann and Uerpmann 1994), a domestic cattle (used in Helmer 1992; Helmer, *in letteris*, see appendix 1), and wild boar (Payne and Bull 1988) were used as reference. Individual measurements of the archaeological specimens will be deposited in the Osteometric Database of Vertebrates in the Caucasus (ODVC)[1].

Sheep (*Ovis* sp.) remains are, on average, smaller at Tepe Hasanlu than at the other sites (Fig. 9). It could indicate a better representation of female domestic sheep in this assemblage while male domestic sheep would be more frequent at the other sites. Larger specimens (LSI value > 0.09) suggest the presence of hunted wild sheep in some of the

1 http://archeozoo-archeobota.mnhn.fr/spip.php?article334.

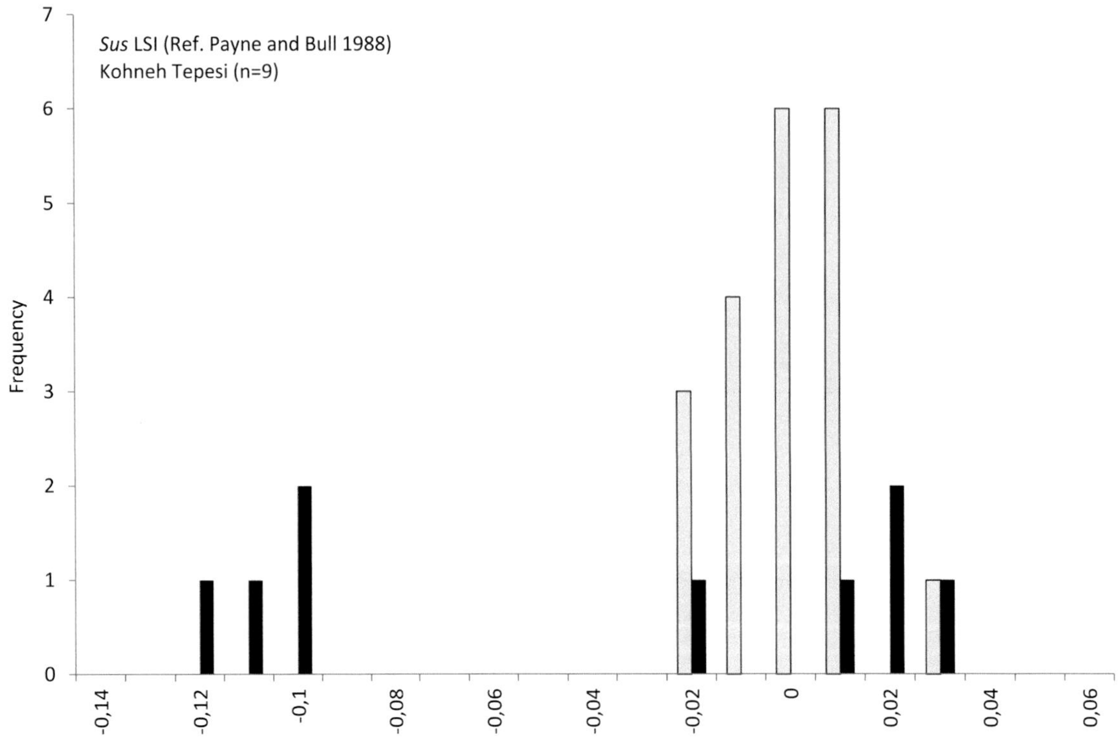

Fig. 12. Size variation in Suidae lower second and third molars from the Kura-Araxes layers at Kohneh Tepesi (black bars) as compared to wild boars from Anatolia (grey bars, data from Payne and Bull 1988).

assemblages. The size of the goat (*Capra* sp.) remains also suggests different sex ratios from one site to another (Fig. 10). Female domestic goats were maybe less slaughtered in Kul Tepe than in Tepe Hasanlu and Kohneh Tepesi. The presence of a few large wild goats is also attested, especially at Kul Tepe. In contrast to sheep and goat, similar average sizes are visible in the cattle assemblages. Looking the size of the cattle (*Bos* sp.) remains, it is more difficult to estimate the sex ratio but both female and male domesticate cattle were slaughtered (Fig. 11). It is interesting to note that smaller specimens are lacking in the rather small assemblage from Haftavan Tepe. Considering a few high LSI values in some of the sites, the presence of aurochs cannot be excluded.

An exploratory analysis of the size of the Suidae remains from Kohneh Tepesi was performed using the lower second and third molars only. Half of the individuals are of comparable size with modern wild boars (*Sus scrofa*) from Anatolia (Payne and Bull 1988) while is other half is composed of smaller specimens that certainly correspond to domestic pigs (Fig. 12).

Discussion and conclusion

The analysis of faunal assemblages from Tepe Hasanlu, Kul Tepe, Haftavan, Ovçular Tepesi, Kohneh Pasgah Tepesi and Kohneh Tepesi offers an important set of new data for this archaeologically under-explored region. It improves our knowledge of the subsistence economy from the second half of the fourth to second half of the third millennia BC. These preliminary analyses indicate that the exploitation of domestic species was always the most important source of animal products. Hunting only played a marginal role at these sites. The wild species indicate the presence of an environmental mosaic around the sites ranging from the steppe to forested areas. The presence of the aurochs is particularly interesting at Kul Tepe and constitutes a precious record for the presence and evolution of this species in the area.

The taxonomic analysis of the Equid remains is still ongoing. The important role of cattle in the subsistence economy and agricultural activities in all the sites indicates favourable environmental conditions for this species that is more demanding in grass and water resources than sheep and goats. The patterns of animal exploitation in north-western Iran and Nakhchivan are comparable to those of contemporaneous settlements in adjacent regions such as Azerbaijan, Armenia, and eastern Anatolia (Fig. 13). For instance, domestic taxa other than cattle, sheep and goats were virtually not exploited at the Kura-Araxes settlement of Sos Höyük in north-eastern Anatolia (Howell-Meurs 2001a: 101-102; Piro 2009). The absence of domestic pig remains is remarkable and constitutes a characteristic of the

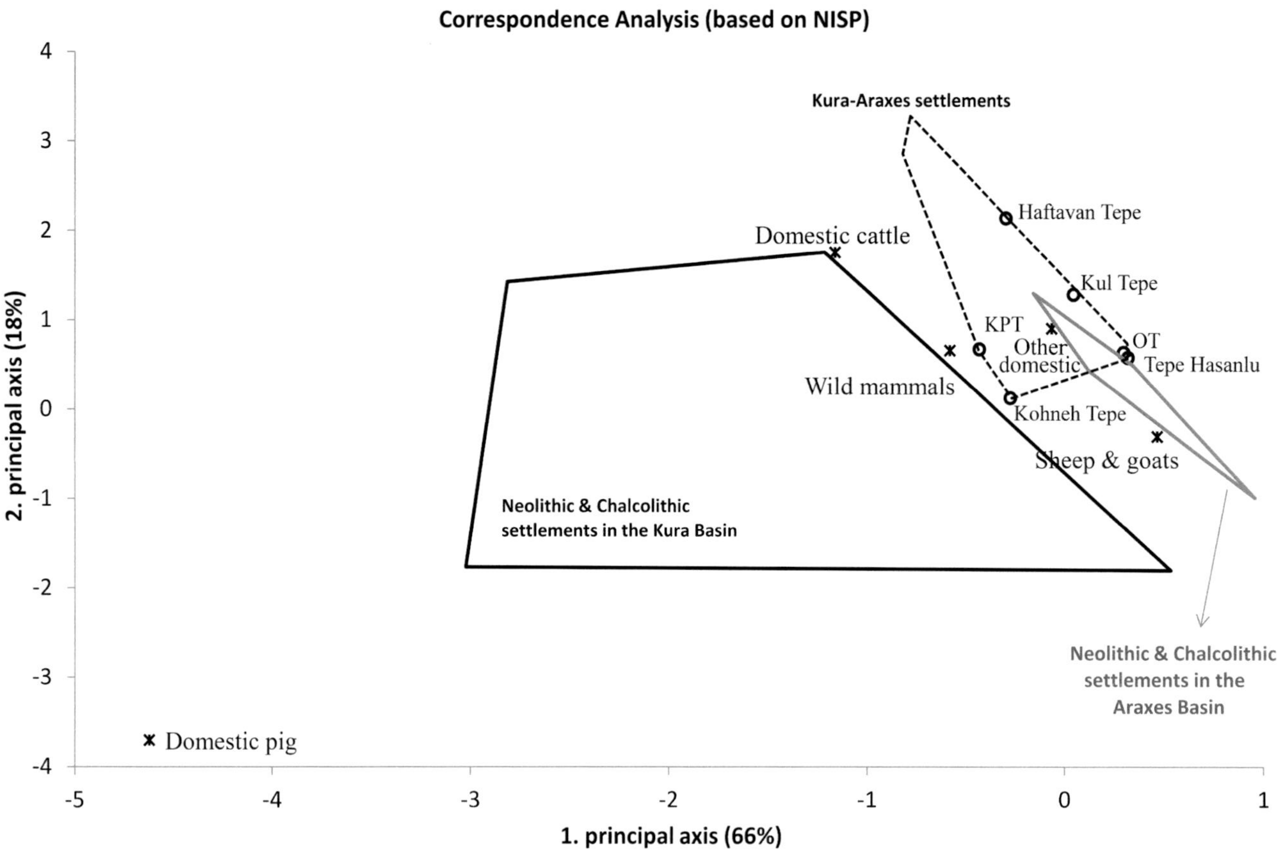

Fig. 13. Correspondence analysis showing the six assemblages presented in this study (OT = Ovçular Tepesi; KTP = Kohneh Pasgah Tepesi) in relation to other Kura-Araxes settlements (from Armenia, Turkey, Georgia and Dagestan, see Decaix *et al.* in press a), and to Neolithic and Chalcolithic settlements from the Kura and Araxes basins (Berthon 2014). Axis I is mainly driven by the abundance of pig remains while Axis 2 is mainly explained by the abundance of cattle remains.

Kura-Araxes subsistence strategies. The few assemblages that yielded pig remains such as Kohneh Pasgah Tepesi, Kohneh Tepesi and Gegharot might belong to the a late phase of the Kura-Araxes culture (Badalyan *et al.* 2014).

Although this study indicates a similar pattern of animal exploitation in the six sites presented here, a deeper understanding of the herding strategies is needed. Further analyses are currently being undertaken in order to establish the kill-off patterns of the different domestic species. We will evaluate the evolution of herding practices in north-western Iran and south Caucasus by comparing our demographic (kill-off patterns and metrical data) with other sites of the region in a diachronic approach. By doing so, we will be able to identify the specificities of the Kura-Araxes pastoral practices.

Acknowledgements

The authors would like to thank the colleagues and friends at the National Museum of Iran who have brought their precious help during the preparation and reconditioning of the very large faunal assemblage of Hasanlu and especially Dr. Fereidoun Biglari, Mrs. Nahid Ghafouri and Mrs. Nayereh Nazari. Also we would like to thank Dr. Michael Danti, Mrs. Kyra Kaercher, and Dr. Stephan Kroll, for their kind helps with the chronology and context analysis of Hasanlu assemblage. The Kul Tepe assemblage was studied in the Archaeozoology of the Archaeometry Laboratory of University of Tehran, directed by Dr. Haeedeh Laleh and Dr. Ahmad Aliyari whom we thank for their kind support. Finally the first author thanks also Roya Khazaeli and Sarieh Amiri for their help in some part of the study of the Hasanlu assemblage. The faunal assemblage from Ovçular Tepesi was made available to study thanks to Dr. Catherine Marro and Pr. Veli Bakhshaliyev. Many thanks to Daniel Helmer who granted us with the permission to publish its reference measurements of modern cattle. Geoffrey Summers and an anonymous reviewer read and provided helpful comments on previous drafts of this paper.

References

Abedi, A., Khatib Shahidi, H., Chataigner, Ch., Niknami, K., Eskandari, N., Kazempour, M., Pirmohammadi, A., Hosseinzadeh, J. and Ebrahimi, Gh. 2014. Excavation at Kul Tepe (Hadishahr),

North-western Iran, 2010: First Preliminary Report. *Ancient Near Eastern Studies* 51: 33-165.

Abedi, A. and Omrani, B. 2015. Kura-Araxes Culture and North-western Iran: New Perspectives from Kul Tepe Jolfa (Hadishahr). *Paléorient* 41(1): 55-68.

Alizadeh, K., Eghbal, H. and Samei, S. 2015. Approaches to Social Complexity in Kura-Araxes Culture: A View from Köhne Shahar (Ravaz) in Chaldran, Iranian Azerbaijan. *Paléorient* 41(1): 37-54.

Badalyan, R., Smith, A.T., Lindsay, I., Harutyunyan, A., Greene, A., Marshall, M., Monahan, B., Hovsepyan, R., Meliksetian, K., Pernicka, E. and Haroutunian, S. 2014. A preliminary report on the 2008, 2010, and 2011 investigations of Project ArAGATS on the Tsaghkahovit Plain, Republic of Armenia. *Archäologische Mitteilungen aus Iran und Turan* 46: 149-222.

Berthon, R. 2014. Past, Current and Future Contribution of Zooarchaeology to the Knowledge of the Neolithic and Chalcolithic Cultures in South Caucasus. *Studies in Caucasian Archaeology* 2: 4-30.

Boessneck, J. 1969. Osteological Differences between Sheep (*Ovis aries*) and Goat (*Capra hircus*). In, Brothwell, D. R. and Higgs, E. S. (eds) *Science in Archaeology: A Survey of Progress and Research*: 331-58. London: Thames & Hudson.

Burney, C. A. 1970. Excavations at Haftavan Tepe 1968: First Preliminary Report. *Iran* 8: 157-72.

Burney, C. A. 1972. Excavations at Haftavan Tepe 1969: Second Preliminary Report. *Iran* 10: 127-50.

Burney, C. A. 1973. Excavations at Haftavan Tepe 1971: Third Preliminary Report. *Iran* 11: 153-72.

Burney, C. A. 1975. Excavations at Haftavan Tepe 1973: Fourth Preliminary Report. *Iran* 13: 149-64.

Clutton-Brock, J., Dennis-Bryan, K., Armitage, P. L. and Jewell, P.A. 1990. Osteology of the Soay Sheep. *Bulletin of the British Museum (National History) Zoology Series* 56(1): 3-56.

Danti, M. D. 2013a. The Late Bronze and Early Iron Age in North-western Iran. In, Potts, D. T. (ed.) *The Oxford Handbook of Ancient Iran*: 327-76. Oxford: Oxford University Press.

Danti, M. D. 2013b. *Hasanlu V: The Late Bronze and Iron I Periods*. Philadelphia: University of Pennsylvania Museum of Archaeology and Anthropology.

Danti, M. D., Voigt, M. M. and Dyson, R. H. 2004. The Search for the Late Chalcolithic / Early Bronze Age Transition in the Ushnu-Solduz Valley, Iran. In, Sagona, A. (ed) *A View from the Highlands: Archaeological Studies in Honour of Charles Burney*: 583-616. Leuven: Peeters.

Davoudi, H., Mashkour, M. and Mohaseb, F. A. 2014. Animal Resource Exploitation and Management at Tepe Hasanlu (North-west Iran) from Early Bronze Age to the Islamic (Ilkhanid) Periods. In, *12nd International Council for Archaeozoology (ICAZ)*, San Rafael, Argentina (poster presented).

Davoudi, H. (in preparation) *Subsistence Economy of Iron Age Societies in North-western Iran Based on Archaeozoological Studies - the Case of Tepe Hasanlu*. PhD. Tarbiat Modares University.

Decaix, A., Berthon, R., Mohaseb, F. A. and Tengberg M. (in press a) Toward a Definition of the Kura-Araxes Agropastoral Systems. In, *Urbanisation, Trade, Subsistence and Production during the Third Millennium B.C. on the Iranian Plateau*. Lyon: Maison de l'Orient et de la Mediterranée.

Decaix, A., Mohaseb, A., Maziar, S., Mashkour, M. and Tengberg, M. (in press b). Subsistence economy in Kohneh Pasgah Tepesi (Eastern Azerbaijan, Iran) during the Late Chalcolithic and the Early Bronze Age based on the faunal and botanical remains. In, *Urbanisation, Trade, Subsistence and Production during the Third Millennium B.C. on the Iranian Plateau*. Lyon: Maison de l'Orient et de la Mediterranée.

Dyson, R. H. 1958. Iran 1957: Iron Age Hasanlu. *University Museum Bulletin* 22(2): 25-32.

Dyson, R. H. 1973. Hasanlu. *Iran* 11: 195-96.

Dyson, R. H. 1989. The Iron Age Architecture at Hasanlu: An Essay. *Expedition* 31(2-3): 107-27.

Dyson, R. H. and Muscarella, O. W. 1989. Constructing the Chronology and Historical Implications at Hasanlu IV. *Iran* 27: 1-27.

Dyson, R. H. and Pigott, V. C. 1975. Hasanlu. *Iran* 13: 182-85.

Edwards, M. 1983. *Haftavan, Period VI. Excavations in Azerbaijan (North-western Iran)*. Oxford: British Archaeological Reports.

Helmer, D. 1992. *La domestication des Animaux par les Hommes Préhistoriques*. Paris: Masson.

Helmer, D. 2000. Etude de la faune mammalienne d'El Kowm 2. In Stordeur, D. (ed) *El Kowm 2, Une Île dans le Désert. La fin du Néolithique Précéramique dans la Steppe Syrienne*: 233-264. Paris: CNRS Editions.

Howell-Meurs, S. 2001a. *Early Bronze and Iron Age Animal Exploitation in Northeastern Anatolia: The Faunal Remains from Sos Höyük and Büyüktepe Höyük*. Oxford: British Archaeological Reports.

Howell-Meurs, S. 2001b. Archaeozoological Evidence for Pastoral Systems and Herd Mobility: The Remains from Sos Höyük and Büyüktepe Höyük. *International Journal of Osteoarchaeology* 11: 321-28.

Khaksar, A., Hemati, E. and Nourozi, A. 2015. The Analysis of Yaniq Culture at Tappeh Gourab of Malayer, Based on Stratigraphical Excavation (in Persian). *Journal of Archaeological Researches of Iran* 4(7): 47-66.

Kohl, P. 2007 *The Making of Bronze Age Eurasia*. Cambridge: Cambridge University Press.

Kushnareva, K. Kh. 1997. *The Southern Caucasus in Prehistory: Stages of Cultural and Socioeconomic Development from the Eighth to the Second Millennium BC*. Translated from Russian by H. N. Michael. Philadelphia: University of Pennsylvania Museum.

Marro, C., Bakhshaliyev, V. and Ashurov, S. 2009. Excavations at Ovçular Tepesi (Nakhichevan, Azerbaijan). First Preliminary Report: The 2006–2008 Seasons. *Anatolia Antiqua* 17: 31-87.

Marro, C., Bakhshaliyev, V. and Ashurov, S. 2011. Excavations at Ovçular Tepesi (Nakhichevan, Azerbaijan). Second Preliminary Report: The 2009–2010 Seasons. *Anatolia Antiqua* 19: 53-100.

Marro, C., Bakhshaliyev, V. and Berthon, R. 2014. On the Genesis of the Kura-Araxes Phenomenon: New Evidence from Nakhchivan (Azerbaijan). *Paléorient* 40(2): 131-54.

Mashkour, M., Biglari, F. and Ghafouri, N. 2012. The Osteoarchaeological Project of the National Museum of Iran: An Interim Report. *Iranian Archaeology* 3: 72-76.

Maziar S. 2010. Excavations at Kohneh Pasgah Tepesi, the Araxes Valley, NW Iran: First Preliminary Report. *Journal of Ancient Near Eastern Studies* 47: 165-93.

Maziar, S. 2015. Settlement Dynamics of the Kura-Araxes Culture: An Overview of the Late Chalcolithic and Early Bronze Age in the Khoda Afarin Plain, North-western Iran. *Paléorient* 41(1): 25-36.

Meadow, R. H. 1999. The Use of Size Index Scaling Techniques for Research on Archaeozoological Collections from the Middle East. In, Becker, C., Manhart, H., Peters, J. and Schibler, J. (eds) *Historia Animalium ex Ossibus. Beiträge zur Paläoanatomie, Archäologie, Ägyptologie, Ethnologie und Geschichte der Tiermedizin. Festschrift für Angela Von den Driesch*: 285-300. Rahden/Westf.: Verlag Marie Leidorf.

Mohaseb, F. A. 2012. *Exploitation des Animaux durant l'Age du Bronze au début de la Période Islamique dans le Nord-ouest de l'Iran: étude Archéozoologique de Haftavan Tepe*. PhD. Université Panthéon-Sorbonne.

Mohaseb, F. A. 2013. Animal Exploitation from the Bronze Age to the Early Islamic Period at Haftavan Tepe (Western Azerbaijan, Iran) (in Persian). *Iranian Archaeology* 4: 120-49.

Motarjem, A. and Niknami, K. 2011. Early Bronze Age at Central Zagros, Iran (in Persian). *Journal of Archaeological Studies* 4: 35-54.

Muscarella, O. W. 2006. The Excavation of Hasanlu: An Archaeological Evaluation. *Bulletin of the American Schools of Oriental Research* 342: 69-94.

Payne, S. and Bull, G. 1988. Components of variation in measurements of pig bones and teeth, and the use of measurements to distinguish wild from domestic pig remains. *Archaeozoologia* 2: 27-65.

Piro, J. 2009. *Pastoralism in the Early Transcaucasian Culture: The Faunal Remains from Sos Höyük*. PhD. New York University.

Rothman, M. S. 2003 Ripples in Stream: Transcaucasia-Anatolian Interaction in the Murat / Euphrates Basin at the Beginning of the Third Millennium. In, Smith, A. T. and Rubinson,K. S. (eds) *Archaeology of the Borderlands: Investigations in Caucasia and Beyond*: 95-110. Los Angeles: Cotsen Institute of Archaeology at UCLA.

Sagona, A. G. 1984. *The Caucasian Region in the Early Bronze Age*. Oxford: Archaeopress.

Summers, G. D. 1982. *A Study of the Architecture, Pottery and Other Material from Yanik Tepe, Haftavan VIII and Related Sites*. PhD. University of Manchester.

Summers, G. D. 2004. Yanik Tepe and the Early Trans-Caucasian Culture: Problems and Perspectives. In, Sagona, A. (ed) *A view from the Highlands: Archaeological Studies in Honour of Charles Burney, Ancient Near Eastern Studies:* 617-43. Leuven: Peeters.

Summers, G. D. 2014. The Early Trans-Caucasian Culture in Iran: Perspectives and Problems. *Paléorient* 40(2): 155-68.

Uerpmann, H.-P. 1979. *Probleme der Neolithisierung des Mittelmeerraums. Beihefte zum Tübinger Atlas des Vorderen Orients.* Wiesbaden: Dr. Ludwig Reichert Verlag.

Uerpmann, H.-P. 1987. *The Ancient Distribution of Ungulate Mammals in the Middle East. Fauna and Archaeological Sites in Southwest Asia and Northwest Africa.* Wiesbaden: Dr. Ludwig Reichert Verlag.

Uerpmann, M. and Uerpmann, H.-P. 1994. Animal Bone Finds from Excavation 520 at Qala'at al Bahrain. In, Hojlund, F. and Andersen, H. (eds) *Qala'at al Bahrain, Volume I. The Northern City Wall and the Islamic Fortress*: 227-37. Aarhus: Aarhus University Press.

Von den Driesch, A. 1976. A *Guide to the Measurement of Animal Bones from Archaeological Sites*. Cambridge: Peabody Museum of Archaeology and Ethnology, Harvard University.

Appendix I

Measurements of a modern cattle from Camargue (France). Courtesy of Daniel Helmer (Helmer 1992). Definition of the measurements from Von den Driesch (1976), except for "H" (Helmer 2000).

Reference measurements of ***Bos taurus*** *used for the LSI*							
	Bp	**Bd**	**GLl**	**GL**	**GB**	**DLS**	**H**
Humerus	-	73.55	-	-	-	-	40.25
Radius	74.4	63.3	-	-	-	-	-
Metacarpus	52.3	51.1	-	-	-	-	-
Patella	-	-	-	-	47.4	-	-
Tibia	-	58.9	-	-	-	-	-
Astragalus	-	-	62.1	-	-	-	-
Calcaneum	-	-	-	122.75	-	-	-
Metatarsus	46.1	49.15	-	-	-	-	-
Phalanx 1	27.08	-	-	-	-	-	-
Phalanx 2	25.28	-	-	-	-	-	-
Phalanx 3	-	-	-	-	-	61.35	-

9 Unravelling the meaning of faunal assemblages in an early urban domestic neighbourhood: worked bone frequencies from the Early Bronze Age III neighbourhood at eṣ-Ṣâfi/Gath, Israel

Haskel J. Greenfield[1], Annie Brown[2], Itzhaq Shai[3], Aren M. Maeir[4]

1 University of Manitoba, Department of Anthropology and St. Paul's College, Winnipeg, MB, R3T5V5, Canada; Haskel.greenfield@umanitoba.ca
2 University of Manitoba, Department of Anthropology and St. Paul's College, Winnipeg, MB, R3T5V5, Canada; umbro392@myumanitoba.ca
3 Israel Heritage Department, Ariel University, P.O.B. 3, Ariel 40700 Israel; shai.itzick@gmail.com
4 The Martin (Szusz) Department of Land of Israel Studies and Archaeology, Bar-Ilan University, Ramat Gan 52900, Israel; arenmaeir@gmail.com

Abstract
Most studies of the faunal remains found within and around structures assume that they relate to the subsistence activities of the occupants. From this basic assumption, the data are used to reconstruct diet and economy. More recently, zooarchaeologists have to come understand that faunal remains may also represent other types of activity areas, such as worked bone production, use, and discard. In this paper, we present newly collected data on the worked bone frequency (tool/ornament) remains recovered from the early urban centre of Tell eṣ-Ṣâfi/Gath, Israel that challenge this assumption. Detailed microscopic and taphonomic examination of the zooarchaeological assemblage collected during the excavation of several structures from an Early Bronze (EB) III (2800-2500 BC) non-elite neighbourhood at the site suggests that formal and informal worked bone are present in far greater numbers than normally described in most contemporary assemblages. This suggests that the distribution of bones in residential areas may be related more to use and fragmentation of modified bone rather than to subsistence. Based on these data and analyses, we discuss the significance of identifying use wear polish on even fragmentary remains in order to understand faunal osteological frequencies in early urban settings.

Keywords
Early Bronze Age, Israel, Worked bone, Use wear polish, Zooarchaeology

Introduction
In recent years, the collection and analysis of animal remains from early urban Near Eastern contexts has become more and more common and systematic (e.g. Allentuck 2004, 2014; Atici 2004; Berthon 2011; De Cupere *et al.* 2006; Greenfield and Allentuck 2011; Hesse and Perkins 1974; Allentuck and Greenfield 2010; Horwitz 1997, 2003; Kansa 2004; O'Connor 2003; Sasson 2010; Zeder 1988, 1990, 2003). In spite of this positive trend, it is often simplistically assumed by most excavators (as well as analysts) that the animal remains recovered from early urban domestic contexts are direct by-products of diet and subsistence activities of the site's inhabitants. From this basic assumption, the data are often used to reconstruct diet and economy. In contrast to the larger archaeological community, zooarchaeologists understand that the distribution of faunal remains may also represent other types of activities. Bone, as the *plastic of antiquity*, would have been used for a variety of functions. In particular, faunal remains may represent a variety of non-food related activities, such as worked bone (tool and ornament) production, use, and discard, as well as residual materials from earlier phases (e.g. Allentuck 2011; Choyke 2000; Marom and Zukerman 2011; Nishimura 2008, 2012, 2014). Hence, the ancient inhabitants of a site do not have to be envisioned as the quintessential 'slobs', living amidst the discards of their last meal. In reality, this assumption makes little sense given the fact that ethnoarchaeological studies have cross-culturally demonstrated that the inhabitants of sedentary urban communities continuously clean up their immediate living space. While they may discard

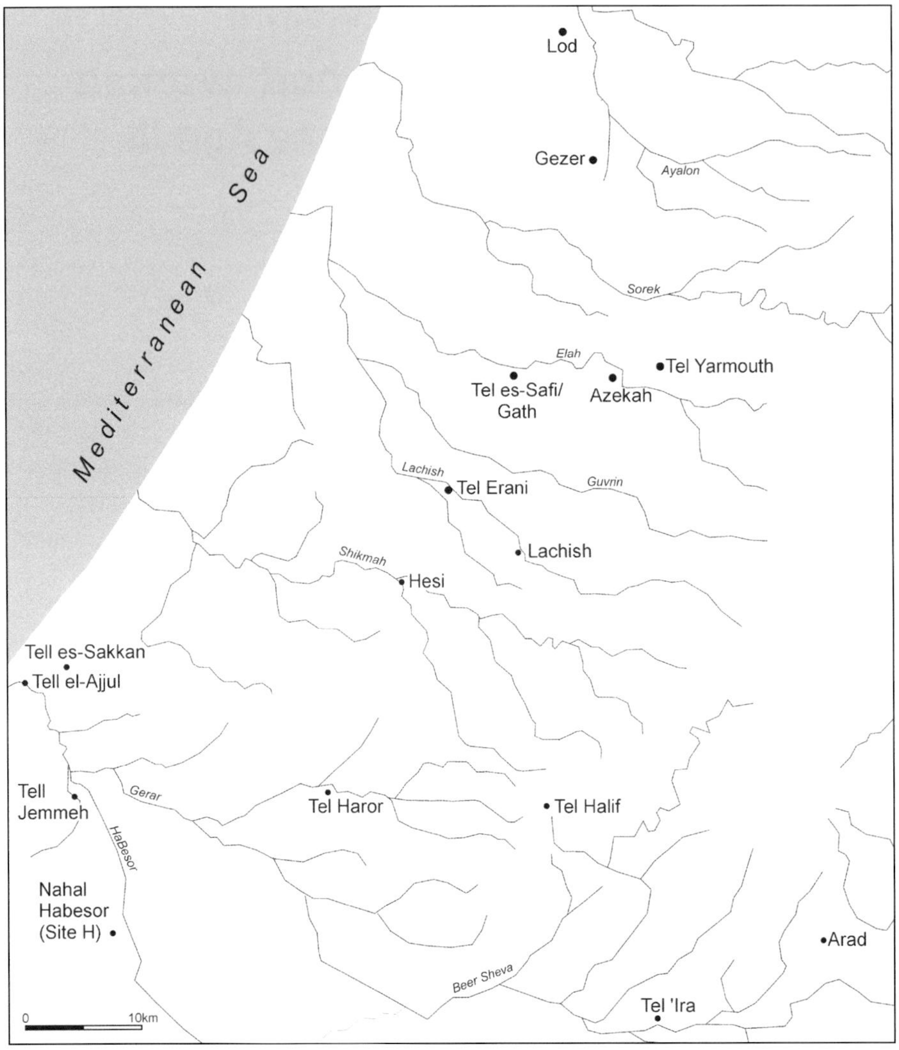

Fig. 1. Map showing location of nearby sites with EB occupations. Copyright Tell eṣ-Ṣâfi/Gath Archaeological Project.

it over the next wall into their neighbour's yard or an abandoned house, they are continuously cleaning up their domestic environments (e.g. David and Kramer 2001; Kramer 1982).

As a result of these findings, we have begun to question the usual assumption that there is a direct relationship between animal remains in early urban sites and the traditional assumption that they represent ancient food production, consumption and/or disposal behaviour. As the research described in this paper will demonstrate, this might not always be the case, since a substantial proportion of the bone debris recovered in domestic urban contexts from a Near Eastern tell site reflect other cultural uses, such as bone tool or ornamental use. This leads us to question the nature of bone tool assemblages in early urban assemblages. By employing a microscopic approach to bone analysis (with the use of even a simple light optical binocular microscope), one can often see patterns of use (e.g. polish) that are not usually visible to the naked eye (Killick 1996; Maeir *et al.* 2009). the bones are examined more slowly and individually. It is important to use the full panoply of technical resources that are at our disposal. The faunal remains recently recovered from the Early Bronze Age (EB) layers at the southern Levantine site of Tell eṣ-Ṣâfi/Gath (Israel) will be used to demonstrate that a taphonomic approach to the interpretation of faunal remains is essential.

The Context – Tell eṣ-Ṣâfi/Gath

Geographically, Tell eṣ-Ṣâfi/Gath is located halfway between the coastline and the pinnacle of the central hill spine where Jerusalem sits (Fig. 1). It is on the edge of the foothills (*Shephelah*) and the coastal plain.

The site is currently under excavation and analysis is ongoing. It is being excavated in a modern scientific fashion, with a tightly controlled radiocarbon and typochronology, strict excavation and recovery controls, systematic sampling of all types of deposits, and incorporation of scientific specialists into fieldwork. Chronologically, the occupation at the site extends from the modern era to the Chalcolithic of the region (Maeir 2012).

Systematic surface collection and excavations across the site shows that the EB II/III occupation covers the entire Tell, when it becomes one of the largest EB settlements (c. 24 ha) in this region (Shai *et al.* 2014; Uziel and Maeir 2005, 2012). During this period, it emerges as a major urban centre for the

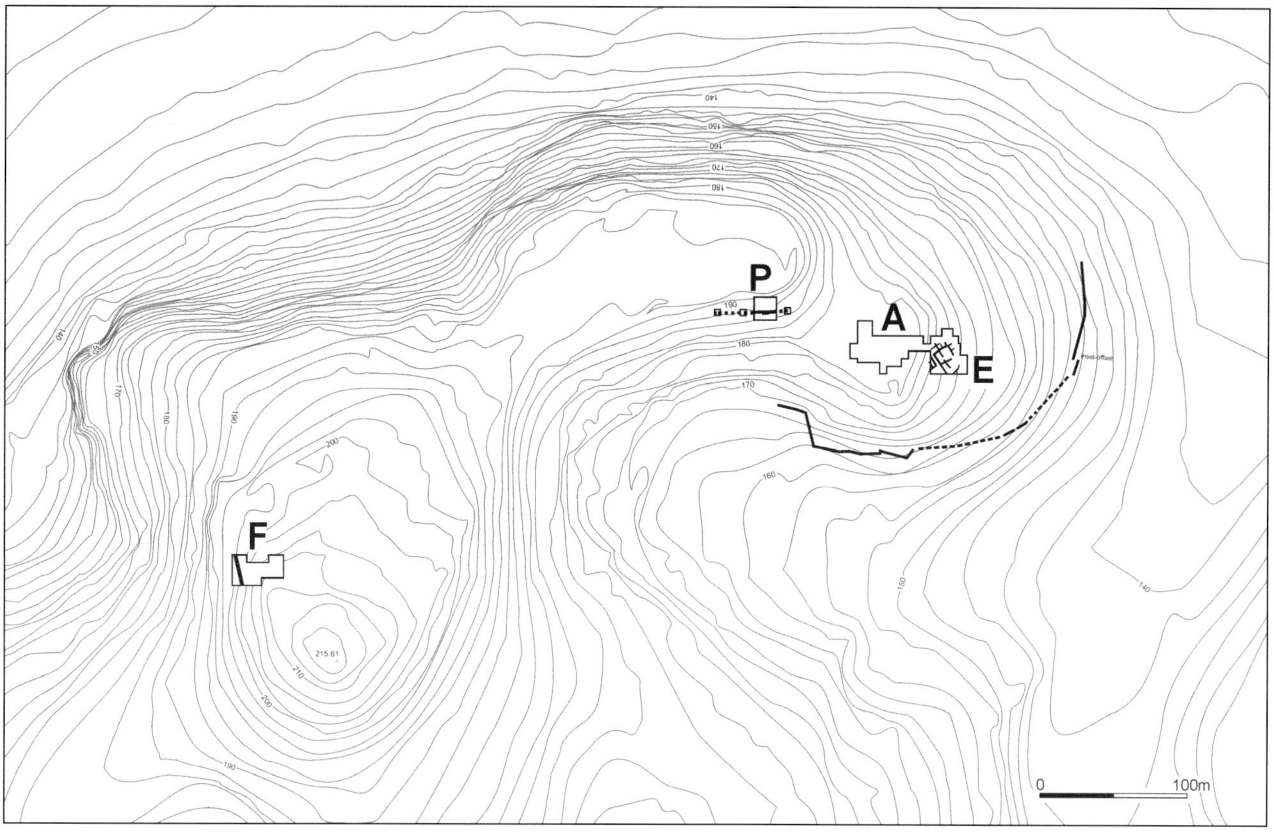

Fig. 2. Map of Tell eṣ-Ṣâfi/Gath showing the location of excavated EB features and fortification. Copyright Tell eṣ-Ṣâfi/Gath Archaeological Project.

region, and is surrounded by an extensive fortification system (Shai *et al.* 2016). It is positioned near the top of the regional settlement hierarchy in terms of size and becomes a major urban centre for this region. The regional political system is best characterised as one of competing peer-polities (cf. Renfrew and Cherry 2009), rather than a single integrated political unit (e.g. kingdom) (de Miroschedji 2009).

Area E – an EB III 'lower stratum' neighbourhood

Our research has focused on the east end of the Tell (Excavation Area E) where a tongue of the EB occupation sticks out beyond the later occupations at the eastern end of the site (Fig. 2). In this part of the site, the EB domestic occupation in is not deeply buried beneath thick later deposits (Maeir 2012). The excavations of Area E uncovered the remains of a non-elite domestic neighborhood from the later EB III (IIIB) (Fig. 3). The remains of at least four building complexes from the latest EB III stratum (Stratum E5) have been partially or totally uncovered to date. One is on the west side of an alley, while the other three are on the east side. Even though the three buildings on the east side have adjoining walls and no clear doorways, they are separated into different building complexes based on a number of criteria, including differential wall thickness and nature of construction, and the distribution features (e.g. hearth) associated with food preparation. Each complex includes courtyards, rooms (for cooking, sleeping, and storage), and various installations (Shai *et al.* 2014).

Radiocarbon dates for the later EB III occupation have been obtained through high precision dating and Bayesian modeling. It relies upon the use of short-lived organics, such as olive pits that derive only from very secure depositional contexts that reflect their use and/or discard. In general, the 1 sigma calibrated range for the EB III occupation at the site is c. 2850-2550 BC (Regev 2013; Shai *et al.* 2014).

The EB occupation is built on a gently sloping lower terrace of the Tell and follows the natural slope of the terrain. The structures descend in elevation from NW to SE. This becomes important when we examine the bone remains from an alleyway that divides the two building complexes. The remains in the alleyway were likely exposed to torrential rain since the runoff made a V- or U-shaped depression through the centre of the alley. To counter the effects of erosion from rain and likely to make the alleyway a more comfortable walking surface, the ruts were periodically and purposefully filled in with small and medium-sized stream pebbles, broken ceramics, and other artefacts and ecofacts (bones and carbonised plant remains).

Fauna

The fauna described here derive from Stratum E5 (later EB III) as a whole (including the various phases in this stratum) and are treated as a single

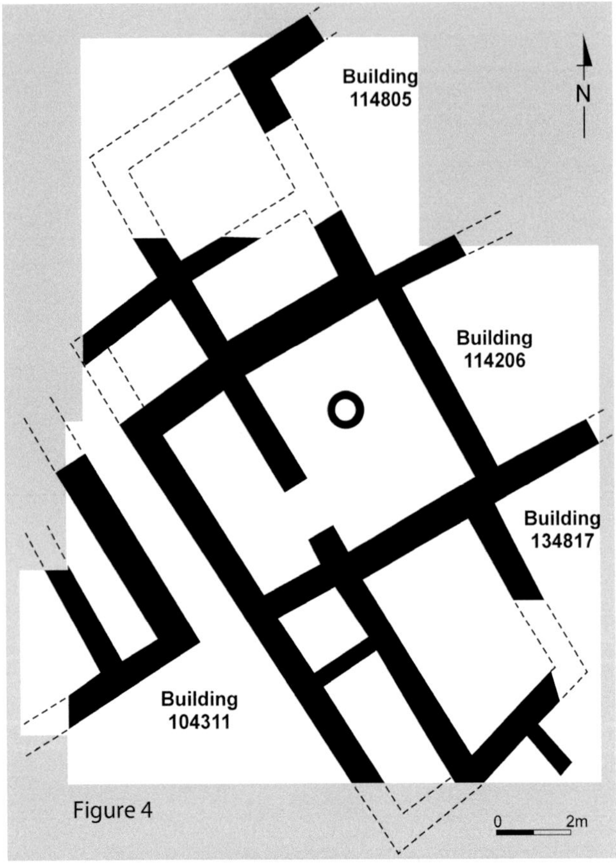

Fig. 3. Plan of Tell eṣ-Ṣâfi/Gath Area E buildings from Stratum 5a-c. Copyright Tell eṣ-Ṣâfi/Gath Archaeological Project.

analytical unit to ensure chronological integrity and avoid later intrusions. It includes the fauna collected and analysed during the 2004-2010 excavations.

Bone taphonomy

Faunal recovery at the site is extremely systematic. All excavated sediments from primary contexts (floors, accumulations immediately above floors) are dry sieved with a 1 cm^2 mesh or subjected to systematic floatation. Those from secondary deposits (fill from collapsed upper stories of houses) are selectively sieved and/or floated. Tertiary deposits (fill brought in from elsewhere for construction purposes) are not sieved. Hence, excavator bias in recovery is not a factor. The samples discussed here all derive from primary and secondary contexts.

Similarly, weathering does not appear to have been a large-scale cause of bone attrition. The vast majority of remains exhibited only light weathering (93.76%), with very few medium (6.05%) and even fewer with heavy weathering (0.18%).

Bone integrity does not appear to be impacted by soil acidity since a relatively base pH has been recorded for most sediment and for the site in general (c. 6-8 pH). A testament to the fact that soil pH is not a variable in bone preservation is that the deposits are also very rich in carbonised plant remains (Frumin and Weiss 2014).

While the soil from the EB III horizon can be described as a sandy loam, the amount of sand in the sediment does not appear to have heavily impacted the assemblage since most bones still have their outer surface (*substantia compacta* for diaphyses and *substantia corticalis* for epiphyses) intact.

Even so, it is difficult to conduct many types of zooarchaeological analyses on the bulk of specimens since the bones readily fragment into small pieces when uncovered. This is evident from the high number of fragments with modern breaks (75%). Given the high level of preservation of the outer surface of bones, the cause of fragmentation is suspected to be a result of the very chalky nature of the soil. The chalky components creates retains moisture, despite the fact that the deposits lie atop an elevated rise. The chalk component in the EB sediments, with its water retention capability, seems to have created a situation where the soil conditions are less conducive to bone preservation in comparison to the later deposits at the site. In consequence, the high breakage level is not a result of careless recovery, curation, etc. It is a function of the changes wreaked on the specimens by the very chalky soil conditions.

Culturally worked bones at Tell eṣ-Ṣâfi/Gath

In most faunal reports from the region for this period, morphologically modified or worked bone (tool/ornaments) are usually a small component of faunal assemblages (1-5%) (e.g. Davis 1988; Horwitz 1997, 2003; Horwitz *et al.* 1996; Horwitz *et al.* 2002). In contrast, in the EB deposits from Tell eṣ-Ṣâfi/Gath, more than a quarter (28.5% of total assemblage; NISP = 637) are probably fragments of culturally modified bones in the form of tools, ornaments, decorations, gaming pieces, etc. This frequency is substantially higher than is normally reported from sites in the region. This is probably a result of the pressure to analyse the bone assemblage as rapidly as possible in order to remain within budget and time guidelines. As a result, there are very few faunal reports that systematically record the intensity of use wear polish as opposed to morphologically identifiable worked bone specimens.

Two basic types of culturally worked bone remains were recovered from the EB layers at Tell eṣ-Ṣâfi/Gath. These include the better known and more easily objects that are morphologically identifiable as tools or ornaments, or fragments of their manufacture. The second type of culturally modified bone is the very fragmentary, and not morphologically identifiable as tools/ornaments. These are more difficult to photograph and describe. They are often only identifiable based on microscopic use wear polish that is differentially distributed across the bone. We will describe each of these types next and their

Table 1. Frequency distribution of cultural modified osteological remains (NISP).

Type of modification	*NISP*	*%*
Morphologically modified with polish	92	4.1%
Polished only	490	21.9%
Total cultural modified	**582**	
Not modified	1654	74.0%
Grand Total	**2236**	**100.0%**

Table 2. Frequency of culturally modified objects by general type (NISP).

Type of modification	*NISP*	*% of total*
Morphological subtotal	**92**	**15.8%**
Decoration	3	0.5%
Decoration?	1	0.2%
Decoration/ornament	1	0.2%
Ornament	3	0.5%
Tool	57	9.8%
Tool?	2	0.3%
Tool/ornament	3	0.5%
Tool/ornament/decoration	2	0.3%
Toy	9	1.5%
Toy?	11	1.9%
Polish subtotal	**490**	**84.2%**
Tool	419	72.0%
Tool?	20	3.4%
Toy?	1	0.2%
Unknown	50	8.6%
Grand Total	**582**	**100.0%**

implication for understanding of the interpretation of bone remains from EB domestic contexts.

Morphologically identifiable worked bone

The number of bones that are morphologically identifiable as culturally modified objects, such as tools/ornaments, at Tell eṣ-Ṣâfi/Gath is quite small overall, but still represents a significant part of the assemblage (n=92; 4.1%; Table 1).

The frequency distribution of culturally modified objects (e.g. tool types) can be divided into general categories, such as decorations (part of the decoration of something larger), ornaments (jewellery), tools, toys, etc. Some objects could not be assigned to such categories (Table 2). The vast majority of morphologically identifiable culturally modified objects by general type, including possible ones, are tools and possible tools, including a cylinder seal (N=59; 64% of all morphological tools). This is followed by various toys (e.g. gaming pieces) and possible toy pieces (N=20; 22%), after which there is a significant drop in frequency to the very few ornaments (N=3, 3.2%), possible decorations (N=4; 4.4%), and decoration/ornament fragments (N=1;

Table 3. Frequency of culturally modified remains identified by morphology. polish. and type (NISP).

	Identified primarily based on:			
Type of modification and functional category	***Morphology*** **NISP**	***Polish only*** **NISP**	***Morphology*** **%**	***Polish only*** **%**
Unknown		49	0.0%	10.0%
Decoration				
Unknown	2		2.2%	0.0%
Cap of something?	1		1.1%	0.0%
Decoration?	1		1.1%	0.0%
Decoration/ornament	1		1.1%	0.0%
Ornament				
Unknown	1		1.1%	0.0%
Ornament	2		2.2%	0.0%
Tool				
Unknown	21	419	22.8%	85.5%
Awl	5		5.4%	0.0%
Awl/needle	1		1.1%	0.0%
Awl?	3		3.3%	0.0%
Flat tool	1		1.1%	0.0%
Handle?	11		12.0%	0.0%
Point	1		1.1%	0.0%
Punch?	2		2.2%	0.0%
Rubber	2		2.2%	0.0%
Rubber. scraper	1		1.1%	0.0%
Scoop. scraper	1		1.1%	0.0%
Scraper?	1		1.1%	0.0%
Shovel	4		4.3%	0.0%
Spatula	2		2.2%	0.0%
Cylinder seal	1		1.1%	0.0%
Tool?	2	20	2.2%	4.1%
Tool/ornament	1		1.1%	0.0%
Tool/ornament	2		2.2%	0.0%
Tool/ornament/ decoration	2		2.2%	0.0%
Toy				
Toy		1	0.0%	0.2%
Game piece	6	1	6.5%	0.2%
Game piece?	11		14.2%	0.0%
Dice	1		1.1%	0.0%
Grand Total	**92**	**490**	**100.0%**	**100.0%**

1%). The remainder are even more ambiguously labelled as tool/ornament/decorations (N=5; 2.3%).

Tools

Most of the morphologically identifiable tools are assigned 'functions' based on their shape and use wear, which may not directly correspond with their entire range of possible functions. The relatively common types include awls, spatulas, handles, scraper, punch, rubber, and shovel/scraper/dust pan (Table 3). Most of the tool pieces could not be assigned to a particular function since they were too fragmentary (N=21; 23%). The next most common category was possible handles (N=11; 12% - Fig. 4), followed by fragments of awls/needles, (Fig. 5) and points (N=10; 11%

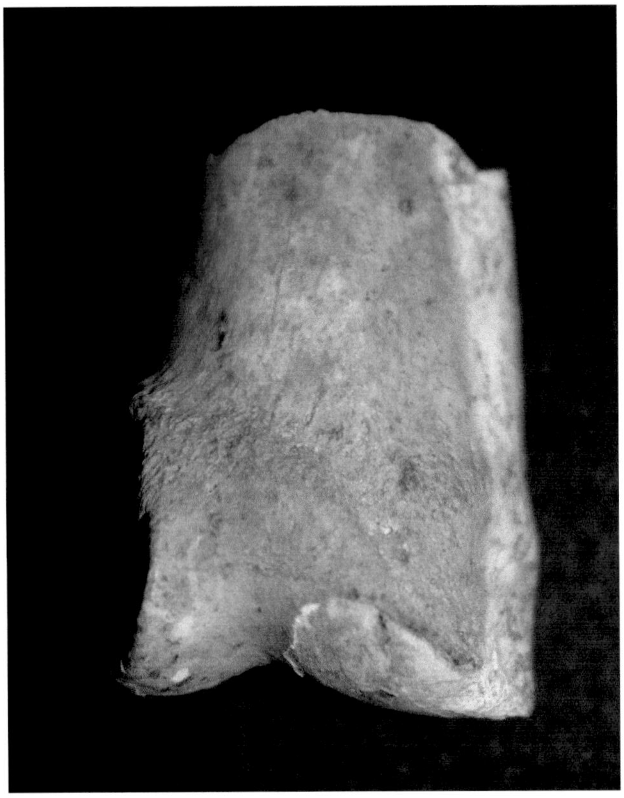

Fig. 4. Possible handle - *Bos taurus*, phalange I with surface polish (Area E, L114203, B1142019, bone # 1, Stratum E5a), Photo 3336. Copyright Tell eṣ-Ṣâfi/Gath Archaeological Project.

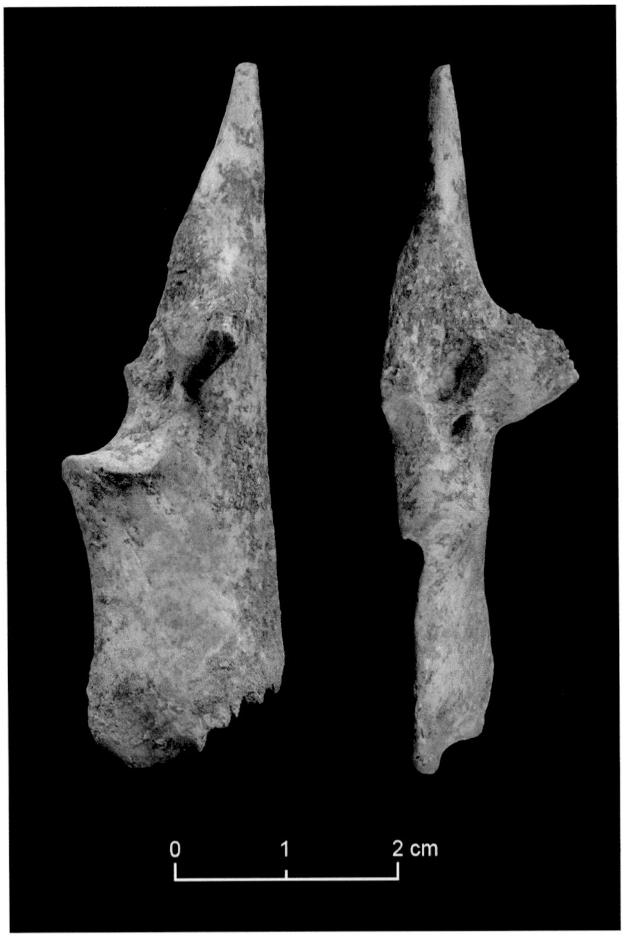

Fig. 5. Awl – *Gazella gazella*, ulna modified and with surface polish (Area E, L134304, B1343012, bone # 1), Photo 1343012. Copyright Tell eṣ-Ṣâfi/Gath Archaeological Project.

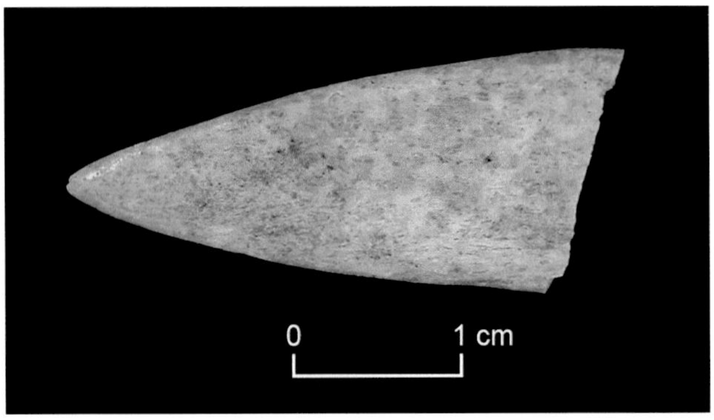

Fig. 6. Point - Large mammal long bone shaft (Area E, L 104108, B 1041034, bone #13, Stratum E5a), Photo 3204. Copyright Tell eṣ-Ṣâfi/Gath Archaeological Project.

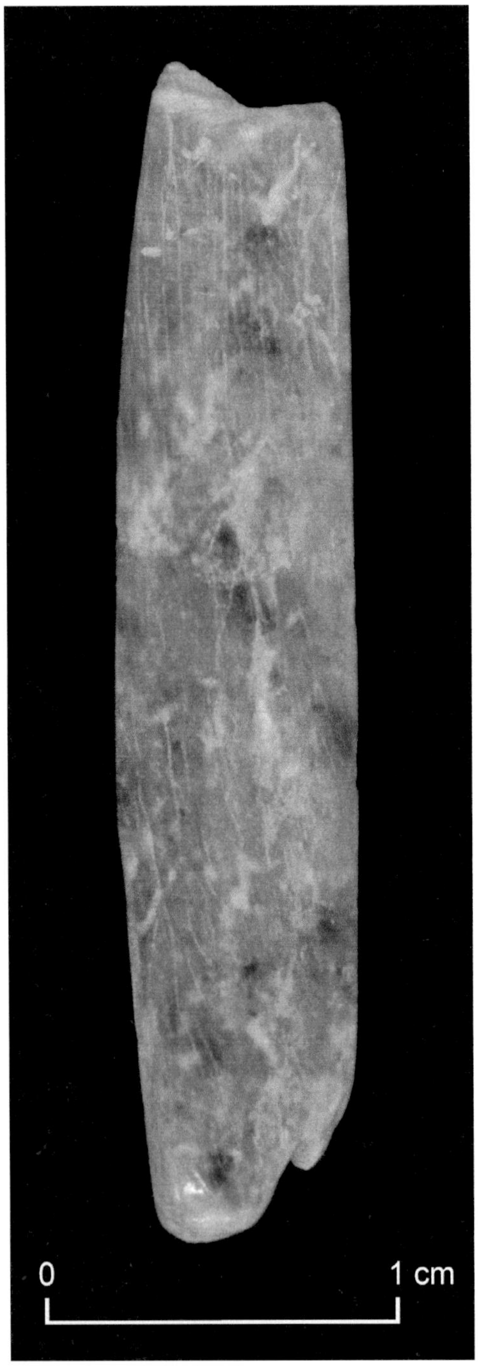

Fig. 7. Medium intensity polished specimen - Medium mammal (probably *Ovis/Capra*) rib shaft (Area E, L114105, B1141015, bone #5), Photo 3201. Copyright Tell eṣ-Ṣâfi/Gath Archaeological Project.

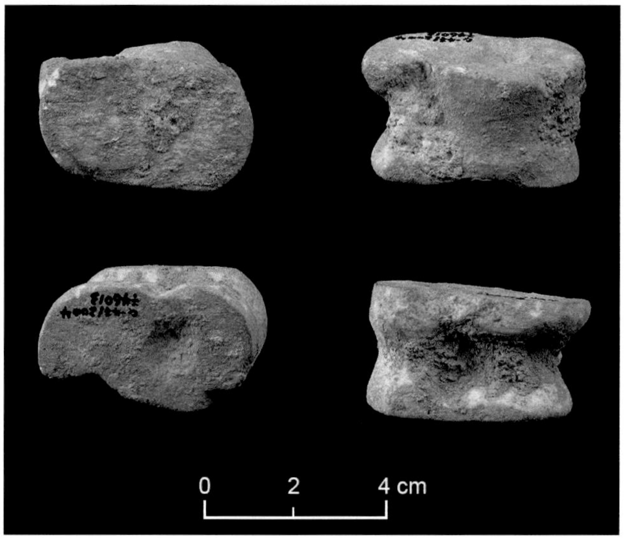

Fig. 8. Possible gaming specimen – culturally modified and polished *Bos taurus* astragalus (Area E, L 746013, B 1141015, bone #5), Photo 746013. Copyright Tell eṣ-Ṣâfi/Gath Archaeological Project.

- Fig. 6). No needles were found in the assemblage. Given this, we suspect that all the indeterminate awl/needle fragments actually belong to awls.

Several fragments that probably belong to a difficult to distinguish group that includes spatulae, scoops, scrapers and shovels/dust pans (N=8; 8.5%) were also observed. Tools showing evidence of having been used in a rubbing manner (possibly for ceramic production or leather manufacture) were also observed (N=3; 3.2% - Fig. 7).

Game pieces

There are several small and hard bone elements that were modified and/or exhibit use wear polish, particularly calcanei and astragali, carpals and tarsals, and phalanges (Fig. 8). Some are intentionally shaped, such as the astragali that alter their dimensions. The size, shape and use-wear polish of these bones suggest that they were used as gaming pieces (Gilmour 1997; Kisos 2014) or for divination (Minniti and Peyronel 2005; Plug 1987).

At least three game board fragments have been found thus far. They are made of chalk, probably locally produced. They have vertical and horizontal lines incised on them. In addition, there are a few small chalk pieces that may be the gaming pieces. The shape of the board is reminiscent of the well-known Egyptian game called *Senet*. It is the most common type of gaming board known from the EB II-III in the southern Levant (Sebbane 1991: 233–236). Recreation was clearly part of life during the EB of Tell eṣ-Ṣâfi/Gath.

Ivory

Two types of ivory have been found in the EB excavations. First, there is an exquisitely carved hippopotamus ivory cylinder seal. This is a very rare find for the EB of the southern Levant. It was found *in situ* in one of the residential complexes (Building 114805, Phase E5c – Fig. 9) (Maeir *et al.* 2011). It depicts a lion standing in dense vegetation. The skill level to make it suggests that the seal was made by a highly proficient artist. The quality of the raw material and the drawing also suggest that the owner was of relatively high status. The presence of such items hints that the socio-economic organization of the local inhabitants was relatively complex and would have included administrative or exchange specialists.

The cylinder seal was discovered within a nearly complete holemouth jar with four beads (three were made of faience and one was made of carnelian), which may indicate that they were part of a necklace. It is similar to one found at Megiddo (Brandl 2013). It is important to note the rarity of such items, particularly with their absence at nearby Tel Yarmuth, often considered the primary site in this region (de Miroschedji 1999: 14, 2009).

Another piece of ivory was also found in the excavation area. They are found in a different house (Locus E15AT03, Basket E15AT020 - Building 104311 – Fig. 10). It was also made of hippopotamus ivory, but its function could not be determined. It is sawn into a small well-defined piece, rounded in shape, and drilled through the centre. It almost seems as if it was the cap at the end of a rod or stick. It was clearly part of something larger.

Hippopotamus ivory appears to be a highly desired non-local trade good. It did not derive from the immediate vicinity around the site. Yet, it could have been from hippopotamus populations in the larger region since local Levantine hippopotamus populations were present in the region until the Iron Age (Horwitz and Tchernov 1990). Alternatively, the ivory may have derived from much farther away, such as the Nile Valley. Only isotopic studies may eventually answer whether the ivory came from the hippopotamus coastal populations or much further afield.

The ivory finds are best interpreted as evidence for the participation of the inhabitants of the site in intra- or inter-regional exchange systems. The location of the site is probably a reflection of this since it is strategically positioned to monitor movement along the two major routes – the north-south Inland Coastal Plains Highway that runs along the edge of the boundary between the coastal plain and the *Shephelah* (foothills). This highway is drier and easier to traverse, the rivers and streams are easier to cross, and has access to clean water, shade and pasture. It skirts that poor water and swampy coastline of the region with its better known *Via Maris* (Way of the Sea). The second highway that runs past the site is the route from the coast into the central hills,

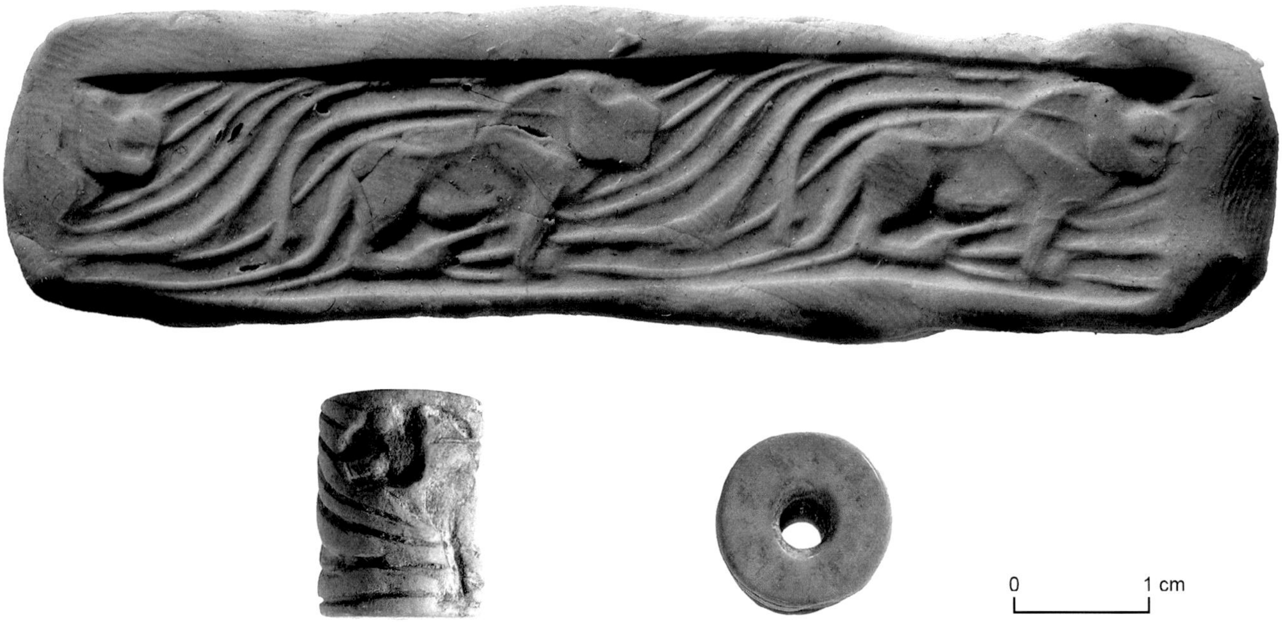

Fig. 9. *Hippopotamus amphibius* ivory cylinder seal (Area E, L 84301, B 843005, bone #1). Copyright Tell eṣ-Ṣâfi/Gath Archaeological Project.

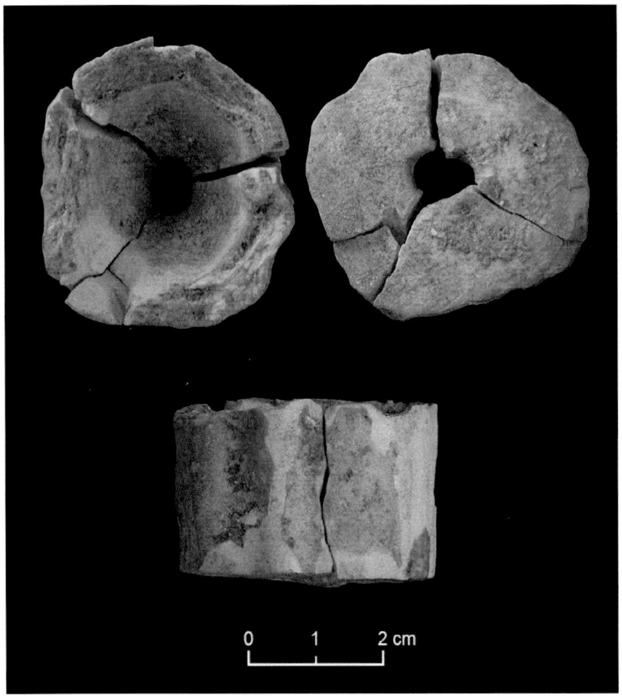

Fig. 10. *Hippopotamus amphibius* ivory rounded cap (Area E, L E15AT020, B E15AT020, bone #1). Copyright Tell eṣ-Ṣâfi/Gath Archaeological Project.

that follows the Elah River valley up to the general region of Bethlehem (Rainey and Notley 2006).

The presence of non-local bone remains, such as ivory, is usually indicative of the wealth and importance of a site's inhabitants. Yet, the material culture in this part of the site suggests that they were not residences of the elite, but possibly merchants, which would explain the presence of a rare administrative objects such as a cylinder seal. Clearly, expensive and possibly exotic trade goods were being brought into the site from outside of the immediate region. Their discovery in a commoner's house is indicative that such goods were not simply signatures of elite trade networks.

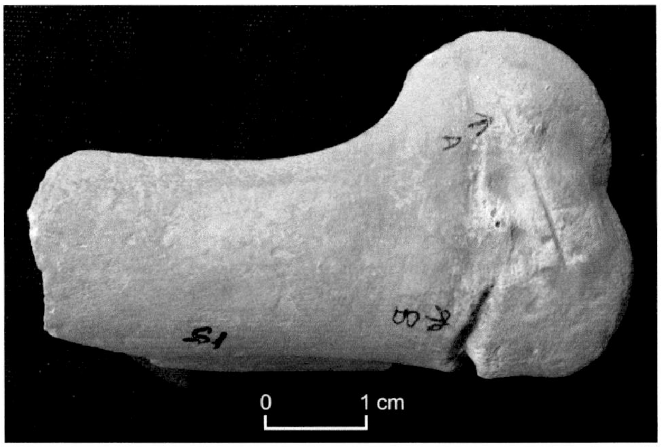

Fig. 11. *Ovis aries* (Sheep) humerus with butchering marks, but no evidence of cultural use wear polish (Area E, L 74505, Basket 745020, bone # 15) Photo 1700. Copyright Tell eṣ-Ṣâfi/Gath Archaeological Project.

Unidentifiable fragments

Most worked bone is recognisable only through the presence of use wear polish since they are often too fragmentary to retain the tool or ornament morphology. Yet the literature often ignores the importance of identifying and recording use wear polish as part of the regular training (Jin and Shipman 2010). A reflection of this is that it is not mentioned in any of the commonly cited textbooks on zooarchaeology (Davis 1987; Hillson 1996; O'Connor 2000; Schmid 1972). Only recently is polish acknowledged as important, but is largely attributed to the effects of canid digestion or the sedimentary matrix on the bone (Reitz and Wing 2008: 46, 136).

Table 4. Frequency distribution of cultural modified osteological remains (NISP).

Type of modification	*NISP*	*%*
Morphologically modified with polish	92	4.1%
Polished only	490	21.9%
Total cultural modified	582	
Not modified	1654	74.0%
Grand Total	**2236**	**100.0%**

Cultural polish is best identifiable by observing the differential distribution and intensity of polish across all surfaces of the bone. Most bones in an assemblage do not have any polish (Fig. 11). For example, if the polish is equally present on all (or most) surfaces, the polish might be simply from exposure to natural processes, such as running water; if the polish is generally restricted to the exterior bone surfaces, this would signal that it is cultural polish (the object may be part of a tool or ornament) since both surfaces of the bone are not equally modified through use; or if there is a higher intensity of polish where use wear marks exist, this may signal the function and location of most intensive use of the object.

Cultural or use wear polish appears almost as a polished marbled or glassy surface. It is usually smooth to the touch and very shiny (Jin and Shipman 2010; Shipman 1989). Such polish was identified on over a quarter of the total assemblage (N=582; 26% - Table 4, Fig. 12 – see gloss on shaft).

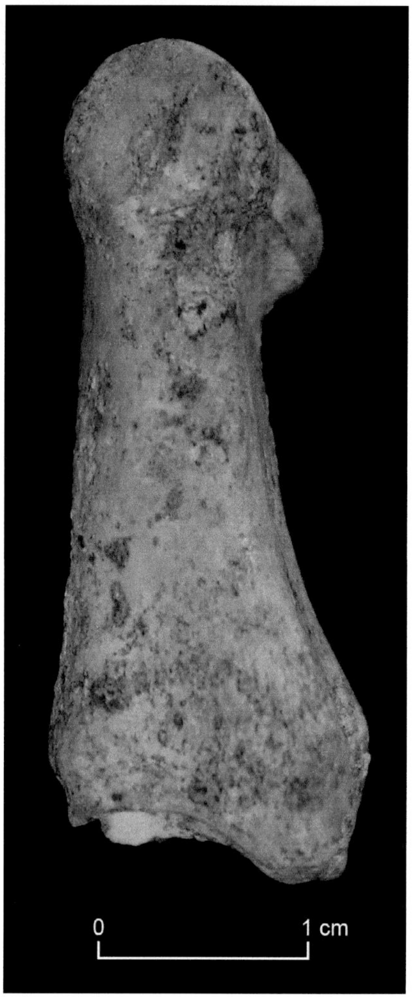

Fig. 12. *Capra hircus* phalange 1 with surface polish (Area E, L 104308, B1043043, bone # 1) Photo 3322. Copyright Tell eṣ-Ṣâfi/Gath Archaeological Project.

Morphological identifiable tools versus only identifiable through use wear polish

In most assemblages, the frequency of morphologically recognisable bone tools tends to be quite small. But, if the fragmentary polished remains are included in the analysis, the sample size grows dramatically. As noted before, the number of bones that are morphologically identifiable as tools at Tell eṣ-Ṣâfi/Gath is quite small (n=92 – Table 1). In contrast, 5x as many are distinguishable through use wear polish patterns (N=490, which represent 84% of the culturally modified assemblage – Table 2).

Polish and function

All specimens that were morphologically modified by shaping also exhibited polish. Two had very little polish (i.e. astragali), and were probably only briefly used (probably as toys or possibly for ritual practices; Gilmour 1997; Minniti and Peyronel 2005). The vast majority (N=418; 85% of polished specimens; Table 4) were part of some kind of tool/ornament, but their overall function could not be discerned (plus another twenty specimens as probably parts of tools/ornaments).

Polish implication

The frequency of polished remains at Tell eṣ-Ṣâfi/Gath may suggest that the reason we find bones in houses may be different than normally assumed – that they are not all directly related to food processing and consumption. It is suggested here that they represent the remains of bone tool use, particularly of those bones that fragmented during use and/or the final abandonment of houses.

Conclusions - What do these bones mean?

In conclusion, our excavations in Area E of Tell eṣ-Ṣâfi/Gath have demonstrated the presence of an EB III domestic quarter with several houses and a small street. There is no evidence of industrial activities. The houses were built and maintained over a long period of time. They are typical multi-room structures characteristic of the period, with 1-2 rooms associated with a courtyard, and in few cases also a small storage room. The courtyard was used for domestic activities, which included grain storage and other working platforms (stone built circular installations). This is a common layout in EB urban centers.

It is clear from the differential distribution of remains across the neighborhood that the inhabitants of this neighbourhood did not discard their garbage into their long-lived and well-maintained solid multi-room structures. For example, there is no evidence of middens in the houses, nor significant clusters of bone remains that might represent purposeful dumping episodes. All the structures in the neighbourhood were inhabited contemporaneously until the entire site (including this area) was abandoned c. 2500 BC. It would appear that they discarded most of the food remains either in the nearby alleyway or elsewhere - on or off the site. Only further excavation of this occupational horizon in other parts of the site may answer this question more fully.

It is also clear that the inhabitants of EB III Tell eṣ-Ṣâfi/Gath did not employ metal tools for processing of animal remains, or modifying bone remains into tools, ornaments or other functions. Such activities were restricted to stone tools. Chipped stone tools were used primarily for butchering and cleaning of bones for tool/ornament manufacture. Grinding was used for shaping of bone tools.

When there is a significant disparity between the frequencies of what are traditionally recognised as bone tools/ornaments and the much greater number of bones with evidence of use wear polish, the nature of interpretations of the material must change. We must recognise that not all bone accumulations in and around houses are reflections of subsistence. The bones may have a secondary use, such as tools/ornaments/gaming pieces. In our case study, there are far fewer formal tools in the assemblage in comparison to those identified through use wear polish. If such a large number of bones with evidence of such cultural modification are present in houses, this suggests that a large proportion of the bones recovered in houses were probably from fragmentation resulting from these secondary uses, and not necessarily from food preparation and consumption as is commonly assumed. Hence, a more substantial proportion of assemblages might be related to bone tool/ornament creation, use, fragmentation, loss, and discard. The importance of bone in the pre-modern world cannot be underestimated. It is the 'plastic' of antiquity, since it could be modeled and shaped for many functions and roles.

The very high frequency of use wear polish at Tell eṣ-Ṣâfi/Gath is a far larger number (26%) than normally reported in the comparative literature and may suggest that we need to consider several important questions in future studies. Some of these questions include:

- 'Why is there so much garbage littering the floors of houses?'
- 'Do they really represent the remains of food discarded randomly of people while they are still occupying the structure or could they represent the remains of bone tools that fragmented or were abandoned during use in the house'?
- Why would people clean their living environment from food consumption refuse, but not from the remains of fragmented bone tools?
- In other words, 'what do these bones mean?'

Such issues are rarely addressed and can dramatically affect the interpretation of faunal and other remains from urban settlements.

In conclusion, we propose that the nature of zooarchaeological practice taking place throughout the region (which does not systematically include microscopic use wear analysis) and the often simplistic interpretation of osteological remains as food residues commonly found in the literature (*mea culpa*, included) must change. It is necessary to conduct detailed microscopic studies in order to be able to identify and interpret the remains. We must begin to reconsider the nature of animal remains recovered, particularly from urban domestic contexts, since they may represent a very different set of behavioural actions than simply the left-overs of food consumption. The 'food' residues found on and above the floors may instead represent early examples of recycling of food residues into bone tool and ornaments and their ultimate discard or loss.

Acknowledgements

We would like to acknowledge the following organizations for generously funding our research: The Tell eṣ-Ṣâfi/Gath Archaeological Project and its staff and team members; Bar-Ilan University and the Department of Land of Israel Studies and Archaeology at Bar-Ilan University; the Social Science and Humanities Research Council of Canada (Grant #895-2011-1005); Faculty of Arts, St. Paul's College and Near Eastern and Biblical Archaeology Laboratory of the University of Manitoba; St. John's College and the McDonald Institute of the University of Cambridge; and Jewish and Catholic Foundations of Manitoba. Of course, we gratefully thank the host of colleagues and students for all their efforts. Without their unstinting support, this research would not have been successfully carried out. Any errors are the responsibility of the authors.

The overall excavations of the site are directed by Prof. Aren Maeir, while the excavations of the Early Bronze Age occupation atTell eṣ-Ṣâfi/Gath are conducted as a joint operation by the University of Manitoba and Bar-Ilan University. The directorship is shared by Professors Haskel Greenfield and Aren Maeir, while Prof. Itzhaq Shai served as Area Supervisor.

References

Allentuck, A. 2004. *Production, Distribution and Consumption of Animals at Early Bronze Age Titriş Höyük, Southeastern Turkey: A Zooarchaeological Approach.* Winnipeg: University of Manitoba.

Allentuck, A. 2011. Raw material availability and technological choice: modified metapodia from an Early Bronze Age Site in central Israel. *International Journal of Osteoarchaeology* 23: 379-394

Allentuck, A. 2014. Temporalities of human–livestock relationships in the late prehistory of the southern Levant. *Journal of Social Archaeology* 15(1): 94-115.

Allentuck, A. and Greenfield, H.J. 2010. The organization of animal production an early urban center: the zooarchaeological evidence from Early Bronze Age Titriş Höyük, southeast Turkey. In, Campana, D. V., Crabtree, P. J., DeFrance, S. D., Lev-Tov, J. and Choyke, A. M. (eds) *Anthropological Approaches to Zooarchaeology: Colonialism, Complexity and Animal Transformations*: 12-29. Oxford: Oxbow Press.

Atici, L. 2004. Centralized or decentralized: the mode of pastoral economy at Early Bronze Age Kaman-Kalehöyük. *Anatolian Archaeological Studies* 14: 119-127.

Berthon, R. 2011. *Animal exploitation in the Upper Tigris River valley Turkey between the 3rd and the 1st millennia.* PhD. Ur- und Frühgeschichte Institüt, Christian-Albrechts University, Kiel, Germany.

Brandl, B. 2013. Cylinder seals. In, Finkelstein, I., Ussishkin, D. andCline, E.H. (eds) *Megiddo V: The 2004–2008 Seasons,* Vol 30: 993–1010. Tel Aviv: The Institute of Archaeology of Tel Aviv University.

Choyke, A. M. 2000. Bronze Age antler and bone manufacturing at Arslantepe (Anatolia). In, Mashkour, M., Choyke, A. M. and Buitenhuis H. (eds) *Archaeozoology of the Near East,* Vol 32: 170-183. Groningen: ARC Publication.

David, N. and Kramer, C. 2001. *Ethnoarchaeology in Action.* Cambridge: Cambridge University Press.

Davis, S. J. 1987. *The Archaeology of Animals.* New Haven: CT, Yale University Press.

Davis, S. J. 1988. The mammal bones from Tel Yarmuth 1980-1983. In de Miroschedji, P. (ed.) *Yarmouth 1, Rapport sur les trois Premières Campagnes de Fouilles à Tel Yarmouth (Israel), 1980-1982*: 143-149. Paris: Association Pour la Diffusion de la Pensée Française, Editions Recherche sur les Civilisations.

De Cupere, B., Duru, R., Umurtak, G., Choyke, A. and Buitenhuis, H. 2006. Animal husbandry at the Early Neolithic to Early Bronze Age site of Bademagacì (Antalya province, SW Turkey): evidence from the faunal remains. In, Vila, E., Gourichon, L., Choyke, A. M. and Buitenhuis, H. (eds) *Archaeozoology of the Near East VIII: Proceedings of the 8th International Symposium on the Archaeozoology of Southwestern Asia and Adjacent Areas, vol. 2*: 367-406. Lyon: Archéorient, Maison de l'Orient et de la Méditerranée.

de Miroschedji, P. 1999. Yarmuth: the dawn of city-states in southern Canaan. *Near Eastern Archaeology* 62 (1): 2-19.

de Miroschedji, P. 2009. Rise and collapse in the southern Levant in the Early Bronze Age. *Scienze dell'antichità. Storia Archeologia Antropologia* 15: 101-129.

Frumin, S. and Weiss, E. 2014. Diet and Environment at Early Bronze Age Tell eṣ-Ṣâfi/Gath: The botanical evidence. *Paper presented at the ASOR 2014 Conference, San Diego.*

Gilmour, G. H. 1997. The nature and function of astragalus bones from archaeological contexts in the Levant and eastern Mediterranean. *Oxford Journal of Archaeology* 16 (2): 167-175.

Greenfield, H. J. and Allentuck, A. 2011. Neighbourhood differences in animal exploitation and consumption patterns in an early (Early Bronze Age) urban center: the zooarchaeology of Titriş Höyük, SE Turkey. In, Morton, S., Butler, D., and Reese-Taylor, K. (eds) *It's Good to Be King: The Archaeology of Power and Authority (Proceedings of the 41st Chacmool Conference, November 7-10, 2008)*: 171-180. Calgary: University of Calgary, Chacmool Archaeological Association.

Hesse, B. and Perkins, D. J. 1974. Faunal remains from Karataş-Semayük in southwest Anatolia. *Journal of Field Archaeology* 1: 149-160.

Hillson, S. 1996. *Mammal, Bones and Teeth: An Introductory Guide to Methods of Identification, 2nd edition.* London: Institute of Archaeology, University College London, Reprinted with corrections Ed.

Horwitz, L. K. 1997. Faunal remains. In E. Braun (ed.) *Yiftaḥ'el: Salvage and Rescue Excavations at a Prehistoric Village in the Lower Galilee, Israel. Israel Archaeological Antiquity Reports, No. 2*: 155-172. Jerusalem: Israel Antiquities Authority.

Horwitz, L. K. 2003. Early Bronze Age animal exploitation at Qiryat Ata. In A. Golani (ed.) *Salvage Excavations at the Early Bronze Age Site of Qiryat Ata. Israel Antiquity Authority Reports 18*: 225-241. Jerusalem: Israel Antiquities Authority.

Horwitz, L. K. and Tchernov, E. 1990. Cultural and environmental implications of Hippopotamus bone remains in archaeological contexts in the Levant. *Bulletin of the American Schools of Oriental Research* 280: 67-76.

Horwitz, L. K., Hellwing, S., and Tchernov, E. 1996. Patterns of animal exploitation. In R. Gophna (ed.) *Excavations at Tel Dalit*: 193-216. Tel Aviv: Tel Aviv University and Ramot Publishing.

Horwitz, L. K., Tchernov, E., Mienis, H. K., Hakker-

Orion, D., and Bar-Yosef Mayer, D. 2002. The archaeozoology of three Early Bronze Age sites in Nahal Besor, north-western Negev. In, Van den Brink, E. C. M. and Yannai, E. (eds) *Quest of Ancient Settlements and Landscapes. Archaeological Studies in Honour of Ram Gophna*: 107-133. Tel Aviv: Tel Aviv University, Institute of Archaeology.

Jin, J. J. H. and Shipman, P. 2010. Documenting natural wear on antlers: A first step in identifying use-wear on purported antler tools. *Quaternary International* 211: 91-102.

Kansa, S. W. 2004. Animal exploitation at Early Bronze Age Ashqelon, Afridar: What the bones tell us - initial analysis of the animal bones from Areas E, F and G. *'Atiqot* 45: 279-297.

Killick, D. 1996. Optical and electron microscopy in material culture studies. In Kingery, W. D. (ed.) *Learning from Things*: 204-230. Washington D.C.: Smithsonian Institution Press.

Kisos, S. 2014. Leisure activity in the EB III: game boards as a reflection of social interaction at Tell eṣ-Ṣâfi/Gath. *Paper presented at the 2014 ASOR conference*, San Diego.

Kramer, C. 1982. *Village Ethnoarchaeology: Rural Iran in Archaeological Perspective.* New York: Academic Press.

Maeir, A. M. (ed.) 2012. *Tell eṣ-Ṣâfi/Gath I: Report on the 1996–2005 Seasons. Ägypten und Altes Testament 69 (vol. 1/text).* Wiesbaden, Harrassowitz Verlag in Komnmission, Vol 1: Text.

Maeir, A. M., Shai, I., and Horwitz, L. K. 2011. 'Like a Lion in Cover': a cylinder seal from Early Bronze Age III Tell eṣ-Ṣâfi/Gath, Israel. *Israel Exploration Journal* 61 (1): 12-31.

Maeir, A. M., Greenfield, H. J., Lev-Tov, J., and Horwitz, L. K. 2009. Macro- and microscopic aspects of bone tool manufacture and technology in the Levantine Iron Age: A 9th century BCE workshop from Tell eṣ-Ṣâfi/Gath, Israel. In, Rosen, S. A. and Roux ,V. (eds) *Techniques and People: Anthropological Perspectives on Technology in the Archaeology of the Proto-Historic and Early Historic Periods in the Southern Levant*: 41-68. Paris: De Boccard.

Marom, N. and Zukerman, S. 2011. Applying on-site analysis of faunal analysis assemblages from domestic contexts: a case study from the lower city of Hazor. In, Landau-Yasur, A., Ebeling, J. R., and Mazow, L. B. (eds) *Household Archaeology in Ancient Israel and Beyond,* Culture and History of the Ancient Near East, Vol 50: 37-54. Leiden: Brill.

Minniti, C. and Peyronel, L. 2005. Symbolic or functional astragali from Tell Mardikh-Ebla (Syria). *Archaeofauna* 14: 7-26.

Nishimura, Y. 2008. *North Mesopotamian Urban Space: A Reconstruction of Household Activities and City Layout at Titriş Höyük in the Third Millennium B.C.* PhD. University of California.

Nishimura, Y. 2012. The life of the majority: a reconstruction of household activities and residential neighborhoods at the late-Third-Millennium urban settlement at Titriş Höyük in northern Mesopotamia. In, Parker, B. J. and Foster, C. P. (eds) *New Perspectives on Household Archaeology*: 347-372. Winona Lake, IN: Eisenbrauns.

Nishimura, Y. 2014. North Mesopotamian urban neighborhoods at Titriş Höyük in the Third Millennium B.C. In, Creekmore, A. T. I. and Fisher, K. D. (eds) *Making Ancient Cities: Space and Place in Early Urban Societies*: 74-110. New York: Cambridge University Press.

O'Connor, T. P. 2000 *The Archaeology of Animal Bones.* Texas: Texas A&M University Press.

O'Connor, T. P. 2003. *Analysis of Urban Animal Bone Assemblages.* London: Council for British Archaeology.

Plug, I. 1987 Analysis of witchdoctor bones. *Research NATCO Museum* 1 (3): 47-67.

Rainey, A. and Notley, R. S. 2006. *The Sacred Bridge: Carta's Atlas of the Biblical World.* Jerusalem: Carta.

Regev, J. 2013. *Chronology of the Early Bronze Age in the Southern Levant Based on 14C Dates in Relation to Context, Stratigraphy and Cultural Remains, Modeled with Bayesian Analysis.* PhD. Bar-Ilan University.

Reitz, E. J. and Wing, E. S. 2008. *Zooarchaeology, Second Edition.* Cambridge: Cambridge University Press.

Renfrew, C. and Cherry, J. F. (eds) 2009. *Peer Polity Interaction and Socio-political Change.* Cambridge: Cambridge University Press.

Sasson, A. 2010. *Animal Husbandry in Ancient Israel: A Zooarchaeological Perspective on Livestock Exploitation, Herd Management and Economic Strategies.* London: Equinox Publishers.

Schmid, E. 1972. *An Atlas of Animal Bones.* New York: Elsevier Publishing Company.

Sebbane, M. 1991. EB and MB I Board Games in Canaan and the Origin of the Egyptian Senet Game. *EI* 21: 233-238.

Shai, I., Greenfield, H. J., Eliyahu-Behar, A., Regev, J., Boaretto, E., and Maeir, A. M. 2014. The Early Bronze Age remains at Tell eṣ-Ṣâfi/Gath, Israel: an interim report. *Tel Aviv* 41 (1): 20-49.

Shai, I., Chadwick, J. R., Welch, E., Katz, J., Greenfield, H. J., and Maeir, A. M. 2016. The Early Bronze Age fortifications at Tell eṣ-Ṣâfi/Gath, Israel. *Palestine Exploration Quarterly* 148 (1): 42-58.

Shipman, P. 1989. Altered bones from Olduvai Gorge, Tanzania: techniques, problems, and implications of their recognition. In, Bonnichsen R. and Sorg, M. (eds) *Bone Modifications*: 317-334. Orono, Maine: Center for the Study of Early Man.

Uziel, J. and Maeir, A. M. 2005. Scratching the surface at Gath: implication of the Tell eṣ-Ṣâfi/Gath surface survey. *Tel Aviv (Journal of the Institute of Archaeology of Tel Aviv University)* 32 (1): 50-75.

Uziel, J. and Maeir, A. M. 2012. The location, size and periods settlement at Tell eṣ-Ṣâfi/Gath: the surface survey results. In, Maeir, A. M. (ed.) *Tell eṣ-Ṣâfi/Gath I. Report on the 1996–2005 Seasons. Ägypten und Altes Testament 69*: 173-181. Wiesbaden.

Zeder, M. A. 1988. Understanding Urban Process through the Study of Specialized Subsistence Economy in the Near East. *Journal of Anthropological Archaeology* 7: 1-56.

Zeder, M. A. 1990. Animal Exploitation at Tell Halif. in The Bronze Age Settlements at Tell Halif: Phase II Excavations, 1983-1987 by Seger, J. D. *et al*. In, Rast, W. E. (ed.) *Bulletin of the American Schools of Oriental Research. Supplementary Studies, No. 26 (Preliminary Reports of ASOR-Sponsored Excavations 1983-87)*: 1-32. Boston: ASOR.

Zeder, M. A. 2003. Food provisioning in urban societies: a view from northern Mesopotamia. In, Smith, M. (ed.) *The Social Construction of Ancient Cities*: 156-183. Washington: Smithsonian Press.

10

Animal exploitation at Iron Age Kinet Höyük (Turkey) between 700-550 BC[1]

Canan Çakırlar[2], Jort Bosman[3], Salima Ikram[4]

1 This paper was originally presented in ASWA VII (2004, Ankara)

2 University of Groningen, Groningen Institute of Archaeology, the Netherlands; c.cakirlar@rug.nl
3 Stockholm University, Department of Ecology, Environment, and Plant Sciences, Sweden
4 American University in Cairo, Department of Sociology, Egyptology and Anthropology, Egypt

Abstract

Kinet Höyük was a small Mediterranean harbor located on the northeastern corner of the Mediterranean. The advanced maritime activity and rapid acculturation of the settlement between 700-550 BC are manifest in a significant increase in pottery types originating from the Aegean realm. In regional chronology, this period is the Late Iron Age. The assemblage dating to this period reflects an animal economy that depended on animal husbandry, hunting, and aquatic resources. Fish, shellfish and turtle exploitation intensified during the period in question, assuming a commercial nature, as best attested by the abundance of the so-called murex remains used in dye production. The assemblage suggests the presence of both *Dama dama dama* (the Anatolian fallow deer) and *D. dama mesopotamica* (the Mesopotamian fallow deer), just north of the Amanus Mountains, adding information on the complex biogeographical history of the region. Biometric and ageing data are provided in Appendices 1 and 2.

Introduction

The northeastern corner of the Mediterranean is a key region in terms of the archaeology and palaeoenvironments of Southwest Asia. This is the strip of land remaining between the eastern Taurus Mountains and the Amanus Mountains, the northern end of the Great Rift, forming a transitional zone between the Syro-Levantine and the Anatolian zoogeographical regions (Fig. 1). Culturally, the region belongs to Eastern Cilicia, but it is adjacent the Cypro-Aegean cultural orbit via the Mediterranean Sea, while maintaining affinities both with Anatolian and the Syro-Palestinian policies in the adjacent Amuq Plain (Gates 2013; Gates 2015). Now part of the Hatay province of Turkey, the potential role of this region for the prehistoric and protohistoric investigations of the Eastern Mediterranean was recognized early in the early 20th century (Braidwood and Braidwood 1960; Goldman 1963; Seton-Williams 1954; Woolley 1938). Despite extensive excavations, the early 20th century projects did not produce systematic evidence pertaining to subsistence, diet or palaeoenvironments.

Interest in the region's pre- and protohistoric past was renewed in the 1990s, with updated research agendas focusing specifically on exploring subsistence economies through time (Gates 1999). Kinet Höyük was the perfect multi-period mound to achieve this aim because it was clear from its location on the coast and surface pottery that it was occupied at least since the Late Neolithic Period (Halaf Period) until the Middle Ages (the Crusader period).

This paper reports on a small segment of the zooarchaeological evidence from the intensively excavated archaeological deposits from the site, namely (Architectural) Periods 7 and 6 dating roughly to the period between 700 and 550 BC. In regional chronology, this is the Late Iron Age. As such, the paper is part of the diachronic study of animal exploitation at the settlement, which is on-going (Ikram *et al.*, in preparation). One motivation for the study was the rich nature of the zooarchaeological assemblage from Periods 7 and 6: The assemblage is unique because it is visibly richer in aquatic remains than other phases in Kinet, it is well preserved, and contains a high number of vertebrate and invertebrate specimens. What makes the study of this assemblage also interesting is the fact that we know virtually nothing about the subsistence economy of this region at this crucial time that corresponds

Fig. 1. Location of Kinet Höyük in the Eastern Mediterranean, showing the mountain ranges as well as the northern extention of the Great Rift as the Jordan Valley. Map image by Frits Steenhuisen (Groningen Institute of Archaeology).

Fig. 2. Kinet Höyük. View from the Sea. Excavation Operation E/H is visible on the slope. Note also the mound's proximity to the coastline, the breadth of the coastal plain and the moist suspended on the slopes of the Amanus. It's a typical day in late July. Photo by Tuğrul Çakar, courtesy of the Kinet Höyük excavations.

to the period of Greek colonization, as well as the intensification of long-distance, maritime trade networks in the Mediterranean in general, associated with the Phoenician expansions (Hodos 2006). It is a period defined more by entrepreneurship than strict state control. Finally, the study is a contribution to the understanding of the ancient biogeography of this intermediate region between the Levant and Anatolia at its Mediterranean interface.

The site and its environment

The site is situated on a narrow coastal plain (Erzin=Issos Plain) defined by the Amanus Mountains to the east and by the Mediterranean Sea (Iskenderun Gulf) to the west (Fig. 2). It has limited access to the Amuq Plain via the so-called Syrian Gates to the south. According to geomorphological surveys, the mound was originally located on a promontory inside a former estuary just to the south and controlling a natural bay to its north (Ozaner *et al.* 1993); a perfect location for a typical eastern Mediterranean harbor (e.g. Beirut, Tyre and several others). Today, the coastline has a straight profile and deep anchorage. Geomorphological reconstructions indicate that the nearby river has begun its southward movement to the south during the Hellenistic Period, filling the estuary completely and possibly causing its abandonment after the Hellenistic Period (Beach and Luzzader-Beach 2008). The site was re-invested during the Crusader period (Redford *et al.* 2001); the Islamic settlement preceding the Crusader Period occupation on top of the mound was located on the plain, only a few hundred meters away from the mound (Eger 2010).

The excavations at the mound were conducted between 1993 and 2012 by M.-H. and C. Gates of the Bilkent University in Ankara.

Methods

All remains discussed in this study come from hand-collected lots. The light and heavy residue from flotation samples were saved and sorted, but they await further study. The specimens were counted and weighed and identified to the lowest possible taxon using a variety of reference collections and osteomorphological manuals (e.g. Schmid 1972). Fish and turtles were identified using the collections in Tübingen University, for example. Jort Bosman

Table I. Summary table of taxonomic composition: NISP and weight of higher taxa.

Taxonomic description	*NISP*	*%NISP*	*Weight (gr)*	*%Weight*
Unidentified mammals	7248	49	23950	19
Domestic mammals	5524	37	70960	56
Wild mammals	255	2	5829	5
Marine turtles	666	4.5	19807	15
Birds	74	0.5	70	0.1
Fish	495	3.3	1050	0.9
Molluscs (excluding Murex accumulations)	539	3.6	4102	3.3
Total	**14798**	**100**	**125768**	**100**

identified the bird remains in 2014 using the bird collections of Groningen Institute of Archaeology (GIA), University of Groningen.[1] Osteometric measurements were taken following Von den Driesch 1976. Tooth eruption and wear patterns of teeth in mandibles (complete and partial) in sheep, goat, cattle, and pig/boar were recorded using the Grant's scheme (Grant 1982). Age stages provided in the tables and figures summarizing the age-at-death should be taken as estimations, not absolute ages. They are rough determinations based on fusion and tooth eruption and attrition sequences following works such as Grant 1982, Habermehl 1975, Payne 1973, Silver 1970, and Zeder 2006. While quantifying the relative frequencies of culling ages, again instead of 'correcting' the raw data (e.g. Vigne and Helmer 2007), we chose not to manipulate the NISP values. For the interpretation of the analytical results based on ageing data in terms of economic strategies, we take the models suggested by Payne (1973), Vigne and Helmer (2007), and Stein (1998) into consideration, but we apply none of them directly to our data.

Results

The faunal assemblage discussed in this paper comes from the Building Periods 7 and 6 of Kinet Höyük, dated roughly to 700-550 BC. Faunal remains were hand-picked from excavated deposits covering residential and open-air activity areas, disuse levels and pits. The layers were uncovered during the 1995, 1997, and 1998 seasons from two main excavation areas: Operation E/H on the western slope with a direct view of the sea (Fig. 2), and Operations A-AII and D on the eastern slope overlooking the coastal plain. Further details on these excavations can be found in Gates 1999.

A total of 14,798 specimens, weighing about 125 kg were examined using the reference collections of Kinet Höyük, the British Institute in Ankara, and the collection of the Tübingen University Institute of Scientific Archaeology in 2001 and 2002 (Table 1). The NISP (=Number of Identified Specimens) and weight counts were used to assess relative abundance of taxa and relative contribution to meat diet respectively. Quantifications resulting from the calculations of both units are of course taken to be indications of these targeted variables, not at face value. About 49% of the sum of the assemblage could only be identified to different size classes of mammals, due to their fragmentary condition. These make up only about 19% of the weight of the material. Bones and shells were retrieved through hand-collection and wet-sieving after floatation. Here we deal with the hand-collected material. Butchery marks are well visible and abundant. Bone and shell surfaces are mainly smooth with little traces of surface erosion as if they were buried quickly. However, chewing and gnawing marks are abundant.

Molluscs

Molluscs of a wide diversity were used at Late Iron Age Kinet for different purposes. They do not seem represent a significant contribution to the diet in Kinet, in comparison, for example, to the case of some protohistoric coastal sites in the Aegean, where shells make almost half of the faunal remains, such as at Bronze and Iron Age Kastanas, Miletus or Troia (Becker 1986, Peters 1993, Çakırlar 2009). Despite the fact that the majority of shells consist of edible bivalves, such as oysters, cockles and clams, it is interesting that no single species or genus was targeted, unlike most sites where certain species of shellfish seem to be dietary staples (for example cockles at Troia; Çakırlar 2009). If mollusks were gathered alive, the variety of aquatic niches exploited must have been great –from fully marine to freshwater environments, in Kinet's case probably the nearby river. Littoral gastropods living on rocks such as limpets (*Patella* spp.) and topshells (*Monodonta* spp. and *Gibbula* spp.) are common, but so are species living mostly buried inside sandy bottoms such as the bittersweet clams (Glycymeridae), cockles (Cardiidae), and coquinae shells (Donacidae). The habitats of different species are indicative of at least two very different types of coastal waters, one aligned with a rocky shore, and the other shallow with sandy bottom. For some specimens, it is clear that they were not collected alive. For example, two specimens of *Conus mediterraneus* display typical marks of natural abrasion and therefore represent clearly beach picks, not food refuse.

In contrast to the generally diverse taxonomic representation characterized by equilibrium rather

1 A recent list of the bird skeletal specimens held by the Groningen Institute of Archaeology can be found at https://dataverse.nl/dataset.xhtml?persistentId=hdl:10411/20702 (accessed 11-05-2018).

Fig. 3. Fills of crushed murex shells in Period 7, Operation E/H on the seaside of the mound. Photo by M.-H. Gates, Courtesy of Kinet Höyük Project.

Table 2. Summary table of Molluscs except *Murex* shells.

Description	*NISP*	*% of NISP*	*Weight (gr)*	*% of Weight*
Gastropoda				
Patella sp.	16	3.0	29	0.7
Monodanta articulata	2	0.4	16	0.4
Monodonta turbinata	3	0.6	21	0.5
Cyprea lurida	6	1.1	33	0.8
Neverita josephine	7	1.3	39	0.9
Phalium undulatum	18	3.3	126	3.1
Tonna galea	3	0.6	20	0.5
Murex brandaris	10	1.9	105	2.6
Hexaplex trunculus	*	*	*	*
Thais haemastoma	8	1.5	93	2.3
Euthria cornea	6	1.1	16	0.4
Nassarious gibbosulus	68	12.6	58	1.4
Conus mediterraneus	12	2.2	118	2.9
Bivalvia				
Arca tetragona	4	0.7	19	0.5
Glycymeridae	99	18.4	526	12.8
Spondylus gaederopus	9	1.7	226	5.5
Ostrea edulis	12	2.2	869	21.2
Cardiidae	11	2.0	45	1.1
Venus sp.	119	22.1	1607	39.2
Donax sp.	104	19.3	69	1.7
Sepia sp.	12	2.2	17	0.4
Freshwater taxa				
Unionidae	5	0.9	23	0.6
Unidentified molluscs	5	0.9	28	0.7
Total of mollusc remains	**539**	**3.6**	**4102**	**3.3**
Total of studied material	**14798**	**100**	**125768.15**	**100**

than dominance of one or other taxon, there are (almost literally) tons of crushed shells of *Hexaplex* (formerly = *Murex*) *trunculus*, which is unequivocal evidence for purple dye production. Under the plaster floors of Period 7 structures on the western side of the mound, thick layers of crushed murex shells were placed, as levelling fill on top of the ruins of previous building periods (Gates 2003) (Fig. 3). Shallow pits on the eastern side of the mound have been used to store murex shells during Period 6, associated with the kiln excavated in the same area. These are remains of reused residue of shell dye production, which probably took place not on the mound proper, but someplace else by the coast. Calculations estimate that one or two tons of crushed murex shells were deposited in the excavated areas dating to Periods 6 and 7. These were not included in the general quantification of the assemblage presented here.

Table 3. Summary table of fish.

Description	*NISP*	*% NISP*	*Weight*	*% Weight*
Requieum and whaler sharks (Carcharhinidae)	2	0.4	2	0.1
Mackerel sharks (Lamnidae)	2	0.4	4	0.4
Unid. cartilaginous fish (Elasmobranchii indet.)	5	1	9	0.8
Groupers. bass. perches (Serranidae indet.)	6	1.2	9	0.9
Groupers (*Epinephelus* spp.)	143	28.9	545	52.0
Jacks (Carangidae)	15	3	63	6.0
Seabass (*Dicentrarchus* spp.)	4	0.8	3	0.3
Sea breams (Sparidae indet.)	28	5.7	32	3.1
Gilthead sea bream (*Sparus aurata*)	7	1.4	12	1.1
Mullets (Mugilidae)	1	0.2	0	0.0
Grey trigger fish (*Balistes carolinensis*)	11	2.2	13	1.3
North African catfish (*Clarias gariepinus*)	119	24	223	21.3
Unidentified Bony Fish (Osteichthyes indet.)	152	30.7	134	12.8
Total of Fish Remains	**495**	**3.4%[1]**	**1050**	**0.0**

[1] Of the total assemblage.

Fish

The absence of certain smaller fish taxa, the common representation of bigger individuals, and the presence of larger skeletal elements are probably due to recovery techniques. The availability of comparative material as well as the aim of the current study, which was to reach implications of economic significance depending on the habitat preferences of different taxa, determined taxonomic diversity to an extent.

The cartilaginous fish are represented by Elasmobranchii vertebral centra. Only four out of nine specimens could be identified to family level. Two vertebral centra belong to individuals of the Carcharchinidae family, commonly named as requiem and whaler sharks. One of the specimens belongs

Table 4. Representation of the elements of Cheloniidae.

Elements	NISP	% of NISP	Weight	% of Weight
Exoskeleton (Carapace and plastron)	479	71.9	16654	84.1
Ribs	24	3.6	66	0.3
Indet. vertebrae	21	3.2	132	0.7
Cranium	13	2.0	59	0.3
Mandible	2	0.3	29	0.2
Hyoid	2	0.3	3	0.0
Scapula	34	5.1	1013	5.1
Humerus	19	2.9	1041	5.3
Radius	4	0.6	55	0.3
Ulna	7	1.1	158	0.8
Pelvis	2	0.3	32	0.2
Femur	13	2.0	314	1.6
Tibia	3	0.5	52	0.3
Fibula	2	0.3	16	0.1
Acropodium and zeugopodium	27	4.1	73	0.4
Indet. Parts	14	2.1	109	0.6
Total	**666**	**100.0**	**19807**	**100.0**

to an individual estimated 3 meters in TL (=Total Length). The specimen's diameter is 28.2 mm. These remains may represent occasional or chance captures. They can also be remnants of beach pickings.

Groupers, seabasses and perches (Serranidae) are the most common elements of the fish sample. Their economic importance today and in the ancient times is well-attested (Whitehead *et al.* 1986, Desse and Desse-Berset 1996). Their high percentage may have been inflated by differential preservation. Their skeletal elements are more robust as opposed to, for example, Mugilidae bones (Although this would need to be verified with density studies). The second most abundant fish taxon is Sparidae. As in many other sites, most of the Sparid specimens are large teeth bearing bones (i.e. dentary and premaxilla). The only species identified among them with certainty is the guilt-headed sea bream (*Sparus aurata*). The very low representation of mullets (Mugilidae) is noteworthy, for mullets commonly inhabit estuaries and calm bays, which must have been the type of coastal environment immediately available to Kinet in the Late Iron Age according to geomorphological results. Grey trigger fish (*Balistes carolinensis*) lives on rocky bottoms at depths between 10-100 m and are solitary species. They can be fished diving with a harpoon.

The riverine North African catfish (*Clarias gariepinus*) is the only freshwater species identified. Kinet falls within the present and ancient distribution of these species (Arndt *et al.* 2003).

From the habitats and occurrences of the fish represented here, it is clear that the pelagic environments were rarely used for fishing. Continental waters, still and sandy-muddy parts of the river or the rivers were exploited. The proportion of estuarine species is very low to fish of coastal, but not necessarily of estuarine origin. This data falls in disagreement with the geomorphological reconstructions of the coastline at the time, but may be a reflection of recovery or preservation bias, or may indicate a subsistence strategy. To explore this question, wet-sieved samples of fish from this and other periods of the site must be studied.

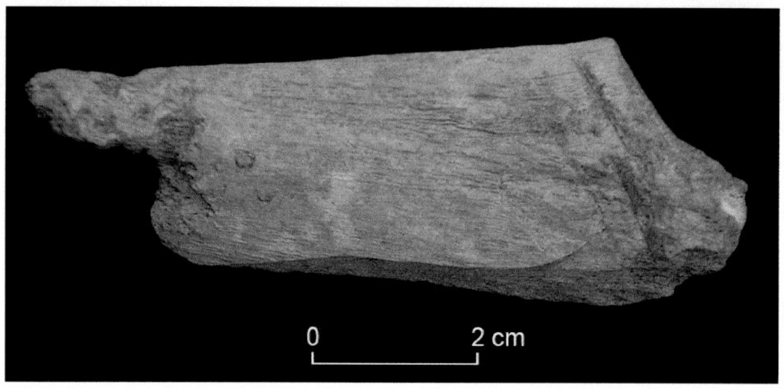

Fig. 4. Chelonid coracoid fragment (KT8779 749) with traces of butchery marks and carnivore gnawing (Photo by Turgut Çakar).

Turtles

One of the phenomenal features of the fauna is the large amount of marine turtles (Table 4). The skeletal and exoskeletal elements of marine turtles are easy to recognize, but identification at the species level is problematic. There are five species occurring in the Mediterranean (Geldiay *et al.* 1995). Only the Loggerhead turtle (*Caretta caretta*), the Green turtle (*Chelonia mydas*), and the Soft-shelled Nile turtles (*Trionyx triunguis*) are commonly encountered in the Iskenderun Bay today (Oruç 2001). The post-cranial elements (including the carapace and the plastron) of the Loggerhead and the Green turtle are virtually identical. The bones of the *Trionyx* are distinguishable from these two. After studying a sample of the material with the archaeobiological and zoological collections of Tübingen University and by induction, it is concluded that the remains belong to either one of the common marine turtle species.

The animals seem to have been brought to the site as a whole and processed there (Fig. 4). Many extremity bones have been subject to chewing and butchering. Only one complete specimen is present. It is a fibula, measuring 9.3 cm (GL) in good morphological agreement with an adult *C. mydas* in the Tübingen collection. Most other exoskeletal elements are also in good morphological agreement with the green turtle.

The remains of marine turtles are virtually absent from contexts dating to other periods represented at Kinet. Marine turtle consumption to this extent has not been attested at other sites of the Holocene Mediterranean either. What makes turtles such a

Table 5. Summary table of birds.

Taxa	*Common name*	*NISP*	*Remarks*
Phalacrocoracidae	Cormorants/shags	2	
Anatidae	Duck, geese and swan family	3	
Anser sp.	Grey and white geese	2	one of them (a femur) with a cut mark
Accipitriformes	Diurnal birds of prey	1	
Galliformes	Chickens, quails etc.	8	
Gallus domesticus	Domestic chicken	4	
Fulica atra	Eurasian coot	6	
Fulica cf. *atra*	Eurasian coot or similar	3	
Charadriiformes	Waders, gulls and auks	1	
Procellariidae	A group of seabirds	1	
Passeriformes	Perching birds/songbirds	2	
Total of identified birds		**33**	

Table 6. Measurements of *Sus scrofa*.

Radius			*Atlas*			
KT Nr.	Bp		KT Nr.	BFcr		
8532	34.1		7478	70		
Metacarpus II			***Metacarpus V***		***Metacarpus V***	
KT Nr.	Bd	GL	KT Nr.	GL	KT Nr.	GL
7478	14.7	77.3	8769	78	8769	78
					7478	73
Tibia			***Metatarsus IV***			
KT Nr.	Bd	Dd	KT Nr.	Bp	Bd	GL
7478	39.2	34.5	10667	20.2	22.1	119.5
7478	38.3	36.8				
8403	37.2	34				
7235	32.5	31.5				

Table 7. Summary table of Cervidae.

Description	*NISP*	*% NISP*	*Weight (gr.)*	*% Weight*
Dama dama dama/ mesopotamica	48	30	1365	30
Cervus elaphus	84	51	2496	55
C. elaphus/D. dama	24	15	565	12
Capreolus capreolus	6	4	144	3
Total	**162**	**100**	**4570**	**100**

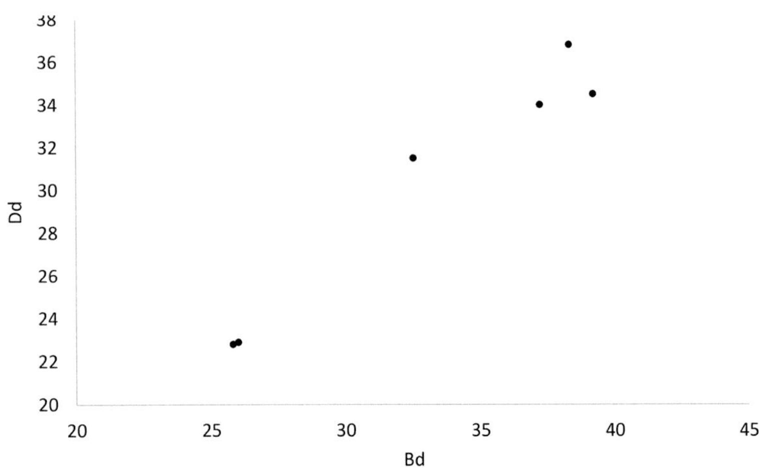

Fig. 5. Size comparison of *Sus* distal tibiae.

basic part of the subsistence economy at this site and at this period? Answering this question will require further research.

Birds

Among 69 bird specimens, 33 were identified (by Jort Bosman) to order (i.e. –formes) or lower taxa. The dominance of aquatic taxa, including coastal and freshwater birds such as coots, geese and cormorants (Table 5) is clear and expected. Noteworthy is the presence of domestic chicken remains, actually as a significant feature of the assemblage, no less than four specimens, perhaps up to 12, representing the gradual, human-mediated migration of domestic chickens from their homeland towards the west via Assyria during the first millennium BC (Perry-Gal *et al.* 2015).

Wild mammals

Wild mammals constitute a small portion of the fauna. Brown Hare (*Lepus capensis*), red fox (*Vulpes vulpes*), stone marten (*Martes foina*), and badger (*Meles meles*) are found in small numbers, indicating catches of those individuals who dared to come close enough to the

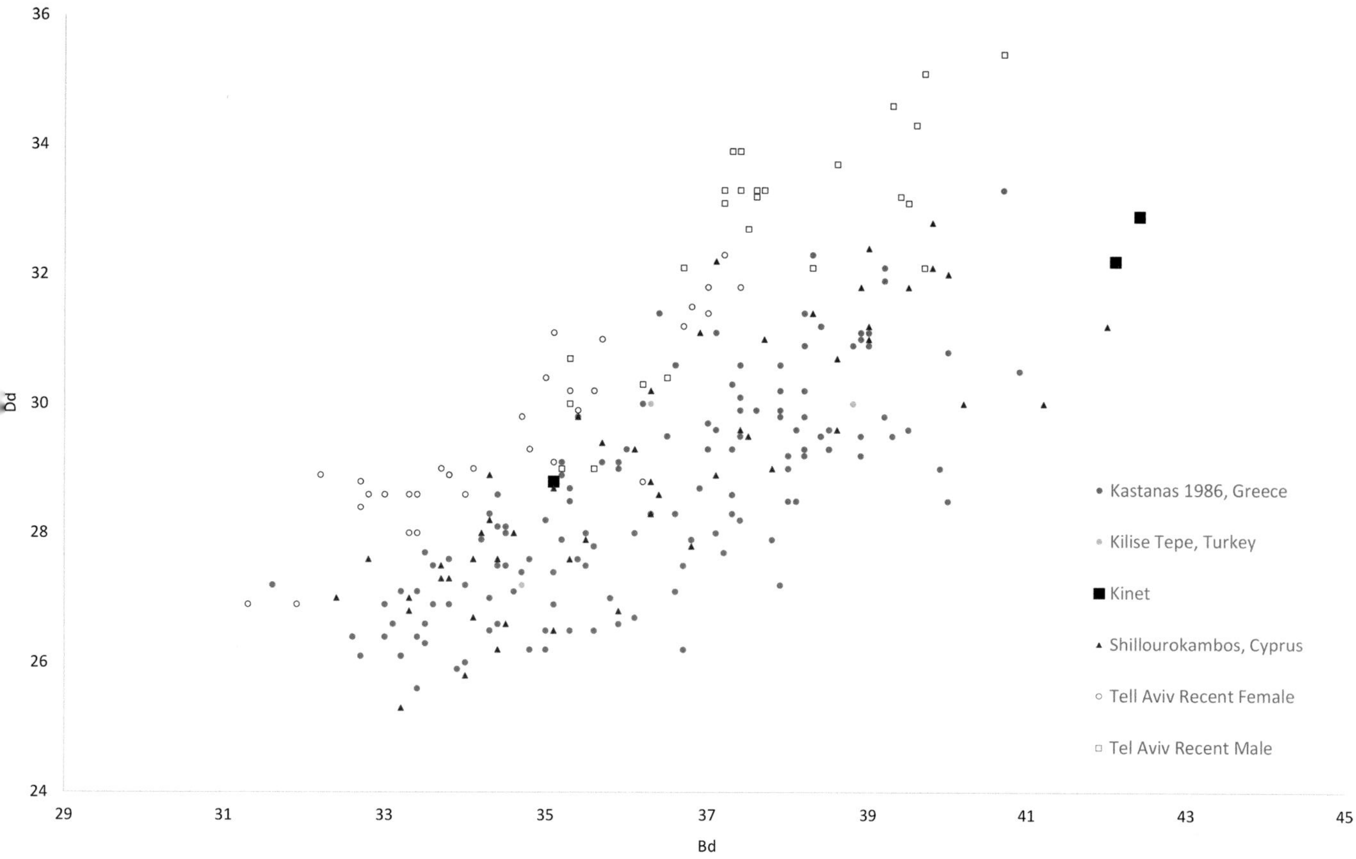

Fig. 6. Scatter distribution of distal tibia measurements Bd and Dd: recent *D. dama mesopotamica* from Tel Aviv, *Dama dama* from Kilisetepe north of Kinet, Kastanas in Greece, and Kinet. All data other than Kinet comes from https://www.nottingham.ac.uk/zooarchaeology/deer_bone/search.php (accessed 1-10-2018).

settlement. One humeral, one fibular, one metatarsal (V), and two ulnar fragments are identified to the brown bear (*Ursus arctos*). All of the *Ursus* fragments are uncovered from the eastern side of the mound, two from the same excavation unit. They probably all belong to the same sub-adult or adult individual. The distal epiphysis of the humerus, and the distal and proximal portions of one of the ulnar shafts bear cut marks. The locations of the cut marks imply that the carcass of the bear was processed to obtain meat.

Wild boar (*Sus scrofa*) are present. Their identification was based on their osteometric measurements. Of the 25 bones of wild boar, 15 come from a refuse pit on the eastern side of the mound.

Deer (Cervidae) remains are fairly common in Kinet's Late Iron Age faunal assemblage. It was sometimes not possible to distinguish the fallow deer (*Dama dama*) remains from those of the red deer (*Cervus elaphus*), therefore these are classified as *C. elaphus/D.dama*.

There are two subspecies of fallow deer in the Near East: The European or the Anatolian fallow deer (*Dama dama dama*) and the Mesopotamian fallow deer (*Dama dama mesopotamica*). The latter was formerly considered a separate species based on its antler form and geographical distribution (Vogler 1997). The Mesopotamian subspecies is also larger than its European counterpart. The region around Kinet is devoid of both subspecies at present. The status of the Mesopotamian fallow deer is endangered. In fact, only a few individuals survive in European zoos and gated herds in Israel. Kinet is located between the well-attested archaeological populations of the Anatolian fallow deer to the northwest and the populations of Mesopotamian fallow deer to the east and south. The fallow deer remains in Sirkeli, just northwest of Kinet on the Seyhan River were attributed to the *D. d. mesopotamica*, thereby dragging the northern border of this subspecies north of the Amanus mountains into Anatolia (Vogler 1997). Based on antler remains and measurements, the fallow deer of Kinet can be both *D. dama dama* and *D. dama mesopotamica*. On the whole, the fallow deer at Kinet are smaller than the population at Sirkeli, but at Kinet there are both very large and very small individuals (Fig. 6).

The red deer (*Cervus elaphus*) appears to be the largest contributor to the protein diet obtained from wild terrestrials. Their populations have been wiped out from the region today. Six well-preserved specimens of *Capreolus capreolus* represent the roe deer.

Of the small wild bovids, gazelle, wild sheep and wild goat are present, but are rare. This must be due to the high humidity of the region. The gazelle were attributed to *G. gazella*, the mountain gazelle, based on a complete horn core with cut marks at its base. The size distribution of the specimens (eight

Table 8. Summary table of domestic animals.

Domestic mammals	*NISP*	*% of NISP*	*Weight (gr)*	*% of weight*
Cattle	816	14.8%	27653	39.0%
Cattle/Red deer	1	0.0%	4	0.0%
Sheep	254	4.6%	3367	4.7%
Goat	149	2.7%	1873	2.6%
Sheep/goat	3555	64.4%	25458	35.9%
Pig	663	12.0%	9302	13.1%
Horse	6	0.1%	474	0.7%
Donkey	36	0.7%	1805	2.5%
Horse/donkey	35	0.6%	924	1.3%
Dog	6	0.1%	36	0.1%
Dog/fox	1	0.0%	1	0.0%
Total	**5522**	**100.0%**	**70897**	**100.0%**

altogether) are in comparable range with those from Sirkeli Iron Age levels (Vogler 1997). The presence of gazelles at both Sirkeli and Kinet is interesting, because there is no Pleistocene or Holocene record of gazelle populations in Anatolia proper (that is Turkey excluding Southeastern Turkey) (Uerpmann 1987). It is worthwhile considering that gazelles at both sites may be imports from regions south and east of the Amuq Mountains, rather than individuals from local populations living on the coastal plain and hilly country surrounding the sites in Mid-Late Holocene. Eight *Capra aegagrus* and two *Ovis orientalis* were identified. The identification of the *C. aegagrus* and *O. orientalis* is based on size differences with their domestic counterparts from the same levels of Kinet.

Domestic mammals

The largest domestic contributor to the meat economy was the sheep and goat group. They make up 43% of the domestic animals in terms of weight. The ratio of sheep to goats is 5:3. Among these is one infant skeleton of a sheep or a goat (KT 10239). Domestic cattle was the second biggest contributor of the meat part of the subsistence economy (with 39% of the weight of domestic animal remains). Pig's contribution to the economy is limited by a 13% of the bone weight of the domestic mammals. Pigs constitute about 4.5% in NIS and 7.5% in WIS of the entire assemblage. It is interesting that both the pigs and their wild ancestors are present in relatively low numbers. One would expect pigs to have thrived in the vegetation and climate around Kinet. Wild boar is abundant today in the oak covered western slopes of the Amanus Mountains. There must be reasons other than environmental adaptation for the low representation of pigs and boar at Kinet. One of these might be the importance of vertical seasonal migrations, which are an important part of the annual life cycle in the region today. Because pigs tend to be less mobile (although they too can be herded), sheep, goat and cattle may have better fitted the seasonally mobile system of animal husbandry. We provide the tooth eruption and wear patterns of domestic sheep, goat, sheep/goat, cattle and pig, as well as their post-cranial measurements in Appendix 1.

There is a good number of cattle that were slaughtered at a rather young age (see Appendix 2), indicating milk, meat and traction were all-important. A much larger sample of mandibles with teeth is available to draw the mortality profiles of sheep and goat, but here too slaughter ages are well-distributed among different age-at-death groups, indicating that a mixed economy was practiced, and maintaining healthy herds was a principle economic goal. Almost all pigs were culled when juvenile or as young individuals, indicating sty-keeping and seasonal culling rather than free-range herding.

Of the 77 specimens attributed to domestic equids, 36 were identified as donkeys, and five as horses. The absence of second and third phalanges among Equid remains is notable. According to the 20 limb bones providing evidence about the age of mortality of the equids, all appear to have died latest when sub-adults. Both donkey and horse specimens had cut marks on them.

Conclusions

Altogether, the remains of aquatic animals make up 19.2% in weight and 11.4% in NISP of the assemblage. This itself and the two maritime innovations of the period, namely capturing of marine turtles and murex dye production, along with the expansion of trade contacts with maritime cultures of the Eastern Mediterranean, confirm that there should be, as it has always been suspected and suggested (Jacobsen 1981, Biagi *et al.* 1984), a relationship between coastal adaptations and sea bound acculturation. We cannot elaborate more on the cause-and-effect nature of this relationship at Kinet at this state of research. Additionally, the habitat preferences of the aquatic species in the record challenges our understanding of the ancient coastline next to the site.

The presence of *Dama dama dama* to the north of the Amanus Mountains is suggested. Game animals of the surrounding mountains and hills are present as well, but to a lesser degree. The virtual absence of the wild boar is not in accordance with our current knowledge of the ancient environment, and less so considering the extent of deer hunt which took place during the period. Was there a political reason why the slopes of the Amanus were not exploited as much as the coast or was it a matter of optimizing subsistence costs that the Late Iron Age inhabitants of Kinet preferred coastal resources to obtain animal resources? Finally, what was the reason/s behind the sudden and short-lived explosion in sea turtle exploitation?

Acknowledgements
This paper is largely summary of Çakırlar's M.A. thesis submitted to the Department of Archaeology and History of the Faculty of Arts and Sciences at the American University of Beirut in March 2003. CÇ would like to thank Dr. Marie-Henriette Gates, director of the Kinet Höyük Project, Prof. Dr. Helga Seeden, her thesis supervisor, D. James Rackham, who helped her learn bones, and Prof. Dr. Hans-Peter Uerpmann. Dr. Margarethe Uerpmann. Dr. Can Gündem (all of Tübingen University at the time) for their invaluable advice, assistance, and support during identifications, analyses, and writing of the presentation for the ASWA 2004 meeting in Ankara and the MA thesis.

References
Arndt, A., Van Neer, W., Hellemans, B., Robben, J., Volckaert F., and Waelkens, M. 2003. Roman trade relationshiops at Sagalassos (Turkey) elucidated by ancient DNA of fish remains. *Journal of Archaeological Science* 30: 1095-1105.

Beach, T. P. and Luzzadder-Beach, S. 2008. Geoarchaeology and aggradation around Kinet Höyük, an archaeological mound in the Eastern Mediterranean, Turkey. *Geomorphology* 101(3): 416-428.

Becker, C. 1986. *Kastanas. Die Tierknochenfunde*. Prähistorische Archäologie in Südosteuropa. Band 5. Berlin: Wissenschaftsverlag Volker Spiess.

Biagi, P., Torke, W., Tosi, M. and Uerpmann, H-P. 1984. Qurum: A case study of coastal archaeology in Northern Oman. *World Archaeology* 16(1): 43-61.

Braidwood, R. J. and Braidwood, L. S. 1960. *Excavation in the Plain of Antioch*. Oriental Institute Publications. no. 61. Chicago: University of Chicago Press.

Çakirlar, C. 2009. *Mollusk shells in Troia. Yenibademli. and Ulucak: an Archaeomalacological Approach to the Environment and Economy of the Aegean*. John and Erica Hedges.

Desse. J. and Desse-Berset, N. 1996. Archaeozoology of groupers (Epinephelinae): Identification. osteometry and keys to interpretation. *Archaeofauna* 5: Proceedings of the eighth meeting of the ICAZ fish remains working group: Ichthyoarchaeology and the archaeological record held in Madrid 3-11. October 1995: 121-127.

Eger, A. 2010. Hisn al-Tinat on the Islamic-Byzantine Frontier: Synthesis and the 2005–2008 Survey and Excavation on the Cilician Plain (Turkey). *Bulletin of American School of Oriental Research* 357: 49-76.

Gates, M-H. 1999. Kinet Höyük in Eastern Cilicia: A case study for acculturation in ancient harbours. In *Olba* 2. Vol 1: Proceedings of the Uluslararası Kilikya Arkeolojisi Sempozyumu held in Mersin 1999: 304-312. Mersin: Mersin Üniversitesi Kilikya Arkeolojisini Araştırma Merkezi Yayınları.

Gates, M-H. 2003. 2001 Season at Kinet Höyük (Yeşil-Dörtyol. Hatay). *Kazı Sonuçları Toplantısı* 24 (1): 283–98.

Gates, M. H. 2013. The Hittite Seaport Izziya at Late Bronze Age Kinet Höyük (Cilicia). *Near Eastern Archaeology* 76(4): 223-234.

Gates, C. 2015. Kinet Höyük (Classical Issos) A Harbor Town in Southeast Cilicia during the Persian and Hellenistic Periods. *Journal of Eastern Mediterranean Archaeology & Heritage Studies* 3(2): 81-104.

Geldiay, R. Koray, T., and Balık S. 1995. Status of sea turtle populations (*Caretta c. caretta* and *Chelonia m. mydas*) in the Northern Mediterranean Sea. Turkey. In (ed.) K.A. Bjorndal. *Biology and Conservation of Sea Turtles*: 425-434. Washington D.C.: Smithsonian Institution Publications.

Goldman, H. 1963. *Excavations at Gözlükule. Tarsus III: The Iron Age*. Princeton: Princeton University Press.

Grant, A. 1982. The use of tooth wear as a guide to the age of domestic ungulates. In, Wilson, B., Grigson, C. and Payne, S. (eds.) *Ageing and Sexing Animal Bones from Archaeological Sites*. British Archaeological Series International Series 109: 91-107. Oxford: Archaeopress.

Habermehl, K-H. 1975. *Die Altersbestimmung bei Haus- und Labortieren*, 2. Berlin.

Hodos, T. 2006. *Local Responses to Colonization in the Iron Age Mediterranean*. Routledge.

Ikram S., Kabatiar R. and Çakırlar, C. in preparation. The zooarchaeology of Kinet Höyük. In, Gates, M.-H. (ed.) *Kinet Höyük Volume 1*.

Jacobsen, T. W. 1981. Franchthi Cave and the Beginning of Settled Village in Greece. *Hesperia* 50(4): 303-319.

Oruç, A. 2001. Trawl fisheries in the eastern Mediterranean and their impact on marine turtles. *Zoology in the Middle East* 24: 119-125.

Ozaner F. S., Gates, M-H. and Özgen, I. 1993. Dating the coastal dunres of Karabasamak district (Iskenderun Bay) by geomorphological and archaeological methods. *Arkeometri Sonuçları Toplantısı* 7: 357-367.

Payne, S. 1973: Kill-off patterns in sheep and goats: the mandibles from Aşvan Kale. *Anatolian Studies* 23: 281-303.

Perry-Gal, L., Erlich, A., Gilboa, A. and Bar-Oz, G. 2015. Earliest economic exploitation of chicken outside East Asia: Evidence from the Hellenistic Southern Levant. *Proceedings of the National Academy of Sciences* 112(32): 9849-9854.

Peters, J. 1993. Archaic Milet: Daily life and religious customs from an archaeozoological perspective. In, Buitenhuis, H. and Clason, A.T. (eds) *Archaeozoology of the Near East* I. Proceedings of the first meeting of the Archaeozoology of Southeast Asia Working Group of ICAZ: 88-96. Leiden: Universal Book Services/Dr. W. Backhuys .

Redford, S., Ikram, S., Parr, E. M. and Beach, T. 2001. Excavations at Medieval Kinet. Turkey: A Preliminary Report. *Ancient Near Eastern Studies* 38: 58-138.

Schmid, E. 1972. *Atlas of Animal Bones for Prehistorians, Archaeologists and Quarternary Geologists*. Amsterdam: Elsevier.

Seton-Williams, V. 1954. Cilician Survey. *Anatolian Studies* 4: 121-74.

Silver, I. A. 1970. The ageing of domestic animals. In, Brothwell, D. R. and Higgs, E. S. (eds) *Science in Archaeology: A Survey of Progress and Research*, 2nd ed.: 283-302. New York: Praeger Publishing.

Stein, G. 1998. Medieval pastoral production system at Gritille. In, Redford, S. (ed.) *The Archaeology of the Frontier in the Medieval Near East: Excavations at Gritille, Turkey*, Archaeological Institute of America Monographs:181-210. New series 3, Boston: Archaeological Institute of America.

Uerpmann, H-P. 1987. *The Ancient Distribution of Ungulate Mammals in the Middle East*. Beihefte zum Tübinger Atlas des Vorderen Orients. Reihe A. Nr. 27. Wiesbaden: Dr. Ludwig Reichert Verlag.

The University of Nottingham. 2018. Deer Bone Database [online] Available at: https://www.nottingham.ac.uk/zooarchaeology/deer_bone/results.php [Accessed 7 May 2018].

Vigne, J. D. and Helmer, D. 2007. Was milk a "secondary product" in the Old World Neolithisation process? Its role in the domestication of cattle, sheep and goats. *Anthropozoologica* 42(2): 9-40.

Vogler, U. 1997. *Faunenhistorische Untersuchungen am Sirkeli Höyük. Adana. Turkei.* PhD. München: Ludwig-Maximillians-Universität.

Von den Driesch, A. 1976. *A Guide to the Measurement of Animal Bones from Archaeological Sites*. Peabody Museum Bulletin, no. 1. Cambridge: Harvard University Press.

Whitehead, P. J. P., Bauchot, M.L., Hureau, J.C., Nielsen, J. and Tortonese, E. 1986. *Fishes of the North-Eastern Atlantic and the Mediterranean*. [CD Rom] Paris: Unesco Publications.

Woolley, C. L. 1938. The Excavations at Al Mina. Sueidia. *Journal of Hellenic Studies* 58: 1-30.

Zeder, M. A. 2006. Reconciling rates of long bone fusion and tooth eruption and wear in sheep (*Ovis*) and goat (*Capra*). In, Rucillo, D. (ed.) *Recent Advances in Ageing and Sexing Animal Bones*: 87-118. Oxford: Oxbow.

Appendix I: Measurements

Measurements: *Cattle*

	KT Nr.	GL/ GLl	SLC/ LA	GLP/ GLm/GLF	LG	BG/ GB	Bp	BFp/ BFcd	Dp	SD/Dl	Bd	BT	BFd	DD/Dm	Dd
Scapula	7629		64.1	75.5	63.4	53									
	8769		54.8	68.8	59.6	48.5									
	8403		52.9	64	53.5	46									
	8180		49.9	62.7	52.3	42.8									
	7227			60.1	57.7	43.2									
	10578			60.1	52.1	40.5									
	8540		41.8	58.9	51.4										
Humerus	8403										74.6	65.6			
Radius	8530						81.5	73.5							
	8136						71.5	66.4							
	6629										67.9		60.9		
	8128										66.8		60.4		
	7909										64.1		57		
	8534										60		54.5		
Metacarpus	10254						48.1		28.7						
	6533						49.8								
	6629						54.3								
	8405						49.8		30.1						
	9948						58.5		34.5						
	8769										62.6				32.6
	7276										62.3				21.7
	7486										59.1				31.2
	0574										53.9				
	7469										52.6				28.6
	8079										51				29.8
Phalanx 1, anterior	7469									28.3	30			20.9	22.6
	8470	60.2					31.5		35.2	25.1	29.1			19.5	22.1
	7204	63.9					30.9		31.2	25.7	28.8			19.4	21.1
	8304	56.5					28.5		29.2	24	26.5			17.6	19.8
	8540	57.7					27.9		30.3	23.9				15.9	
	9460	56.8					27.7		30.4	22.7	26.7				
	8405	50.4					26.5		28.1	20.5	22.5			19.7	17.1
	6629	58.3					26.4		29.4	22	23.8			16.7	18.9
	8789	62.7					26.4			21.4	24			16.4	19.1
	10574	56.1					24		26	20.4	20.8			20.1	22.1
Pelvis	7401		69.7	Female											
Tibia	10995										55				
	10574										60.6				43.4
	7673										53.9				40
	7489										63.5				
	7291						80.8								
Astragalus	7673	67.8		63.7						39.6	42.8			37	
	8512	67.5		62.1						37.3	42.1			37.8	
	10899	66.4		59.4						39.3	42.1			32.2	
	8405	65		58.7						35	43.9			37.1	
	8339	64.4		59.5						38.3	41.5			36.1	
	10574	63.2		57.8						33.9	37.2			32.4	
	7274	60.8		56.5						33.8	39.3				
	9458	60.2		56.5						33.1	37			33.4	
	8530	55.5		50.2						31.2	35.7			32.2	
Metatarsus	8203										55.8				33.6
	8502										49.9				29.9
	8532										47.8				29.1
	6414						44.6								
	8084						43		39.7						
	8302						41.6								

	8502									25.7	29.1			18.5	21.2
Phalanx1, posterior	8547	47.3							25.4	22.1	23.4			15.5	18.1
	10667	61.6					31.7		33.7	28.6	30.4			20	22.3
	5898	58.5					31.1		31.3	26	29.1			18.7	22.2
	7204	61.6					30.3		33.4	24.7	29.4			20.2	21.2
	10574	54.1					30.2		23.2	23.8	25.4			19.1	21.1
	8084	60.3					30		32.3	27.1	30.2			18.4	21.5
	8532	60.8					29.1		20.7	27.8	27.6			20.1	20.7
	8489	57.5					28.5		29.9	22.9	25.5			16.3	18.7
	8532	59					28.2		29.2	22.8	25.8			17.1	19.3
	10574	57.6					27.7		29.8	24.8	26.2			27.3	19.3
	7156	48.7					25.5		24.7	21.4	24.2			16	17.9
	10258	50.2					24.4		24.1	20.5	23.4			15.9	18.2
Atlas	8534	87		74		125.5		83.7							

Measurements: *Sheep*

	KT Nr.	**GL/ GLl**	**SLC/ LA**	**GLP/ GLm/GLF**	**LG/LAR**	**BG/ GB**	**Bp**	**BFp/ BFcd**	**Dp**	**SD/Dl**	**Bd**	**BT**	**BFd**	**DD/Dm**	**Dd**
Scapula	7291					21									
	7235			32.4											
	7272			30.9											
	8779		21.8	30.2	24.6	19.1									
	10992		21.2												
	8534		21.1	30.7	38.1	21.2									
	8528		20.9	33	26	19.8									
	8079		20.3	32.1	27.4	21.2									
	8271		19.3	30.2	27.7	19.6									
	7909		19.2	31.9	24	21.8									
	8713		19.2	30.9	27	22.1									
	7227		18.9	31.9	24.3	20									
Humerus	8413											29.4			
	8769											28.5			
	8405										32	30.8			
	8610										31.6	28.5			
	7486										31.5	29.7			
	8090										30.5	29.9			
	10426										30	28.9			
	10479										29.4	28.7			
	6010										29.1	28.1			
	8343										28.6	26.7			
	7746										28.5	26.5			
	8308										28	27.1			
	10570										28	26.1			
	7727										27.2	26.4			

Radius	8720										31.3		26.4		
	8717										30.2		25.5		
	7601										29.6		24		
	6619										29.3		24.8		
	8000										29.1		23.2		
	8128										28.6		25.4		
	8042										27.3		25.2		
	11105						33.3	30.2							
	8100	150.8					32.8	31.8		20.1	30.5		28.1		
	10574						31.9	29.4							
	7725						31.7								
	8128						31.5	29.1							
	8789						31.5	28.2							
	8517						31.3	27.9							
	8076						31	29.8							
	7221						30.6	28.2							
	8534						30.6	26.9							
	5429						30.3	27.3							
	8000						29.9	28.6							
	8534			37	122.1		29.9	17.6					24		
	8318						29.8								
	8313						29.7	27.3							
	8186						29.2	26.7							
Metacarpus	6954										25.3				
	9957										25.2				
	6954										24.4				16.6
	7227										27.5				16.4
	7465										25.6				16.4
	7286										25.5				16.1
	10437										27.2				15.9
	8534										25.1				15.8
	7272										23.5				14.8
	8403	131.9					25.1								
	8195	109.8					25.4		17.8	17.2					
	7843	104					21.3		15.7	12.3	25				15.3
	8130	98.9					24.2		16.8		25.8				15.3
	7408	94.8									25.8				14.5
Phalanx 1, anterior	8072	32.8					12.8		13.1	11.4	12.9			8.9	11.5
	8139	34.2					12.4		14	10.3	11.1			8.2	10.9
	10431	34.2					12.2		14.3	9.7	11.3			8.3	11.1
	7227	33.4					12.1		9.6	9.5	11.1			7.8	9.8
	6297	34.3					12.1		14.4	9.4	12.3			7.7	9.7
	8306	34					12.1		13	9.3	10.5			7.2	9.5
	7494	34.3					11.6		13.6	9.3	11.1			8.2	9.5
	5451	33.9					11.6		11	9.3	10.9			7.5	10.8
Pelvis					30.2	Male									
Femur	10893										36.7				
Tibia	10899										27.4				20.1
	8534										27				
	7247										26.1				19.4
	6178										26.1				19.1
	10578										26				
	7601										24.4				18.4
	7621										24.3				20
	10899										24.3				19.3
	8323										24.2				
	10247										23.6				17.3
	10992										23.3				19.4
	9460										21.1				18.8

Astragalus	8195	30.6		28.6						17.1	20			17.7	
	9866	30.6		26.2							18.3				
	10987	30.4		28.6						15.2	18.5			16.4	
	7998	30.1		29.8						17.3	20.2			16.9	
	730	30		28.7						17.1	20			17.1	
	8128	29.8		28.4							20.1				
	10254	29.4		27.8						16.1	17.1				
	10473	29.1		28.5						16.5	17.9			16.5	
	7227	28.5		27.1						15.9	17.9			16.5	
	10565	28.5		27						15.7	27.8			15.6	
	8076	28		26.4						15.9	19			15.9	
	8502	27.2		25.5						15.8	17.2			15.9	
	8607	26.7		25.9						15.3	17.8			15.9	
	8534	25.5		25.2						14.7	18.3			15.5	
Calcaneus	10995	59				20.4									
	8034	51.8				18.6									
Metatarsus	8534						19								
	10258										23.9				
	6619	109.5					19.9				23.7				
	10995										22.4				
	7227									11.1	22.4				14.9
Phalanx 1, posterior	6063	38.1					14.1		15.9	10.9	12.9			8.5	11.2
	10992	37.3					13.6		16.1	10.7	12.3			8.8	10.1
	8139	36.8					13.4		15.1	11.1	12.5			9.2	11.6
	6195	36.1					13.2		14	10.2	12.1			8.6	11.8
	8130	34.5					12.6		14.4	10	11.7			8.6	11.5
	7841	34.2					12.5		15.3	11.2	12.2			8.5	10.9

Measurements: *Goat*

	KT Nr.	GL/ GLl	SLC/ LA	GLP/ GLm/GLF	LG/LAR	BG/ GB	Bp	Dp	SD/Dl	Bd	BT	DD/Dm	Dd
Scapula	10992		23.7	34.1	28.6	24.5							
	8534		21.7	35.5	30.6								
Humerus	8100										32.1		
	8079										32.4		
	6187									35.2	34.7		
	7202									33	32.7		
	8405									30.9	28.9		
	7494									30.6			
	8042									30.3	29.5		
	7475									30.2	28.9		
	7428									30	29.7		
	8339									29.9	29.6		
	7632									29.1	28.4		
Metacarpus	8318									32.2			
	10258									31			
	8128	113.2					24.7	17.3	14.2	30.7			17.2
	10258									25.6			
Phalanx 1, anterior	7247	38.1					12.4	13.8	10.9	11.9		9.1	10.6
	7401	35.3					12	13.4	9.6	11.1		7	9.1
	8201	41.2					11.1	17.9	12.8	14.9		9.4	11.6
	8526	34.1					10.8	14.8	9.8	11.1		7.1	9.2
Pelvis	8607				29.7								

	8413											28.3	
	7621											27.2	20.7
	8630											26.9	21
	8130											26.9	20.8
	10992											26.9	19.6
	8403											26.3	20.9
	7621											26.3	19.8
	9652											26.3	18.3
	8769											26.1	20.2
Tibia	7889											26.1	20.1
	10992											26.1	19.2
	10995											26	18.3
	7227											25.7	20.1
	8072											25.5	18.6
	8021											25.4	19.8
	6533											25.2	20.1
	6187											24.4	18.2
	7415											22.4	19.8
	5898											21.8	
	5444											21.4	20.6
	10437	31.8		29.9					17.2			18.2	21.2
	8124	31.6		30					16.8			17.3	20.1
	7736	31.2		28.8					16.3			17	19.9
	8325	30.9		28.4					16.1			17.7	20.2
	7730	29.9		28					16.2			17.3	19.9
Astragalus	10995	29.7		27.6									18.1
	7623	28.5		27.1					16.3			16.3	17.6
	8530	27.3		25.7					15.4				18.8
	10467	27.1		25.1					14.8			15.4	16.8
	8769	26.4		25.5									17.9
	7736	26.2		25.9					14.4			15.8	17.1
	8769	56.7				19.5							
	7641	57.8				19.6							
Calcaneus	7621	58				20.8							
	7469	55.2				20.3							
	7219	58.4				18.4							
Metatarsus	10247									27			18.1
	7918	137.9					19.4	20.4	10.6	23			15.5
	8769	40.8					13.6	14.5	11.9	14		9.4	12.3
	8409	37.1					13.6	15	11.3	12.2		9.2	11.2
	7673	38.1					13.3	14.7	11.8	13.5		9.1	10.7
	8489	37.1					13.2	16	10.5	12		8.2	9.9
	9957	41					12.5	15	10	11.4		8.8	10
Phalanx 1, posterior	9460	36					12.3	14.8	11.2	12.7		8.8	10.2
	10570	38.1					12.3	14.8	11	11.9		7.4	9.8
	8084	38.5					12.1	13.5	9.7	11.4		7.8	10.1
	5882	32.7					11.9	13.6	9.1	10.5		7.4	9.4
	8185	35.5					11.4	13.7	9.8	11.1		7.8	9.4
	7623	37.7					11.4		9.6	10.9		7.5	10
	7748	38.4					11.1	13.7	9.8	11.5		7.8	10.1
Phalanx 1, anterior or	8610	35.7					12.1	13.9	10.9	12.7		8.7	11.4

Measurements: *Sus*

	KT Nr.	GL/ GLl	GLP/ GLm/GLF	LG/LAR	BG/ GB	Bp	BFp/ BFcd	Bd	Dd
Scapula	9468		29.3						
Humerus	7204							33.1	
	8779							31.1	
Radius	8532					34.1			
	8547					27.2			
	8532					25.1			
Metacarpus II	8201	33.1						13.7	
	6006	46						8.8	
	6414							7.9	
Metacarpus III	7907					15.6			
	10149	60.9				14.8		14.5	
	7469	64.9				11.3		15	
Metacarpus IV	8263					19.3			
Pelvis	8534			31.1					
	9941			29.4					
	7219			27.8					
Tibia	8425							26.1	
	7269							26	22.9
	8547							25.8	22.8
	7673							24.6	23
Astragalus	7494	31.8	32.6						
Calcaneus	8339	67.8			21.6				
Metatarsus II	8136	72.9						12.8	
Metatarsus IV	7498	62.3				11		13.5	

Measurements: *Horse*

	KT Nr.	GL/ GLl	SLC/ LA	GLP/ GLm/GLF	LG/LAR	BG/ GB	Bp	Dp	SD/Dl	Bd	BT	BFd	DD/Dm	Dd
Atlas	7181			79										
Humerus	8532						75.4	75.2						

Measurements: *Donkey*

	KT Nr.	GL/ GLl	SLC/ LA	GLP/ GLm/GLF	LG/LAR	BG/ GB	Bp	BFp	Dp	SD	Bd	BT	BFd	DD	Dd
Scapula	8723		46.8	64.6	42.1	34.5									
	8502			59.8	39.8	34.3									
Humerus	7673										52.3	51.5			
	8615											56.6			
Radius	8339						51.3								
	8534	252.8					54.5	50.7		25.8	51		42.2		
	8723						61	55.1							
Metacarpus	6297						36.7		24.8						
Phalanx 1, anterior	7401	65.3					35.9		27.3	21.8	32.1			15.4	17.9
Tibia	7472										50.2				32.7
	8076										47				32.8
Astragalus	8128	GH	45.6	GB	41	LmT	39.1							BFd	30.1
Calcaneus	8413	80.4				39.2									
	9652					40.3									
Metatarsus	8547						41.2		33.8						
Phalanx 1, posterior	8136	51.8					34.3		25.7	22.1	29.7				

Appendix II: Tooth wear and eruption scores following Grant 1982

Cattle

Kinet no	d4	p4	m1	m2	m3
11105	16	0	12	7	0
6006	11	0	3	0	0
10987	11	0	11	0	0
10578	0	0	0	0	15
7673	0	0	0	0	12
6006	0	2	0	0	0
8214	0	11	0	0	0
8540	0	0	12	11	7
8405	0	0	0	2	0

Goat

Kinet no	d4	p4	m1	m2	m3
8403	0	0	19	18	17
8534	9	0	0	0	0
6936	13	0	11	2	0
10473	10	0	2	0	0
7725	9	0	3	2	0
7204	9	0	3	0	0

Sheep

Kinet no	d4	p4	m1	m2	m3
8540	0	16	17	17	15

Sheep/goat

Kinet no	d4	p4	m1	m2	m3
10574	0	0	12	10	0
10247	17	0	12	10	2
10258	0	12	12	12	12
10437	0	16	17	14	12
10143	0	0	10	3	0
9948	10	0	0	4	0
10258	10	0	0	0	0
10143	14	0	7	1	0
10243	0	13	13	12	12
10563	0	6	12	11	4
9866	0	0	0	10	0
10426	13	0	11	0	3
10258	0	14	15	13	12
10893	14	0	12	9	0
10987	10	0	3	0	0
10987	0	0	12	10	0
10987	0	16	17	13	12
10238	9	0	0	3	0
10238	0	0	13	12	0
10893	11	0	3	0	0
10893	11	0	3	0	0
10893	11	0	4	0	0
10426	14	0	0	0	0
10893	12	0	11	0	0
10574	12	0	5	0	0
10254	0	12	14	13	12
10998	0	14	0	12	12
10998	0	0	13	12	11
10998	10	0	0	0	0
10426	0	16	18	13	12
10426	12	0	2	0	0
10437	0	12	13	12	12
10574	0	13	14	12	12
10574	13	0	11	2	0
10893	10	0	0	0	0
10995	0	16	17	13	12
9868	0	16	14	12	12
10995	10	0	0	0	0
10995	13	0	0	0	0
10995	0	13	14	11	10
10995	17	0	12	4	0
10995	9	0	0	0	0
10995	12	0	0	0	0
10570	0	16	17	0	0
10995	18	0	12	7	0
10570	12	0	0	0	0
10995	11	0	6	0	0
10992	11	0	5	1	0
10992	0	14	17	13	13
10992	11	0	0	0	0
10992	11	0	6	0	0
8823	0	0	4	4	0
10992	0	0	12	11	0
8789	0	0	0	0	11
10995	0	0	13	0	0
11105	10	0	0	0	0
10567	0	0	13	12	0
10567	13	0	11	3	0
10567	9	0	0	0	0
9957	0	11	13	12	10
11105	16	0	12	0	0
11105	11	0	2	0	0
11105	11	0	0	0	0
10570	0	0	6	0	0
11105	11	0	0	0	0
10437	12	0	9	2	0
11105	11	0	0	0	0
11105	0	13	14	0	0
11105	11	0	0	0	0
10578	0	13	13	12	8
10578	0	16	17	14	12
10578	14	0	11	0	0
10578	13	0	13	10	0
10578	13	0	12	0	0
11105	17	0	0	0	0
8769	0	0	3	0	0
10473	0	14	17	13	12
8320	10	0	3	0	0
8311	0	13	14	0	12
8313	0	10	12	10	10
8313	13	0	0	0	0
8074	0	13	13	12	10
7843	13	0	9	0	0
8769	0	0	10	8	2
7181	0	11	13	12	11
8769	9	0	0	0	0
8271	13	0	7	0	0
8076	0	3	10	0	0
8076	0	0	0	0	9
7746	0	12	12	11	8
7850	13	0	13	3	0
7560	0	11	12	11	4
7560	6	0	0	0	0
7560	0	0	13	12	0
8079	0	12	0	0	0

8769	0	11	12	0	0
8042	0	0	12	12	0
7437	0	11	11	0	0
7437	16	0	10	5	0
7269	0	7	7	0	0
8827	0	0	0	9	5
8610	0	13	12	12	11
8789	9	0	0	0	0
7345	16	0	3	0	0
7345	16	0	0	0	0
7181	11	0	8	0	0
8302	0	9	11	0	7
8033	10	0	3	0	0
8489	13	0	0	0	0
8779	0	4	11	10	4
8601	0	13	12	10	0
8044	0	11	12	11	10
8044	0	0	13	12	0
8044	13	0	10	0	0
8782	0	13	15	0	0
8271	0	16	18	17	14
8302	0	16	17	17	13
10236	0	0	11	0	2
10899	11	0	0	0	0
10899	11	0	0	0	0
10899	0	0	4	0	0
10899	11	0	0	0	0
9460	14	0	13	0	0
9460	11	0	0	0	0
9460	0	0	0	12	11
9460	0	0	0	11	9
8784	0	12	12	0	0
10664	0	16	18	17	13
10899	14	0	0	0	0
9652	0	13	12	0	0
10467	13	0	11	0	0
10467	0	12	12	0	0
10677	17	0	12	0	0
10675	9	0	3	0	0
10992	13	0	10	2	0
10421	0	13	13	12	12
8789	12	0	8	3	0
9460	0	17	17	0	0
10899	10	0	3	0	0
10897	0	13	13	12	10
8033	0	13	15	12	0
7725	0	13	14	12	11
7641	14	0	11	4	0
8630	14	0	11	5	0
7730	8	0	0	0	0
8604	17	0	0	0	0
8604	0	0	12	10	2
10899	0	0	11	9	0
10899	11	0	8	0	0
10899	11	0	0	0	0
10899	15	0	13	12	0
10899	0	14	15	13	12
10899	10	0	0	0	0
10899	0	16	18	14	12
10899	10	0	0	0	0
10899	10	0	0	0	0
10899	0	14	12	0	0
10899	11	0	6	0	0
8482	13	0	12	0	0
8090	0	12	13	12	10
7855	12	0	0	0	0
6533	0	13	0	0	0
7227	0	10	0	0	0

7494	0	0	0	0	0
8329	11	0	3	0	0
8848	16	0	12	9	0
8528	13	0	11	9	2
8528	11	0	7	0	0
7227	0	0	13	0	0
8139	0	13	13	12	11
7227	0	16	0	0	0
8514	0	15	17	14	0
8339	16	0	11	7	0
8339	0	0	0	0	3
7401	0	12	12	0	0
7401	0	0	0	13	8
7415	0	16	14	13	12
7415	0	16	14	13	12
7415	0	13	12	0	0
8195	0	16	18	0	0
6063	0	0	0	0	0
6278	0	12	13	0	0
6278	0	13	12	12	10
6297	0	14	0	0	0
6187	0	0	17	13	13
6021	0	10	13	12	0
5898	15	0	11	0	0
8789	0	3	0	0	0
7227	0	0	0	0	0
6063	0	0	0	0	12
8130	11	0	0	0	0
7227	0	16	18	14	12
7227	0	14	17	13	0
7227	0	0	0	0	12
7227	0	0	11	11	0
7227	0	13	13	13	0
7227	0	0	0	12	10
7227	11	0	4	0	0
7227	0	0	12	11	4
6531	0	11	12	11	3
8128	0	0	2	0	0
7889	0	0	0	0	4
7486	0	0	12	10	0
7221	0	12	12	12	9
7679	0	0	5	1	0
8209	12	0	11	0	0
8547	0	13	14	13	10
8702	11	0	3	0	0
8130	11	0	0	0	0
8532	7	0	0	0	0
7469	12	0	0	0	0
8128	0	14	17	13	0
8128	12	0	4	0	0
7204	0	10	0	0	0
7276	0	10	12	11	0
7276	0	13	13	0	0
8018	0	0	0	0	12
8018	10	0	0	0	0
10258	0	2	12	9	0
8532	11	0	0	0	0
7839	0	3	12	10	0
5429	0	0	0	0	0
8328	14	0	12	9	0
7235	0	17	18	0	0
7235	0	14	14	13	12
7601	0	2	12	11	0
7601	10	0	3	0	0
7601	0	0	12	10	0
8720	13	0	11	3	0
7889	0	0	17	14	12
8214	18	0	12	8	0

7469	12	0	0	0	0
8403	14	0	7	2	0
8403	0	0	12	11	0
8403	0	0	0	11	0
8403	0	0	0	0	0
7284	10	0	3	0	0
7284	12	0	7	2	0
7284	0	0	3	0	0
8130	0	13	14	12	12
7219	11	0	3	0	0
8534	0	0	11	0	0
8540	0	0	12	0	0
8540	0	0	9	0	0
8540	12	0	3	0	0
8405	18	0	0	0	0
8405	11	0	7	0	0
8405	14	0	11	2	0
8405	0	14	14	13	12
8534	9	0	0	0	0
8534	0	14	13	12	10
8540	12	0	4	0	0
8534	13	0	0	0	0
8534	11	0	0	0	0
8534	0	4	12	11	0
8534	0	16	17	14	13
8534	18	0	12	10	0
8534	0	16	0	13	0
8534	13	0	0	0	0
8405	0	4	12	11	2
8413	0	2	12	0	0
8789	0	13	13	0	0
8413	0	0	0	12	8
8413	13	0	9	0	0
8413	12	0	10	2	0
8413	0	14	14	12	11
8413	11	0	0	0	0
8413	13	0	9	2	0
8540	11	0	0	0	0
8413	10	0	0	0	0
8540	18	0	12	0	0
8413	10	0	0	2	0
8413	0	0	12	0	0
8413	0	0	12	5	0
7467	10	0	0	0	0
8540	0	0	17	13	13
8540	0	13	13	12	11
8540	0	13	13	13	12
8405	7	0	0	0	0
8413	13	0	11	3	0
8405	13	0	10	2	0
7478	0	0	9	0	0
7478	10	0	0	0	0
7478	10	0	0	0	0
7673	0	14	17	14	0
7673	0	16	18	16	13
7673	0	15	14	13	0
7478	17	0	13	0	0
7824	7	0	0	0	0
8201	0	0	18	17	14
7824	0	0	10	0	0
7824	9	0	3	0	0
7824	0	0	0	0	8
7821	12	0	11	4	0
7465	0	9	0	0	0
5425	0	12	0	0	0
5451	0	2	12	10	0
8534	12	0	12	0	0
7478	9	0	2	0	0
8502	0	0	0	3	0
8502	13	0	11	0	0
9941	11	0	0	0	0
8502	0	0	0	13	12
10254	0	0	12	7	1
7821	0	11	0	0	0

Sus

Kinet no	d4	p4	m1	m2	m3
8764	0	6	0	0	0
7478	0	13	13	12	0
8044	0	0	7	3	0
8718	0	0	0	2	0
7629	0	0	0	10	0
7478	0	11	17	0	0
6531	0	0	6	2	1
6297	0	0	0	6	2
8413	0	0	10	6	3
8470	0	0	0	0	3
10987	0	6	14	8	4
8534	0	6	12	8	5
7478	0	0	0	8	5
10238	5	0	0	0	0
8771	7	0	3	0	0
8777	7	0	3	0	0
7286	7	0	2	0	0
8547	8	0	0	0	0
8540	8	0	2	0	0
8076	8	0	0	0	0
7478	8	0	2	0	0
8403	8	0	2	0	0
7240	9	0	3	0	0
8547	9	0	6	0	0
8309	9	0	3	0	0
8547	9	0	6	0	0
10247	9	0	0	0	0
8403	9	0	1	0	0
8723	9	0	7	2	0
8318	10	0	7	0	0
8540	10	0	6	2	0
10677	10	0	0	0	0
10258	10	0	7	0	0
8273	11	0	3	0	0
8547	11	0	7	2	0
8540	11	0	6	0	0
6010	12	0	6	0	0
8789	13	0	5	0	0
10992	13	0	10	3	0
10258	13	0	0	0	0
10992	14	0	11	3	0

11

Hunting hartebeest (*Alcelaphus buselaphus*) in Sinai during Persian times (6th to 4th century BC)

Louis Chaix[1]

1 Dpt of Archaeozoology - Museum of Natural History - Geneva and APA. Archéologie et peuplement de l'Afrique - Université de Genève

Abstract

The site of Tell el-Herr in northern Sinai, is a mound built in different periods from Late Ptolemaic until Roman times. A fortress and habitations delivered numerous animal bones dated from the Persian period (6th to 4th century BC). The fauna is dominated by domestic mammals, mainly caprines. Wild fauna, including birds and fishes, is characterized by bones of hartebeest. This antelope was hunted in the area of the site and prepared locally. A minimum of 15 individuals are present, with a majority of adult males. These animals were exploited for their meat but also for their horns, which often bear technological marks. The presence of hartebeests in northern Sinai around 400 BC bears witness to the ancient distribution of the species, which used to live in the Near-East and is currently restricted to Central and South Africa.

Introduction

The site of Tell el-Herr is situated in the Sinai (Egypt), in the north-eastern part of the Nile Delta (Fig.1). This mound is one of the most important Persian settlements in the area of Peluse and a strategic place on the road between Egypt and the Near East. Since the prehistory, the region delivered remains of various cultures (Caneva 1992; Hassan 1998). The tell itself was excavated first by an Israeli mission and since 1985 by a French-Egyptian team from Sorbonne University (Valbelle 1987; 2007; Valbelle and Abd El-Maksoud 2007; Valbelle and Nogara 2000). The site comprises of different fortresses and a habitation area (Marchi 2014a, 2014b). The Persian levels are dated between the 6th and the 4th century BC. They delivered numerous animal bones reflecting the diet and the way of life of the inhabitants of the tell (Chaix 2002). For the Persian occupation, the fauna of the 5th century consists of four groups: domestic animals (51.3%), wild animals (3.6%), birds (5.3%) and fishes (39.6%). For the 4th century, the percentages are: domestic animals (51.8%), wild animals (0.3%) birds (2.3%) and fishes (45.5%; Table.1). In both assemblages, the livestock is dominated by domestic caprines (38.1% and 43.2%), of which 83% can be attributed to sheep and 17% to goat. Cattle hardly occurs. The domestic mammals represent the main source of meat. The numerous remains of fishes belong to small *Claridae*, between 10 and 15 cm, and only represent a small part of the diet.

The wild fauna, with the exception of birds, reptiles, fishes and molluscs, is characterized by a significant number of bones of hartebeest (*Alcelaphus buselaphus*), mainly in the levels dated from the last part of the 5th century BC (cf. Tab.1; Fig. 2). 106 remains were attributed to this species, representing a minimum of 15 individuals. A majority of them are adult, as shown by the fusion of four proximal tibia and six calcanei. According to the data known for a majority of ruminants, the hartebeests from Tell el-Herr were more than 3.5 years old.

Today, *Alcelaphus buselaphus* is an African antelope, with a distribution limited to some countries south of the Sahara (Namibia, Botswana, Tanzania, Kenya and Ethiopia). Its weight ranges from 75 to 200 kg and its withers height between 150 and 245 cm. Both males and females are horned and, except the weight and stature, there are few morphological differences between the two sexes (Vrba 1979; Kingdon 2013). To this day, eight subspecies have been identified, mainly based on their geographical distribution, some particularities of dimensions, colour and horn shape (Sclater and Thomas 1894; Arctander *et al.* 1999). One frontal from Tell el-Herr is similar to the orientation of horn-cores observed in the division of *A. buselaphus tora* (East 1999; Hildyard 2001; Fig. 3). Osteometrical data were compared with measurements in the literature (Eastern and Central Africa; Peters *et al.* 1997; Van

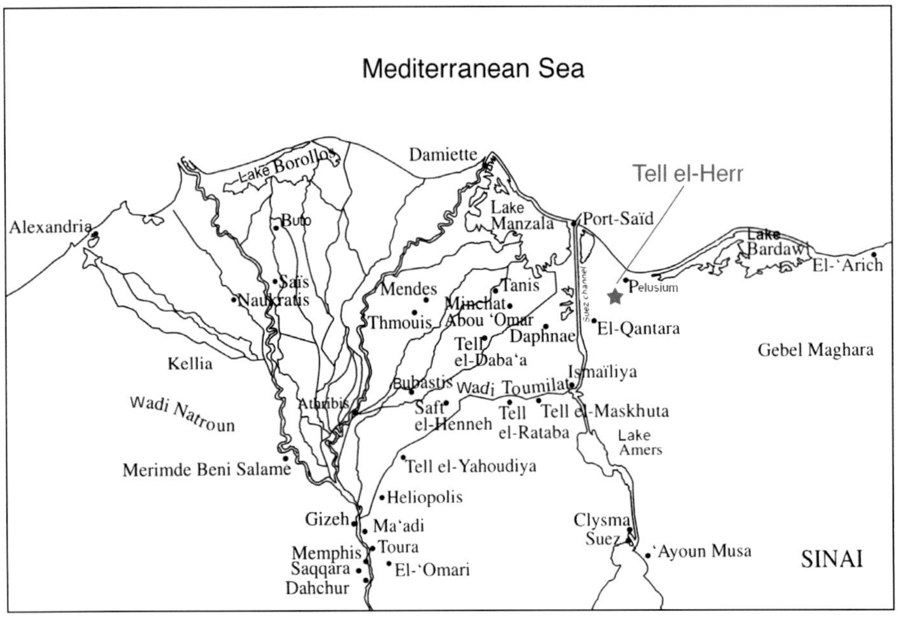

Fig. I. Geographical situation of Tell el-Herr.

Table I. Tell el-Herr faunal spectrum from the Persian levels. (NISP: number of identified bones).

5th Century BC		
	NISP	**%**
cattle	143	4.60
caprines	1183	38.14
dromedary	38	1.23
horse	31	1.00
donkey	169	5.45
pig	26	0.84
dog	3	0.10
hartebeest	**106**	**3.42**
gazelle	1	0.03
wild boar	4	0.13
hippopotamus	1	0.03
birds	166	5.35
fishes	1231	39.68
Total	**3102**	**100**

4th Century BC		
	NISP	**%**
cattle	14	4.65
caprines	130	43.19
dromedary	1	0.33
horse	1	0.33
donkey	7	2.33
pig	2	0.66
dog	1	0.33
hartebeest	**1**	**0.33**
birds	7	2.33
fishes	137	45.52
Total	**301**	**100**

Fig. 2. Hartebeest male (Sclater and Thomas 1894).

Neer 1989) and from skulls from Western Africa (National Museum of Natural History, Paris).[1] All the measurements (after Von den Driesch 1976) are presented in annexes (Annexes I and II). The dimensions of horn-cores are smaller than those of *Alcelaphus b. cokii* and *A. b. liechtensteini* (Fig. 4), and there is a lack of data about *A. b.alcelaphus*. The same observation can be made for the proximal breadth of the radius, significantly smaller than the males and females used for comparison (Fig. 5). The age of this bone could possibly influence this observation, because seven radius for a total of 14 are only represented by their proximal part. The dimensions of the first phalanx are placed in the

1 We are greatly indebted to Dr. Joséphine Lesur from the National Museum of Natural History of Paris for her permission to study the skulls of hartebeest kept in their collections.

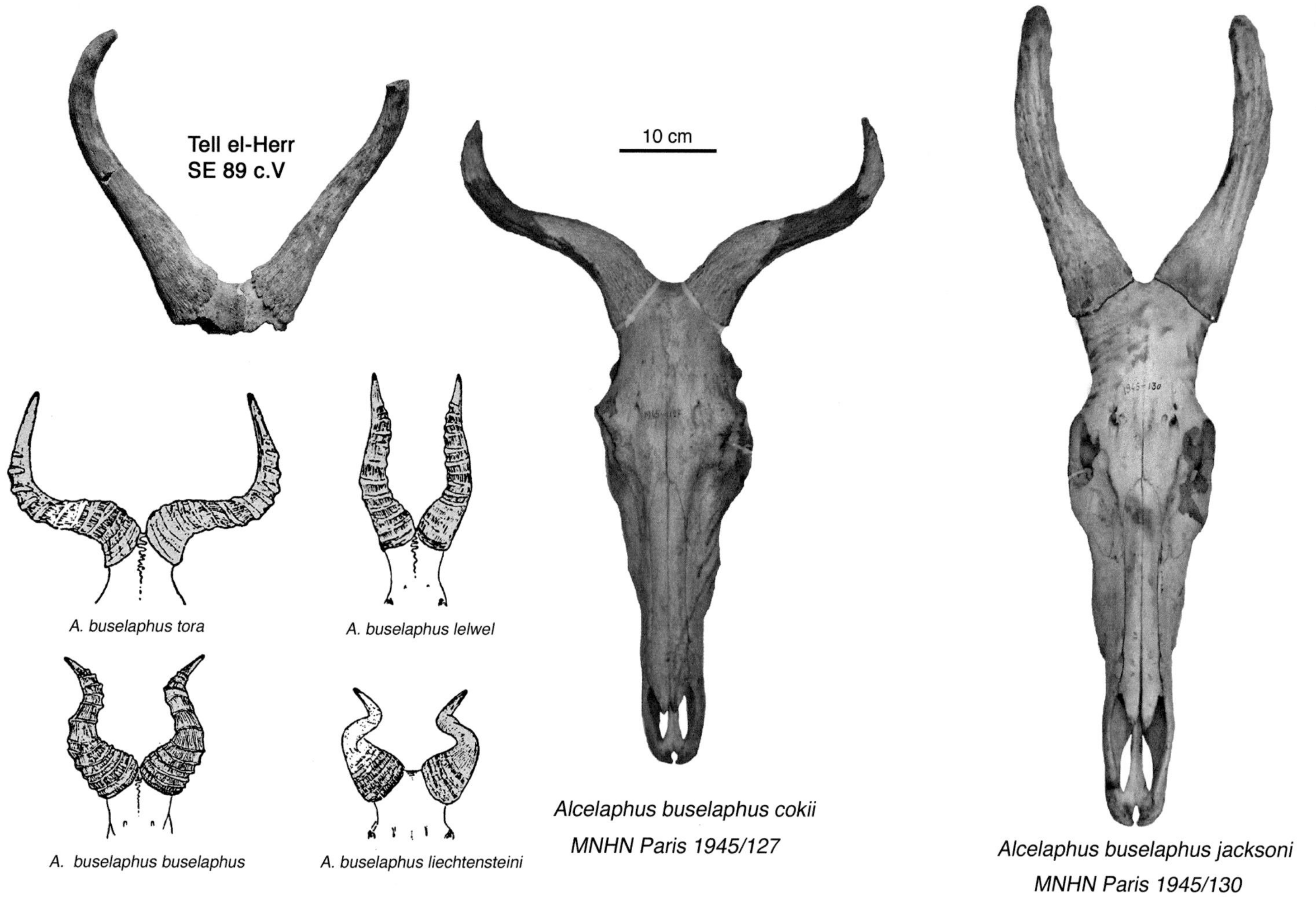

Fig. 3. The frontal of Tell el-Herr compared with skulls of *A. buselaphus cokii* (center), *A. buselaphus jacksoni* (right) from the Museum in Paris (MNHN-ZM-AC-1945-127 et MNHN-ZM-AC-1945-130) and with the figures of various subspecies (after Sclater & Thomas 1894: 125).

upper part of the distribution and probably pertain to male individuals (Fig. 6).

The distribution of skeletal elements suggests the presence of whole animals on the site. This observation shows that the antelopes were hunted not too far from the site, allowing hunters to bring fresh carcasses ready for consumption. Meanwhile, the repartition of the different skeletal parts shows some significant differences with the theoretical distribution: skulls and forelimb are well represented when ribs and vertebrae and extremities are clearly underrepresented. The lack of elements of the backbone can be due to the difficulty of determination of fragments of vertebrae and ribs. This phenomenon can also result from the exploitation of hartebeest carcasses outside the fortress (Fig. 7). Among the fragments of head, horn-cores are dominant and represent 87.5%. The repartition of these pieces displays a concentration in the south-eastern sector, in some storage rooms (Fig. 8). Other elements like radius and phalanxes are also numerous in the same sector. The exploitation of hartebeests is attested by marks linked to different operations: skinning, disarticulating, defleshing and technological use (Figs. 9 and10).

Horn-cores are particularly well represented, with 44.8% of the remains. Most of them present marks, sometimes near the base, probably for horn-sheath extraction. Others are sawed in the middle part or near the tip (Fig. 11). The use of the horn-core seems linked to the production of rings and possibly tubes (Fig. 12).

Tell el-Herr brings new information on the past distribution of this antelope, nowadays found only in Africa, south of the 15th Northern parallel. (Haltenorth and Diller 1977; Heckel 2009; Fig. 13). During the Pleistocene (Middle Palaeolithic and Late Palaeolithic, hartebeests were present in some favourable areas of the Near East, like Syria (Douara cave), Jordan (Azraq) and Israel (Tabun) (Uerpmann 1987). Later on, a strong decrease of wild ungulates occurred during the Iron Age, a period characterized by a clear expansion of human populations. Hartebeest was impacted by this phenomenon and during all periods, the remains of this antelope are found in low number, from the Natufian until the

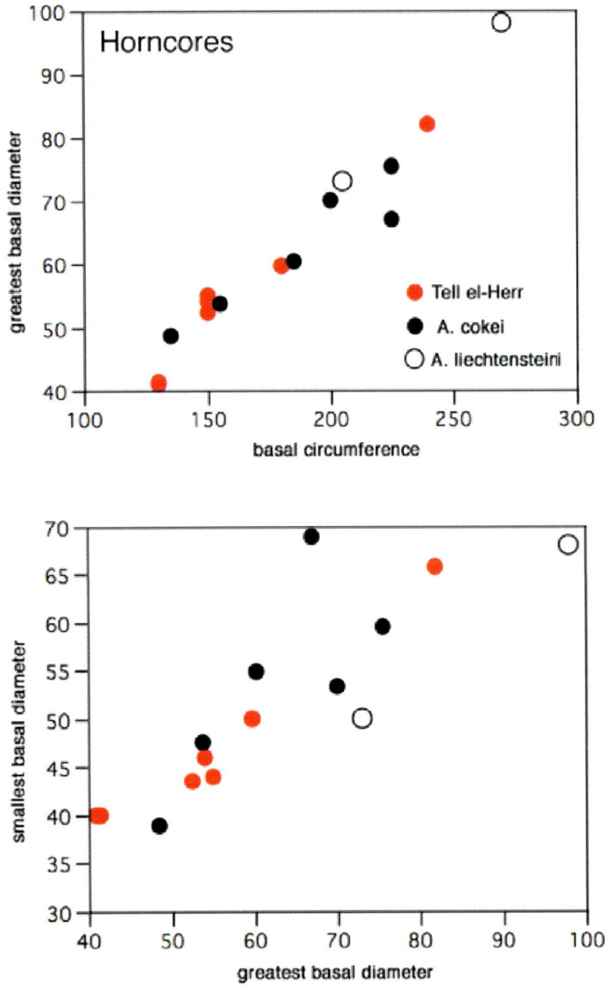

Fig. 4. Diagram showing the dimensions of the horncores from Tell el- Herr compared with those of *A. b. cokii* and *A. b. liechtensteini*.

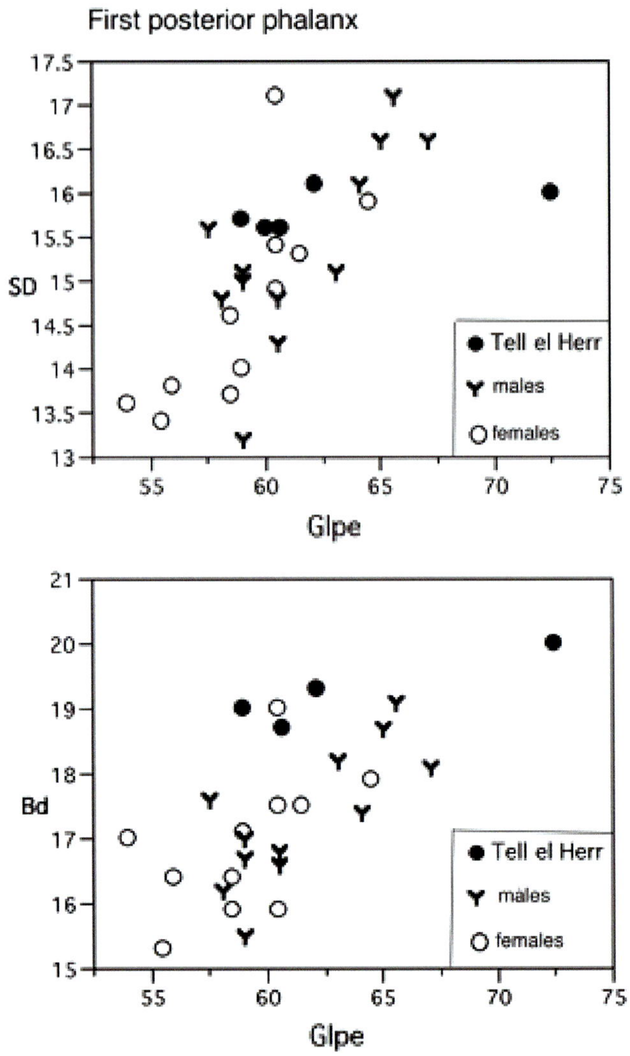

Fig. 6. Diagram showing the dimensions of the first phalanx from Tell el-Herr compared with hartebeest males and females.

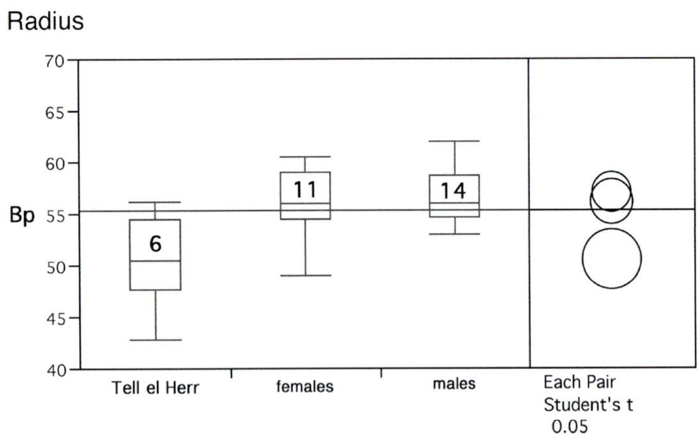

Fig. 5. Diagram showing the dimensions of the radius proximal from Tell el-Herr compared with hartebeest males and females.

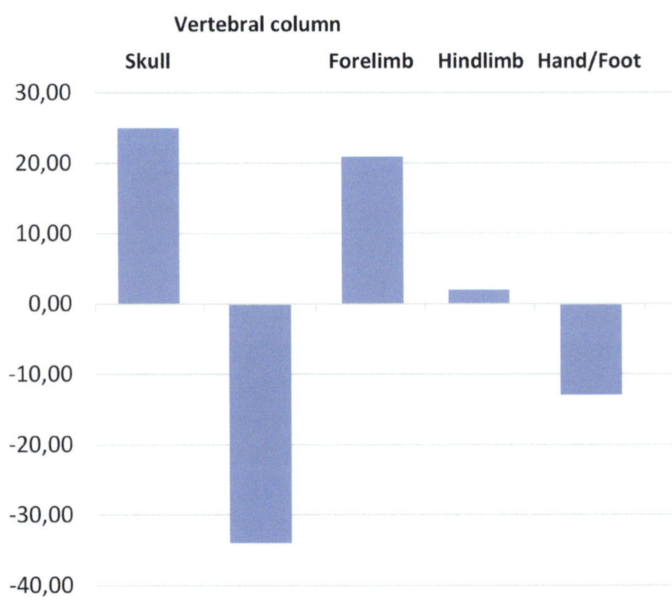

Fig. 7. Distribution of the skeletal elements from hartebeest from Tell el-Herr.

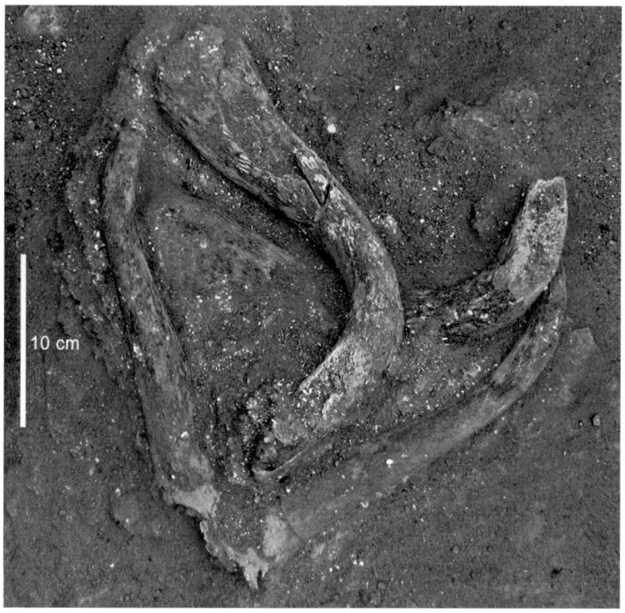

Fig. 8. Deposit of hartebeest's horns found in the dump of Unit C, in the south- eastern part of the site.

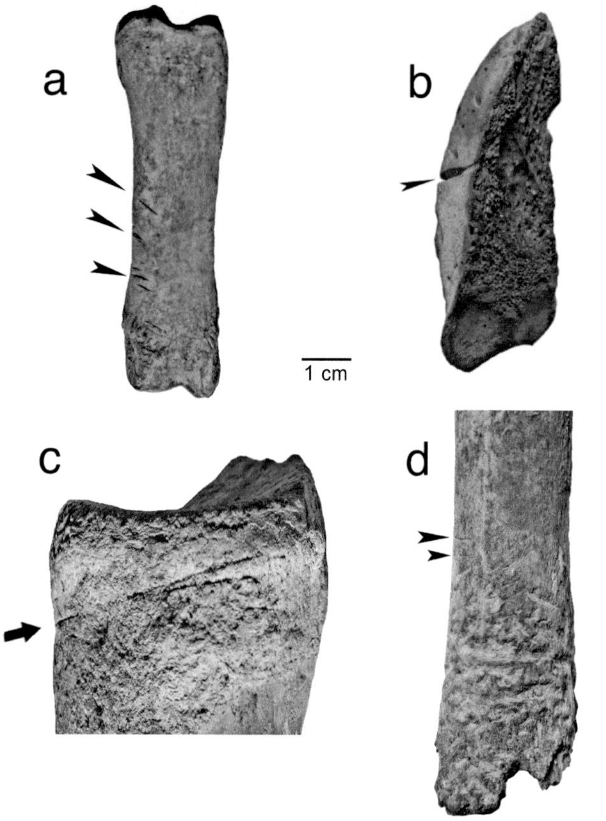

Fig. 10. Various type of traces: a) marks of skinning on a first phalanx; b) marks of skinning ? on a third phalanx; c) tracks of disarticulation on a proximal radius; d) traces of defleshing and biting on a distal metacarpal.

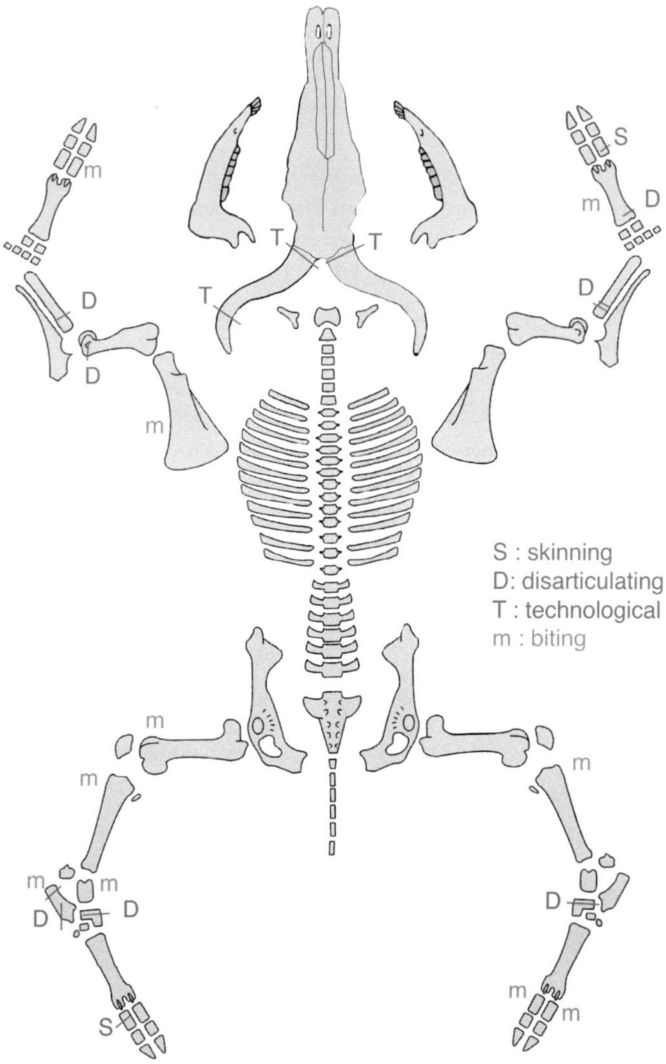

Fig. 9. Synthetic diagram illustrating the various marks observed on the hartebeest's skeleton from Tell el-Herr.

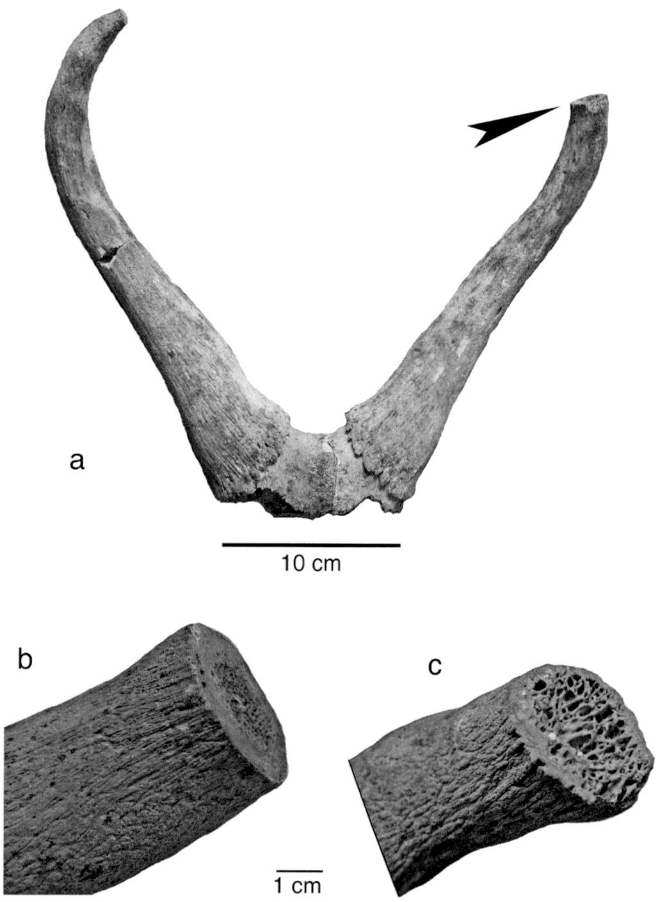

Fig. 11. Technological traces (sawing) on hartebeest's horncores.

Fig. 12. Horncore of hartebeest sawed to prepare rings.

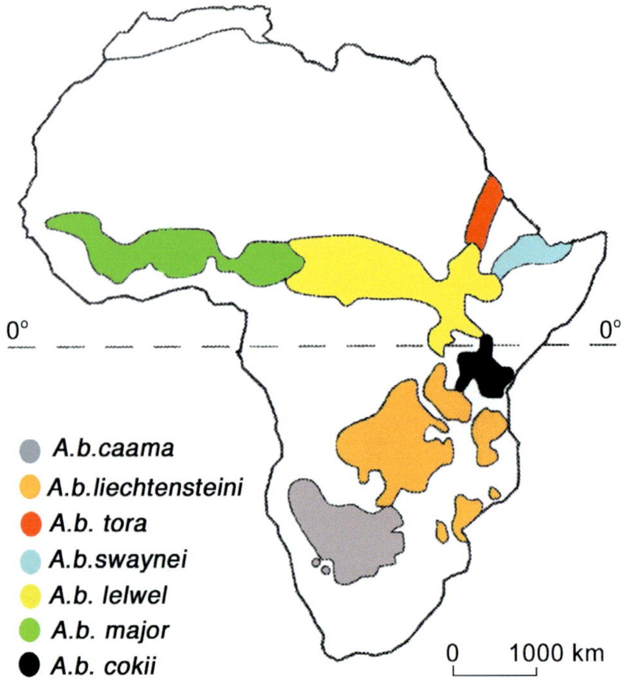

Fig. 13. Map of distribution of actual subspecies of hartebeest (modified after Kingdon 1997).

Iron Age, where the last specimen was found in Lachish (Israel; Tsahar *et al.* 2009; Ussishkin 2004).

The hartebeest remains from Tell el-Herr Persian levels (600 to 350 BC) can be considered as an informative stage in the geographic evolution of this species from its former range to its present African homeland (Uerpmann 1987: 88). In Africa, the history of this antelope shows a clear decrease since the Palaeolithic (Linseele and Van Neer 2009; Pöllath 2009). The subspecies *Alcelaphus buselaphus buselaphus* is now extinct in Northern Africa: the last individuals were killed in Tunisia in 1925 (Harper 1945) and in Egypt at the end of the 20th century (Osborn and Helmy 1980; Anderson 1902). In Egypt, hartebeest was present since prehistoric times. (Osborn and Osbornova 1998). Two whole males were buried in Saqqara during the 18th Dynasty (around 1500 BC; Lortet and Gaillard 1903: 72-77). For these authors, they belong to the subspecies "*buselaphus*". Rare reliefs show the hunting of hartebeest, like the tombs of Amenemhet and Rekhmire in Thebes dated from the 18th Dynasty (Davies *et al.* 1915; Davies and De Garis 1943). Like other antelopes (*Oryx* and *Addax*), hartebeest was tamed, as attested by a scene from the Mastaba of Mereruka, in Saqqara, dated from the 6th Dynasty (Osborn & Osbornova 1998: 172).

To conclude, the hartebeests from Tell el-Herr bring new information about the history of this ruminant, from its largest distribution reaching the Near East to its present status, restricted to a few subspecies distributed from Ethiopia to West Africa and in some countries south of the equator. Hartebeests were hunted in the area of the fortress. Along with their consumption, they provided raw material, especially horns used to produce various artefacts.[2]

2 Thanks to Dr. Pierre Meyrat for the English revision of this text.

References

Anderson, J. 1902. *Zoology of Egypt: Mammalia* (revised and completed by W.E. de Winton). London: Hugh Rees.

Arctander, P., Johansen C. and Coutellec-Vreco, M.A. 1999. Phylogeography of three closely related African bovids (tribe Alcelaphini). *Molecular Biology and Evolution* 16 (12): 1724-1739.

Caneva, I. 1992. Le littoral nord-sinaïtique dans la préhistoire. *CRIPEL* 15: 33-38.

Chaix, L. 2002. Animal exploitation at Tell El-Herr (Sinaï, Egypt) during Persian times: first results. In, Buitenhuis, H., Choyke, A. M., Mashkour, M. and Al-Shyab, A.H. (eds) *Archaeozoology of the Near-East* V: 268-272, Groningen: ARC – Publicaties 62.

Davies, N, De Garis, N. and Gardiner, A.H. 1915. *The Tomb of Amenemhet*. London.

Davies, N. N. and De Garis, N. 1943. The tomb of Rekh-mi-Re at Thebes. *Publications of the Metropolitan Museum of Art Egyptian Expedition* 11.

East, R. 1999. African antelope database 1998. *IUCN Species Survival Commission*: 186-193.

Haltenorth T. and Diller, H. 1977. *A field guide to the mammals of Africa including Madagascar*. London: Collins.

Harper, F. 1945. Extinct and vanishing mammals of the Old World. *American Committee for International Wildlife Protection*: 642-648. New York.

Hassan, F. A. 1998. Relations culturelles entre l'Égypte et ses voisins orientaux durant la Préhistoire récente. In, Valbelle D. and Bonnet, C. (eds) *Le Sinaï durant l'Antiquité et le Moyen-Age*: 12-19. Paris: Ed. Errance.

Heckel, J.O. 2009. The present status of hartebeest (*Alcelaphus buselaphus* ssp.) with special focus on north-east Africa and the Tora hartebeest (*Alcelaphus buselaphus tora)*. In, Riemer, H., Förster, M., Herb, M. and Pöllath, N. (eds) Desert animals in the eastern Sahara: status, economic significance, and cultural reflection in antiquity. *Colloquium Africanum* 4: 141-153.

Hildyard, A. 2001. *Endangered Wildlife and Plants of the World*: 674-675. New York: Marshall Cavendish.

Kingdon, J. 2013. *Mammals of Africa*: London: Bloomsbury.

Linseele, V. and Van Neer, W. 2009. Exploitation of desert and other wild game in ancient Egypt : the archaeozoological evidence from the Nile Valley. In, Riemer, H., Förster, F., Herb, M. and Pöllath, N. (eds) *Desert Animals in the Eastern Sahara: Status, Economic Significance, and Cultural Reflection in Antiquity. Colloquium Africanum* 4: 47-78.

Lortet, L.C. and Gaillard, C. 1903. La faune momifiée de l'Ancienne Egypte. *Archives du Museum d'Histoire Naturelle de Lyon* 8: 1-206.

Marchi, S. 2014a. Construire et vivre à Migdol à l'époque perse. L'habitat en contexte fortifié de Tell el-Herr (Nord-Sinaï). *Egypte, Afrique et Orient* 72: 33-42.

Marchi, S. 2014b. *L'habitat dans les forteresses de Migdol (Tell el-Herr) durant les Ve et IVe siècles av. J.-C. Etude archéologique*. Paris: Presses Universitaires Paris Sorbonne.

Osborn, D.J. and Helmy, I. 1980. The contemporary land mammals of Egypt (including Sinaï). *Fieldiana* (Zoology) 5 (1309): 1-579.

Osborn, D.J. and Osbornová, J. 1998. *The Mammals of Ancient Egypt*. Warminster: Aris & Phillips Ltd.

Peters, J., Van Neer, W. and Plug, I. 1997. Comparative postcranial osteology of hartebeest (*Alcelaphus buselaphus*), scimitar oryx (*Oryx dammah*) and addax (*Addax nasomaculatus*), with notes on the osteometry of gemsbok (*Oryx gazella*) and Arabian oryx (*Oryx leucoryx*). *Annales Sciences Zoologiques*, vol. 280. Tervuren: Musée Royal de l'Afrique Centrale.

Pöllath, N. 2009. The prehistoric gamebag: the archaeozoological record from sites in the Western Desert of Egypt. In, Riemer, H., Förster, F., Herb, M. and Pöllath, N. (eds) Desert animals in the eastern Sahara: status, economic significance, and cultural relection in antiquity. *Colloquium Africanum* 4: 79-108.

Sclater, P. L. and Thomas, O. 1894. *The Book of Antelopes*. Vol. I. London: Porter.

Tsahar, E., Izhaki, I., Lev-Yadun, S. and Bar-Oz, G. 2009. Distribution and extinction of Ungulates during the Holocene of the Southern Levant. *PLOSOne* 4 (4): 1-11.

Uerpmann, H. P. 1987. The ancient distribution of Ungulate mammals in the Middle East. Fauna and archaeological sites in Southwest Asia and Northeast Africa. *Beihefte zum Tübinger Atlas des Vorderen Orients* A 27.

Ussishkin, D. 2004. The renewed archaeological excavations at Lachish (1973-1994). *Monographs of the Institute of Archaeology* 22, Tel Aviv: Tel Aviv University.

Valbelle, D. 1987. Entre l'Égypte et la Palestine, Tell El-Herr. *Bulletin de la Société française d'Égyptologie* 109: 24-38.

Valbelle, D. 2007. Tell el-Herr. *Les Niveaux Hellénistiques et du Haut-Empire*. Paris: Ed. Errance.

Valbelle, D. and Abd El-Maksoud, M. 2007. *Mission franco-égyptienne de Tell el-Herr (Nord-Sinaï)*. Paris: Ed. Errance.

Valbelle, D. and Nogara, G. 2000. La forteresse du IVe siècle avant J.-C. à Tell el-Herr (Nord-Sinaï). *CRIPEL* 21: 53-56.

Vrba, E. 1979. Phylogenetic analysis and classification of fossil and recent Alcelaphini, Mammalia: Bovidae. *Biological Journal of the Linnaean Society* 11 (3): 207-228.

Van Neer, W. 1989. Contribution to the archaeozoology of Central Africa. *Annales Sciences zoologiques*, vol. 259. Tervuren: Musée Royal de l'Afrique Centrale.

Von den Driesch, A. 1976. A guide to the measurement of animal bones from archaeological sites. *Peabody Museum Bulletins* 1: 1-136.

Apendix I

Osteometrical data from Persian hartebeest (after von den Driesch, 1976).

Horn-core

External length	Basal circumference	Greatest diameter	Least diameter
330	240	82	65,7
290	180	59,8	50
265	150	54	46
260	150	55	44,1
	210		
	150	52,4	43,6
	130	41	40
	130	41,5	40

M3 inferior

L

L	B
30	11
28,5	11,6

Atlas

BFcr	BFcd
77	67

Scapula

GLP	LG	BG	SLC
41			25
54	44	35	34,8
58	46	39	38,9
55	42,5	35,3	35
57,6	44,7	37,6	40,1

Humerus

Bd	BT
51,4	46,4
51	48
54	52

Radius

BP	BFp	Bd
51,1	47,2	
49,2		
42,8		42
54	49	
49,9		
56,1		48,1

Ulna

LO	DPA
74	49
75,4	

Metacarpal

Bd	Dd
43	27
44	27,5

Phalanx I anterior

Glpe	Bp	SD	Bd
68,8	20,7	16	20
68	22,8	19	21,5
61,8	20,1	15,3	18,3
59,5	20,3	15,3	19
			20,5
			17,2

Apendix II

Osteometrical data from Persian hartebeest (after von den Driesch, 1976).

Phalanx II anterior

GL	Bp	SD	Bd
40	23,5	17	17
38,8	20,4	17	19

Tibia

Bd	Dd
45,4	35,9
44	35
36	27,1

Talus

GLm
42,3

Calcaneus

GL	DTmax
105,3	
105,1	35,4
104,9	35,3
102	36
102	

Naviculo-cuboid

GB
42
39,2
39

Metatarsal

GL	Bp	Dp	SD	DD	Bd	Dd
254	34,8	37	21	22	40	27
	35	36				
	33	39			43	25
					46	28

Phalanx I posterior

Glpe	Bp	SD	Bd
72,5	23,2	16	20
71	23	15	20
62,2	21,1	16	19
60,7	21,3	16	19
60	20,5	16	
59	19,4	16	19

Phalanx II posterior

GL	Bp	SD	Bd
39,5	18,5	14	17
37,8	18,5	15	16
36,9	17,5	14	15
32,8	20,2	16	

Phalanx III

DLS	Ld
57,9	
56,5	
55,5	48,9
51,6	45,6

12 Persian Period Dog Burials of Beirut: Morphology, Health, Mortality and Mortuary Practices

Yasha Hourani[1]

1 Department of Archaeology -The University of Sheffield, Minalloy House - 10-16 Regent Street, Sheffield S1 3NJ (United Kingdom); yashahourani@hotmail.com

Abstract

Sixteen nearly complete and partial dog burials, in addition to scattered canid remains, were recently recovered from Persian period layers in Beirut. The practice of dog inhumation in Levantine Persian period is widely attested. However, few detailed archaeozoological analyses were so far carried out on the canid remains. This study outlines the archaeological characteristics of the Beirut burials and provides a comprehensive assessment of the biological characteristics of the buried dogs. Morphological and metric analysis show that all the specimens were of mediolinear type. The mortality profile points to an attritional death, whereas pathological conditions show that dogs suffered from traumatic injuries, joint and periodontal diseases. Relying on archaeothanatological observations, a reconstruction of the burial space and mortuary practices is also attempted. The analysis suggests that, aside from stone and amphora fragments, dogs were buried under perishable covers or in perishable containers. Furthermore, the location of the burials reveals new patterns on this tradition of dog inhumation. The burials were laid out at the outskirts of the ancient city, aligned parallel to the shore of a water stream and to the axis of a street. Hence, this study corroborates previous observations on the biological features of the dogs inhumed in Levantine cities during the Persian period and, by addressing previously not tackled issues on mortuary practices, contributes to improve our knowledge of this phenomenon.

Keywords

Dog burials, Persian period Beirut, Archaeothanatology, Morphotype, Pathology

Introduction

The practice of dog burial in Persian period Levant is by now widely attested (Dothan and Porath 1982; Bennett and Schwartz 1989; Bennett and Blakely 1989; Wapnish and Hesse 1993; Kelm and Mazar 1995; Nadelman 1995; Maher 2005; Sapir-Hen and Bar-Oz 2006; Sapir-Hen 2011; Çakirlar *et al.* 2014). This tradition was widespread from the late sixth to early fourth centuries BC along the central and southern Levantine coastal strip, between Beirut in the north and the Gaza Valley in the south (Çakirlar *et al.* 2014). Many explanations of this phenomenon and its origins have been proposed. Ashkelon burials were considered, on one hand, as special ritual deposits (Stager 1991; Heltzer 1998; Halpern 2000; Edrey 2008); on the other hand as arising from a local custom of burying dogs, while the burials themselves were not associated with any ritual (Wapnish and Hesse 1993). Osteological analyses showed that most of the buried dogs were of medium size and mesocephalic skull type, with limb proportions lying in the range of modern shepherd dogs (Wapnish and Hesse 1993, Çakirlar *et al.* 2014).

Within the borders of modern Lebanon, three sites yielded intentional dog burials dating to the Iron Age. At Khaldé, Saidah (1967) mentions the presence of eight dog burials near a human burial site. Nonetheless, the brief description of their stratigraphic position makes it unclear whether they are contemporary with the Iron Age II human cemetery, hence earlier than the Persian period (Dixon 2013). Tell el-Burak yielded two burials along with six scattered canid remains dating to the fourth century BC (Kamlah and Sader 2003; Sader and Kamlah 2010; Çakirlar *et al.* 2014). Beirut (lot BEY 020) yielded ten burials, dating to the fifth century BC, out of which seven were excavated (Finkbeiner and Sader 1997). The canid remains from Khaldé and lot

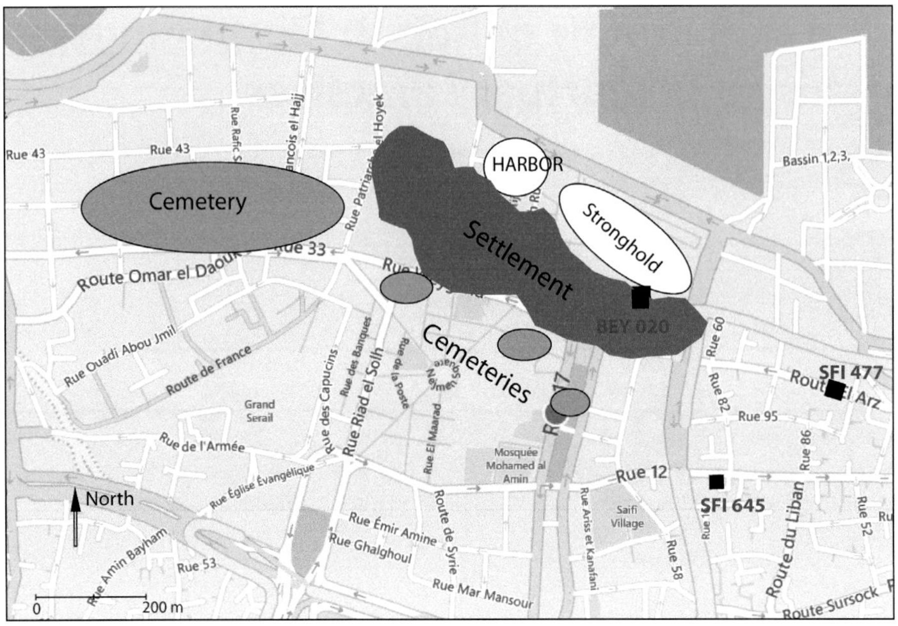

Fig. 1. Location of the three lots (BEY 020, SFI 477 and SFI 645) that yielded Persian period dog burials in Beirut. Superimposed over a recent map of modern Beirut (google map), are depicted the Persian period settlement in the ancient Tell area, the human cemeteries and the Phoenician harbor (modified after Curvers 2013 with his kind permission). Lot BEY 020, located at the edge of the settlement yielded ten dog burials uncovered in debris layers deposited on top of the Iron Age glacis. Lot SFI 477 and SFI 645, located to the east, yielded sixteen burials that are analyzed in this paper.

BEY 020 in Beirut have not been analyzed from an archaeozoological viewpoint. On the other hand, Tell el-Burak burials were thoroughly analyzed based on archaeological and osteological approach. Still, the sample was too small to be representative of the dog population that was once present at Burak. Hence, an accurate picture of the dog population involved in the practice of dog inhumation in the central Levant is still missing.

Recent excavations held in 2013 and 2014 at Beirut yielded sixteen additional Persian burials. This paper is focused on those recently excavated canid remains. Contextual location of the burials and their stratigraphic position in relation to urban development revealed that they were laid at the outskirts of the ancient city. Herein, a reconstruction of the burial space and mortuary practices is attempted, relying on archaeological and archaeothanatological observations at hand. Further, a comprehensive assessment of the morphological traits, health status and mortality profile of the buried dogs is exposed. The outlined biological and archaeological characteristics would define whether the Beirut burials fit in the features known for the large-scale dog inhumation tradition in the Levantine Persian period, and to shed more light on this phenomenon in the central Levant.

Historical Background and Site Location

During the Persian period (538 – 332 BC), Beirut was a small port-city known under the name of *B'rt* (Markoe 2008), an active and prosperous cosmopolitan milieu where local traditions met external cultural inputs, especially from the Egyptian and Greek cultures (Elayi 2001; Elayi and Sayegh 2000). The city was under the political control of the city-state of Sidon whose territory stretched out north to Nahr el-Kelb and south to Jaffa, with the exception of Tyre and its dependencies. Furthermore, Sidon itself was under the Persian suzerainty (Elayi 2001; Elayi and Sayegh 2000).

The organization of the ancient city was brought forward through archaeological research undertaken in the 1950's and during the reconstruction of war-devastated downtown Beirut in the 1990's. With regard to the Late Iron Age, a residential quarter located to the west of the ancient Tell and dating to the Persian period was uncovered in the 1990's. The sector was built at the end of the sixth century BC, under the reign of king Eshmunazor or Bodashtart and during the expansion period of Sidon in southern Phoenicia. An earthquake likely destroyed it at the end of the fourth / early third century AD (Elayi 2001; Elayi and Sayegh 2000). The residential area, laid out in an orthogonal plan adapted to the local topography, comprised houses, shops and workshops as well as a small temple seemingly dedicated to Astarte. The design of the edifices fits in the typical Phoenician architectural style (Elayi 2001; Elayi and Sayegh 2000). Another lot on the Tell area, BEY 020, yielded layers of debris deposited above the Iron Age glacis and dating to the Persian period. In these layers, the Persian period dog inhumation phenomenon was first observed, with ten uncovered burials (Finkbeiner and Sader 1997).

In the last decade, more rescue excavations took place at the eastern fringes of the ancient city. Two lots, SFI 645[1] and, 250 meters further to the north, SFI 477[2], both located in the Saifi quarter of modern Beirut (Fig. 1), were excavated in 2013 and 2014. Both recorded a slight Persian period activity linked

1 Rescue excavations of the Directorate General of Antiquities (Lebanon) under the supervision of Dr. Assaad Seif
2 Rescue excavations of the French Institute for the Near-East (Beirut) under the supervision of Dr. Joyce Nassar

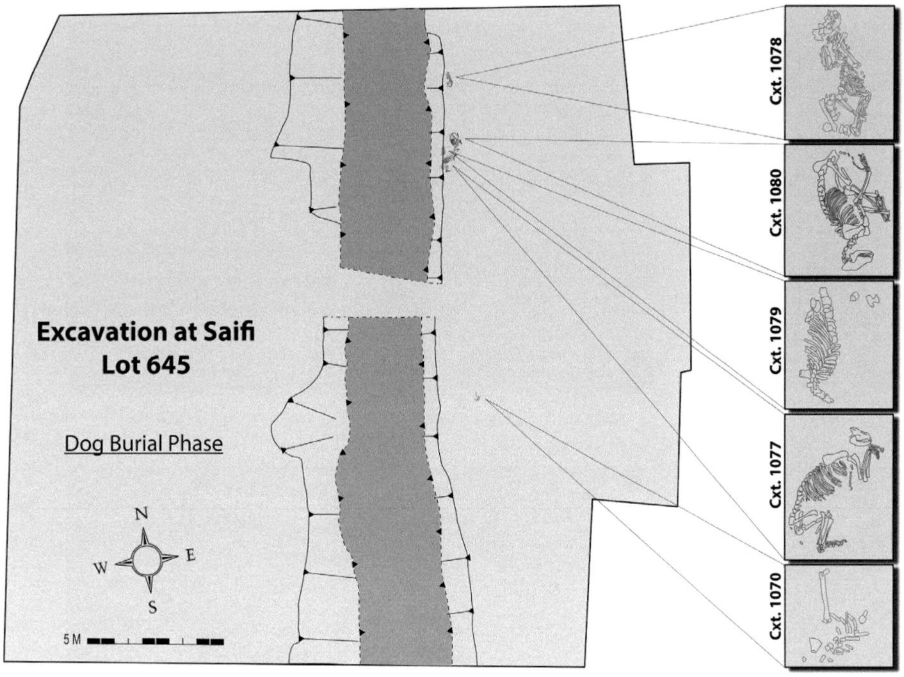

Fig. 2. Location of the dog burials along the shore of the water stream at lot SFI 645. The bottom of the stream is shown in dark green and its stepped edges by the black lines. Canid remains lay on the eastern side of the stream (Courtesy of SFI 645 excavation project, Directorate General of Antiquities, Lebanon).

to a burial-ground. At SFI 645, the area was exclusively devoted for dog burials (Seif and Choueiri, in prep.). Meanwhile, at SFI 477 human and dog burials were laid out on one side of a street dating to the Persian period; the stratigraphic relationship between the human and the dog burials will be more finely defined with the progress of current archaeological research (Nassar and Chanteau 2018).

Material

Sixteen nearly complete and partial dog burials, in addition to some scattered dog remains, have so far been recovered from the Persian layers uncovered in the Saifi quarter.

At the time of Late Iron Age occupation, a seasonal stream dissected lot SFI 645 on a south-north axis and showed many phases of extension and regression (Seif and Choueiri, in prep.). In this sector, dog burials have been laid out in accumulation layers deposited by the river activity. Articulated nearly-complete and partial skeletons of four dogs (Canid 1077, 1078, 1079, 1080) and a group of bones partially in anatomical connection (Canid 1070) were uncovered at the eastern edge of the stream, laid parallel to its axis (Figs. 2 and 3). The partial skeleton (1079) was truncated by the natural cut of the water stream during one of its extension phases. Five scattered remains (atlas, axis, mandible, humerus and tibia) belonging to at least two more individuals were also recovered from the same layers.

Lot SFI 477 has highlighted the presence of a street, oriented along an east-west axis, during the Persian period (Nassar and Chanteau 2018). Remains of twelve canids were recovered at the southern side, aligned parallel to the street. Eight specimens were nearly-complete skeletons (DA1, DA3, DA4, DA7, DA8, DA9, DA10, DA11,). Two specimens (DA6, DA12) consisted of articulated partial skeletons truncated by modern foundation-trenches and one specimen (DA5) truncated by an Ottoman-era trench consisted of a cranium articulated with the mandible (Fig. 3). In addition, isolated elements of a right hind paw found next to specimen DA9 do not match any of the partial specimens based on size.

At SFI 477, an amphora fragment covering one of the skeletons made the burials datable to the Late Iron Age. At SFI 645, pottery sherds recovered from the layer superimposed upon the skeletons and from the layer they lie above yielded the *terminus ante* and *post quem*.

Methods

Archaeothanatology

Despite the effort made to delineate the burial pits during the excavations, none were found on either site. Therefore, to understand whether the carcasses decomposed in open air, were buried in earth-grave or placed in a perishable container, an attempt was made to characterize the burial space. Hence, post-depositional displacements of skeletal remains linked to mortuary practices were recorded (Duday 1990, 2006; Duday *et al.* 2009, 2014). The position of skeletal elements indicates whether the decomposition happened in a filled environment or in a void (Duday 1990, 2006; Duday *et al.* 2009, 2014; Hoogland and Hofman 2013). The chronology of natural dislocation of the skeletal joints (Hill and Behrensmeyer 1984; Lyman 1994: 313; Méniel 2008) was taken into account when interpreting the displacements of skeletal elements. This might give useful information about the corpse during the initial phase of

SFI 645. Canid 1070 — SFI 645. Canid 1077 — SFI 645. Canid 1078 — SFI 645. Canid 1079

SFI 645. Canid 1080 — SFI 477. Canid DA1 — SFI 477. Canid DA3 — SFI 477. Canid DA4

SFI 477. Canid DA5 — SFI 477. Canid DA6 — SFI 477. Canid DA7 — SFI 477. Canid DA8

SFI 477. Canid DA9 — SFI 477. Canid DA10 — SFI 477. Canid DA11 — SFI 477. Canid DA12

Fig. 3. Photographs of the five articulated canid specimens at lot SFI 645 and the eleven articulated specimens at SFI 477. Specimen 1070 consists of several bone finds partially in anatomical connection. Specimen 1079 is truncated by the natural cut of the water stream. Specimen DA5 is truncated by an ottoman trench and only the cranium remains in connection with the mandible. Specimens DA6 and DA12 are truncated by modern foundation-trenches. Specimen 1078 was covered by a large cobble that caused the collapse of the thoracic region during decomposition process and specimen DA9 was covered by Late Iron Age amphora fragments (Courtesy of the SFI 645 excavation project – Directorate General of Antiquities, Lebanon and the SFI 477 excavation project – French Institute for the Near-East, Beirut).

decomposition (Duday *et al.* 2009). Based on archaeothanotological observations, suggestions on the structure of the tomb were thus made.

Age and Sex

Age-at-death estimation was based on the fusion of the cranial (Barone 1976) and postcranial skeleton (Barone 1976; Evans and Christensen 1979; Watson *et al.* 1986), dental eruption (Silver 1969) and tooth wear state (Horard-Herbin 2000).

Ages of fusion in cranial and postcranial skeleton, defined on eumorphic dogs, could be applied to the sample under study since none of the specimens showed features of a chondrodystrophic brachymel hound, whose bones differ from those of dogs of a normal build in the earlier fusion of limb epiphyses and cranial base (Hilzheimer 1926: 97; Baxter 2006, 2010). Nevertheless, some discrepancy was observed between the state of cranial and postcranial fusion. This prompted the choice to prevail postcranial fusion data in the age-at-death estimation.

Two types of mortality profile were constructed. The first profile suggests estimation of absolute ages. It is based on ages of postcranial epiphyseal fusion in subadults and adults and on dental eruption ages in juveniles, as some lacked the postcranial skeleton. The second profile is based on dental eruption and wear state of the first lower molar. It is constructed following Horard-Herbin's (2000) scheme which is divided in three age groups, young (stage A,B,C), intermediate (stage D,E,F) and old (stage G); moreover, a stage P, designating Puppies, was added in order to include individuals showing deciduous dentition.

The sex was determined by the presence or the absence of the *baculum* (Evans and Christensen 1979). In the case of partial or poorly preserved skeletons, sex could not be determined.

Dog morphotypology

Regarding ancient dogs, aspects of the physical characteristics as size, shape of the skull and the body proportions can be reconstructed based on skeletal remains. In an attempt to assign the archaeological specimens to morphotypes, classifications of modern dogs were used. The latter are divided in distinct types based on morphological differences in the general appearance or particular criteria like the shape of the skull (Galy 2004).

The typology based on body proportions (height and length of the body) divides dogs in three groups (Regnault 1903; Galy 2004: 8; Belhaoues 2011a: 97): longilinear (body rather high than long with cursorial adaptations), mediolinear (harmonious proportions between body height and length) and brevilinear (body rather massive and large than high). The typology based on the general size divides dogs in four groups: hypermetric (standing more than 60cm at the withers), eumetric (between 40 and 60 cm at the withers), hypometric (between 20 and 40 cm at the withers) and ellipometric (less than 20 cm at the withers; DeGrossiMazzorin and Tagliacozzo 2000: Fig. 10; Belhaoues 2011a: 98). The cranial typology based on the shape of the cranium, mainly the allometric variation of the face (Lignereux *et al.* 1991), also divides dogs in three groups (Stockard 1941; Ellenberger and Baum 1943; Evans and Christensen 1979; Brehm *et al.* 1985; Lignereux *et al.* 1991; 1992): dolichocephalic (narrow cranium, thin and long-snouted, almost-inexistent stop), mesocephalic (medium-snouted, larger snout than in the dolichocephalic type) and brachycephalic (broad and short-snouted cranium). The typology based on the general appearance divides dogs in four groups (Megnin 1897; Galy 2004: 9-10; Belhaoues 2011a: 96-97): lupoid (primitive appearance, mediolinear and sometimes hypermetric, medium-snouted cranium), braccoid (often mediolinear, snout as large at its base as at its extremity, well-defined stop), molossoid (massive body and massive short-snouted head) and graioid (high and slender-longilinear-, dolichocephalic head showing no stop).

These classifications are to some extent appropriate to be applied in the description of archaeological remains (Lignereux *et al.* 1992; De Grossi Mazzorin and Tagliacozzo 2000; Belhaoues 2011a), in contrary to breeds whose concept is based on standards that cannot be readily identified from archaeological remains (coat, color, ears' shape, temperament, *etc.* or genetic disorders to which they are prone; Cruz *et al.* 2008). By the means of morphological description and metric analyses, archaeological specimens can be assigned to the above-mentioned morphotypes. This approach is herein based (following Belhaoues 2011a) on the withers height, cranial shape, conformation of the lower jaw and the limbs, *i.e.* the ratio between the total length and the slenderness index of the mandible and the major limb elements. Throughout these analyses, metric comparisons with modern breeds were used to aid in the description of morphotypes. Metric analyses of modern breeds have been carried out and are described in previous reports, using bivariate or multivariate analysis, and proved to be useful in providing an indication of the dog morphotype (Wapnish and Hesse 1993; De Grossi Mazzorin and Tagliacozzo 2000; Daróczi-Szabó 2006; Baxter 2006; Philips *et al.* 2009; Belhaoues 2011a, 2011b).

The withers height based on the multiplication factors of Harcourt (1974) was estimated for each individual. Various classifications based on the withers height exist (Mégnin 1897; Udrescu 1989); this study adopted the one recently established by Belhaoues (2011a: Fig. 51; reproduced here in Table 1). Belhaoues' classes are finely divided, with intervals

Table 1. Classification of dogs based on the withers height (reproduced after Belhaoues (2011a) with his kind permission).

Withers Height	*Size - Group*
< 25 cm	Dwarf
25 to 30 cm	Very Small
30 to 40 cm	Small
40 to 50 cm	Medium
50 to 60 cm	Medium large
60 to 70 cm	Large
> 70 cm	Very Large

of 5 or 10cm; improved classes are meant to point out the presence of dwarf and very large-sized dogs.

Regarding cranial typology, first, morphological traits related to the concavity of the stop, and the development of the sagittal crest and the external occipital protuberance were described to give an indication of the cranium type. Second, craniometric indices reporting the proportions of the cranium (Stockard 1941, Ellenberger and Baum 1943, Brehm *et al.* 1985 *in* Lignereux *et al.* 1991) were calculated. The cranial index (Stockard 1941) is commonly used to assign specimens to a morphotype. It reflects on the general configuration of the cranium by relating its maximum width to its maximum length. Nevertheless, the limits between the three cranial types, fixed on the basis of the cranial index, vary according to scholars (Stockard 1941; Sisson 1969; Komeyli 1984; for an overview of the literature see Lignereux *et al.* 1991). This study refers to a recent classification, dividing dogs in types and sub-types (Belhaoues 2011a). Sub-types intent to lessen the complexity of classification due to the great cranial shape variability and individual variation which make it difficult to assign a cranial type to a breed (Belhaoues 2011a). The dolichocephalic dogs which have a cranial index less than 55 are divided in three sub-types: Ultra-dolichocephalic, dolichocephalic and sub-dolichocephalic. Mesocephalic breeds have a cranial index between 55 and 65; they are divided in two sub-types, sub- and ultra-mesocephalic. Lastly, brachycephalic dogs have a cranial index more than 65 and are divided in sub-brachycephalic, brachycephalic and ultra-brachycephalic (Belhaoues 2011a: Fig. 48; reproduced here in Table 2 with modifications in the classification of modern breeds according to author's measurements of modern specimens). However, it is important to note that there is a continuum without any natural breaks between dolichocephalic and brachycephalic crania (Helton 2016), therefore any classification remains arbitrary (Belhaoues 2011a: 93).

The craniometric indices (cranial index and skull index) and the related length of the cranium (total and basal length), were compared to modern dog breeds and recent wild canid specimens (wolf, fox and jackal), held in the osteological collections of the National Museum of Natural History of Paris, using bivariate analysis (Table 3 a and b; measurements were taken by the author following Von den Driesch 1976 and Lignereux *et al.* 1991). Two variables (selected following Belhaoues 2011a) were plotted against each other in a scatter diagram in order to outline the trends in modern breeds and archaeological specimens. Linear regression equations were used in quantifying the trends in modern dog breeds and coefficients of determination (R^2) were calculated to show the degree to which the two variables were related to each other (Daróczi-Szabó 2006). The comparative specimens of modern breeds include gazehounds (Barzoï, Sloughi, Whippet), Pointing dogs (German short-haired pointer, Brittany spaniel), Flushing dogs (Cocker spaniel), Herding dogs (German shepherd, Pyrenean shepherd), molossoid breeds (Great Dane, Boxer), Dachshund, Spitz type (Pomeranian, Eskimo dog) and companion dogs (Poodle, Maltese). In addition, further craniometric indices derived from the archaeological specimens were compared to those of modern breeds from published data (Stockard 1941, Ellenberger and Baum 1943, Seiferle 1966, Komeyli 1984, Betti 1990 *in* Lignereux *et al.* 1991; Belhaoues 2011a).

The ratio between the total length and the slenderness index of the limbs and the mandible was also compared to modern breeds. Limb slenderness index (De Grossi Mazzorin and Tagliacozzo 2000) consists of the smallest depth of the diaphysis expressed as a percentage of the greatest length for the major long bones (humerus, radius, femur and tibia: SD x 100 / GL; measurements following Von den Driesch 1976). It is based on Hartcourt's (1974) mid-shaft diameter index, using the smallest diameter of the diaphysis as numerator and as an alternative for the mid-shaft diameter. Mandible slenderness index (Belhaoues 2011a, 2011b; De Grossi Mazzorin and Tagliacozzo (2000) mandibular index 2) consists of the height of the mandible behind the first molar expressed as a percentage of the total length (Ht behind M1 x 100 / TL; measurements following Von den Driesch 1976). The ratio between the total length and the slenderness index aims to assess, to some extent, the relationship between the size and the build of the animal (Belhaoues 2011a, 2011b). The slenderness index proved to be useful in separating modern and archaeological dogs of different breeds as suggested by De Grossi Mazzorin and Tagliacozzo's (2000) ratio between withers height and limb slenderness index.

Pathologies

Assessing skeletal pathologies allows exploring the health status and life conditions of ancient animals (Binois 2012). The skeletal elements were therefore

Table 2. Definition of cranial types and sub-types based on the cranial index (modified after Belhaoues (2011a) with his kind permission).

Cranial type	*Cranial sub-type*	*Cranial index*	*Modern dog breeds and recent wild canids*
Dolichocephalic	Ultra-dolichocephalic	< 45	Barzoï, Sloughi
	Dolichocephalic	45 < IC < 50	Dobermann
	Sub-dolichocephalic	50 < IC < 55	Eskimo dog, German shepherd, Great Dane, Whippet, **Fox**, **Jackal**, **Wolf**
Mesocephalic	Sub-mesocephalic	55 < IC < 60	Brittany spaniel, Cocker spaniel, Dachshund, German shorthaired pointer, Poodle, Pyrenean shepherd
	Ultra-mesocephalic	60 < IC < 65	Maltese, Pomeranian
Brachycephalic	Sub-brachycephalic	65 < IC < 75	Boxer
	Bracycephalic	75 < IC < 85	Bulldog
	Ultra-brachycephalic	> 85	Pekinese, King-Charles spaniel

examined macroscopically for bone surface modifications (Baker and Brothwell 1980; Lovell 1997; Varga *et al.* 1999; Csippán and Daróczi-Szabó 2008). The observed modifications were classified into seven groups of pathological conditions including dental anomalies (periodontal disease, irregular wear and tooth breakage), linear enamel hypoplasia, traumatic injuries, inflammation and infectious diseases, enthesophytes and ossification of ligaments, joint diseases, and congenital abnormalities. Estimated age-at-death of the individuals was taken into consideration when interpreting skeletal pathologies.

Results

Mortuary practices

At SFI 645, the orientation of the corpses follows a south-north axis, the head situated either at the north (2 individuals) or at the south (2 individuals). At SFI 477, the orientation of the skeletons follows an east-west axis, the head situated either at the east (3 individuals) or at the west (7 individuals). Hence, the burials seem to have aligned with the stream and the street. However the position of the bodies is random. The dogs were buried on their right or left side with no correlation in the head-tail orientation (Table 4). The body is generally extended with the limbs either extended or flexed (Fig. 3). Only specimen DA1 is in a curled position and specimen DA3 is laid in a contracted position, the head and the limbs bent towards the thorax.

Boundaries of burial pits were not found for any of the deposits on both sites. This might be due to the difficulty of separating the sediments (sediment matrix v/s burial fill), particularly in a flooded area at SFI 645. In any case, the preservation of the anatomical connections indicates that the decomposition happened in a confined space and all the nearly-complete and partial skeletons appear to be primary, individual deposits. Only specimen 1070 could present a distinct case. It is only partially preserved, consisting of an articulated half of the vertebral column and rib cage with some other scattered cervical and thoracic vertebrae, left femur, fragments of left and right scapulae and cranium fragments. The absence of a pit makes it unclear whether the loose and displaced anatomical connections are due to decomposition in an open space, to a secondary deposit where the remains were moved before the complete decomposition of the ligaments maintaining the vertebrae and ribs in connection, or to post-depositional disturbance. On the other hand, specimen DA6, DA9, DA12 and 1079 were truncated by post-depositional activities (Fig. 3).

Specimen DA9 was covered with a Late Iron Age amphora fragments. Breakage and collapse of the fragments following the decomposition process disturbed the last three cervical and the first thoracic vertebrae, some ribs and the left scapula. Specimen 1078 was covered with a stone (c. 30x40cm) at the thorax; consequently a disjunction occurred in the vertebral column and the thoracic region collapsed under the weight of the stone to a lower level than the rest of the body. Many displacements followed the collapse of the thorax. The connection between the skull and the atlas and between the axis and the third cervical vertebra was lost. The front limbs were also disturbed; the right scapula toppled over to the back and interphalangeal connections were displaced.

The remaining burials did not show any imperishable cover or container. Despite the general preservation of the anatomical connections, displacements of skeletal elements that are not related to later reworking of the burials occurred in all deposits. Some displacements are limited to the original volume occupied by the body. Frequently, the rib cage was flattened indicating the differential in-filling of the internal volume of the cadaver; this situation was accompanied by the disjunction of costo-vertebral joints. Furthermore, the rotation of the femur was recorded in several cases. On the other hand, displacements outside the original volume of the corpse also occurred. These displacements suggest the existence of original voids around the corpses at the initial phase of decomposition.

Table 3a. Measurements of long bones (humerus, radius, femur, tibia), cranium and mandible of modern dog breeds. The specimens are held in the osteological collections 'Mammifères et Oiseaux – Anatomie comparée' of the National Museum of Natural History (Paris) and the measurements were taken by the author following Von den Driesch (1976). Craniometric indices and slenderness Index

Modern breeds	*Barzoi ♀ (13 yrs)*	*Boxer ♀*	*Boxer ♂*	*Brittany spaniel ♀*	*Brittany spaniel ♂*	*Brittany spaniel ♂*	*Cocker spaniel ♀*	*Dachshund ♀*	*Eskimo dog*	*German pointer ♀*	*German shepherd ♀*
MNHN reference	1995-237	1985-1263	1985-1319	1985-1258	1985-1259	1985-1231	1985-1219	1985-1282	1857-20	1985-1297	1985-1337
Humerus GL	228.9	172.2		148.1	152.9		115.9	64.4		172.7	182.5
Humerus SD	14.0	15.1		12.4	11.5		9.2	8.0		13.1	14.2
Humerus slenderness index	6.1	8.8		8.4	7.5		7.9	12.4		7.6	7.8
Radius GL	224.5	167.1		153.5	150.7			54.2			
Radius SD	13.8	13.9		10.9	10.9			10.5			
Radius slenderness index	6.1	8.3		7.1	7.2			19.4			
Femur GL	243.1	184.4		159.4	158.5		125.2	70.7		185.3	206.1
Femur SD	13.4	13.6		12.3	12.5		9.9	11.2		13.1	14.6
Femur slenderness index	5.5	7.4		7.7	7.9		7.9	15.8		7.0	7.1
Tibia GL	251.5	188.9		168.2		178.6		62.8		192.7	206.0
Tibia SD	14.1	15.0				12.5		10.1		13.4	
Tibia slenderness index	5.6	8.0				7.0		16.0		7.0	
Cranium Total length	238.4	173.3		172.4	171.8		157.4	142.7	212.5	185.1	210.9
Cranium Basal length	210.0	151.0		153.6	157.9		141.2	126.0	187.9	171.8	193.2
Cranium Zygomatic breadth	95.9	122.3		103.7	98.0		86.8	81.1	110.4	106.9	107.3
Cranial index	40.2	70.5		60.1	57.1		55.2	56.8	52.0	57.7	50.9
Skull index	45.6	81.0		67.5	62.1		61.5	64.4	58.8	62.2	55.5
Mandible Total Length	177.5	142.5	151.8	132.2	124.8		116.6	104.9	156.2	141.3	157.2
Height of the mandible behind M1	23.4	20.3	23.9	18.2	20.1		17.7	16.1	27.1	23.2	25.9
Mandible slenderness index	13.2	14.3	15.8	13.8	16.1		15.1	15.3	17.3	16.4	16.5

Specimen DA6 is the most dislocated skeleton. This is due to the fact that the surface on which it rests is not flat. The skull and the fore paws are more elevated than the rest of the body, therefore it might have been resting on the edge of a pit that was not discriminated or on a perishable container's wall. Consequently, the articulations of the left forelimb have amply moved. The scapula toppled over to the back of the trunk and was found on the edge. The elbow joint has been dislocated. The phalanges have reversed over the metacarpals. Also, the atlanto-occipital joint was dissociated and the right mandible slightly dropped. In addition, the spinal segments were dissociated and rotated. All these displacements might be indications that the decomposition process did not happen in a filled environment but in a void (Duday *et al.* 2009, 2014). The individual might have been covered with a cover made from perishable material that isolated the corpse from the sediment filling the burial pit.

Specimen 1077 has also undergone many displacements at the joints (Fig. 3). Disjunction of spine segments and costo-vertebral joints occurred. Further displacements occurred at a larger scale like the disjunction of the shoulder and hip joints. Interphalangeal connections were also displaced and some phalanges were reversed. These movements might be indications of decay in a void.

Another example of displacement of skeletal elements is shown in specimen 1080. Apart from the rotation of the femur, flattening of the thorax, vertebral (9-10th thoracic vertebrae) and costo-vertebral (7th left rib) joint disjunction, specimen 1080 showed a displacement of the right wrist joint (Fig. 3). Metacarpals and phalanges toppled over as if they were abutting against a pit edge or any perishable wall that disappeared. For instance, a linear delimitation effect is noticed at the eastern side of the burial along the fore limbs, and the left carpal bones and metacarpals were maintained in equilibrium by a wall. As no pit was delineated, it is not clear whether the linear delimitation corresponds to a pit wall or to a wall of a perishable container.

Furthermore, the absence of burial pits delimitations limited the analysis of the chronological deposition of the burials. It is unclear at both sites whether the burials were concomitant or successive in a short period of time. At lot SFI 645, only specimen 1070 is clearly earlier than the other burials; a sedimentary deposit related to the seasonal flow of of the long bones and mandible were estimated according to these measurements. The highlighted Tibia SD measure is approximate as the diaphyses of the tibia and fibula were still united.

	German shepherd ♀ (4 yrs)	*German shepherd ♀*	*German shepherd ♂*	*German shepherd ♂*	*Great Dane*	*Maltese ♀*	*Pomeranian ♀ (6 yrs)*	*Pomeranian ♂*	*Poodle ♂*	*Pyrenean shepherd*	*Sloughi ♀*	*Whippet ♂*
	1985-1276	1985-1274	1985-1242	1985-1211	1985-1328	1985-1203	1985-1226	1985-1209	1985-1217	1985-1313	1987-258	1996-2468
	173.9	190.0	203.2		247.5	93.1	108.7	113.0	132.7	136.1	187.9	163.5
	16.4	15.6	14.6		19.7	7.8	8.4	8.4	10.4	10.5	11.8	10.3
	9.4	8.2	7.2		8.0	8.3	7.7	7.4	7.9	7.7	6.3	6.3
	180.7	184.9	206.2		232.4	87.7	101.5	113.9			194.8	162.8
	15.3	13.9	13.7		18.2	7.6	8.3	8.7			11.8	10.8
	8.5	7.5	6.6		7.8	8.7	8.2	7.7			6.1	6.6
	198.7	204.0	218.1		255.9	95.3	113.6	125.2	144.2	144.5	206.6	171.2
	15.4	15.6	15.7		19.4	8.4	9.2	8.8	11.0	11.0	13.8	11.2
	7.7	7.6	7.2		7.6	8.8	8.1	7.0	7.6	7.6	6.7	6.5
	205.5			239.0	269.0	94.2	117.7	124.7	149.6	150.9	215.0	183.5
				16.4	19.5						12.9	10.3
				6.9	7.3						6.0	5.6
	206.4	205.9	223.4		248.7	120.9	139.3	139.7	153.6	169.2	203.5	178.5
	186.4	186.8	193.7		224.2	109.2	122.8	124.1	134.5	150.3	178.9	160.6
	108.3	109.9	115.0		136.9	78.1	82.9	85.6	89.7	99.7	98.5	93.7
	52.5	53.4	51.5		55.0	64.6	59.5	61.3	58.4	58.9	48.4	52.5
	58.1	58.8	59.4		61.0	71.6	67.5	68.9	66.7	66.4	55.1	58.3
	160.7	161.5	166.8		186.9	92.5	103.2	104.4	113.6		150.4	134.1
	25.0	26.3	26.0		32.6	12.0	15.2	14.2	15.8		21.8	18.9
	15.5	16.3	15.6		17.5	12.9	14.7	13.6	13.9		14.5	14.1

Table 3b. Measurements of long bones (humerus, radius, femur, tibia), cranium and mandible of recent wild canids. The specimens are held in the osteological collections 'Mammifères et Oiseaux – Anatomie comparée' of the National Museum of Natural History (Paris) and the measurements were taken by the author following Von den Driesch (1976). Craniometric indices and slenderness index of the long bones and mandible were estimated according to these measurements. The highlighted Tibia SD measures are approximate as the diaphyses of the tibia and fibula were still united.

Recent wild canids	*Canis aureus ♀*	*Canis aureus ♂*	*Canis lupus ♀ (5-8 yrs)*	*Canis lupus*	*Vulpes vulpes ♂*	*Vulpes vulpes ♂*	*Vulpes vulpes*
MNHN reference	1992-1503	1912-547	1984-08	1984-036	1922-139	1982-1060	1910-138
Humerus GL	136.7	136.9	196.8	193.9	118.6	124.2	108.0
Humerus SD	9.7	9.4	15.8	16.9	7.1	7.6	7.9
Humerus slenderness index	7.1	6.9	8.0	8.7	6.0	6.1	7.4
Radius GL	140.4	142.0	195.6	191.2	109.5	114.7	98.1
Radius SD	9.5	9.7	14.5	14.8	6.8	7.6	8.5
Radius slenderness index	6.8	6.9	7.4	7.7	6.2	6.6	8.7
Femur GL	151.8	147.1	209.2	207.2	122.5	129.6	115.8
Femur SD	9.8	9.7	15.7	17.2	7.7	8.6	9.1
Femur slenderness index	6.5	6.6	7.5	8.3	6.3	6.7	7.9
Tibia GL	156.1	152.9	221.2	213.9	130.8	141.0	121.0
Tibia SD	9.5	9.8	14.7	15.8	6.9	7.4	8.7
Tibia slenderness index	6.1	6.4	6.6	7.4	5.2	5.2	7.2
Cranium Total length	168.1	173.8	239.7	241.3		144.6	144.2
Cranium Basal length	154.2	157.3	211.3	213.8		133.2	133.0
Cranium Zygomatic breadth	86.4	87.4	128.7	132.7		77.4	75.5
Cranial index	51.4	50.3	53.7	55.0		53.5	52.3
Skull index	56.01	55.54	60.89	62.07		58.09	56.74
Mandible Total Length	121.7	127.7	177.1	181.7	100.7	106.7	108.4
Height of the mandible behind M1	18.6	18.5	28.9	34.3	12.7	14.2	14.3
Mandible slenderness index	15.3	14.5	16.3	18.9	12.7	13.3	13.1

Table 4. Orientation of the dog burials at lot SFI 477 and SFI 645.

Lot	Specimen	Orientation	Side
SFI 477	DA1	West - East	Right
	DA3	West - East	Right
	DA4	West - East	Right
	DA6	West - East	Right
	DA7	West - East	Left
	DA8	West - East	Left
	DA9	East - West	Right
	DA10	East - West	Left
	DA11	West - East	Left
	DA12	East - West	Right
SFI 645	1070		Left
	1077	North/North East - South/South-West	Left
	1078	North - South	Left
	1079	South - North	Left
	1080	South - North	Right

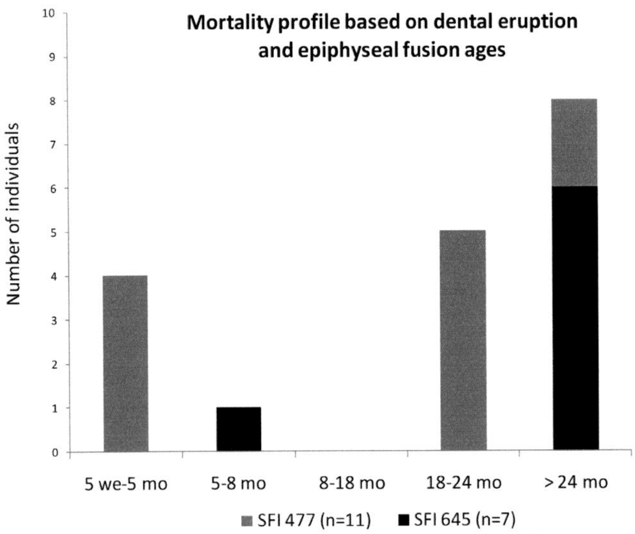

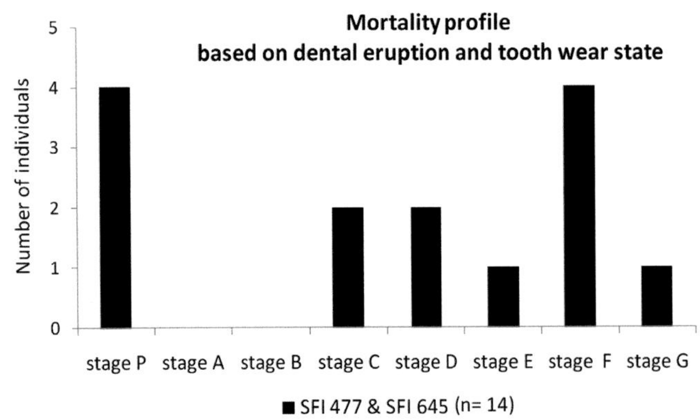

Fig. 4. Mortality profile of Persian period dogs in lot SFI 477 and SFI 645; (a) based on combined data from dental eruption and postcranial epiphyseal fusion; (b) based on dental eruption and tooth wear state (following Silver 1969 and Horard-Herbin 2000).

the water stream separates this specimen chronologically from the others. Further, the presence of an isolated atlas and axis between specimen 1079 and 1080 might be a sign of truncation of earlier deposits. At lot SFI 477, the presence of isolated paw next to specimen DA9 might also indicate truncation of earlier burials.

Demography

Two specimens had their deciduous dentition fully erupted and the alveolus for the first lower molar perforated, making them between 5 weeks and 5 months at the time of death. Two more specimens showed the same state of dental eruption, in addition an unfused atlas suggests an age of less than 3.5 months. These four puppies clustered at SFI 477. An isolated humerus with unfused distal epiphysis and a tibia from lot SFI 645 suggested an age of less than 8 months for at least one individual. Five specimens aged between 18 and 24 months were all buried at SFI 477. Eight individuals were aged more than 24 months at the time of death, out of which six were buried at SFI 645 and two at SFI 477 (Fig. 4a, Table 5).

As reflected by tooth wear state (Fig. 4b), individuals older than 24 months are not considered very old; only one individual (specimen 1080) was categorized as an old individual (stage G). Therefore, this specimen was likely more than 4 years at the time of death (Horard-Herbin 2000: Fig. 4; stage G correlates with age-classes 48-72 months and more than 72 months). On the other hand, seven specimens from both sites belonged to the intermediate-age group (stages D, E and F) and two specimens were classified as young (stage C).

A case of discrepancy between the state of cranial and postcranial fusion and the level of attrition on the first lower molar is illustrated by specimen DA12. This individual had the inter-frontal suture closed and a wear state for the lower first molar corresponding to the stage 'F(b)',suggesting an age of 3-4 years (Barone 1976, Horard-Herbin 2000: Fig. 4). Conversely, the epiphyses of the vertebral centrum were not all fused making the individual aged less than 2 years at the time of death (Barone 1976).

Despite the fact that this sample is not large enough to make statistically significant claims, some general tendencies of the mortality patterns can still be discussed. The distribution of the specimens by age group showed that there was no selection based on age. A single group, 8-18 months, is not represented. Hence, the mortality profile does not indicate a regulation of the dog population. Conversely, it suggests attritional death where the youngest and the oldest individuals are the most affected. The age group 8-18 months might represent the most resistant individuals, those who passed the vulnerable early months of life; a fact that would explain their absence.

The sex of only six specimens could be determined conclusively and consisted of male individuals (Table 5). The sex of specimen DA5, DA6, DA12

Table 5. Overview chart summarizing the biological features of the Persian period dogs at lot SFI 477 and SFI 645.

Specimen	*Sex*	*Age*	*State of conservation and Morphological traits*	*Withers height*	*Pathologies*	*Dental formula and position of teeth*
1070		> 2 yrs	Group of bones partially articulated Medium-large size and slender	56.2 cm	Joint disease (intervertebral and knee joints)	
1077	Male	> 2 yrs	Articulated skeleton Medium-large size, sub-dolicho/sub-mesocephalic, marked robustness of lower jaw	57.7 cm	Dental anomalies (periodontal disease, irregular wear, tooth breakage) Traumatic injury (rib fracture) Congenital anomalies (undersized vertebral articular process, distortion of spinous process) Joint disease (intervertebral, costo-vertebral, interphalangeal joints) and ossification of supraspinous ligaments	Complete dentition, PM and M spaced
1078	indet	> 2 yrs	Articulated skeleton Medium size and slender	47.7 cm	Congenital anomaly (distortion of spinous process) Ossification of tendons (MC 4)	Complete lower dentition, slight crowding of lower P4 and M1
1079	indet	> 2 yrs	Articulated partial skeleton		Traumatic injuries (rib fractures) Joint disease (costo-vertebral joints) and ossification of vertebral ligaments Enthesophytes (trochanter major of the femur, acromion of the scapula)	
1080	Male	> 2 yrs	Articulated skeleton Medium-large size, sub-dolicho/sub-mesocephalic	57.8 cm	Dental anomaly (irregular wear) Congenital anomaly (distortion of spinous process) Joint disease (wrist, costo-vertebral and facet joints) Ossification of ligaments (tibio-fibular, supraspinous) Inflammation (periostitis -microtrauma- or infectious disease)	Complete dentition, slight crowding of lower P4 and M1
1072	indet	< 8 mo	Isolated humerus and tibia (epiphysis unfused)			
1076	indet	> 2 yrs	Isolated atlas and axis (caudal epihysis of axis centrum fused)			
1076	indet		Isolated right mandible			Supernumerary tooth, PM spaced, encroachment of M3 on the vertical ramus of the mandible
DA1	Male	1.5 – 2 yrs	Articulated skeleton Medium size and slender	49.2 cm	Periodontal disease (infection of the mandibular periosteal bone) Ossification of ligaments (supraspinous)	Supernumerary lower tooth, spaced lower PM, slight crowding of lower P4 and M1, encroachment of M3 on the vertical ramus of the mandible
DA3	Male	1.5 – 2 yrs	Articulated skeleton Medium size	46.8 cm	Periodontal disease Physiological stress (LEH)	Persistence of upper deciduous incisors, crowding of lower incisors, slight crowding of lower P4 and M1, encroachment of M3 on the vertical ramus of the mandible
DA4	indet	5 we – 5 mo	Articulated skeleton			

Table 5. (cont.)

Specimen	Sex	Age	State of conservation and Morphological traits	Withers height	Pathologies	Dental formula and position of teeth
DA5	indet	5 we – 5 mo	Articulated cranium and mandible			
DA6	indet	> 2 yrs	Articulated partial skeleton Medium-large size, sub-dolichocephalic	55.1 cm	Traumatic injuries (crush fracture of the frontal bone, greenstick fracture of a spinous process) Congenital anomaly (distortion of spinous process)	Oligodontia (lower M3 absent), slight crowding of lower P4 and M1
DA7	indet	5 we – 3.5 mo	Articulated skeleton			
DA8	indet	> 2 yrs	Articulated skeleton Medium-large size and slender, sub-dolichocephalic	51.8 cm	Congenital anomaly (distortion of spinous process , malformation of the distal part of two ribs) Ossification of ligaments (supraspinous)	Complete dentition, lower PM spaced
DA9	Male	1.5 – 2 yrs	Articulated skeleton Medium-large size and slender	53.2 cm	Traumatic injury (rib fracture) Congenital anomaly (distortion of spinous process)	Complete dentition
DA10	indet	5 we – 3.5 mo	Articulated skeleton			
DA11	Male	1.5 – 2 yrs	Articulated skeleton Medium-large size and slender	56.3 cm	Joint disease (facet joint arthropathy : axis)	Complete dentition, lower PM spaced, slight crowding of lower P4 and M1, crowding of upper P2 and P3
DA12	indet	1.5 – 2 yrs	Articulated partial skeleton Medium size and slender, sub-meso/ sub-dolichocephalic	49.6 cm	Traumatic injuries (cranium, ribs, unhealed vertebrae) Ossification of ligaments (supraspinous)	Complete dentition, telescoping of the lower teeth, encroachment of M3 on the vertical ramus of the mandible, position of upper P1 close to the canine
	indet	> 1 yr	Isolated right hind paw (MT4, MT5, calcaneus, cuboid, phalanx I (x2), phalanx II)			

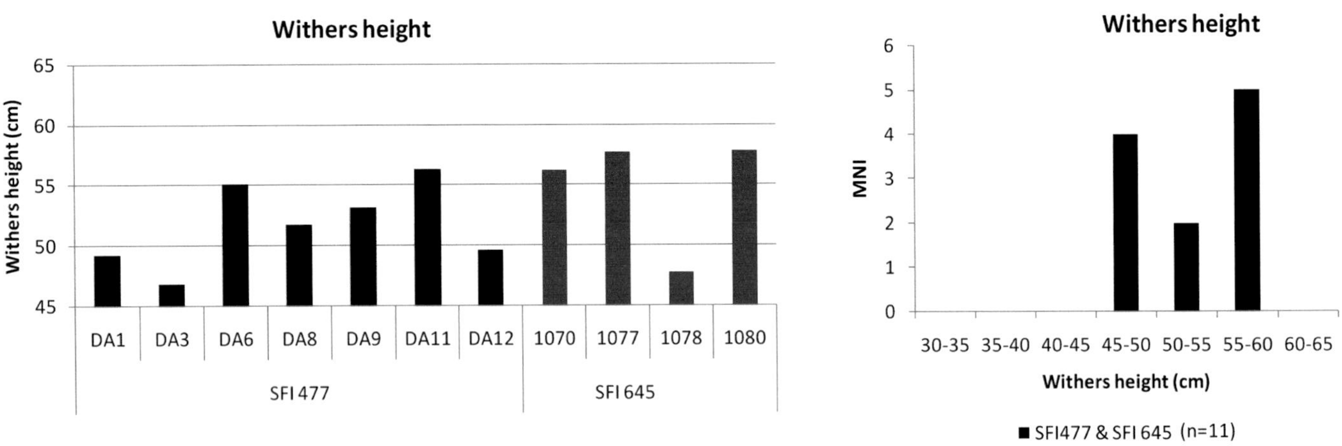

Fig. 5. Estimation of the withers height (following Harcourt 1974); (a) estimated withers height per individual; (b) distribution of the specimens according to their height at the withers in categories of 5cm each.

and 1079, truncated by later trenches, could not be determined. Neither could the sex be determined on individuals represented by scattered bones partially articulated and by isolated bones. Regarding specimens DA4, DA7 and DA10, which are juveniles, taphonomic factors may be responsible for the loss of the *baculum*, as juvenile bones are more prone to taphonomic destruction than adult ones. Concerning specimens DA8 and 1078, it is likely that in the absence of a recovered *baculum* they were females, though fresh breaks at the pelvic region and the absence of several caudal vertebrae make it

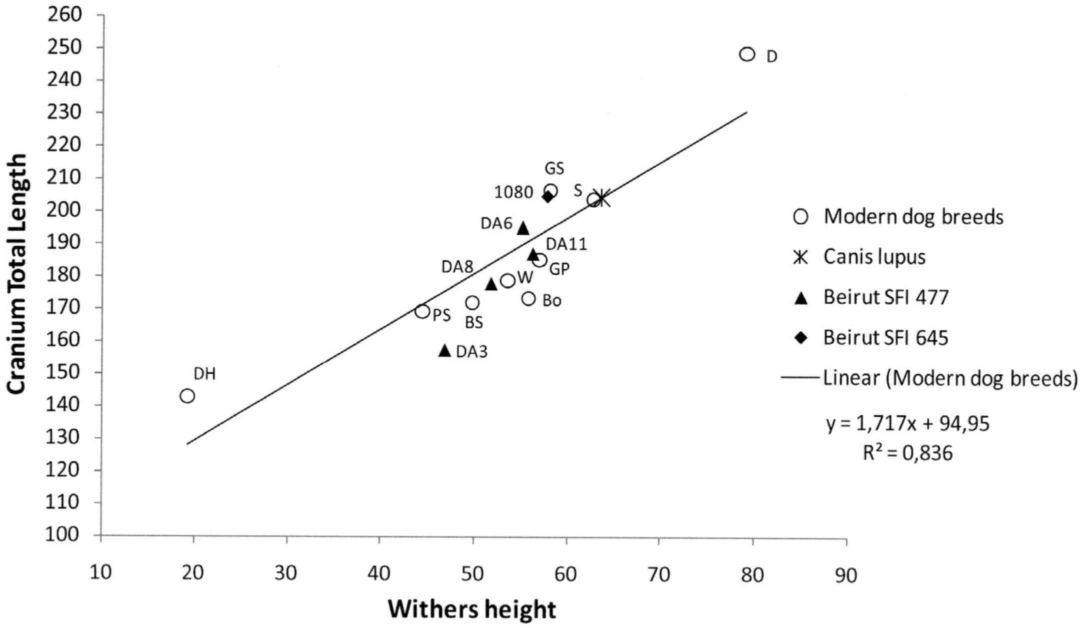

Fig. 6. Scatter plot illustrating the relationship between the total length of the cranium and the height at the withers in recent wolf, modern dog breeds (Bo: Boxer. BS: Brittany spaniel. D: Great Dane. DH: Dachshund. GS: German shepherd. GP: German shorthaired pointer. PS: Pyrenean shepherd. S: Sloughi. W: Whippet) and Beirut specimens. The linear trendline is based on modern dog breeds data.

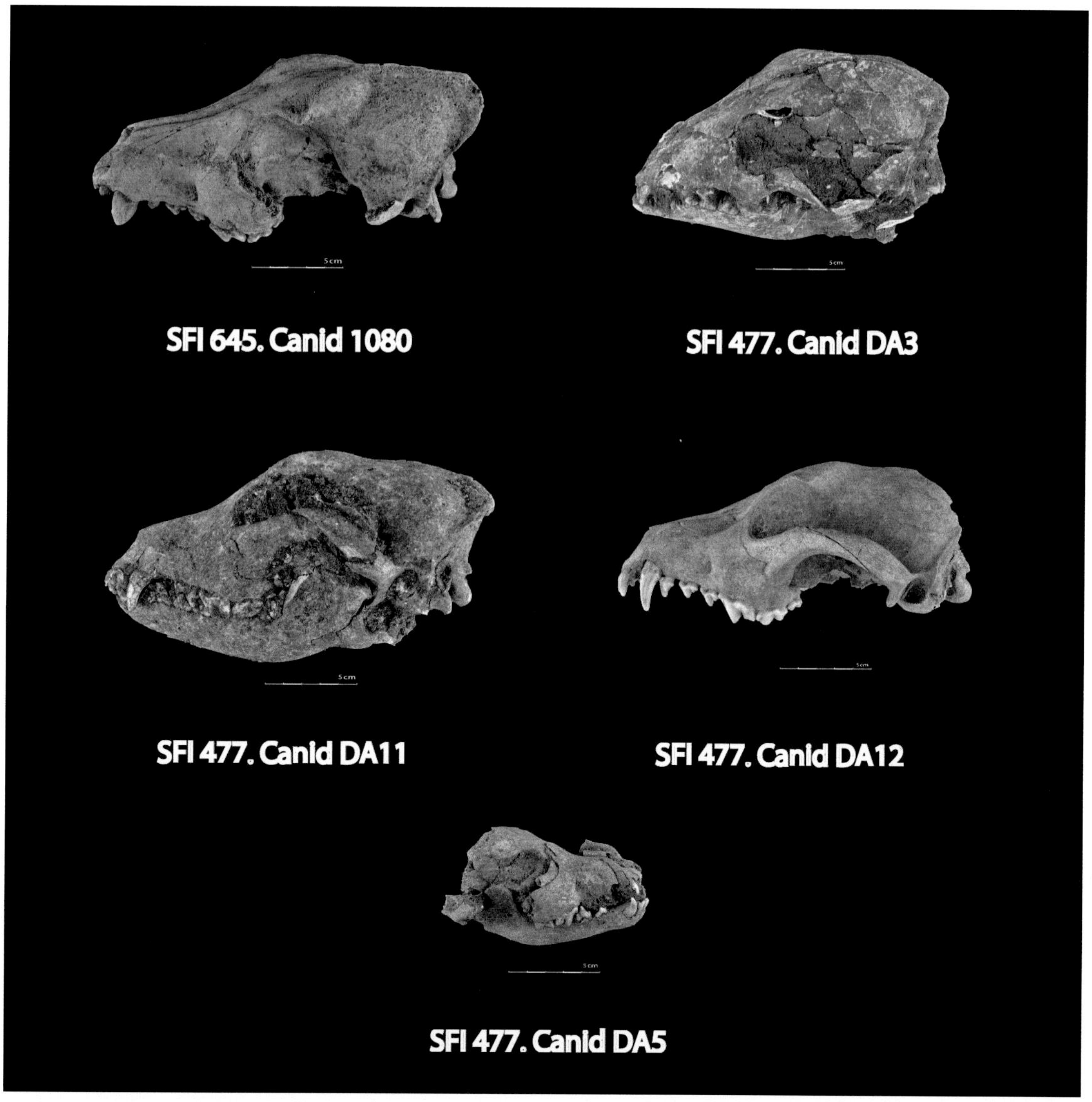

Fig. 7. Crania from lot SFI 477 and SFI 645 in lateral view. The cranium of specimen DA3, DA5 and DAll is shown in connection with the mandible. The cranium of specimen DA5 and DAI2 is incomplete. DA5 is a juvenile individual.

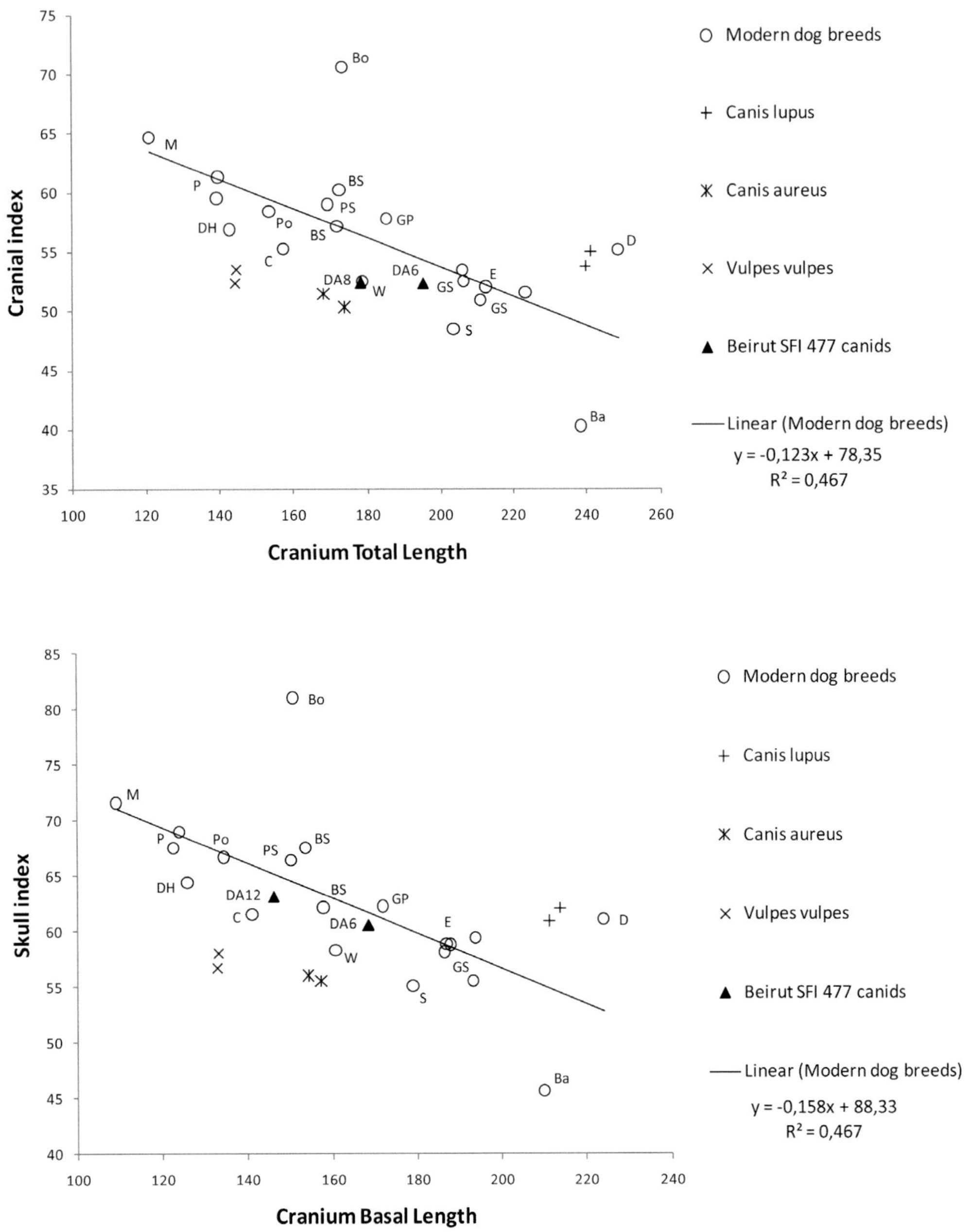

Fig. 8. Scatter plot illustrating the relationship between (a) the cranial index and the total length of the cranium and between (b) the skull index and the basal length of the cranium in recent wild canids (Wolf, Fox, Jackal) and modern dog breeds (Ba: Barzoi. Bo: Boxer. BS: Brittany spaniel. C: Cocker spaniel. D: Great Dane. DH: Dachshund. E: Eskimo dog. GS: German shepherd. GP: German shorthaired pointer. M: Maltese. P: Pomeranian. Po: Poodle. PS: Pyrenean shepherd. S: Sloughi. W: Whippet). The linear trendline is based on modern dog breeds data. Beirut specimens lie closest to the sub-dolichocephalic and sub-mesocephalic breeds.

unclear whether the skeletons were partially disturbed when they were initially uncovered.

Morphotypes

The estimated withers heights vary between 46 and 57 cm (Fig. 5, Table 5). The smallest specimens (DA1, DA3, DA12, 1078) belong to the medium size group and the largest ones (DA6, DA8, DA9, DA11, 1070, 1077, 1080) to the medium-large size group (following the classification of Belhaoues 2011a). They are all eumetric.

The total length of the crania varies between 157 and 204 mm. As expected, the cranial length varies according to the withers height (Fig. 6). In fact, the cranial and postcranial dimensions are correlated in dogs (Alpak *et al.* 2004; Galy 2004), with the exception of Basset Hounds in which the allometry between the skull and the body is linked to chondrodystrophic dwarfism (Crockford 2000; Baxter 2006, 2010). Beside size variation, cranial shape showed also a certain range of variation. The largest specimens (DA6, DA11 and 1080) had a cranium showing a well-developed external occipital protuberance, sagittal and nuchal crest, and a moderate stop. Specimen 1080 had the most pronounced muscle attachments (Fig. 7). Smaller specimens, of the medium size group, had less pronounced crests. Specimen DA12 showed a moderate sagittal crest,

Table 6. Cranial indices of Beirut Persian period canids. The indices were calculated according to Lignereux *et al.* (1991; following Stockard (1941), Ellenberger and Baum (1943), Brehm *et al.* (1985)). The measurements on which the indices are based follow the standard measurements technique divided by Von den Driesch (1976; listed here in numbers) and additional measurements proposed by Lignereux *et al.* (1991; listed here in letters as used by the authors). The most discriminant indices as defined by Lignereux *et al.* (1991) are highlighted in bold. The index values highlighted in grey are approximate, since the measurements were conditioned by the state of preservation of the crania.

Cranial indices (following Lignereux et al. 1991)	*Measurements following von den Driesch (1976) and Lignereux et al. (1991)*		*DA3*	*DA6*	*DA8*	*DA11*	*DA12*	*Cxt 1077*	*Cxt 1078*	*Cxt 1080*
1. Cranial index (Length-Breadth index 1)	30 / 1	Zygomatic breadth / Total length		52.3	52.4					
2. Skull index	30 / 3	Zygomatic breadth / Basal length		60.6			63.2			
3. Length-Breadth index 2	30 / ᾱ	Zygomatic breadth / Distance prosthion-asterion								
4. Snout index	36 / J	Breadth at the canine alveoli / Snout length (prosthion - infraorbital)					67	63.3		69
5. Upper facial index	34 / K	Greatest palatal breadth / Nasal length					97.1	101.5		102.4
6. Palatal index	34 / 13	Greatest palatal breadth / Medial palatal length			62.5		70.9			
7. Palato - basal index	13 / 3	Medial palatal length / Basal length					54			
8. Palato - palatal index	14 /13	Length of the horizontal part of the palatine / Medial palatal length			27		35.4			
9. Premolar index	P / Q	Length of maxillary premolars / Length of the maxillary premolar alveolar process	77.7		75.7		97.9	82.3		87.6
10. Orbito - facial index	N / O	Distance orbit - infraorbit / Distance orbit - prosthion	22.7	27.6		31.1	28.9	31.1		32
11. Length-Breadth index 3	29 / E	Greatest neurocranium breadth / Distance nasion - inion		52.7						50.6
12. Length-Breadth index 4	29 / ß	Greatest neurocranium breadth / Distance nasion - asterion		57.3						57.5
13. Breadth - height index	29 / M	Greatest neurocranium breadth / Height of cranium		83.9						88.4
14. Cranio-facial index 1	E / 8	Distance nasion - inion / Viscerocranium		103.5						115
15. Cranio-facial index 2	ß / 8	Distance nasion - asterion / Viscerocranium		95.1						101.2
16. Cranio-facial index 3	E / K	Distance nasion - inion / Nasal length		163.4						169.8
17. Cranio-facial index 4	ß / K	Distance nasion - asterion / Nasal length		150.2						149.5
18. Zygomatic index	30 / 8	Zygomatic breadth / Viscerocranium					111.9			
20. Mandibular premolar index	U / V	Length of mandibular premolars / Length of the mandibular premolar alveolar process	84.4	73.8	72	81.4	86.3	63.6	85.1	73.6

although its cranium is incomplete. Specimen DA3 also showed a moderate sagittal crest and external occipital protuberance; though this could be linked to its unfused cranial sutures (Fig. 7).

The cranial index is estimated at 52 (Fig. 8a) for two specimens, DA6 and DA8, and therefore fall in the sub-dolichocephalic group. Other discriminant craniometric indices (Fig. 8b, Table 6: index 2, 4, 5, 6, 9, 14, 15, 16, and 17, following Lignereux *et al.* 1991) suggest that the position of specimens DA6, DA8, DA12, 1077 and 1080 lies closest to the sub-dolichocephalic and sub-mesocephalic dogs. The snout, upper facial and palatal indices involving the width of the palate express the ratio between the width and the length of the face. The cranio-facial indices express the relationship between the length of the cranium and the length of the face. According to these indices, none of the complete specimens showed an elongated nasal region as in the gazehound type, neither a short-snouted face as in brachycephalic dogs. Hence, cranial dimensions and shape suggest that the Beirut dogs' affinities lie in the vicinity of

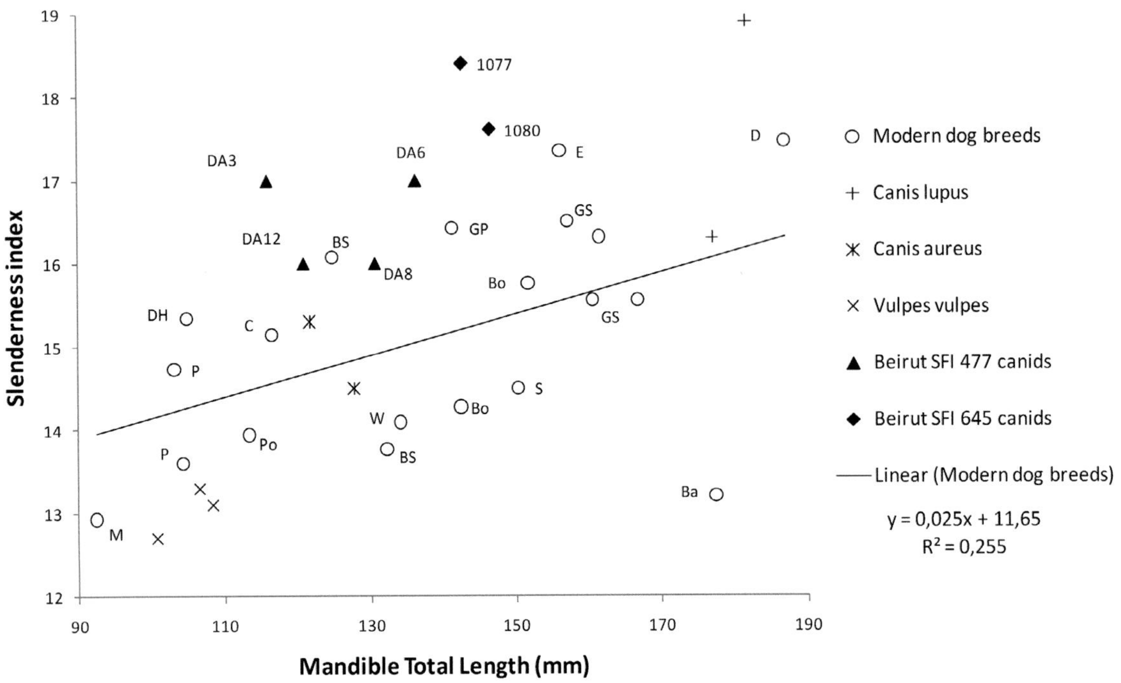

Fig. 9. Scatter plot illustrating the relationship between the slenderness index of the mandible and its total length in recent wild canids (Wolf, Fox, Jackal) and modern dog breeds (Ba: Barzoi. Bo: Boxer. BS: Brittany spaniel. C: Cocker spaniel. D: Great Dane. DH: Dachshund. E: Eskimo dog. GS: German shepherd. GP: German shorthaired pointer. M: Maltese. P: Pomeranian. Po: Poodle. S: Sloughi. W: Whippet). Compared to the linear trendline (based on modern dog breeds data), Beirut specimens show a relative tendency to robustness, clearly away from the gazehound group (Whippet, Sloughi, Barzoi).

sub-dolichocephalic and sub-mesocephalic type of modern dog breeds (Table 5).

Compared to modern breeds, the mandible showed a tendency towards robustness in all specimens (Fig. 9). DA8 and DA12 are close in size and robustness to the Brittany spaniel, a braccoid hunting dog. Specimens DA6, 1077 and 1080 are closer to lupoid dogs (German shepherd, Eskimo dog). The mandible of specimen DA3 is robust and smaller than the other individuals.

Furthermore, the dental formula and the position of the teeth on the mandible showed variability amongst specimens (Fig. 10, Table 5). The lower third molar is absent in specimen DA6 while specimen DA1 exhibits a supernumerary tooth between the canine and the lower first premolar. The medium-large sized specimen 1077 showed spaced teeth on the lower jaw whereas some individuals exhibited telescoping of teeth like in specimen DA12 or a slight crowding of the fourth premolar and the first molar like in specimen DA1, DA3, DA6, DA11, 1078 and 1080. Specimen DA3 also showed crowding of the lower incisors, which could be due to the upper deciduous incisors overlapping the permanent ones. The third lower molar encroaches on the vertical ramus of the mandible in specimen DA1, DA3 and DA12. Changes in the position of teeth were also noted in the upper jaw. For instance, the position of the upper first premolar is close to the canine in specimen DA12, in which the telescoping of the lower teeth followed the diminution of the maxillary diastema's length. In specimen DA11, the upper second and third premolars are crowded and the third premolar is skewed, the anterior cusp being oriented towards the lingual side.

The length of the first lower molar varies between 19.7 and 22.8 mm (Fig. 11). The length of the lower carnassials is used in metric analysis, though it has been shown in studies on modern dogs to be of limited use in defining size (Clark 1996, 2000). This statement is corroborated through the archaeological sample analyzed here where the length of the carnassials was not correlated in all cases with the total length of the mandible neither with the overall size of the individual. For instance, the first lower molar of specimen 1077 is smaller than that of specimen DA1, DA3 and DA12 although it belonged to a larger individual.

Measurements of the major limb elements were also compared to modern breeds (Fig. 12). The ratio between the total length and the slenderness index of the humerus showed that three individuals, DA6, 1077 and 1080, lie between braccoid (German shorthaired pointer, Brittany spaniel) and lupoïd (German shepherd) modern breeds. Specimen 1078 is close to the Pyrenean shepherd based on its size and slenderness. Four specimens, DA8, DA9, DA11 and DA12 are more slender than the others, though they show great robustness compared to the gazehound dogs (Whippet, Sloughi, Barzoi).

The conformation of the radius showed that specimen 1080 stands out of the lot by its robustness,

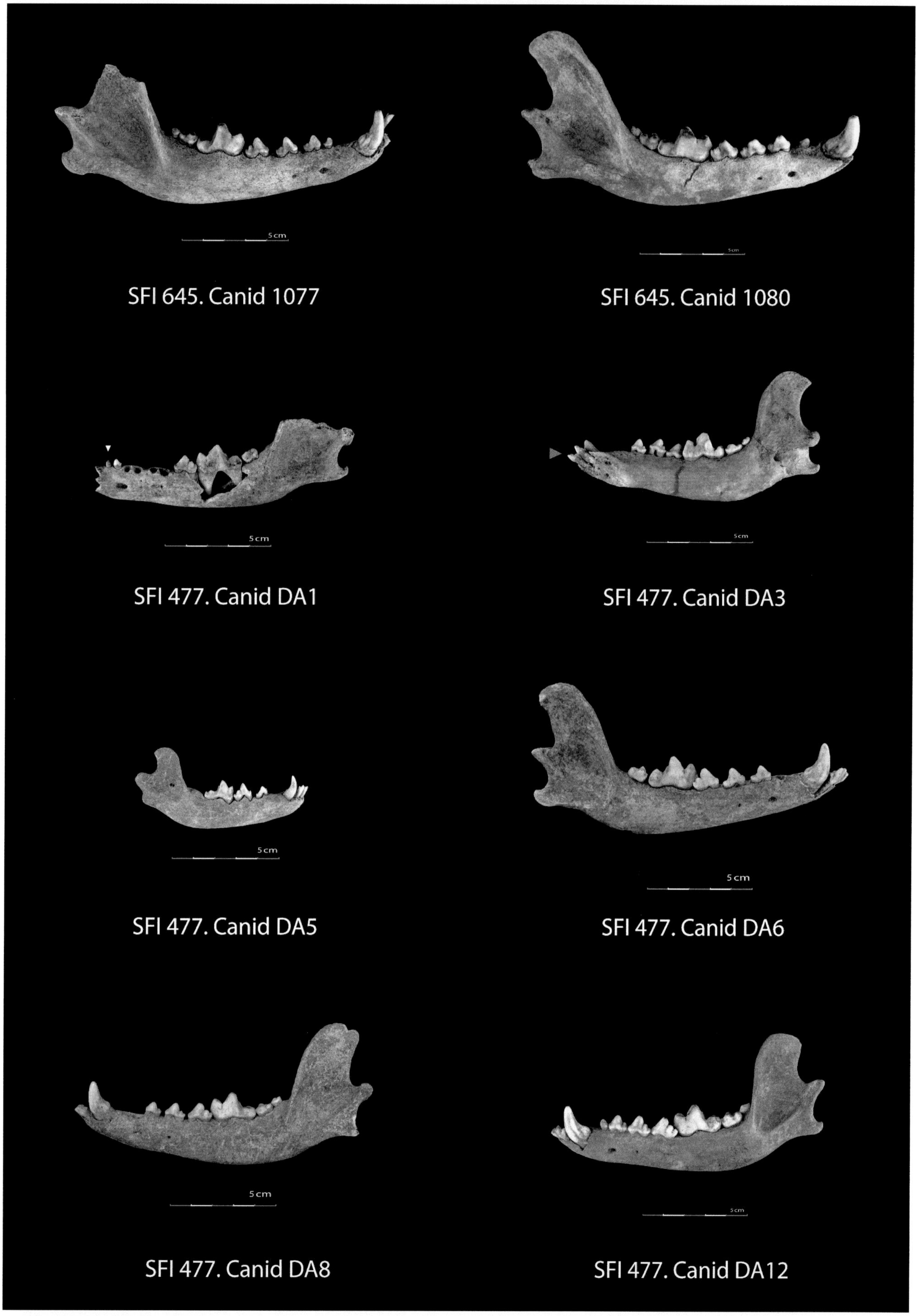

Fig. 10. Mandibles from lot SFI 477 and SFI 645. The right mandible of specimen 1077, 1080, DA5, DA6 is in lateral view. The right mandible of DA3 is in medial view showing crowding of the incisors indicated by the red arrow. The left mandible of DA1, DA8, DA12 is in lateral view. DA5 is a juvenile individual showing deciduous teeth. DA1 displays a supernumerary tooth indicated by the white arrow. DA6 shows a case of oligodontia with the absence of the third molar.

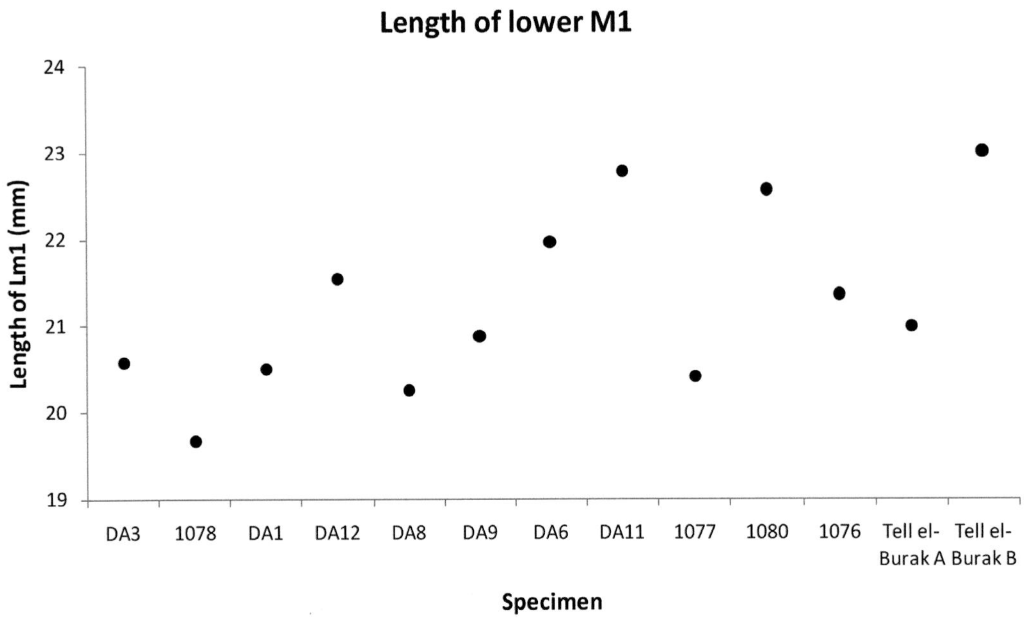

Fig. 11. Length of the mandibular first molar in the Beirut specimens (lot SFI 477 and SFI 645) and the Persian period specimens from Tell el-Burak (Cakirlar *et al.* 2014). Beirut specimens are arranged in ascending order of the withers height with the exception of the isolated mandible 1076. It is apparent that the length of the lower MI is not correlated in all cases with the overall size of the individuals.

12a

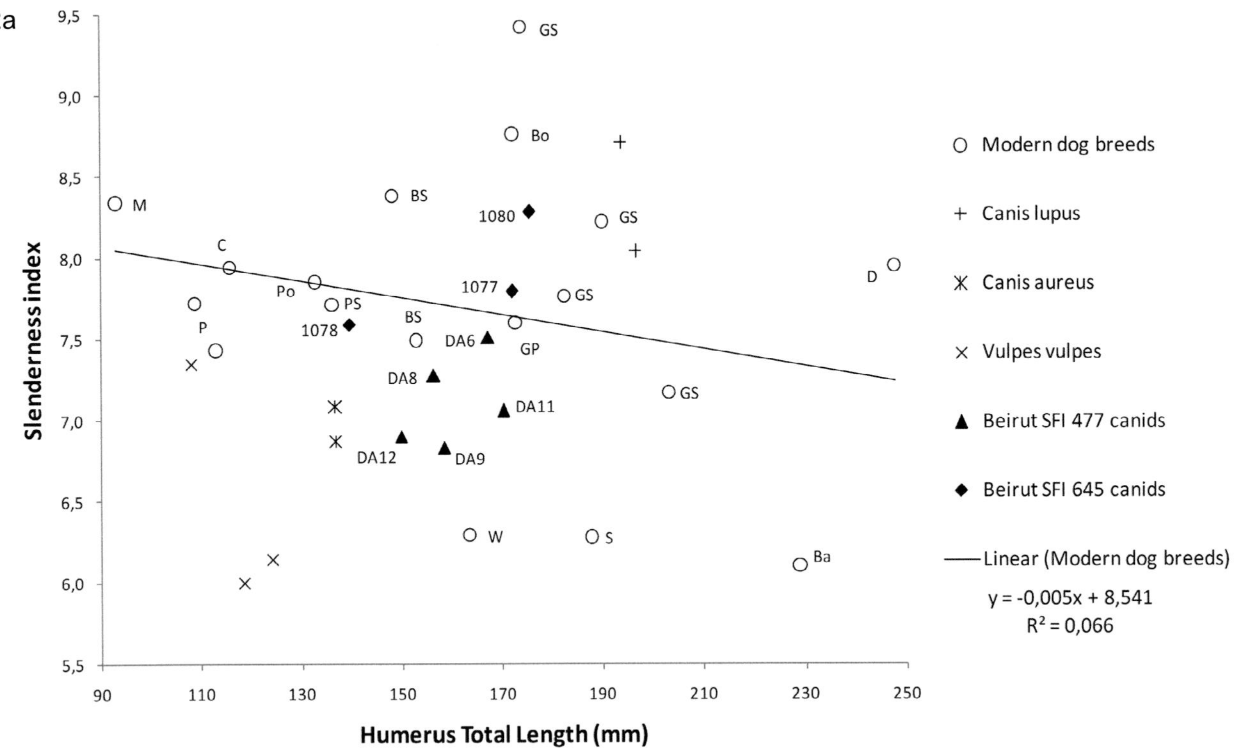

Fig. 12. Scatter plot illustrating the relationship between the slenderness index and the total length of the long bones (humerus (a), radius (b), femur (c), tibia (d)) in recent wild canids (Wolf, Fox, Jackal), modern dog breeds (Modern breeds: Ba: Barzoi. Bo: Boxer. BS: Brittany spaniel. C: Cocker spaniel. D: Great Dane. GS: German shepherd. GP: German shorthaired pointer. M: Maltese. P: Pomeranian. Po: Poodle. PS: Pyrenean shepherd. S: Sloughi. W: Whippet) and Beirut specimens. The linear trendline is based on modern dog breeds data.

with an index occurring in the highest limit of the German shepherd's variation range and a size slightly smaller than this breed. Specimens DA6, DA8 and 1077 are in the variation range of the Brittany spaniel and the German shepherd regarding the slenderness index, and their size is intermediate between these two breeds, especially DA6 and 1077. Three specimens, DA11, DA12 and 1078 are slenderer than the previous group. DA1 appears to be a smaller and slenderer individual; its slenderness index equals that of the Whippet but it is smaller than this breed. Meanwhile, specimen DA3 of a nearly similar size shows a greater robustness.

As for the femur, specimen 1080 lies in the upper limit of the German shepherd's robustness variation. The femur of specimen 1077 lies between

12b

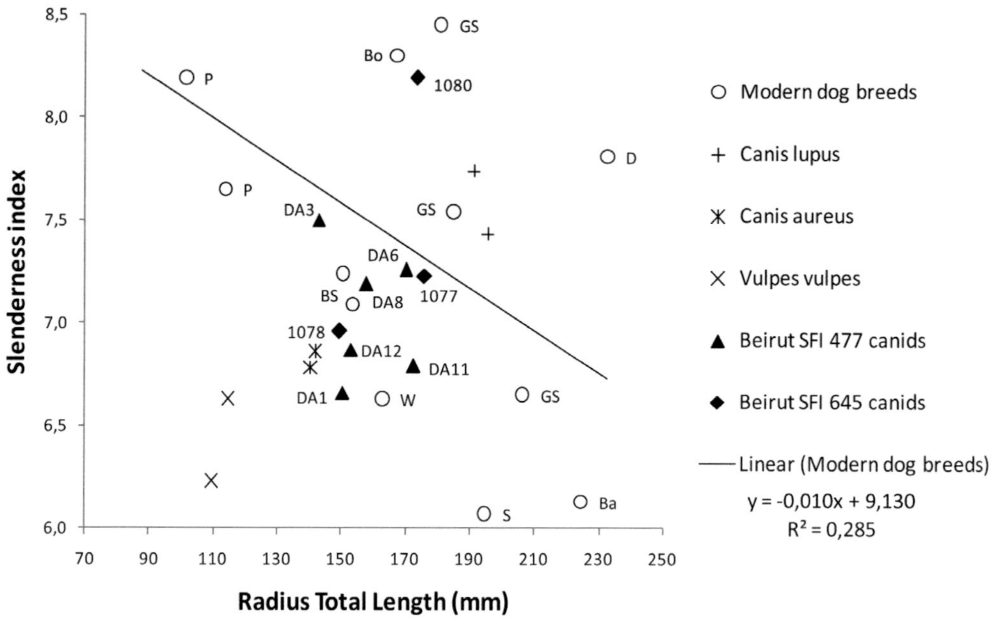
Slenderness index
Radius Total Length (mm)
Modern dog breeds
Canis lupus
Canis aureus
Vulpes vulpes
Beirut SFI 477 canids
Beirut SFI 645 canids
Linear (Modern dog breeds)
y = -0,010x + 9,130
R² = 0,285

12c

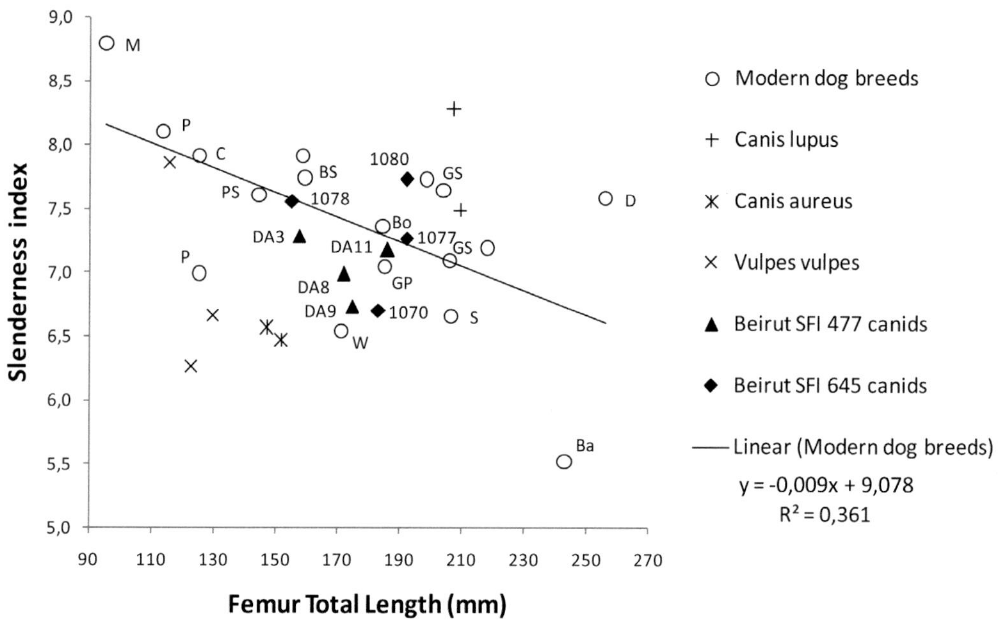
Slenderness index
Femur Total Length (mm)
Modern dog breeds
Canis lupus
Canis aureus
Vulpes vulpes
Beirut SFI 477 canids
Beirut SFI 645 canids
Linear (Modern dog breeds)
y = -0,009x + 9,078
R² = 0,361

12d

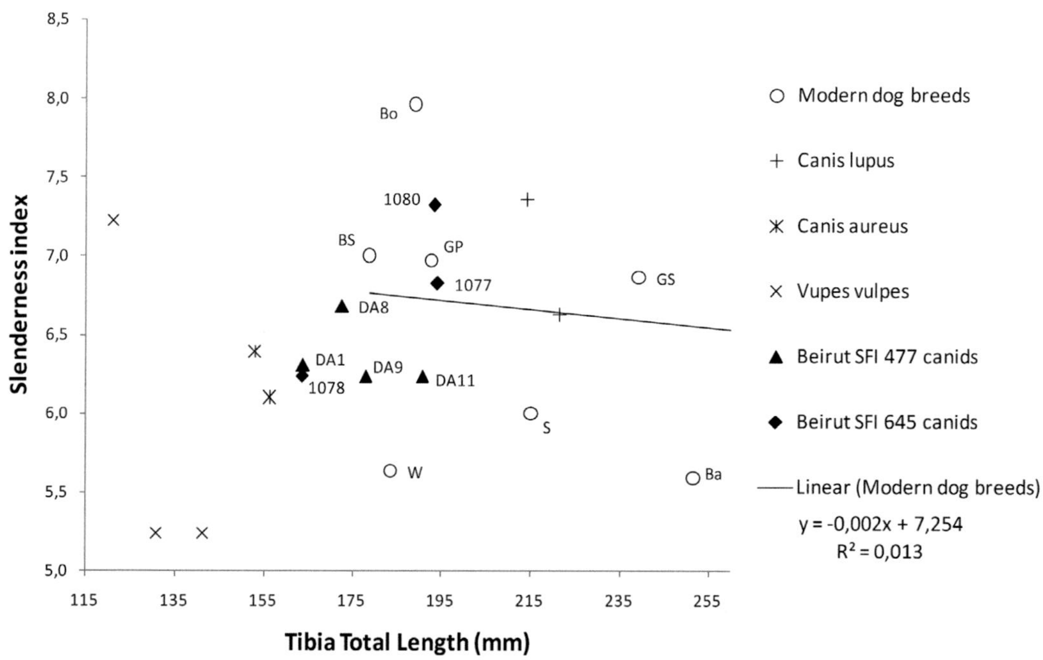
Slenderness index
Tibia Total Length (mm)
Modern dog breeds
Canis lupus
Canis aureus
Vupes vulpes
Beirut SFI 477 canids
Beirut SFI 645 canids
Linear (Modern dog breeds)
y = -0,002x + 7,254
R² = 0,013

Table 7. Long bone measurements. R. GL: range of Greatest Length (mm), following Von den Driesch (1976). R. SI: range of Limb slenderness index, following De Grossi Mazzorin and Tagliacozzo (2000).

	R. GL (mm)	*R. SI*
Humerus	139 - 175	6.8 - 8.3
Radius	142 - 175	6.7 - 8.2
Femur	155 - 192	6.7 - 7.7
Tibia	163 - 194	6.2 - 7.3

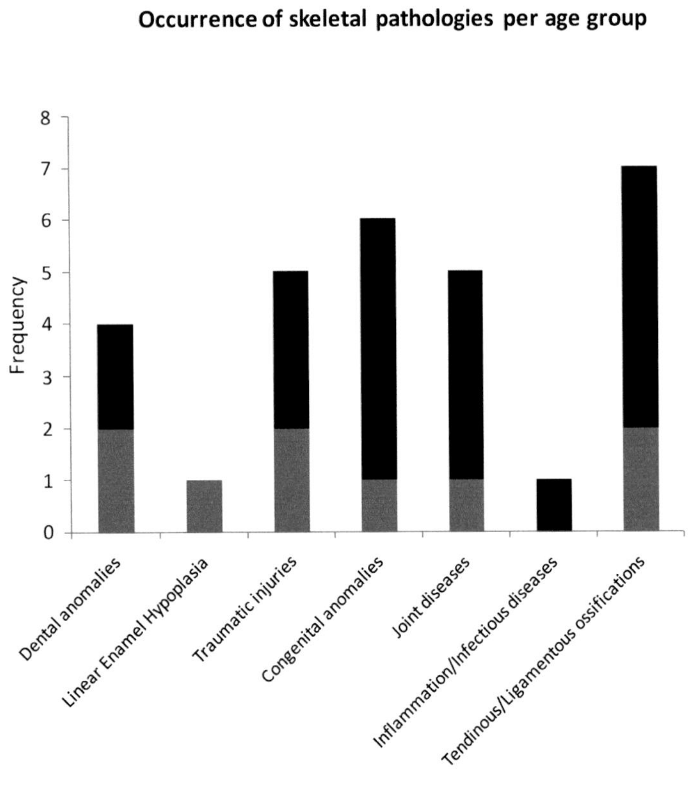

Fig. 13. Occurrence of pathological changes per age group.

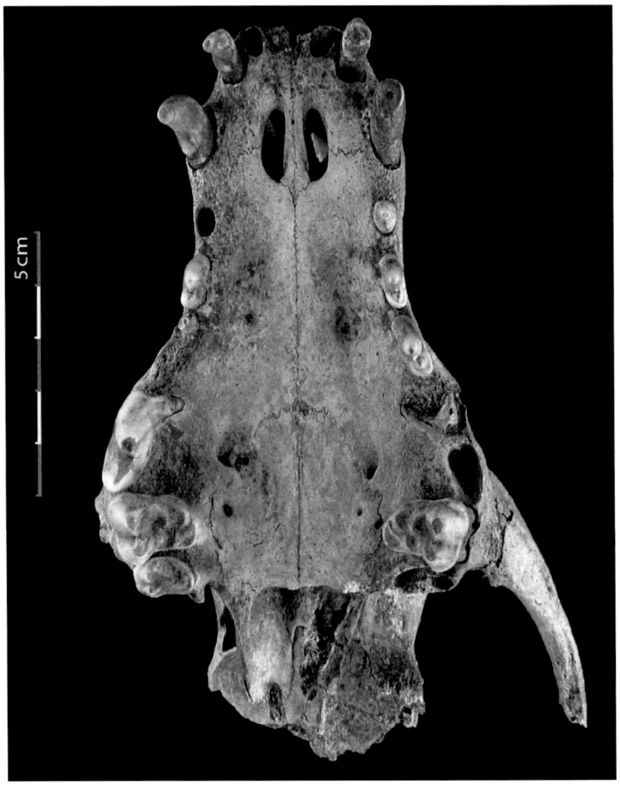

Fig. 14. Cranium of specimen 1077 in ventral aspect showing *ante mortem* tooth loss of the right and left first incisor and the right third premolar, breakage of the left fourth premolar and irregular wear of the left third incisor and canine tooth. Resorption of bone occurred in the incisive bone.

the Boxer and the German shepherd. Specimen DA11 is less robust and its dimensions are closer to the German shorthaired pointer. DA8 has also a slenderness index close to this breed although it is smaller. Specimens DA9 and 1070 are slenderer, though their slenderness does not equal that of the Whippet. Specimen 1078 equals the slenderness index of the Pyrenean shepherd and its size is comprised between this breed and the Brittany spaniel. Specimen DA3 is slenderer than specimen 1078.

Finally, the tibia also showed a high robustness in specimen 1080 and, to a lesser degree, in specimen 1077. Both specimens have a tibial length close to that of the German shorthaired pointer. Specimens DA1, DA8, DA9, DA11, and1078 are slenderer but their slenderness does not equal that of the gazehound group.

Morphological and metric analysis pointed out the variation in the general aspect of the cranium and mandible amongst Beirut specimens. The ratio between the slenderness index and the total length of the limbs showed variation in size and robustness, especially among the medium-large specimens. Nevertheless, measurements on modern skeletons showed that within a breed there is an inter-gender variation as well as sexual dimorphism-related variation (Table 3a). Regarding the archaeological sample, at least two groups are separated by size (medium and medium-large group) and a variation of the slenderness index is noticed within each group. The sex of the archaeological specimens could not be documented in all cases, though sexual dimorphism does not appear to be the main parameter related to allometric and robusticity variation in this sample. Male individuals are identified in the two size groups. In the medium-large size group, specimens DA11, 1077 and 1080 are all considered male individuals of similar size but they show differences in limb robustness. Therefore, specimen DA11 is slenderer than the other specimens; meanwhile specimen 1080 is characterized by the robustness of its cranial and postcranial bones coupled with well-developed muscle insertions. The same fact can be noticed in the medium size group where specimens DA1 and

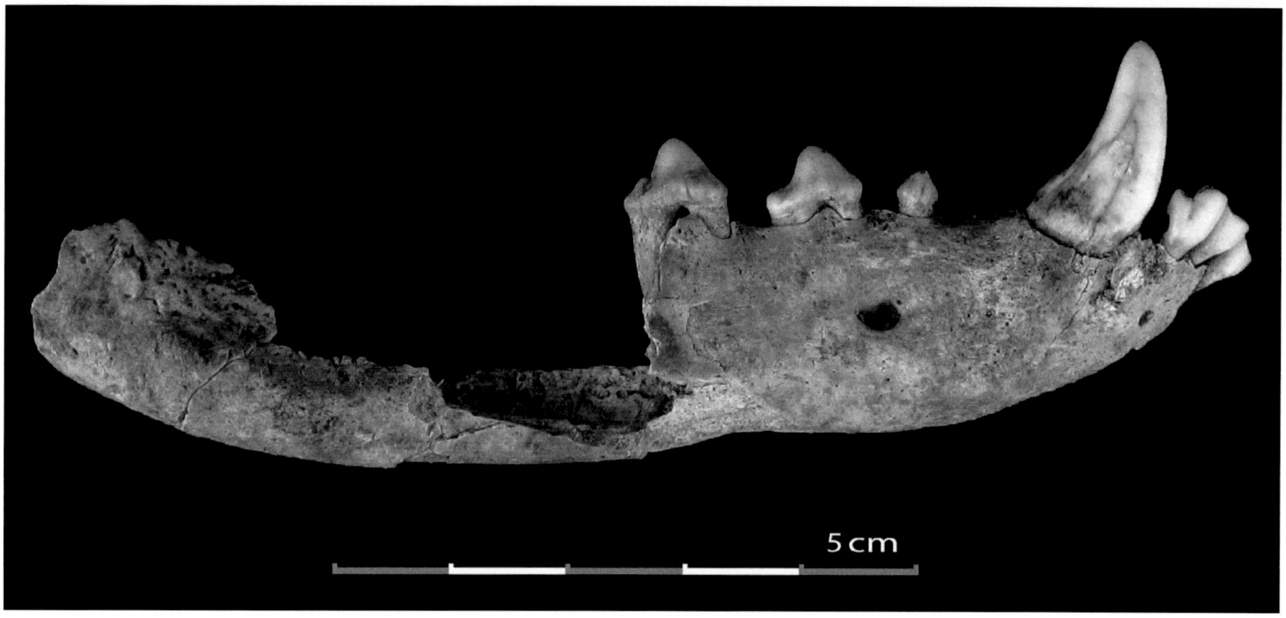

Fig. 15. Right mandible of specimen DA1 in lateral aspect showing periosteal infection.

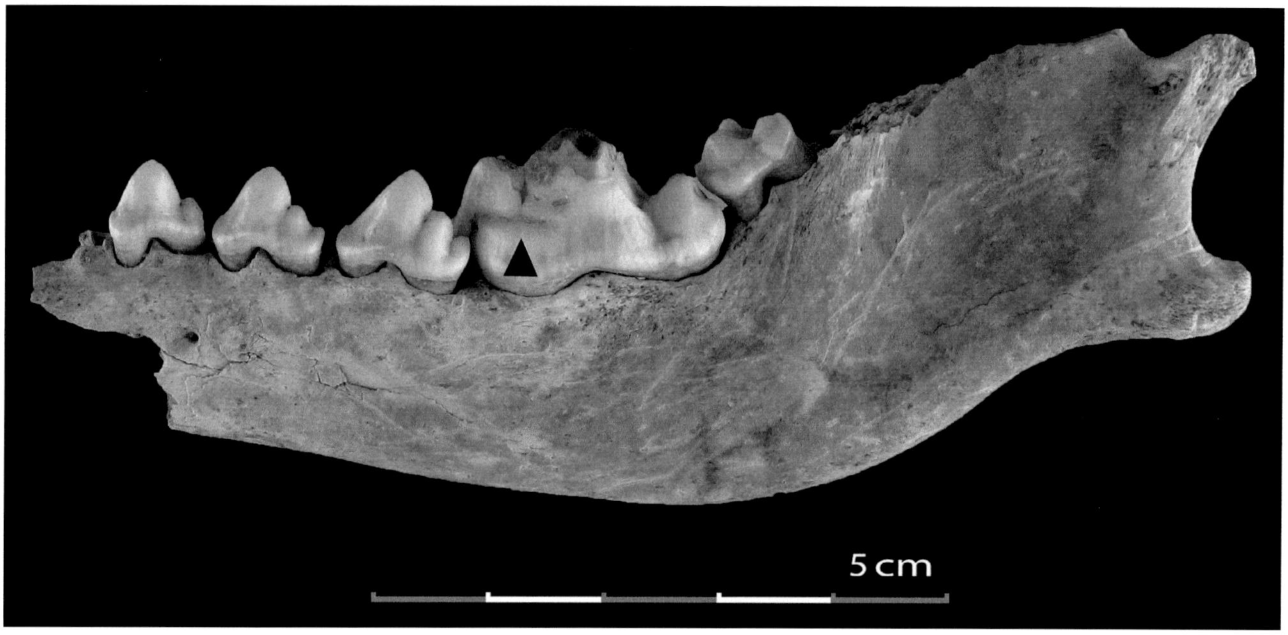

Fig. 16. Left mandible of specimen DA3 in lateral aspect showing linear enamel hypoplasia in the first cusp of the carnassial tooth (indicated by the black arrow).

DA3 are both males showing a different degree of robustness regarding the radius.

Further, the analysis indicated that the gazehound and the molossoid type are not present in the Beirut sample. The combined metric and morphological traits of the skull and the limbs define the specimens as mediolinear dogs (Regnault 1903). They are of medium bodily proportions and show a medium-snouted cranium along with a moderate stop. In modern mediolinear breeds, the lupoid and the braccoid types are generally difficult to discriminate (Belhaoues 2011a). The lupoid type is actually represented by the Spitz, the Nordic dog breeds and the majority of the shepherds. The majority of the hunting dogs represents the braccoid type. Regarding the archaeological sample, the specimens showed a gradual variation in size and robustness between the braccoid and the lupoid dogs; although specimen 1080 reveals characteristics more similar to large lupoid-type breeds, in particular the cranium that exhibits well-developed crests.

The extent of size variability amongst the specimens from the Persian period in Beirut correspond to that of pre-Roman archaeological samples that are generally considered as originating from a relatively homogenous-size population. For instance, the overall size (46-57cm at the withers) and long bone measurements (Greatest length and slenderness index; Table 7) of Beirut dogs are included in the range of variation of Bronze Age specimens of Britain, which are considered as resulting from a single freely interbreeding population (Harcourt 1974, Clark 2000). Interestingly, on the other hand,

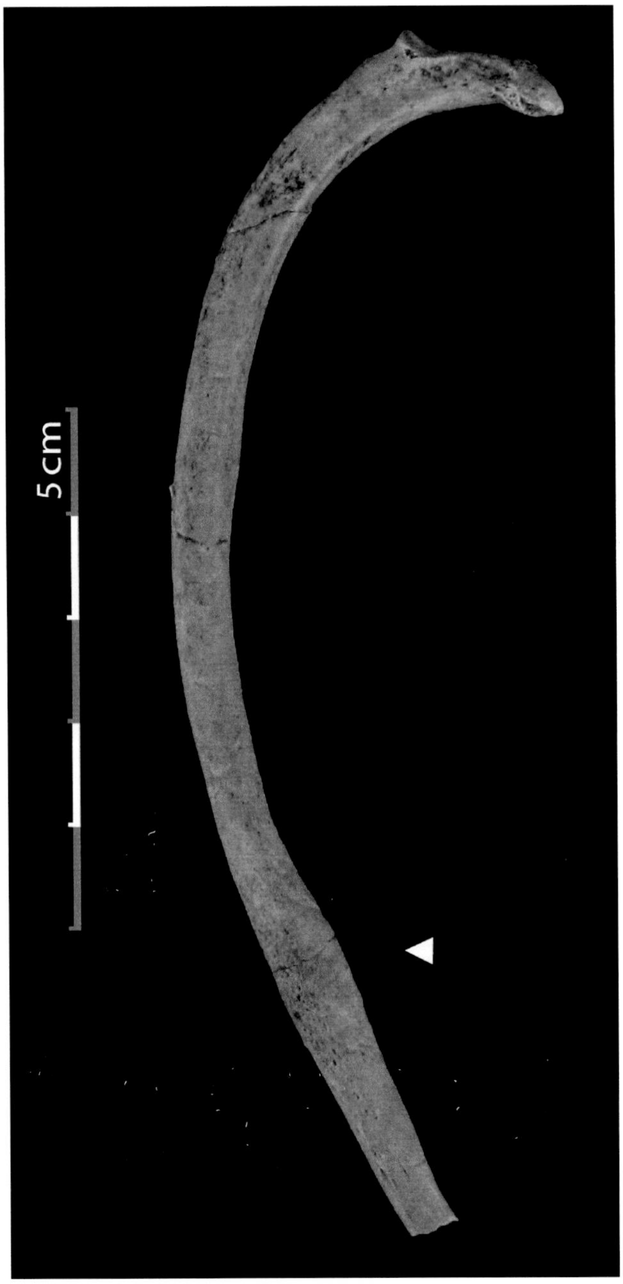

Fig. 17. The 7th left rib of specimen 1079 showing a callus formation on the corpus (indicated by the white arrow).

the absence of the mandibular third molar is commonly not detected in Pre-Roman material (Clark 2000). This condition, along with tooth crowding, results from diminution of jaw length (Clark 2000).

Then again, specimens of the medium size group, DA1, DA12 and 1078 are not much larger than the jackal and show a similar slenderness index regarding the limbs. Commonly, the Levantine jackal is close in size to domestic dogs (Wapnish and Hesse 1993). Furthermore, Daróczi-Szabó (2006) pointed out the 'jackal-look' of pariah dogs, stating that the head shape of many feral dogs breeding freely appears more similar to jackals rather than wolves. Morphologically, the Beirut specimens showed different cranial traits from the jackal, such as the well-developed sagittal crest and the moderate stop;

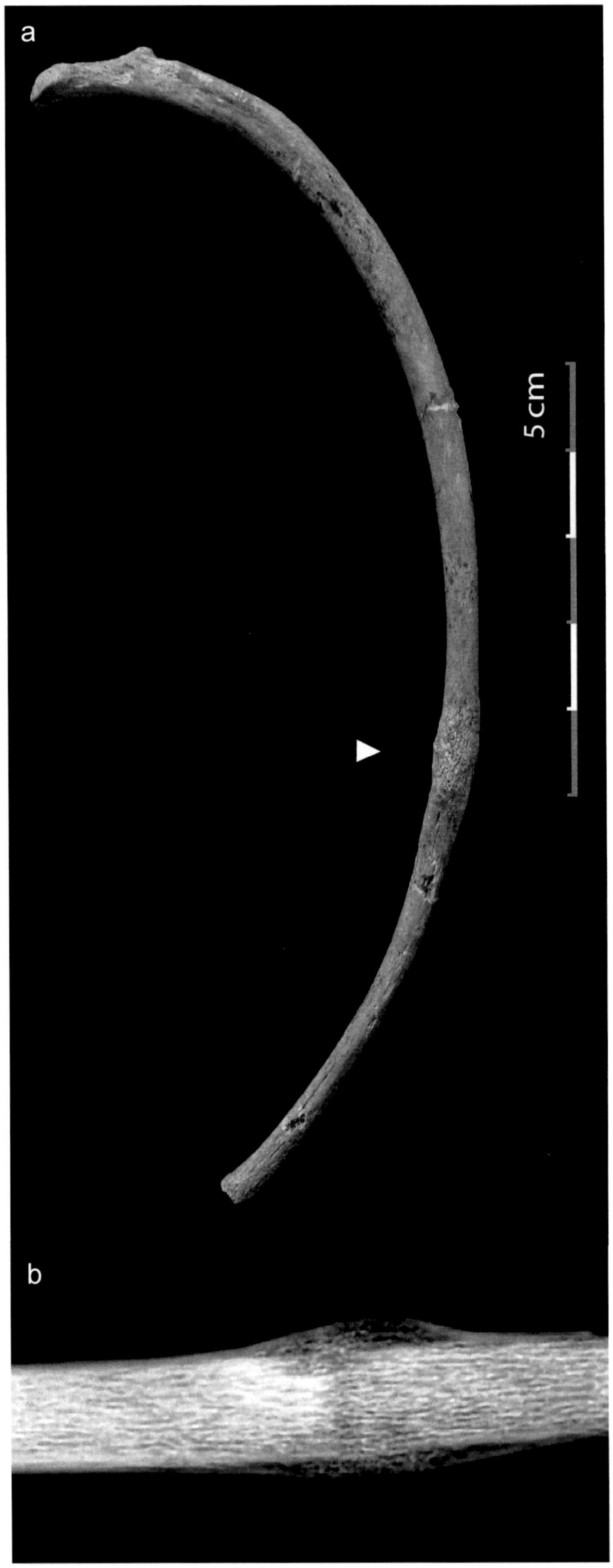

Fig. 18. The 11th left rib of specimen DA9 showing callus formation on the corpus (indicated by the white arrow - a). The X-ray image shows the fracture site (b).

nonetheless, only the cranium of specimen DA12 from the medium size group is preserved.

Health status

While juvenile individuals did not show any sign of pathological modifications, all of the twelve

a

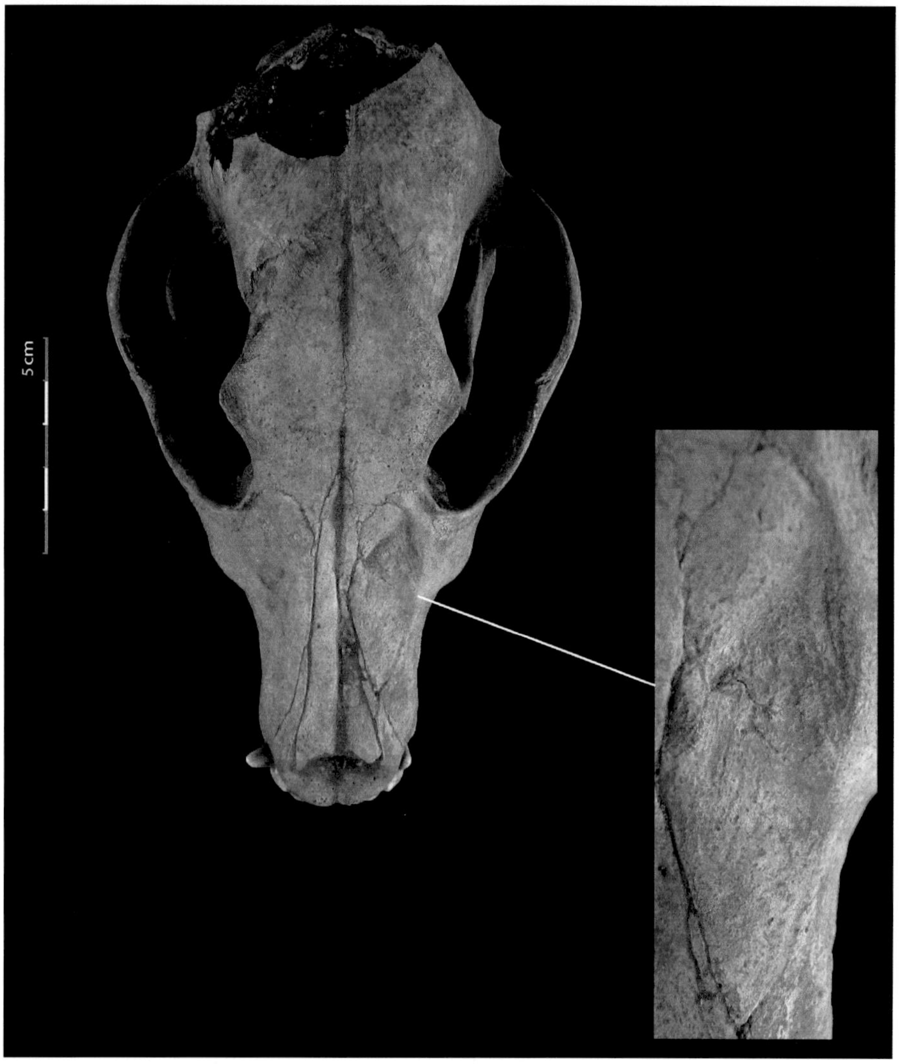

b

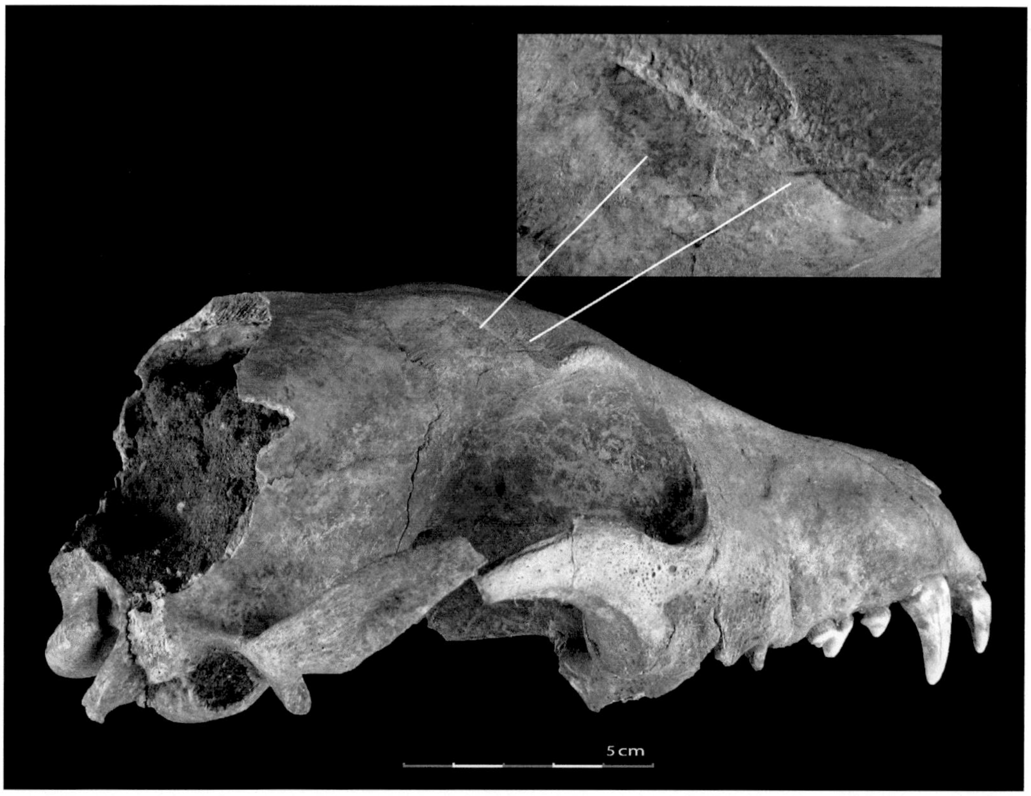

Fig. 19. Cranium of specimen DAI2 showing a crush fracture in the nasal bone (a) and a penetrating injury in the supraorbital area (b).

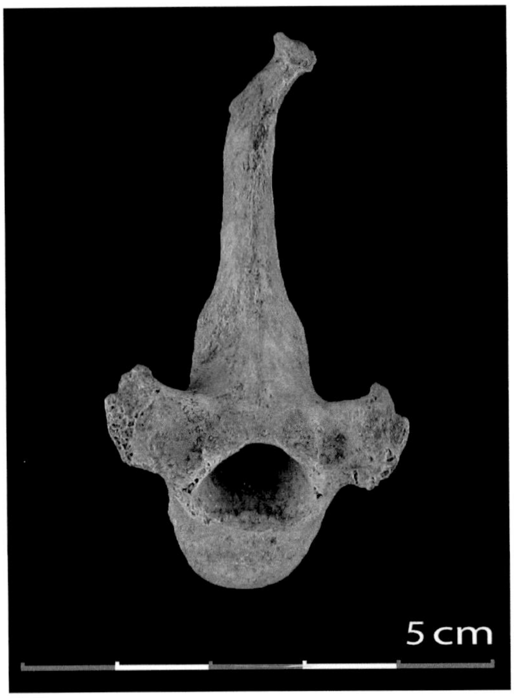

Fig. 20. The 8th thoracic vertebra of specimen DA8 showing a distortion of the spinous process.

fully- and partially-articulated skeletons of older individuals showed evidence of skeletal pathologies (Fig. 13, Table 5).

Periodontal disease was recorded in three individuals, two of which exhibited *ante-mortem* tooth loss with remodeling of the empty alveolus (Fig. 14). One individual showed periosteal infection in the horizontal ramus of the mandible (Fig. 15). Dental anomalies also consisted in irregular wear and tooth breakage. A case of linear enamel hypoplasia was recorded and indicates a physiological stress (Fig. 16).

Traumatic injuries were noted in five individuals. These lesions were typically located on the trunk and cranium. Transverse fractures of ribs were observed in four specimens (Fig. 17 & 18); the callus appears to have remodeled indicating an advanced healing process and that the individuals survived for at least several weeks after they were injured (Aufderheide and Rodríguez-Martín 1998: 21; Ortner 2003: 126 - 128). Conversely, one of these specimens had two thoracic vertebrae showing fracture of the spinous process while the apex of the

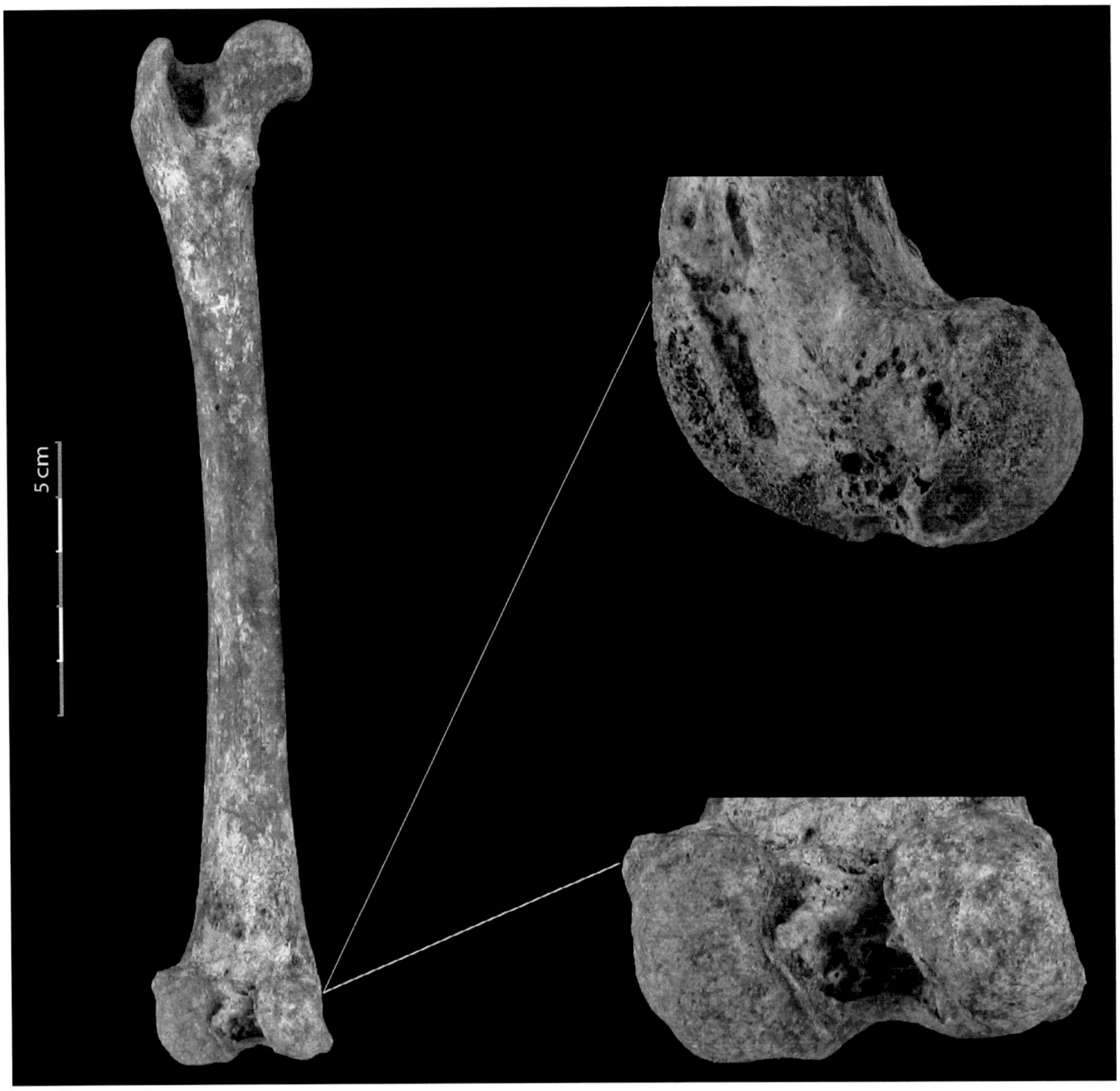

Fig. 21. Left femur of specimen I070 showing osteophytes, marginal lipping and pitting at the distal end.

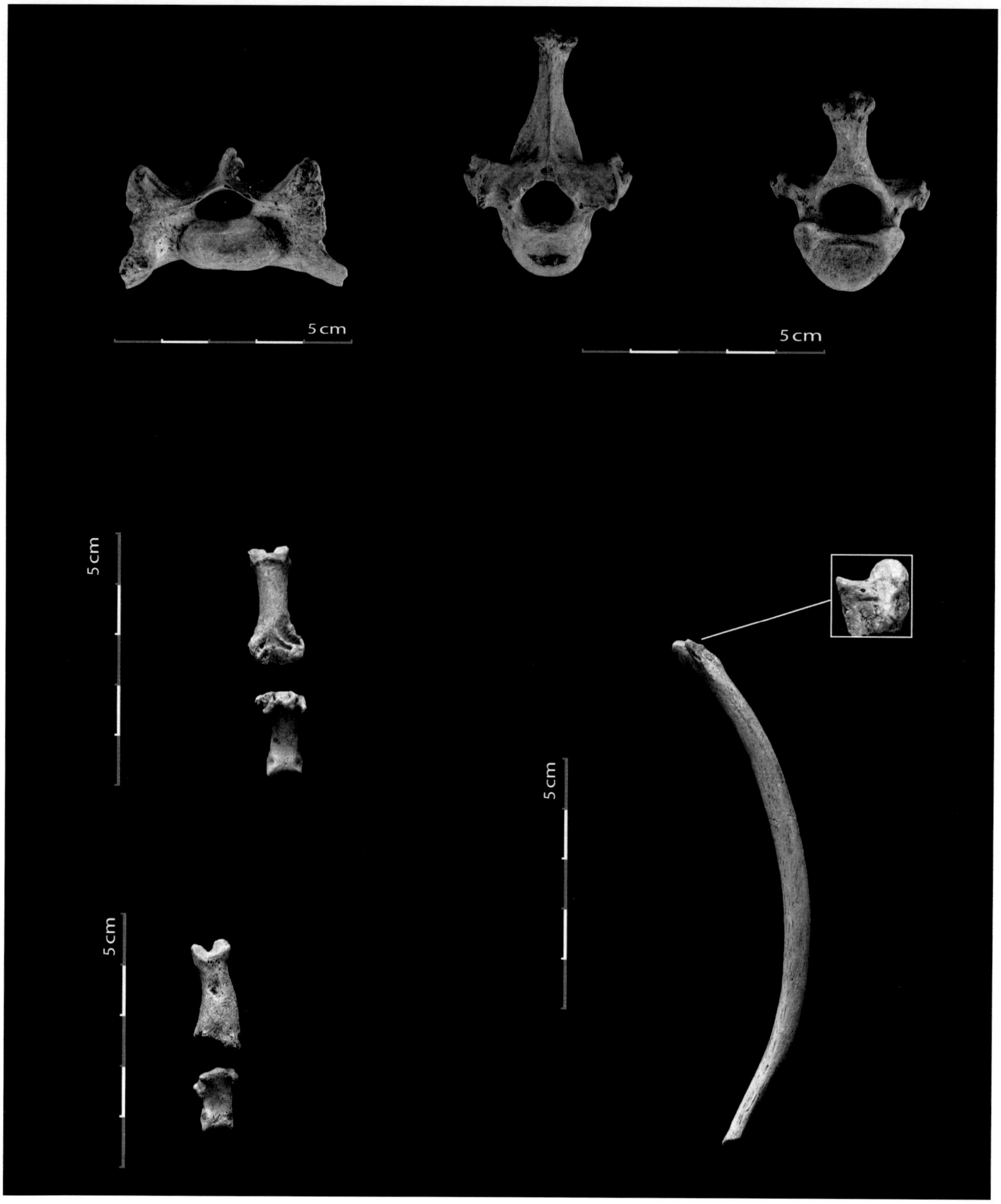

Fig. 22. Bone changes in specimen 1077. Fracture of the right ala of the sacrum and distortion of the spinous processes; transversal lesion on the body endplates of the 9th thoracic vertebra; osteophytes at the proximal and distal end of phalanges of the fore and hind paws; marginal lipping at the head of the 13th right rib.

processes had not re-fused with the vertebrae at the time of death. Another specimen showed a green-stick fracture (Lovell 1997) of a thoracic spinous process. The injuries inflicted to two crania appeared as crush fractures (Lovell 1997) in the frontal and nasal bone. Moreover, one of the crania displayed a penetrating injury in the supraorbital area (Fig. 19).

Bone abnormalities *a priori* of congenital origin were observed. Undersized vertebral anterior and/or posterior articular processes were found in one specimen. The curvature of the vertebral spinous process was observed in six specimens in the thoracic, lumbar and sacral vertebrae (Fig. 20). These modifications are unlikely to have affected the animals.

Five individuals exhibited signs of joint diseases. Eburnation and porosity aspects were present on articular surfaces and osteophytes surrounded the periphery of the joint margins. Lesions were observed in one case at the knee (Fig. 21), wrist (marginal lipping at a carpal bone surface) and interphalangeal joints (Fig. 22). The highest prevalence was observed in the costo-vertebral joints (3 cases; Fig. 22),

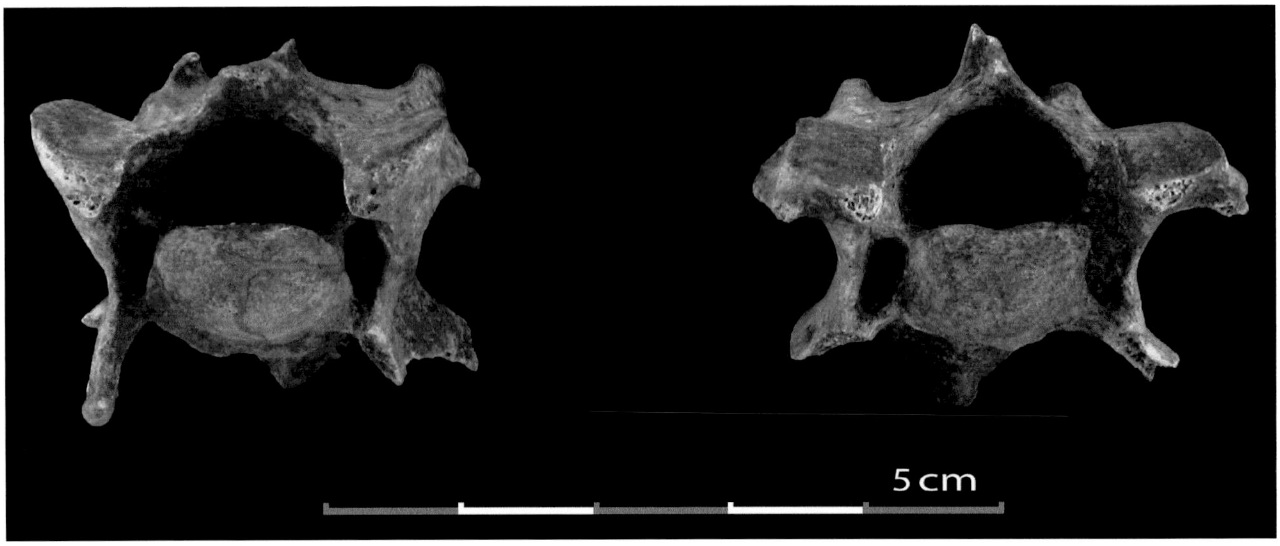

Fig. 23. Cranial aspect of the 3rd and 4th cervical vertebrae of specimen 1070 showing marginal osteophytes at the vertebral body endplates.

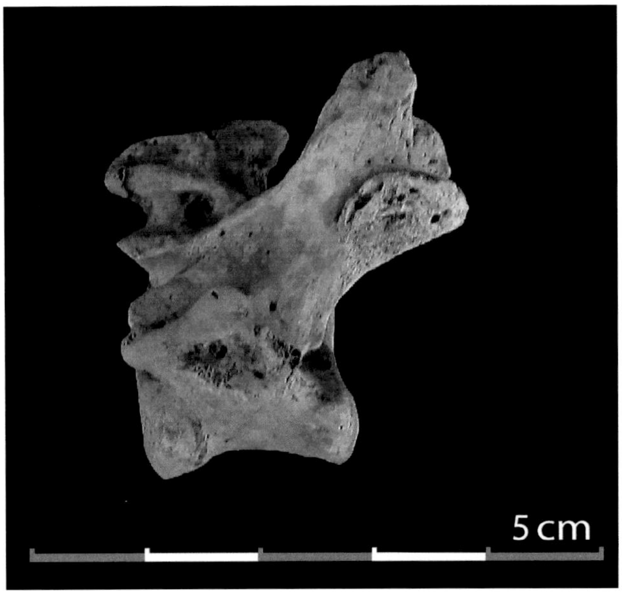

Fig. 24. Lateral aspect of the 10th thoracic vertebra of specimen 1080 showing pitting and widening of the contour of the caudal facet joints.

intervertebral and vertebral facet joints (four cases; Figs. 22, 23 and 24).

Ligamentous ossifications led to exostoses formation at the apex of vertebral spinous processes in six specimens and to the fusion of vertebrae in one of these cases with syndesmophyte formation between vertebral bodies (Fig. 25; Bartosiewicz 2013: 139). A case of ligament ossification is also the reason behind the fusion of the distal diaphyses of the tibia and the fibula (Fig. 26). Another case of tendon ossification in a metacarpal was recorded. Enthesophytes were found in one specimen in which bone spurs formed at the surface of the greater trochanter of the femur and at the acromion process of the scapula. Conditions such as enthesophytes and ligamentous ossifications advance with age but can also be linked to environmental factors or inherited disorders (Bendrey 2007; Bartosiewicz 2013: 139).

One case of inflammation was recorded. The two scapulae of specimen 1080 showed cloacae and proliferation of new bone that are likely to originate from periostitis related to microtrauma, or from an infection transmitted from an adjacent organ (Fig. 27).

Discussion

The burials at SFI 477 and SFI 645 fit in most of the archaeological characteristics of the dog inhumation custom in southern Levant outlined by Çakırlar *et al.* (2014). For instance, the Beirut case affirms the tendency to place the burials in clusters rather than in isolation. As at other Late Iron Age settlements, the deposits consisted of individual burials and the dogs were buried with care on their sides. Imperishable grave goods were absent. However, the Beirut case differs from other Levantine sites by some aspects of the mortuary practices. While no repetitive body position occurred, a regular orientation of the corpse is noticed and seems to follow the axis of the water-stream and the street. Further, unlike the other sites were the corpses were deposited in shallow pits (Ashkelon, Ashdod, Tell el-Burak); no graves were found for Beirut dogs. Hence, it is not clear whether the indices of linear delimitations observed on some skeletons correspond to the edges of a pit or to those of a perishable container or cover. The latter hypothesis might explain the presence of a void around the corpses at the initial stage of the decay process. Currently, the issue of the absence of burial pits cannot be attributed to a technical difficulty of separating the sediments (sediment matrix v/s burial fill) or to a particular burial practice. However, of particular interest is the absence of actual grave-pits at lot BEY 020 (Finkbeiner and

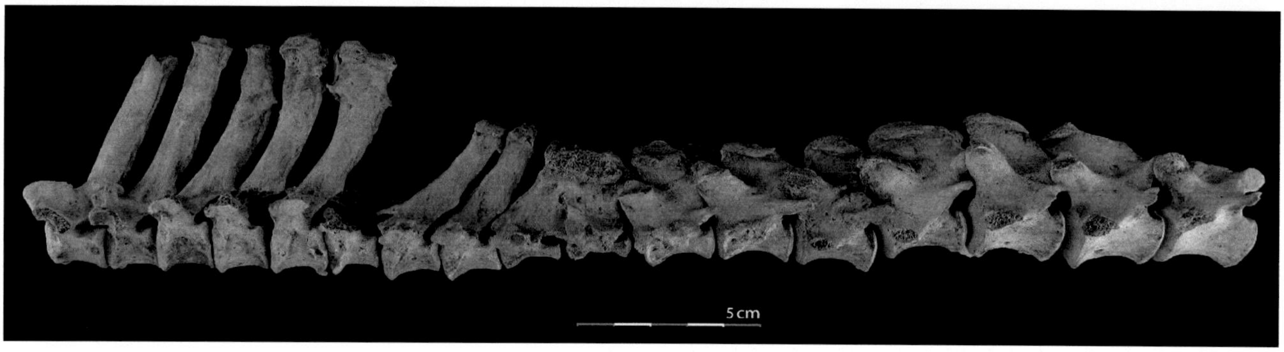

Fig. 25. Left lateral aspect of a spine segment (1st thoracic – 4th lumbar vertebra) of specimen 1079 showing syndesmophyte formation between vertebral bodies (T5, T6, L1, L2), ossification of supraspinous (T3, T4, T5, T7, T9, T10) and short ligaments of the apophyseal joints leading to the fusion of the spinous processes and the apophyseal articulations in two vertebrae (T9, T10).

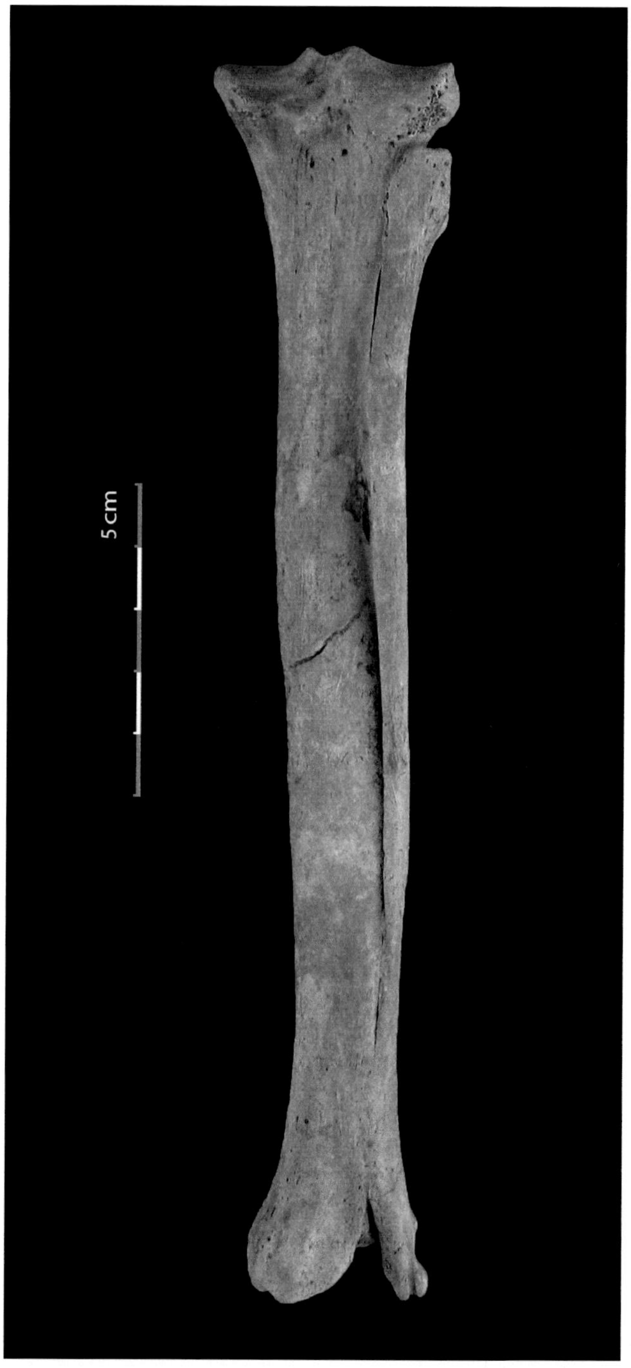

Fig. 26. Ossification of the tibio-fibular ligaments, leading to the fusion of the distal diaphysis of the right tibia and fibula of specimen 1080.

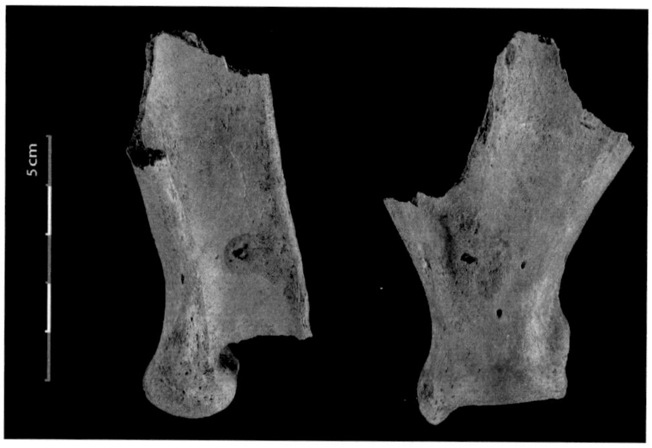

Fig. 27. Caudal and medial aspects of the right scapula of specimen 1080 showing cloacae and proliferation of new bone.

Sader 1997)[3] where the sediment matrix is also different from that at SFI 477 and SFI 645. At any rate, skeletal displacements suggest that some dogs were not lying on bare earth but were placed in a perishable container or under a perishable cover isolating the corpse from the sediment. Likewise, a specimen was covered with amphora fragments and another one with a large cobble at the thoracic region. This practice was previously attested at Beirut at lot BEY 020 and at Tell el-Burak where respectively three and one dog were covered with fragments of Late Iron Age amphorae.

At Ashkelon, Tel Dor and Tell el-Burak, the dogs were placed in empty spaces left between buildings and streets with no clear demarcations of the limits of the dog concentrations. Further, at Ashkelon, Tell el-Burak and Shoham the burials were concentrated at the edge of the settlement (Wapnish and Hesse 1993, Kamlah and Sader 2003, Çakirlar *et al.* 2014, Nadelman 1995). The dog burials uncovered

3 Currently a rescue excavation held by the Directorate General of Antiquities is taking place in the extension of lot BEY 020 and an additional dog burial was recovered in the debris overlaying the Iron Age glacis. There was also no grave-pit for the dog (L. Sheikho, personal communication, may 2016)

previously in Beirut at lot BEY 020 (Finkbeiner and Sader 1997), showed a similar contextual pattern; the dogs were interred in the Persian period debris deposited above the Iron Age glacis. The newly excavated sites in the Saifi quarter add to the understanding of the contextual location of the dog burials. Those cluster outside the urbanized area, on the eastern outer limits of the city. Furthermore, many intra-site indices at Beirut indicate that the deposits were successive and accumulated intermittently. Development of current archaeological research would allow a more thorough comparison of the stratigraphic position of the burials between the different lots in the future and would add information about the time span in which the dogs were interred. Likewise, at Ashdod the dogs were inhumed over a length of time rather than all at once and at Ashkelon the burials accumulated over about eighty years and many burials disturbed earlier inhumations.

These features indicate that the Beirut burials are rather related to a tradition of burying dogs than to a punctual event as an epizootic outbreak or a sacrificial act. In addition, the skeletons showed no butchery marks. The available data does not allow any speculation about the intention behind this practice. In this regard, Çakırlar *et al.* (2014) state that 'the accumulated evidence on Persian period dog burials in the Levant accommodates a basic definition of the archaeological characteristics of the phenomenon, yet it serves poorly towards finding its meaning'.

Nevertheless, the biological features of the dogs involved in this practice are more explicit and the traits of Beirut dogs appear to fall into what has been described for the Levantine sites. At Batash, the withers height of a dog was estimated at 48.5 cm, and Ashkelon dogs stood between 48 and 61 cm at the withers (Wapnish and Hesse 1993). This variation range includes most of the dogs from Beirut which are slightly smaller, standing 46-57 cm at the shoulder. Further, the specimens from Ashkelon and Tell el-Burak showed a mesocephalic skull type and a lack of specialized osteomorphological traits (Wapnish and Hesse 1993; Çakirlar *et al.* 2014). The dogs from Beirut appeared to be sub-dolichocephalic or sub-mesocephalic. Variation in the overall size, skull shape and dimensions, occur in the Beirut sample but the skeletons did not show specialized adaptation that allows the identification of distinct morphotypes resulting from 'anthropogenic selective pressures' (Bartosiewicz 2000: 186).

This having been said, accidental interbreeds are likely to occur among watchdogs, herding dogs and pariah dogs in an urban settlement (Bökönyi 1984; Bartosiewicz 1995) and the disparate distribution of size involves at times uncontrolled reproductions (Gardeisen *et al.* 2011; Belhaoues 2011b). Hence, the variation of the specimens from Beirut are either the result of the non-control of the reproduction, or rather the opposite from the production of different '*breeds*' targeting only mediolinear type dogs that do not exhibit hyper-specialized adaptations perceptible from the skeletal morphology as in gazehound-type or the modern Basset hound. *A fortiori* modern mediolinear dogs can serve a variety of uses, mainly hunting and guarding. By all means, the range of variation is limited to medium sized and medium-large dogs. Small and large dogs are not present. Various faunal analyses argued that such range of variation in the stature corresponds to a low diversity (Harcourt 1974) and is likely the result of accidental panmixis (De Grossi Mazzorin and Tagliacozzo 2000). Selective pressures resulting in greater variability in size and morphology of dogs, indicative of '*breed*' formation, is argued to become noticeable in archaeological samples beyond the borders of Italy at the advent of the Roman era (Harcourt 1974; Lepetz 1996: 68; Bartosiewicz 2000). At Ashkelon, the variability of the dogs was interpreted as originating from a naturally occurring canid population, upon comparison with the unmanaged pariah dogs of Palestine (Wapnish and Hesse 1993).

The demography of Beirut sample seems to reflect a natural mortality where the youngest and the oldest individuals are the most affected. Mortality profiles of modern dogs indicate that 88% reach or go beyond five years of age and that 62% go beyond ten years (Egenvall *et al.* 2000) but these patterns cannot be applied to archaeological dogs given the amount of care given nowadays to pet animals. Conversely, in a population of free-ranging urban dogs, fifty percent die at juvenile age (Beck 1973 *in* Wapnish and Hesse 1993). Likewise, the distribution of mortality at Ashkelon with 62% of young individuals/infants (0-6 months), 5% of subadults (6-18 months) and 33% of adults (18 months and older) indicates that buried individuals were not pre-selected based on age but show a parallel with mortality patterns in free urban dogs (Wapnish and Hesse 1993). The non-selective nature of the burials was also noticed at Tell el-Burak where one of the individuals was slightly less than 6 months at the time of death and the other one was an adult of 3-4 year-old (Çakırlar *et al.* 2014). Similarly, at Tel Dor twelve individuals were assigned to age groups which consisted of seven juveniles and five adults that all died of natural causes[4] (Sapir-Hen 2011). The frequency of juveniles is, however, lower in the Beirut sample in comparison with the sample from Ashkelon and Tel Dor and consists of only a quarter of the sample.

4 One of these dogs is dated to the early Hellenistic period

With regard to the health status, Beirut dogs display traces of pathological conditions having different etiologies. Periodontal disease originates from alveolar infections (Bartosiewicz 2008) and may be related to oral injury caused by hard food. *Lato sensu* arthritis originates from environmental and nutrition conditions, inheritance or congenital predisposition, as well as age-related decrease in bone vascularisation (Chaplin 1971; Von den Driesch 1975; Holmberg and Reiland 1984; Bartosiewicz 2008). In Beirut dogs, joint diseases and ligamentous ossifications may be prevalently related to inherited disorders or environmental factors considering the relatively young age of the individuals. While a few pathologies reflect physiological stress and illness, in particular infectious disease, none of these appears to be directly implicated in the animal's death. Also, most of the traumatic fractures seem to have healed at least partially at the time of death.

Similarly, at Ashkelon c. 5% of the subadult and adult individuals showed pathological traces remnants of both disease and injuries like damaged paws, knitted breaks (including a case of broken and healed first cervical vertebra), dislocations, parasite infections and dental anomalies (Wapnish and Hesse 1993).

The patterning of the trauma injuries recorded in Beirut dogs does not concord with intra-species aggression. In general, bite wounds in dogs are often directed at the head and shoulder region (Baranyiová *et al.* 2003) while in the analyzed sample, fractures occurred most frequently in the vertebrae and ribs. Moreover, none of the fractures seem to be caused by puncture wounds (Churcher 1993: 53; Groot 2008). These injuries may have been inflicted by humans as a result of a blow to the head and to the body and therefore may reflect cultural conditions under which the dogs were kept. On the other hand, the injuries may have been sustained during work if the individuals were working dogs, as being kicked or gored by cows when herding or having encountered aggressive episodes when guarding. At any rate, these dogs have received a special treatment in death, be it symbolic in nature or purely practical.

Conclusion

The practice of dog burial in the Levantine Persian period has been documented on several sites but not all the findings were analyzed in a detailed archaeozoological and archaeothanatological perspective. Dog burials recently discovered on the eastern outskirts of ancient Beirut brought further light on this tradition and highlighted trends specific to the Persian period in Beirut. Areas at the outskirts of the city were devoted to dog inhumation, along the shore of a water stream and the axis of a street, as well as in a dumping zone above the Iron Age fortification system. A glimpse on the biological identity of the buried dogs has been attempted through morphological and metric analyses and examination of skeletal pathologies. The dogs appeared to be of medium size and medium bodily proportions. They suffered mostly from conditions linked to environmental factors and inherited susceptibility. Nonetheless, several signs of traumatic injuries were apparent on the skeletons, showing that the concerned individuals were not pampered in life but were poorly treated or were plain work-dogs (guard, herd or hunting dogs). However, after their natural death they received a careful treatment in the form of an individual burial. The reconstructed biological features (stature, cranial shape, health conditions and mortality patterns) fit into what has been previously described for the Levantine dogs.

Yet, many aspects of the dog burial phenomenon in Beirut remain to be explored. Investigation of the sediment samples collected from inside the burials at different regions of the skeletons may, on one hand, provide more information about possible concealed aspects of the burial space such as the existence of a perishable container or grave goods and, on the other hand, reveal more biological data reflecting on the health state of the dogs. Analysis of faunal assemblages from other contexts dating to the Persian period shall offer a better understanding towards the relation between dogs and people in Beirut. In addition, a thorough comparison with pre- and post-Persian period dog burials will add to our understanding of this long-standing tradition.

Acknowledgements

I am grateful to Dr. J. Nassar and Dr. J. Chanteau, Dr. A. Seif and Mr. H. Choueri for having entrusted me the study of this material. I am thankful also to Mr. S. El-Khoury, Director General of Antiquities (Lebanon), for his gracious support.

Prof. L. Bartosiewicz and Dr. P. Csippán generously provided helpful comments on skeletal pathologies. Anthropologist colleagues, in particular Ms. V. Kalenderian, offered valuable advice on archaeothanatological observations. Dr. C. Çakırlar reviewed and improved the first draft. I gratefully acknowledge their support. Any error of identification or interpretation remains mine.

References

Alpak, H., Mutus, R. and Onar, V. 2004. Correlation analysis of the skull and long bones measurements of the dog. *Annals of Anatomy* 186: 323–330.

Aufderheide, A. C. and Rodríguez-Martín, C. 1998. *The Cambridge Encyclopedia of Human Palaeopathology*. Cambridge: Cambridge University Press.

Baker, J. R. and Brothwell, D. 1980. *Animal Diseases in Archaeology*. London: Academic Press.

Baranyiová, E., Holub, A., Martiníkova, M., Nečas,

A. and Zatloukal, J. 2003. Epidemiology of intra-species bite wounds in dogs in the Czech Republic. *Acta Veterinaria Brno* 72: 55–62.

Barone, R. 1976. *Anatomie comparée des Mammifères Domestiques.* Tome 1, Ostéologie. Paris: Vigot.

Bartosiewicz, L. 1995. *Animals in the Urban Landscape in the Wake of the Middle Ages*: Tempus Reparatum. Oxford: British Archaeological Reports International Series 609.

Bartosiewicz, L. 2000. Metric variability in Roman period dogs in Pannonia province and the Barbaricum (Hungary). In, Crockford, S. J. (ed.) *Dogs Through Time: An Archaeological Perspective. Proceedings of the 1st ICAZ Symposium on the History of the Domestic Dog. Eighth Congress of the International Council for Archaeozoology (ICAZ 98) August 23-29, 1998, Victoria, B.C., Canada*: 181–189. Oxford: BAR International Series 889.

Bartosiewicz, L. 2008. Environmental stress in early domestic sheep. In, Miklíková, Z. and Thomas, R. (eds) *Current Research in Animal Palaeopathology. Proceedings of the Second ICAZ Animal Palaeopathology Working Group Conference*: 3–13. Oxford: BAR International Series 1844.

Bartosiewicz, L. 2013 (with a contribution by Erika Gál). *Shuffling Nags, Lame Ducks: the Archaeology of Animal Disease*. Oxford, Oxbow Books.

Baxter, I. L. 2006. A Dwarf Hound Skeleton from a Romano-British Grave at York Road, Leicester, England, U.K., with a discussion of other Roman small dog types and speculation regarding their respective aetiologies. In, Snyder, L. M. and Moore, E .A. (eds) *Dogs and People in Social, Working, Economic or Symbolic Interaction. Proceedings of the 9th Conference of the International Council of Archaeozoology, Durham, August 2002*: 12–23. Oxford: Oxbow Books.

Baxter, I. L. 2010. *Small Roman Dogs*. [online] Available at: <http://alexandriaarchive.org/bonecommons/archive/files/baxter_2010_small_roman_dogs_6dc7d64928.pdf>.

Beck, A. M. 1973. *The Ecology of Stray Dogs*. Baltimore: York Press.

Belhaoues, F. 2011a. Etude d'un assemblage canin inédit issu du puits 103 à Ambrussum (Hérault). *Mémoire de Master 2 Archéologie, Université Montpellier III – Paul Valéry*.

Belhaoues, F. 2011b. Variabilité des chiens antiques. L'apport inédit du puits PT103 à Ambrussum (Hé-rault). *Publication des « Cinquièmes Rencontres archéozoologiques de Lattes, 17 juin 2011:* 1–11.

Bennett, W. J. and Blakely, J. A. 1989. *Tell el-Hesi: The Persian Period (Stratum V)*. Winona Lake, Indiana: Eisenbrauns.

Bennett, W. J. and Schwartz, J. H. 1989. Faunal Remains. In, Bennett, W.J. and Blakely, J.A. (eds) *Tell el-Hesi: The Persian Period (Stratum V)*. Winona Lake, Indiana: Eisenbrauns.

Betti, E. 1990. Céphalométrie comparée chez le Chien. *Thèse vét*. Nantes.

Binois, A. 2012. Une vie de chien... Traumatismes multiples et maltraitance potentielles sur un chien médiéval, un cas paléopathologique. *Bull. Soc. Méd. Sci. Vét*. 12: 37–50.

Bökönyi, S. 1984. *Animal Husbandry and Hunting in Tác-Gorsium*. Budapest: Akadémiai Kiadó.

Brehm, H., Loeffler, K. and Komeyli, H. 1985. Schädelformen beim Hund. *Zbl. Vet. Med. C. Anat. Histol. Embryol.* 14: 324–331.

Çakırlar, C., Amer, V., Kamlah, J. and Sader, H. 2014. Persian Period Dog Burials in the Levant: New Evidence From Tell El-Burak (Lebanon) and a Reconsideration of the Phenomenon. In, De Cupere, B., Linseele, V. and Hamilton-Dyer, S. (eds) *Archaeozoology of the Near East X. Proceedings of the Tenth International Symposium on the Archaeozoology of South-Western Asia and Adjacent Areas:* 243–264.

Chaplin, R. E. 1971. *The Study of Animal Bones from Archaeological Sites*. London and New York Seminar Press.

Churcher, C. S. 1993. Dogs from Ein Tirghi Cemetery, Balat, Dakleh Oasis, Western Desert of Egypt. In, Clason, A. T., Payne, S. and Uerpmann, H.-P. (eds) *Skeletons in Her Cupboard*: 39-59. Oxford: Oxbow Monograph, 34.

Clark, K. M. 1996. Neolithic dogs: a reappraisal based on evidence from the remains of a large canid deposited in a ritual feature. *International Journal of Osteoarchaeology* 6: 211–219.

Clark, K. M. 2000. Dogged persistence: The Phenomenon of Canine Skeletal Uniformity in British Prehistory. In, Crockford, S.J. (ed.) *Dogs Through Time: An Archaeological Perspective*: 163–169. Oxford: BAR International Series 889.

Crockford, S. J. 2000. Dog evolution: a role for thyroid hormone physiology in domestication changes. In, Crockford, S.J. (ed.) *Dogs Through Time: An Archaeological Perspective*: 3–20. Oxford: BAR International Series 889.

Cruz, F., Vilá, C. and Webster, M. T. 2008. The legacy of domestication: accumulation of deleterious mutations in the dog genome. *Molecular Biology and Evolution* 25: 2331–2336.

Csippán, P. and Daróczi-Szabó, L. 2008. Animal diseases from medieval Buda. In, Miklikovà, Z. and Thomas, R. (eds) *Current Research in Animal Palaeopathology. Proceedings of the Second ICAZ Animal Palaeopathology Working Group Conference*: 74–79. Oxford: BAR International Series 1844.

Curvers, H. 2013. *Burial Customs of Beirut* (Amman September 9, 2013). Oral Talk, Round-Table Conference Burial Customs of Bilad ash-Sham (University of Jordan / DAAD).

Daróczi-Szabó, M. 2006. Variability in Medieval Dogs from Hungary. In, Snyder, L.M. and Moore, E.A. (eds) *Dogs and People in Social, Working, Economic or Symbolic Interaction. Proceedings of the 9th Conference of the International Council of Archaeozoology, Durham, August 2002*: 85–95. Oxford: Oxbow Books.

Dixon, H.M. 2013. *Phoenician Mortuary Practice in the Iron Age I – III (ca. 1200 – ca. 3000 BCE) Levantine "Homeland"*. PhD. University of Michigan.

Dothan, M. and Porath, Y. 1982. Ashdod IV. Excavation of Area M. The Fortifications of the Lower City. *Atiqot* 15. Jerusalem: Publications of the Israel Antiquities Authority.

Duday, H. 1990. Observations ostéologiques et décomposition du cadavre: sépulture colmatée ou en espace vide. *Revue Archéologique du Centre de la France* 29(2): 193–196.

Duday, H. 2006. L'archéothanatologie ou l'archéologie de la mort (Archaeothanatology or the Archaeology of death). In, Gowland, R. and Knüssel, C. (eds) *Social Archaeology of funerary remains*: 30–56. Oxford: Oxbow Books.

Duday, H., Cipriani, A.-M. and Pearce, J. 2009. *The Archaeology of the Dead: Lectures in Archaeothanatology*. Oxford: Oxbow Books.

Duday, H., Le Mort, F. and Tillier, A.-M. 2014. Archaeothanatology and Funeral Archaeology, Application to the Study of Primary Single Burials. *Anthropologie* LII(3): 235– 246.

Edrey, M. 2008. The Dog Burials at Achaemenid Ashkelon Revisited. *Tel Aviv* 35(2): 267–282.

Egenvall, A., Bonnett, B., Shoukrim, M., Olson, P., Hedhammar, A. and Dohho, I. 2000. Age pattern of mortality in eight breeds of insured dogs in Sweden. *Preventive Veterinary Medicine* 44: 1–14.

Elayi, J. 2001. *L'antique Beyrouth à la lumière des dernières fouilles archéologiques*. [online] Available at: <http://www.clio.fr/BIBLIOTHEQUE/l_antique_beyrouth_a_la_lumiere_des_dernieres_fouilles_archeologiques.asp> [Accessed on 10 May 2016].

Elayi, J. and Sayegh, H. 2000. *Un quartier du port phénicien de Beyrouth au Fer III/Perse. Archéologie et histoire*. Paris: Éditions Gabalda.

Ellenberger, W. and Baum, H. 1943. *Handbuch der Vergleichenden Anatomie der Haustiere*. 18th ed. Berlin: Springer.

Evans, H. E. and Christensen, G. C. 1979. *Miller's Anatomy of the Dog*. Philadelphia: W.B. Saunders Co.

Finkbeiner, U. and Sader H. 1997. BEY 020: Preliminary Report on the Excavations 1995. *Bulletin d'Archéologie et d'Architecture Libanaises* 2: 114–166.

Galy, G. 2004. Analyse ostéopathique de la variété des morphologies canines. *Mémoire de l'European School of Animal Osteopathy (E.S.A.O)*. Brighton.

Gardeisen, A., Valenzuela-Lamas, S. and Belhaoues, F. 2011. Variabilité des morphotypes canins archéologiques et implications sur le statut des chiens dans l'Antiquité. In, Brugal, J.-P., Gardeisen, A. and Zucker, A. (eds) *Prédateurs dans tous leurs états: Evolution, Biodiversité, Interaction, Mythes, Symboles. Actes des XXXIe Rencontres internationales d'Archéologie et d'Histoire d'Antibes*. Antibes: Editions APDCA.

Groot, M. 2008. Understanding past human-animal relationships through the analysis of fractures: a case study from a Roman site in The Netherlands. In, Miklíková, Z. and Thomas, R. (eds) *Current Research in Animal Palaeopathology. Proceedings of the Second ICAZ Animal Palaeopathology Working Group Conference*: 40–50. Oxford: BAR International Series 1844.

Halpern, B. 2000. The Canine Conundrum of Ashkelon: A Classical Connection? In, Stager, L. E. Greene, J. A. and Coogan, M. D. (eds) *The Archaeology of Jordan and Beyond: Essays in Honor of James A. Sauer*: 133–144. Winona Lake, Indiana: Eisenbrauns.

Harcourt, R. 1974. The dog in prehistoric and early historic Britain. *Journal of Archaeological Science* 1: 151–175.

Helton, W. S. 2016. Development of expertise in working dogs: The role of social affordances. *Proceedings of the 5th Canine Science Forum*, 1st ed., Vol. 1: 89 – 96. Padova University Press.

Heltzer, M. 1998. On the Vth Century BCE Dogs from Ashkelon. *Transeuphratène* 15: 149–152.

Hill, A. and Behrensmeyer, A. K. 1984. Disarticulation patterns of some modern east African mammals. *Paleobiology* 10 (3): 366–376.

Hilzheimer, M. 1926. *Naturaliche Rassengechichte der Haussaugtiere*. Berlin.

Holmberg, T. and Reiland, S. 1984. The influence of age, breed, rearing intensity and exercise on the incidence of spavin in Swedish dairy cattle. A clinical and morphological investigation. *Acta Veterinaria Scandinavica* 25(1): 113–127.

Hoogland, M. L. P. and Hofman, C .L. 2013. From corpse taphonomy to Mortuary Behavior in the Caribbean. A case study from the Lesser Antilles. In, Keegan, W. F., Hofman, C. L. and Rodríguez Ramos, R. (eds) *The Oxford Handbook of Caribbean Archaeology*. Oxford: University Press.

Horard-Herbin, M.-P. 2000. Dog management and use in the Late Iron Age: the evidence from the Gallic site of Levroux (France). In, Crockford, S. J. (ed.) *Dogs through Time: An Archaeological Perspective*: 115–121. Oxford: BAR International Series 889.

Kamlah, J. and Sader, H. 2003. The Tell el-Burak archaeological project: preliminary report on the 2002 and 2003 seasons. *Bulletin d'Archéologie et d'Architecture Libanaises* 7: 145–173.

Kelm, G. L. and Mazar, A. 1995. *Tinnah. A Biblical*

City in the Sorek Valley. Winona Lake, Indiana: Eisenbrauns.

Komeyli, H. 1984. Nasennebenhöhlen bei dolicho-, meso und brachyzephalen Hunden unter besonderer Berücksichtigung der rassesspezifischen Schädelformen. *Inaug. Diss.* Giessen, 151 pp.

Lepetz, S. 1996. L'animal dans la société gallo-romaine de la France du Nord. *Revue Archéologique de Picardie*, Numéro Spécial 12.

Lignereux, Y., Regodon, S. and Pavaux, Cl. 1991. Typologie céphalique canine. *Revue de Médecine Vétérinaire* 142: 469–480.

Lignereux, Y., Regodon, S., Personnaz, B. and Pavaux Cl. 1992. Typologie céphalique du chien et ostéo-archéologie : A propos d'une population canine du XVIIe siècle toulousain. *Revue de Médecine Véérinaire* 143(2): 139–149.

Lovell, N. 1997. Trauma analysis in paleopathology. *Yearbook of Physical Anthropology* 40: 139–170.

Lyman, R. L. 1994. *Vertebrate Taphonomy*. Cambridge Manuals in Archaeology. Cambridge University Press.

Maher, E. F. 2005. Chapter 8: the faunal remains. In, Dothan, M. and Ben-Shlomo, D. (eds) Ashdod VI. The Excavations of Area H and K (1968–1969). *Israel Antiquities Authority Reports* 24: 238–290. Jerusalem: Israel Antiquities Authority.

Markoe, G. 2008. Mediterranean: Phoenicia. In, Pearsall, D. M. (ed.) *Encyclopedia of Archaeology*: 875-887. Elsevier Academic Press.

De Grossi Mazzorin, J. and Tagliacozzo, A. 2000. Morphological and Osteological Changes in the Dog from the Neolithic to the Roman Period in Italy. In, Crockford, S. J. (ed.) *Dogs Through Time: An Archaeological Perspective. Proceedings of the 1st ICAZ Symposium on the History of the Domestic Dog. Eighth Congress of the International Council for Archaeozoology (ICAZ 98) August 23-29, 1998, Victoria, B.C., Canada*: 141–161. Oxford: BAR International Series 889.

Mégnin, P. 1897. *Le chien et ses races*. Deuxième édition, Tome premier. Broché.

Méniel, P. 2008. *Manuel d'Archéozoologie Funéraire et Sacrificielle. Âge du Fer*. Editions Infolio.

Nadelman, Y. 1995. Shoham. *Excavations and Surveys in Israel* 14: 80–81.

Nassar, J. and Chanteau, J. 2018 New discoveries in the Ancient Beirut Necropolis. Preliminary Results of the Saifi 477 Excavations. *Bulletin d'Archéologie et d'Architecture Libanaises* 18: 27-44.

Ortner, D.J. 2003. *Identification of Pathological Conditions in Human Skeletal Remains*. 2nd ed. San Diego: Academic Press.

Philips, C., Baxter, I. L. and Nussbaumer, M. 2009. The Application of Discriminant Function Analysis to Archaeological Dog Remains as an Aid to the Elucidation of Possible Affinities with Modern Breeds. *Archaeofauna* 18: 51–64.

Regnault, F. 1903. Essai sur les proportions du corps. *Bulletins et Mémoires de la Société d'Anthropologie de Paris*. Ve Série, 4: 276–291.

Sader, H. and Kamlah, J. 2010. Tell el-Burak : A new Middle Bronze Age Site from Lebanon. *Near Eastern Archaeology* 73(2-3): 129–141.

Saidah, R. 1967. Chronique. *Bulletin du Musée de Beyrouth* 20: 155–180.

Sapir-Hen, L. 2011. Dog Burials at Dor. In, Nitschke, J. L., Martin, S. R. and Shalev, Y. (eds) *Between Carmel and the Sea. Tel Dor: The Late Periods*: 138.

Sapir-Hen, L. and Bar-Oz, G. 2006. *Archaeozoological Analysis of the Faunal Remains from Tel Megadim*. Report submitted to Sam Wolff, November 2006.

Seif, A. and Choueri, H. In preparation. Urban Excavations at SFI 645 – A preliminary Report. *Bulletin d'Archéologie et d'Architecture Libanaises.*

Seiferle, E. 1966. Zur Topographie des Gehirns bei lang- und kurzköpfigen Hunderassen. *Acta anat.* 63: 346-362.

Silver, A. 1969. The Ageing of Domestic Animals. In, Brothwell, D.R., Higgs, E.S. and Clark, G. (eds) *Science in Archaeology: A comprehensive Survey of Progress and Research*: 283–302. London: Thames and Hudson.

Sisson, S. 1969. *The Anatomy of the Domestic Animals*. 4th ed. Philadelphia: Saunders.

Stager, L. E. 1991. Why Were Hundreds of Dogs Buried at Ashkelon? *Biblical Archaeology Review* 17(3): 26–42.

Stockard, C. R. 1941. The genetic and endocrinic basis for differences in form and behavior. *Amer. Anat. Mem.* 19. Philadelphia: Wistar Institute of Anatomy and Biology

Udrescu, M. 1989. Les chiens de l'habitat civil romain de Stolniceni-Vilcea; données zooarchéologiques. *Ann. Roum. Anthropol. Bucarest* 27: 3–8.

Varga, J., Tuboly, S. and Mészarós, J. 1999. *A Háziállatok Fertőző Betegségei. Állatorvosi Járványtan* II. Budapest: Mezőgazda kiadó.

Von den Driesch, A. 1975. Die Bewertung pathologisch-anatomischer Veränderungen an vor- und frühgeschichtlichen Tierknochen. In, Clason, A.T. (ed.) *Archaeozoological Studies*: 413–425. Amsterdam: North Holland Publishing Company.

Von den Driesch, A. 1976. A guide to the measurement of animal bones from archaeological sites. *Peabody Museum Bulletin* 1. Massachusetts: Harvard University, Peabody Museum of Archeology and Ethnology.

Wapnish, P. and Hesse, B. 1993. Pampered pooches or plain pariah? The Ashkelon dog burials. *Biblical Archaeologist* 56: 55–80.

Watson, A. G., Evans, H. E. and de Lahunta, A. 1986. Ossification of the atlas–axis complex in the dog. *Anatomical Histological Embryology* 15: 122–138.

13 Subsistence and ritual: a note on the Achaemenid faunal remains from the site of Kinik Höyük, Southern Cappadocia, Turkey

Pam J. Crabtree[1], Douglas V. Campana, Andrea Trameri[2], Nancy Highcock[3], Lorenzo d'Alfonso[4]

1 Center for the Study of Human Origins, Anthropology Department, New York University
2 Institute for the Study of the Ancient World, New York University
3 Program in Ancient Near Eastern and Egyptian Studies, New York University
4 Institute for the Study of the Ancient World, New York University

Abstract
Kinik Höyük, a multi-period urban site in southern Cappadocia, has been excavated by a joint University of Pavia—New York University archaeological mission since 2011. One of the goals of the excavation has been to understand the site in relation to its environment, including studying how animals were used for both subsistence and ritual purposes. This paper describes the faunal remains recovered from an Achaemenid deposit that may represent some form of ritual feasting.

Keywords
Kinik Höyük, Southern Cappadocia, zooarchaeology, Achaemenid Period, ritual feasting

Introduction
Kinik Höyük (Fig. 1) is a multi-period site located in Southern Cappadocia, Turkey. Until recently, the region around Kinik was poorly known archaeologically. A survey of the area carried out by the University of Pavia under the direction of Lorenzo d'Alfonso between 2006 and 2009 identified Kinik as a major urban centre in the region (d'Alfonso 2010). The site sits along a major east-west trade route that connects Europe and northern Anatolia with Mesopotamia and the Levant. Unlike many other urban centres in Asia Minor, Kinik seems to have been continuously occupied from the Late Bronze Age through the Iron Age and Hellenistic periods (15th through 1st centuries BC). In addition, the uppermost levels of the mound include a Seljuk occupation dating to the 12th—14th centuries AD.

Excavations at Kinik Höyük began in 2011. The project is a joint mission of the University of Pavia (Italy) and the Institute for the Study of the Ancient World at New York University (USA), and eight seasons of excavation have been completed to date (see Highcock *et al.* 2015 for a summary of the first four seasons of excavation at Kinik Höyük). One of the main goals of the project was to understand the urban centre of Kinik Höyük in relation to its biological and geological environment. Particular attention has been paid to the recovery of faunal, archaeobotanical, and other environmental evidence that can inform us about animal husbandry practices, agriculture, and past diets. When these data are combined with the zooarchaeological record for the nearby Neolithic and Chalcolithic site of Tepecik (Fig. 1), we will have a nearly complete record of animal and animal bone use from the earliest aceramic Neolithic through the late Middle Ages in the Melendiz Valley region of southern Cappadocia (see Crabtree, in press; Campana, in press).

This paper will briefly review the zooarchaeological evidence for animal use at Kinik Höyük during the Hellenistic, Iron Age, and Medieval periods based on faunal materials recovered between 2011 and 2012. The Medieval material has been described in detail elsewhere (Crabtree and Campana 2014), and a short summary of the fauna recovered from the first two seasons of excavation has recently been published (Crabtree and Campana 2016). Since the Hellenistic assemblage from the first two seasons of excavation is quite small, we have also included some information on the fauna from the large Hellenistic pit that was excavated in Area B (the top of the mound) during the 2015 season. A more detailed study of these fauna remains is currently under review (Crabtree and Campana, under review). This paper will then focus on the faunal

Fig. 1. Map of Turkey showing the location of Kinik Höyük in southern Cappadocia.

Fig. 2. Hawk statue recovered from Area A.

remains recovered from Achaemenid-Period deposits from Area A that were recovered during the 2013 excavation season.

The Achaemenid Period in Turkey is dated to the 6th to 4th centuries BC when Turkey was part of the Persian Empire. While Kinik Höyük has produced Iron Age deposits dating to the Persian Period in both the upper and lower towns, the area A deposits are unique because they include terracotta figurines and zoomorphic vessels that appear to be associated with ritual and cultic activities. These deposits accumulated specifically in one single room or courtyard (room Ar5) and date to the Late Achemenid and early Hellenistic period. They were deposited through time between the 4th and 2nd C. BC (Trameri, personal communication). The figurative artefacts from this deposit appear as fragments and discards, but identical types of materials came to light in earlier occupation levels, namely from a sounding excavated in room Ar3 within the same building. There the figurative artefacts were entirely preserved and *in situ*. One of the most interesting finds is a stone statue of a hawk, found together with three terracotta cattle figurines (Fig. 2). The room in which the hawk statue was found is part of a building now identified as a temple, thus the deposits can be safely connected with this space and its functions. This is the first Achaemenid-Period temple that is known from Central Anatolia (Trameri 2016). In this paper we have also included some faunal material that was recovered from the lower town during the 2015 and 2016 seasons in order to show how unique the faunal material from Area A is.

Materials and methods

Following the Turkish Ministry of Antiquities protocol, our faunal remains are studied at the site during the excavation season. As a result, we rely on pictures, identification guides (e.g., Schmid 1972;

Table I. Animal bones identified from the Late Iron Age, Achaemenid (Areas A and D), Hellenistic (2011-12), Hellenistic Area B, and Medieval features at Kinik Höyük.

	Late Iron Age	*Area A*	*Area D*	*Hellenistic*	*Hellenistic B*	*Medieval*
Domestic Mammals						
Domestic Cattle; *Bos taurus*	69	69	398	6	336	611
Sheep; *Ovis aries*	48	112	276	10	260	335
Goat; *Capra hircus*	22	36	102	4	110	141
Sheep or Goat; Caprinae	311	878	1075	1	1145	1713
Pig; *Sus scrofa*	20	0	41		7	5
Horse; *Equus caballus*	4	1	79		66	35
Donkey; *Equus asinus*		2	39		8	16
Equid; *Equus* sp.	7		19	2	26	56
Camel; *Camelus dromedarius*						8
Dog; *Canis familiaris*	9		22		5	35
Cat; *Felis catus*			2			
Wild Mammals						
Red Deer; *Cervus elaphus*			1		3	1
Roe Deer; *Capreolus capreolus*						1
Cf. Fallow Deer; *Dama dama*	1					
Hare; *Lepus europaeus*	5		6	4	14	12
Fox; *Vulpes vulpes*			4		23	41
Large artiodactyls						4
Small artiodactyls	169	1785	821	50	910	1226
Large ungulate	57	161	311	3	266	435
Carnivore	1					2
Rodent			1		5	2
Small mammal	7		5	4	3	22
Undetermined mammal	1153	2619	3315	132	2607	5438
Domestic Birds						
Domestic chicken; *Gallus domesticus*	1	3	22	3	18	22
Domestic goose; *Anser anser*					4	5
Mallard/Domestic duck; *Anas platyrhynchos*				1	1	4
Pigeon or dive; *Columba* sp.		1				
Goose-sized bird	1			1	4	9
Domestic fowl-sized bird		2	24		11	18
Wild Birds						
Crane; *Grus grus*						3
Partridge; *Alectoris chukar*	1			2		14
Duck family; Anatidae	1					
Unidentified bird	5	2			11	19
Unidentified fish			1			3
Turtle	15		1		1	167
Undetermined species	1		1			2
Human; *Homo sapiens*	1			1	2	20
Total	**1909**	**5673**	**6640**	**265**	**5846**	**10425**

Cohen and Serjeantson 1996), and our collective experience to analyse the Kinik faunal remains. Any bones that we cannot identify are photographed for further research. In addition to species and skeletal part, we record the side and the portion of the bone present, the degree of fragmentation, and taphonomic information (evidence of burning, weathering, and staining) for each bone or bone fragment. Ageing data based on dental eruption and wear (Payne 1973; Grant 1982) and epiphyseal fusion (Silver 1969) are recorded, as are measurements following Von den Driesch (1976) and any traces of butchery or bone working. The data are entered into a specialized database manager for zooarchaeology, FAUNA (Campana 2010), which allows us to export our data in a range of formats including MS Access and Excel. All our data are recorded on both paper and computer in the field. This allows us to leave a complete record of our data at the site.

Table 2. Species ratios (percentage based on NISP) for cattle, caprines, and pigs from Late Iron Age, Achaemenid Period (Areas A and D), Hellenistic (Area B), and Medieval features from Kinik Höyük.

	Late Iron Age	*Hellenistic B*	*Ach. Area A*	*Ach. Area D*	*Medieval*
Cattle	14.7	18.1	6.3	21.0	21.8
Sheep/goat	81.1	81.5	93.7	76.8	78.0
Pig	4.2	0.4	0.0	2.2	0.2

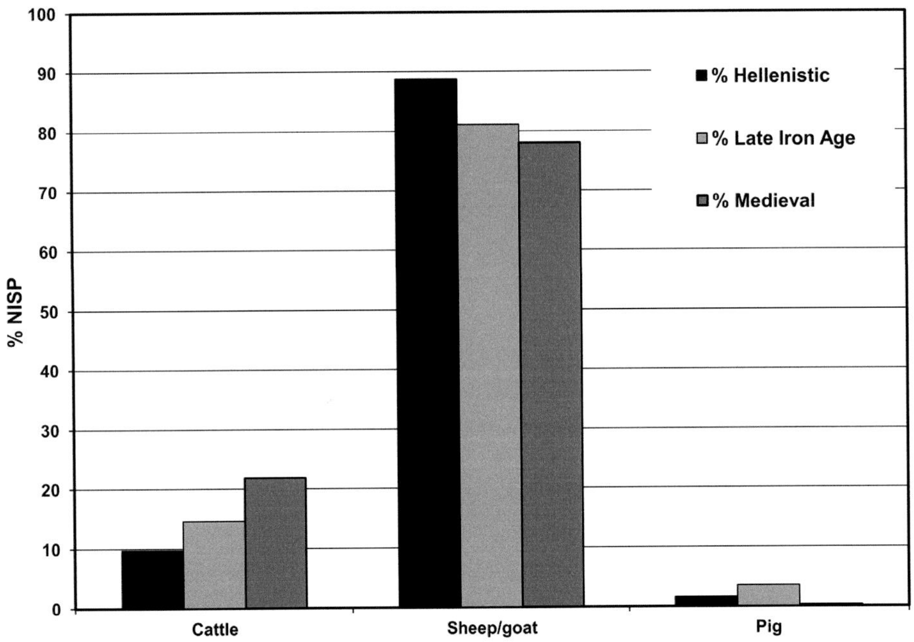

Fig. 3. Species ratios based on NISP for Late Iron Age, Hellenistic, and Medieval features at Kinik.

Faunal remains from the Iron Age, Hellenistic, and Medieval features at Kinik Höyük

The faunal remains recovered from the Late Iron Age (7th-6th centuries BC), Hellenistic, and Medieval features at Kinik Höyük are shown in Table 1. The Hellenistic faunal assemblage from the first two seasons was quite small; most of the faunal material recovered in the early seasons was Late Iron Age or Medieval in date. Since the 2011-12 Hellenistic assemblage was so small, we have also included the animal bone remains recovered from a large Hellenistic trash pit that were excavated and analysed during the 2015 season. All these assemblages are dominated by caprines, with sheep remains consistently outnumbering those of goats. Species ratios for all three periods (Table 2 and Fig. 3) show that caprines make up roughly 80% of the faunal remains, followed by cattle, and very few pigs. This pattern is typical of faunal assemblages from Central Anatolia from the Neolithic onward (see, for example, Steiner *et al.* 2014; Crabtree, in press). The Medieval assemblage is the most diverse. Some of this diversity is certainly a product of the large size of the assemblage, but it may also reflect the less urban and more agropastoral nature of the Medieval settlement at the site. The Medieval occupation has been described as a "modest rural settlement lacking major public buildings" (Highcock *et al.* 2015: 112). Ageing evidence from the Medieval assemblage (Fig. 4) shows that caprines of all ages are present, from neonates (aged 0-2 months) to elderly individuals (aged 8-10 years). The faunal remains from the Late Iron Age and Hellenistic features are what might be expected from a small urban centre in rural central Turkey.

Achaemenid feature from Operation A

In 2013, we encountered a rather unique deposit in Operation A1. The contexts of the area including alternating deposits of material that was rich in ashes, pottery, and animal bone remains. The ceramic remains include fragments of both fine tableware and kitchen ware, in addition to a number of fragmentary terracotta animal figurines, zoomorphic vessels, and a bird statuette (Trameri in Highcock *et al.*, 2015: 106 and Figs. 5 and 6). The material likely dates to the Late Achaemenid period, the second half of the 5th to the beginning of the 3rd century BC.

The faunal material recovered from this deposit differed from the assemblage recovered from the other Late Iron Age contexts. As can be seen from Table 1, this large assemblage is lacking in diversity. Nearly all the bones were recovered from sheep and goats, with smaller numbers of cattle. Species ratios for the large domesticates indicate that the assemblage is made up of 94% caprines, about 6% cattle, and no

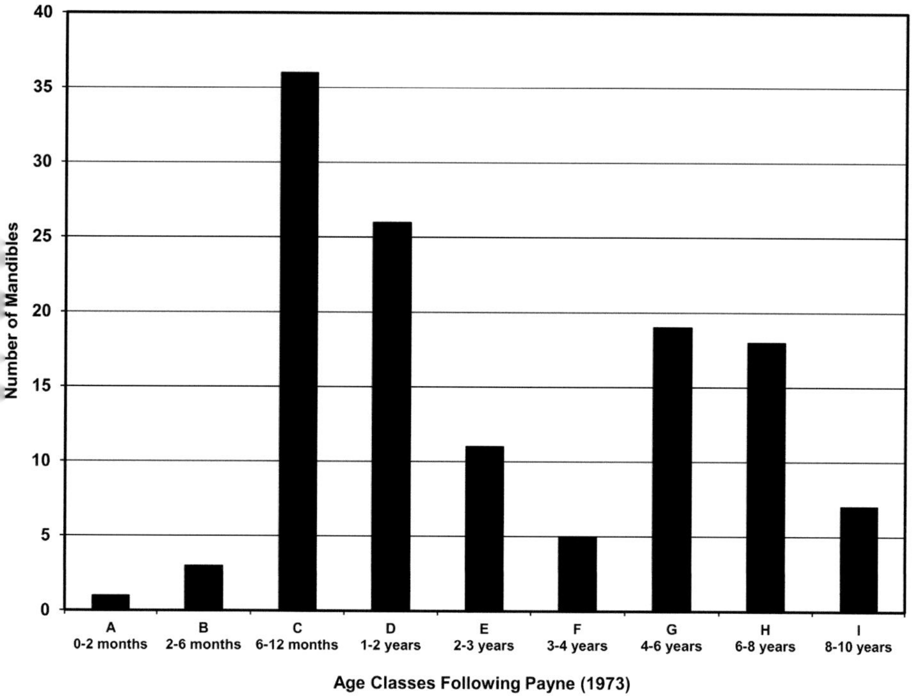

Fig. 4. Age profile for sheep and goats from the medieval (12th to 14th century CE) deposits. The age classes follow Payne (1973).

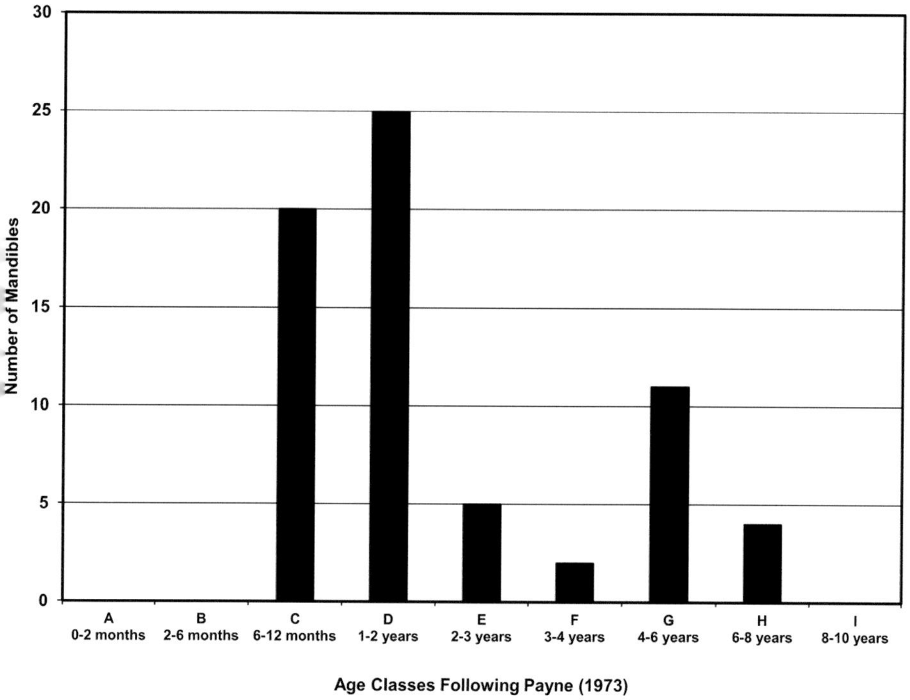

Fig. 5. Age profile for sheep and goat mandibles from the ash deposits in Operation A, Kinik Höyük.

pigs (Table 2). Ageing data show that no very young (under 6 months of age) or very elderly sheep and goat were recovered from this deposit (Fig. 5). The vast majority of the animals recovered from this deposit were between 6 months and two years of age. The assemblage also includes many meaty body parts, including vertebrae, ribs, and upper limb elements, as well as a large number of rib fragments that were identified as small artiodactyl. When we combine the faunal evidence for large numbers of young caprines with the archaeological evidence for high quality tableware and kitchenware, ash layers, and the zoomorphic figurines and bird statuettes, this deposit appears to be something out of the ordinary. We have tentatively identified this deposit as an example of ritual feasting based on the association with ash deposits, the high proportion of sheep and goats, the limited species diversity, and the association with unusual and non-domestic artefacts such as the bird statuettes.

The contrast between the Area A Achaemenid assemblage and the broadly contemporary material from the lower town (Area D) is striking. The Area D assemblage (Table 1) includes higher numbers of cattle, pigs, equids, and wild animals, including red deer, hare and fox. The species ratios based on NISP (Table 2) show that the lower town has a higher proportion of cattle and pigs. Caprines make up only about 77% of the large domestic food mammals in the Area D assemblage. The Area D assemblage is consistent with the faunal assemblages from other Late Iron Age features at the site.

Future Research

This is just a first look at the animal bone remains from Kinik Höyük. Additional animal bone remains have been recovered from Operation A in the 2015 and 2016 seasons, and these may allow us to determine whether our feasting hypothesis is correct. Historic sources suggest that the Kinik area was a centre for specialized horse rearing from the Late Bronze Age through the Roman period (Balza 2013), but this has never been documented archaeologically. We plan to carry out a detailed study of the horse remains from the Bronze Age and Iron Age to shed light on the nature of horse breeding and equine husbandry in the Kinik area. Our long-term goal is to trace the changes in animal husbandry practices, hunting patterns, and diet from the earliest Neolithic through the Medieval period in the Melendiz Valley region of southern Cappadocia by combining the faunal evidence from Kinik with the data from the nearby Neolithic and Chalcolithic site of Çiftlik-Tepecik.

Acknowledgements

Our participation in the 2013 excavation season at Kinik Höyük was supported by a grant from the University Research Challenge Fund (URCF) from New York University. Our participation in the 2014 season was supported by a grant from the American Research Institute in Turkey (ARIT). Our participation in the 2016 season was supported by a URCF grant to Lorenzo d'Alfonso (PI) and Pam Crabtree (co-PI) entitled "Horse Breeding in Ancient Cappadocia, Turkey."

References

Balza, M. E. 2013. Horses and horse husbandry in Central Anatolia during the Hittite and Neo-Hittite Period. *Res Antiquae* 10: 1-14.

Campana, D. V. 2010. FAUNA: Database and Analysis Software for Faunal Analysis. Poster presented at the meeting of the International Council for Archaeozoology, Paris, August 23-26th, 2010.

Campana, D. V. in press. Tepecik Bone Tools Assemblage 2013 Season, Analyzed 2015-16. In, Godon, M. and Biçakçi, E. (eds) *Melendiz Project and Tepecik-Çiftlik's Excavation: Preliminary Outcomes and Ongoing Researches*.

Cohen, A. and Serjeantson, D. 1996. *A Manual for the Identification of Bird Bones from Archaeological Sites*. 2nd ed. London: Archetype.

Crabtree, P. J. in press. Chalcolithic animal bones from Tepecik: Faunal remains recovered from the 2013 season. In, Godon, M. and Biçakçi, E. (eds) *Melendiz Project and Tepecik-Çiftlik's Excavation: Preliminary Outcomes and Ongoing Researches*.

Crabtree, P. J. and Campana, D. V. 2014. Animal use at Medieval Kinik Höyük, a 12th-13th century site in Southern Cappadocia, Turkey. In, Stull, S. D. (ed.) *From West to East: Current Approaches to Medieval Archaeology*: 162-169. Newcastle upon Tyne: Cambridge Scholars.

Crabtree, P. J. and Campana, D. V. 2016. Animal Bone Remains from the Site of Kınık Höyük, Southern Cappadocia, Turkey: Animal Husbandry and Hunting Practices during the Iron Age, Hellenistic, and Medieval periods. In, Stucky, R. A., Kaelin, O., and Mathys, H.-P. (eds) *Proceedings of the 9th International congress on the Archaeology of the Ancient Near East, vol. 2*: 133-139. Wiesbaden: Harrassowitz Verlag.

Crabtree, P. J. and Campana, D. V. under review. *Animal bone remains from the Hellenistic features at Kinik Höyük*.

D'Alfonso, L. 2010. Geo-archaeological Survey in Northern Tyanitis and Ancient History in Southern Cappadocia. In, d'Alfonso, L. (ed.) *Geo-archaeological Activities in Southern Cappadocia, Turkey*: 27-52. Pavia: Italian University Press.

Grant, A., 1982. The use of tooth wear as a guide to the age of domestic ungulates. In, Wilson, B., Grigson, C. and Payne, S. (eds) *Ageing and Sexing Animal Bones from Archaeological Sites*. BAR British Series 109: 91-108. Oxford: British Archaeological Reports.

Highcock, N., Crabtree, P., Campana, D., Capardoni, M., Lanaro, A., Matessi, A., Miller, N., Strosdahl, P., Trameri, A. and d'Alfonso, L. 2015. In, Steadman, S. and McMahon, G. (eds) *The Archaeology of Anatolia: Recent Discoveries (2011-14) Volume I*: 98-117. Newcastle: Cambridge Scholars Press.

Payne, S., 1973. Kill-off patterns in sheep and goats: the mandibles from Aşvan Kale. *Anatolian Studies* 23: 281-303.

Schmid, E. 1972. *Atlas of Animal Bones*. Amsterdam: Elsevier.

Silver, I.A. 1969. The ageing of the domestic mammals. In, Brothwell, D. and Higgs, E.S. (eds) *Science in Archaeology*: 283-302. London: Thames and Hudson.

Steiner, M. C., Buitenhuis, H., Duru, G., Kuhn, S. L., Mentzer, S. M., Munro, N. D., Pöllanth, N., Quade, J., Tsartsidou, G. and Özbaşaran, M. 2014. A forager–herder trade-off, from broad-spectrum hunting to sheep management at Aşıklı Höyük, Turkey. *Proceedings of the National Academy of Sciences* 111 (23): 8404-8409.

Trameri, A. 2016. *A religious building from Kinik Höyük: New finds from the 2016 campaign*. [online] Available at: <http://isaw.nyu.edu/news/new-finds-from-the-2016-campaign-kinik-hoyuk>.

Von den Driesch, A. 1976. *A Guide to the Measurement of Animal Bones from Archaeological Sites*. Peabody Museum Bulletin 1. Cambridge: Peabody Museum of Archaeology and Ethnology, Harvard University.

14

Ottoman Turkish influences on animal exploitation in 16th-17th century Hungary

László Bartosiewicz[1], Erika Gál[2]

1 Osteoarchaeological Research Laboratory, Stockholm University, Lilla Frescativägen 7, 106 91 Stockholm, Sweden; laszlo.bartosiewicz@ofl.su.se
2 Institute of Archaeology, Research Centre for the Humanities, Hungarian Academy of Sciences, Tóth Kálmán utca 4, 10 Budapest, Hungary; gal.erika@btk.mta.hu

Abstract
Between middle of the 16th and the end of the 17th century, the medieval Kingdom of Hungary was divided between the remains of the Catholic Kingdom of Hungary in the West, Protestant Transylvania in the East and the northern tip of the Ottoman Empire wedged in-between from the South. This geopolitical setup, however, should not only be seen in terms of military history and political conflict. It was also a time of vivid cultural interaction, when the exchange of goods and ideas between East and West intensified. Under Southwest Asian influence, not only new domestic species (camel, water buffalo, possibly turkey) appeared in the area, but the known stocks of sheep, horse, poultry, and cattle have also been altered. This means that changes in animal exploitation took place in almost all spheres of everyday life, including common meat diets (declining pork consumption, increasing importance of poultry), animal trade (cattle droves), military transport (new types of mounts and draught animals). As a result of political/ideological competition the use of animals and animal products in self-representation or as markers of social identity became likewise important. These formidable developments will be reviewed on the basis of archaeozoological evidence from Ottoman Period sites in Hungary.

Keywords
animal husbandry, exotic animals, livestock trade, identities, Ottoman Period Hungary

Introduction
In 1536, François I Valois king of France, fearful of Habsburg hegemony in Europe, concluded an agreement with the Sultan Suleiman I, the Magnificent, to lay siege to their imperial city, Vienna. For a Christian king this was a controversial alliance at the time as it clearly fell in line with evident ambitions of European expansion by the Sublime Porte in Istanbul (Takács 1991). During its northward thrust the Ottoman Empire had already invaded the Balkans and the occupation of Hungary began with the decisive battle of Mohács in 1526 in the South. At the time the country had already been divided by an internal war among nobility for the royal throne and further weakened by a failed crusade whose recruits staged a massive peasant uprising in 1514. After the succession debacle of 1540 that followed the death of king John I Szapolyai of Hungary an Ottoman-style government was established and a Pasha was installed in the former royal capital, Buda. The area of medieval Hungary thus became the 'Northwest Passage' on the way to Vienna. In a series of mid 16th century sieges Hungarian forts fell under Ottoman military pressure and a network of Ottoman-Turkish palisade forts also began to be established.

Over half a century later, the series of brutal wars in Central Europe, the Thirty Years' War (1618–1648), also took place against the backdrop of Ottoman expansion at the Southeastern edge of Europe. Fighting between Catholic and Protestant states in the crisis-ridden Holy Roman Empire gradually ignited a general conflict involving most European great powers, fuelled by the rivalry between France and the Habsburgs for full political control. (It was already during the reign of François I that divisions in the Christian religion created permanent rifts as the Protestant movement was becoming increasingly influential through much of Europe). Fearful of the Catholic policies of Ferdinand II, Gabriel Bethlen, the Protestant

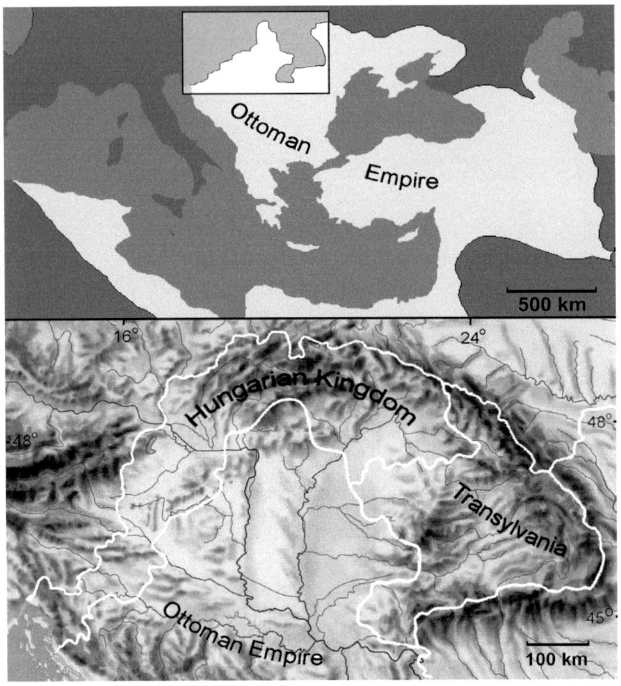

Fig. 1. The European section of the Ottoman Empire and the study area (top; insert) and the mid 16th century geopolitical division of the Carpathian Basin (bottom).

Hungarian Prince of Transylvania, requested a protectorate by the then Ottoman Sultan, Osman II.

During the time-period under discussion here the medieval Kingdom of Hungary, covering the Carpathian Basin, was thus divided into three parts. The Ottoman Empire was wedged into its open, central section from the South along the Danube Valley, occupying most of the plains and hillier Transdanubia that largely correspond to modern-day central Hungary (Fig. 1, bottom). A relatively narrow, sickle-shaped band along its border to the west and north was annexed by Princess Maria von Habsburg and remained nominally the Kingdom of Hungary. Ottoman-Turkish rule lasted until 1699 when the Karlóca Peace Treaty brought an end to the long Austro-Ottoman conflict and resulted in the withdrawal of Ottoman forces from the territory of Hungary.

This hectic time period, however, should not only be seen in terms of political animosity and military developments. The almost 150 years, that is six human generations, were not simply marked by recurring hostility: they also represented a period of intense cultural interaction. Animal remains represent past material culture. They show changing food habits, military tactics and often figure as means of self-representation. Domestic animals themselves should be considered artefacts as their evolution is influenced by humans, reflecting values and aspirations of each community involved with animal husbandry. In this paper, archaeozoological evidence from Ottoman Period sites in Hungary will be reviewed with regard to these fascinating aspects of animal-human relationships in a dynamically changing political landscape.

Material and methods

In an effort to offer a broader picture, Ottoman Period animal remains from Hungary have frequently been presented together with both earlier medieval and later, Early Modern Age assemblages (Bartosiewicz 1999a; Vörös 2003; Bartosiewicz and Gál 2003). In reality, complex stratigraphies at multi-period, especially urban, sites can make precise chronological distinctions difficult. In this study, special emphasis is laid on the number of identifiable specimens (NISP) from 17 deposits assigned to the Ottoman Period within the borders of the empire (Table 1). Archaeological sites in Hungary are identified using a binomial system. The first name is geographical identification, that is the name of the relevant present-day settlement. The second is a specific name referring to the location in relation to the main settlement. Translations of the latter are provided in Table 1.

With one exception, Teleki palota, all material was hand-collected. Water-sieved samples from this site yielded 36 bones of rat (Daróczi-Szabó 2004) which had not been included in the quantitative analyses used in this study. Subsequently these remains were identified as representing black rat (*Rattus rattus* L. 1958; Kovács 2012). In addition to this assemblage, sieved samples were also available from the site of Budapest–Ganz utca (Ganz street). Aside from barely a dozen caprine bones, however, those samples yielded only the remains of plants and commensal microfauna (Tóth *et al.* 2010). Therefore they were not included in this analysis focussing on domestic animals.

Animal remains recovered at the discussed settlements are summarized in Appendices 1 and 2.

The list in Table 1 does not include sites with no detailed faunal lists but with published special finds (especially camel and crested hen remains). Sporadically occurring human bones and worked exotica are not counted in the NISP values.

Eleven contemporaneous settlements from the Kingdom of Hungary under Christian rule (Vörös 2003) were used as a point of reference but not further detailed here. The Ottoman Period assemblages varied strongly in terms of assemblage size (NISP) and closely related taxonomic richness (the number of taxa represented by NISP). Previous calculations using these two parameters showed a monotonous meat supply at forts, while even smaller assemblages from urban and rural sites were indicative of a relatively great diversity of species (Bartosiewicz and Gál 2003: 366, Table 1). This time no such *a priori* distinctions were made between Ottoman Period settlement types for the purposes

Table 1. Ottoman Period settlements available for study.

	Site name	*English translation*	*Century*	*NISP*	*Type*	*Source*
1	Barcs–Palánkvár	Barcs–palisade fort	16th–17th	9591	Fort	Gál and Bartosiewicz 2016
2	Békés–Kastélyzug	Békés–Castle quarters	16th	1814	Fort	Vörös 1980
3	Buda–Pasa palota	Buda–Pasha's palace	17th	1460	Elite	Bökönyi 1974
4	Buda–Teleki palota	Buda–Teleki palace	16th –17th	2068	Elite	Daróczi-Szabó 2004
5	Érd–Palánkvár	Érd–palisade fort	16th –17th	345	Fort	Vörös 2003
6	Gyula–Vár	Gyula–fort	15th –17th	58	Fort	Bökönyi 1974
7	Nagykanizsa–Vár	Nagykanizsa–fort	15th–17th	2049	Fort	Bökönyi 1974
8	Segesd–Pékóföld	Segesd–Pékó estate	16th–17th	1051	Rural	Bartosiewicz 1996a
9	Székesfehérvár–Koch utca	Székesfehérvár–Koch street	15th–16th	69	Urban	Bartosiewicz 1997
10	Szekszárd–Palánk	Szekszárd–palisade fort	16th–17th	6102	Fort	Bartosiewicz 1995c
11	Szolnok–Vár	Szolnok–fort	15th–17th	147	Fort	Vörös 2003
12	Vác–Műszaki Főiskola	Vác–Engineering college	15th–17th	374	Urban	Bartosiewicz 1995b
13	Vác–Kossuth tér	Vác–Kossuth square	15th–17th	96	Urban	Bartosiewicz 1995b
14	Vác–Zeneiskola	Vác–Music school	15th–17th	41	Urban	Bartosiewicz 1995b
15	Vác–Szentháromság tér	Vác–Trinity square	16th–17th	876	Urban	Vörös 2003
16	Visegrád–Alsóvár	Visegrád–Lower fort	16th–17th	2236	Rural	Twigg 2012
17	Visegrád–Salamon torony	Visegrád–Salamon tower	15th–17th	8339	Rural	Bökönyi 1974

Table 2. Summary table of pooled NISP values for livestock.

Location of sites	*n*	*Cattle*	*Sheep*	*Goat*	*Caprine*	*Pig*	*Poultry*	*Total*
Ottoman territory	17	17967	1050	271	8775	1622	4544	34229
Christian territory	11	15270	329	33	2302	4023	2038	23995

of analysis. Classifying settlements by type is a complex task during a long time of dynamic societal change. While Ottoman-Turkish palisade forts of a clear function and more-or-less homogeneous history are relatively easy to identify, urban and elite sites changed throughout the Ottoman Period. The Ottoman administrative centre, Buda, is described as a combination between an overstaffed military camp and a neglected market town. J. Bocatius, a traveller from Košice (outside the Ottoman Empire), visited Buda in 1605, over half a century after it had been taken over by the Ottomans. He refers to chaotic conditions, even corpses abandoned in the streets. Periodic outbursts of warfare, natural disasters and frequent fires in the crowded quarters all contributed to the decline of what was once a Renaissance royal capital (Fekete and Nagy 1975: 345). Even if some accounts have overtones of contemporaneous Christian propaganda, it is likely that medieval urban centres such as the former queen's seat of Segesd (Bartosiewicz 1996a), parts of royal Visegrád and the episcopal centre of Vác (Bartosiewicz 1995a) would have no longer matched the criteria of cities in a central European sense. The archaeological evidence of randomly distributed refuse pits and even semi-subterranean huts amidst the ruins of medieval stone architecture often corroborates this image of urban deterioration (Bartosiewicz 1995a; Twigg 2012; Daróczi-Szabó 2004).

Due to limitations of space in this paper the representation of animal remains was discussed using NISP values, a non-derived measure and in terms of presence and absence in the case of rare species. Neither anatomical, age and sex distributions nor estimations of size could be detailed here. Individual site reports are referred to for this information.

Meat exploitation

The bulk of animal bones brought to light during excavations represent food refuse. A special aspect of Ottoman-Turkish occupation was the presence of Moslem populations who avoided pork on religious grounds. Therefore settlement materials are well worth reviewing in terms of meat consumption. In order to assess differences in meat consumption, pooled NISP values for 17 settlements in the area under Ottoman rule were compared to those from 11 sites in the territory of the Kingdom of Hungary with undisrupted Christian tradition. The thus resulting large numbers of identifiable bones (Table 2) showed that the two groups of sites were quantitatively similar to each other, yielding on average the same number of identifiable bone specimens and displaying similar standard errors ($t=0.199$; $p=0.840$; degrees of freedom=26; Fig. 2). Meanwhile, differences in taxonomic composition also became evident.

The frequencies of animal remains showed a statistically significant difference ($Chi^2= 4892.405$; $p=0.000$; degrees of freedom=5) between Ottoman

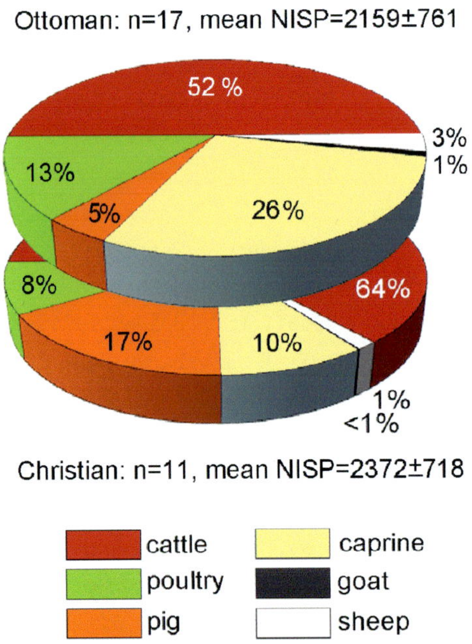

Fig. 2. Relative frequencies of the main livestock remains in the pooled 17 Ottoman (top) and 11 Christian (bottom) assemblages.

Fig. 3. Transversal cut mark on the ventral surface of a sheep axis from Visegrád–Alsóvár (Photo Katalin Boruzs).

and Christian settlements. Under Moslem influence there was a marked drop in 16th–17th century pork consumption, but inter-site variation remained large in the occupied areas. A lot of pork was replaced by mutton: the proportion of caprine (sheep and goat) bones more than doubled in the area under Ottoman occupation whereas the consumption of poultry (mostly chicken) also became more common.

In addition to the avoidance of pork, a rare signature of probably religious meaning was noted at the site of Visegrád–Alsóvár, where the second cervical vertebra (*axis*) of a sheep bore a single transversal cut mark on its ventral surface (Fig. 3). Cuts of this type fall in the proximity of the animals' throat. They have widely been hypothesized to be *perimortem*, i. e. inflicted at the time of death. Since at least Biblical times, this type of injury was observed on atlases of both a sheep and a goat (Bartosiewicz *et al.* 2008). Among others it occurred at Late Bronze Age Kamid el-Loz in Lebanon (Bökönyi 1985: Taf. 85/1-2). There is also a possibility that such cuts are inflicted during the *post mortem* removal of the head. Gilbert (1988: 85, Fig. 5, Pl. XIV/1-4) thus distinguished between two types of knife marks on sheep atlases. The first is inflicted during slaughter (cuts running across the ventral surface after having penetrated the soft tissue of the throat). The second is caused by *post mortem* decapitation (with cuts on the anterior edge of the atlas). In spite of the difficulty of precise identification, the cut mark shown in Fig. 3 is potentially consistent with religious *halal* slaughter whereby the animal's throat must be cut with a swift and deep single incision using a very sharp knife.

Naturally, Christians had adopted cutting the animal's throat as well, however, it may have been far more randomly executed in the absence of religious regulation (Bartosiewicz *et al.* 2008). As mentioned previously the overall shift from pork to mutton consumption in the occupied areas of Hungary varied strongly among the 17 Ottoman assemblages under discussion (Fig. 4). The remains of poultry were not included in this comparison in spite of the percentual increase in their remains during Ottoman times. Differences between the skeletal anatomy, body sizes and taphonomic properties of birds and domestic ungulates (Bartosiewicz and Gál 2007) make it difficult to quantitatively compare their remains in terms of dietary contribution.

The near-complete absence of pig bones was noted only at the high status settlement of Buda–Pasa palota, an evident centre of Moslem power representation. Another assemblage lacking in pig (although ten times smaller) is known from the fort of Szolnok–Vár, where an Ottoman garrison was stationed. Interestingly, the latter site also yielded dozens of sheep astragalus bones, some perforated and filled with lead inlay (Kovács 1989). In addition to the widely known parallels in Classical Antiquity, caprine astragali have also been popular as gaming pieces across Central Asia. The Buda–Teleki palota material, characterized by a likewise high proportion of caprine remains in the food refuse also yielded several thus worked sheep astragali (Daróczi-Szabó 2004: 164). Due to the absence of pig remains, these assemblages may be considered most typical of the presence of Moslem occupants.

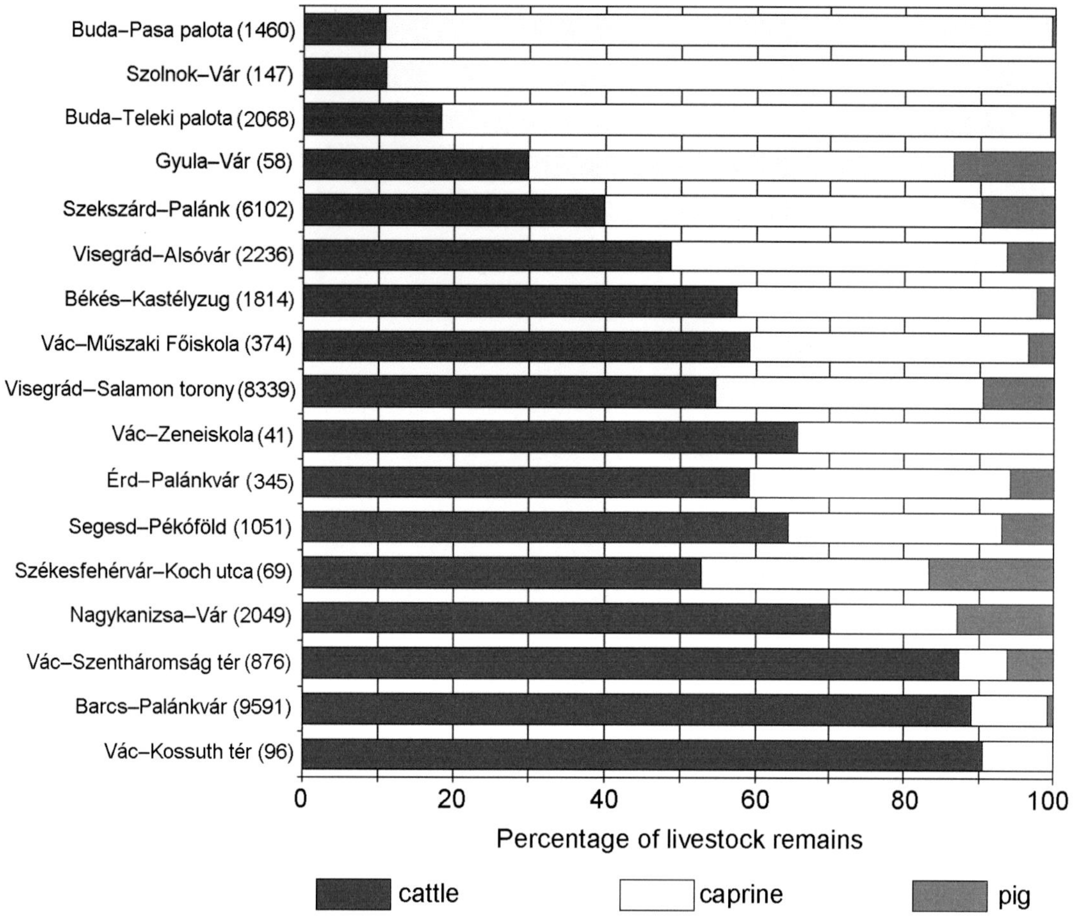

Fig. 4. Relative frequencies of the main ungulate remains in the 17 Ottoman assemblages. Parenthesized numbers indicate the number of identifiable specimens (NISP).

Caprine remains were found in relatively high proportions at several Ottoman palisade forts (Békés–Kastélyzug, Szekszárd–Palánk, Gyula–Vár, the only exception being Érd–Palánkvár, further down the graph). Beef consumption was relatively higher at these settlements (30–60% of NISP) and traces of pork consumption were also sporadically found (3–12% of NISP). Since much of the material originates from Ottoman-Turkish palisade forts, the evidence of pigs seems to indicate the presence of non-Moslem mercenaries of the Sublime Porte. Recruited in the Balkans, such soldiers came from various ethnic/religious backgrounds and thus did not observe Moslem dietary restrictions.

The next group of sites with even higher contributions of beef and pork to the diet consists of urban settlements (Segesd–Pékóföld, Vác–Műszaki Főiskola and Zeneiskola, Székesfehérvár–Koch utca and Visegrád–Salamon torony). Some medieval stratigraphic contamination may occur at such complex sites, as shown by the slight difference between the purely Ottoman Period site of Visegrád–Alsóvár (Twigg 2012), located next to Visegrád–Salamon torony which, however, also had a pre-Ottoman, 13th–15th century component (Bökönyi 1974: 429). Meanwhile the consistent presence of pig bones at urban sites also reflects the situation that cities indeed had more mixed populations, original inhabitants, occupants and travellers of diverse religious and ethnic backgrounds oftentimes living side by side or in tight succession during times of political instability.

Finally, in spite of the order of magnitude difference between their sizes, the assemblages from Barcs–Palánkvár (a palisade fort along the Drava river) and two smaller assemblages from Vác (Kossuth and Szentháromság tér, urban deposits from a market town on the Danube river) represent special settlements located along important cattle-herding routes (Fig. 4, bottom). As illustrated by detailed historical data (Bartosiewicz and Gál 2003: 366) cattle droves originated from redistribution centres near the Danube Bend, trekked past Bratislava, Vienna and St. Pölten, crossing the Inn at Schärding rivers, moving on to Regensburg, and continued toward Nuremberg or Augsburg (Bartosiewicz 1995b). Herds bound Southwest from Hungary advanced toward Croatia crossing the Adriatic by boat between Zadar and Venice (Gaál 1966: 475) or on a land route along the Drava river with a key crossing point at Barcs–Palánkvár (Bartosiewicz 1999b: 48, Fig. 1). Massive concentrations of cattle bone at settlements along such trading routes suggest that locals benefited from the increased supply of live cattle during seasonal cattle

Fig. 5. Military pack camels. Detail from the 1617 map of Szolnok, central Hungary (Library of the Hungarian Central Statistical Office, Budapest).

droves (Bartosiewicz 1995b): even if their bones could not be morphologically distinguished from those of local stock, some animals must have been sold or culled along the way.

The anatomical distribution of cattle bones from Ottoman-Turkish palisade forts is indicative of military meat provisioning based on pre-portioned rations of beef (Vörös 1980; Bartosiewicz and Gál 2003: 368, Ill. 2). Forts of peripheral locations relative to cattle trading routes (Szekszárd, Szolnok, Gyula), however, yielded noticeably smaller percentages of cattle remains.

Exotic beasts of burden

Unless their meat was consumed, draught animals were often disposed of off-site, died along the roadside or in battlefields. Therefore their bones tend to be underrepresented in settlements of focal archaeological importance. In spite of this taphonomic hindrance, until now eight Ottoman Period sites have yielded various numbers of sometimes articulated camel remains in Hungary. Five of them are located in or near the Buda Castle district (Daróczi-Szabó *et al.* 2015). Their discovery may not simply be a product of intense rescue excavations in the present-day municipal area but could also be related to military activity around the capital city of strategic importance. Two individual camel finds were found in forts located right outside the border of the Ottoman Empire (Bajcsa /Bajcsavár/ near the one-time Kanizsa castle in the West and the Northern site of Diósgyőr). Among other possibilities, camels may have ended up as war booty in Christian-controlled areas (Gy. Kovács pers. comm.).

Camels were the most visible exotica introduced by the Ottoman army. They played a key role in the terrestrial distribution of artillery supplies transported by boats on the Danube and were driven to battles as well (Bartosiewicz 2014). They may also have been important in civilian long-distance

Fig. 6. Turks shoeing water buffaloes in the 1571 etching by Melchior Lorch (The State Hermitage Museum, St. Petersburg).

trade. Consistently recovered osteological evidence thus confirms documentary sources that mention thousands of camels deployed in the area under Ottoman occupation (Bartosiewicz 1996b). By the 17th century these animals became a common sight in the Hungarian landscape (Fig. 5).

In most cases the gracile bones are indicative of dromedaries of cross-continental distribution inherent to their function in long-distance communication (Almathen *et al.* 2016). However, conspicuously large finds from Buda–Teleki palota and Budapest–Kacsa utca (Kacsa Street) seem to originate from Bactrian camels (Daróczi-Szabó *et al.* 2015). The skeleton of a robust male hybrid dated to the 17th century, was recovered in Tulln (Austria). Ancient DNA analyses indicated that the animal descended from a dromedary in the maternal line and a Bactrian camel in the paternal line (Galik *et al.* 2015).

According to Vörös (2003), water buffaloes arrived to Hungary as draught animals of the Ottoman army. Water buffaloes in late medieval Europe have widely been associated with Turks (Fig. 6). Two mandible fragments and a loose tooth identified from the presumably 14th–15th century northern courtyard of the Buda palace by Matolcsi (1977: 195) seem to indicate that this species was known in Hungary already prior to Ottoman occupation. Vörös (2003: 356), however, considered these 'early' finds of uncertain stratigraphic date. Unfortunately, most bones of the water buffalo are difficult to distinguish from those of cattle which further exacerbates the aforementioned general taphonomic under-representation of beasts of burden among the food refuse. Taxonomically diagnostic horn cores were identified at two sites in present-day Hungary

Fig. 7. The sedan chair of Sultan Mehmed IV carried by four hinnies (Nordiska Museet, Stockholm. Photo László Bartosiewicz).

(Bökönyi 1958; Vörös 2003), one from Nagykanizsa, from the site of the fort (Bökönyi 1974) and one from the urban settlement of Vác–Szentháromság tér (Vörös 2003). Vörös (2003: 357) summarized numerous references to water buffaloes in 16th–17th century documents from the Kingdom of Hungary and Transylvania. They were mentioned not only as working animals but also as curiosities in game parks and even as dairy animals.

Similarly to the Middle Ages, oxen were probably most commonly used as heavy-duty draught animals during this period. A corroded ox shoe, similar in shape to those shown in the lower section of the shoeing scene in Fig. 6, was found at the fort of Szekszárd–Palánk (István Takács, pers. comm.).

Equids

Equine meat was not regularly consumed either by Christians or Moslems, although no religious ban comparable to that of pork exists concerning horses in either religion. Still, due to the dominant secondary exploitation of equids the osteological evidence of horses and donkeys is sparse in Ottoman Period Hungary. Their bones tend to occur in peripheral areas, often associated with skinning. Much of horse bone working at Christian medieval sites seems to have been associated with the manufacturing of skates and sled runners (e.g. Bartosiewicz 2009), but these artefacts are virtually unknown from Ottoman Turkish settlements.

Horses of great value and importance are thus far better represented in the written record than among food remains. They are usually mentioned under geographical or ethnic labels (e.g. Transylvanian, Polish, Arab or Tartar; Vörös 2003). The few measurable remains reveal variability in withers heights, consonant with the high number of breeds mentioned in documentary sources. Bökönyi (1974: 296) identified traits of the Arab horse on the skeletal remains of a gracile horse recovered from an Ottoman Period refuse pit in the Buda Castle.

Donkeys are popular working animals across the Mediterranean region. Although following their introduction during Roman Period donkeys became known in Hungary, even after this relatively late date they never became widely spread. Their importance possibly increased during the Ottoman Period. These hardy, non-demanding animals were efficient means of short distance transport as is shown by their presence in urban deposits (Vác–Szentháromság tér, Buda–Sándor palota, Eszter Kovács pers. comm.). In addition, donkey bones are also known from the fort of Békés–Kastélyzug and Visegrád–Salamon torony.

Iconographic evidence reveals that the Ottoman army also used pack mules in Europe (Bartosiewicz and Gál 2003: 370), although their bones are yet to be identified from Hungary. One may wonder whether some of the small 'horses' within the 120–156 cm range of withers height mentioned by Vörös (2003: 359) may actually have been mules or hinnies. In sharp contrast with Southwest Asia mules have always been looked down upon in Hungary (Bartosiewicz and Gyöngyössy 2006). The dramatic difference between culturally idiosyncratic attitudes towards animals is illustrated by a painting commissioned by Claes Rålanb, Swedish envoy to the Sublime Porte in Istanbul. It shows the 1657 inaugural procession of Sultan Mehmed IV in which the sultan himself is carried in a sedan chair powered by four hinnies (Fig. 7). A special zoological aspect of this image is that hinnies in particular were easily trained the *tölt*, a special, greatly valued form of gait (Bartosiewicz and Gyöngyössy 2006), clearly recognizable in the painting. In medieval English and Icelandic horses such 'gaitedness' has recently been linked to a mutation in the DMRT3 gene (Wutke *et al.* 2016). As of today, the exact genetic background is unknown in mules and hinnies.

Effects on local animal breeding

In addition to the increasing visibility of newly introduced animal species, traditional domesticates in the Carpathian Basin also underwent gradual change. The Ottoman invasion of the Balkans started a domino effect of human population displacements, some by force, others by further reinforcing traditional movements. Some of the latter were related to animal keeping practices. Christian Vlachs still pursued a mobile pastoral way of life at the time. Some Orthodox Christian and part of the Catholic Vlachs fled the

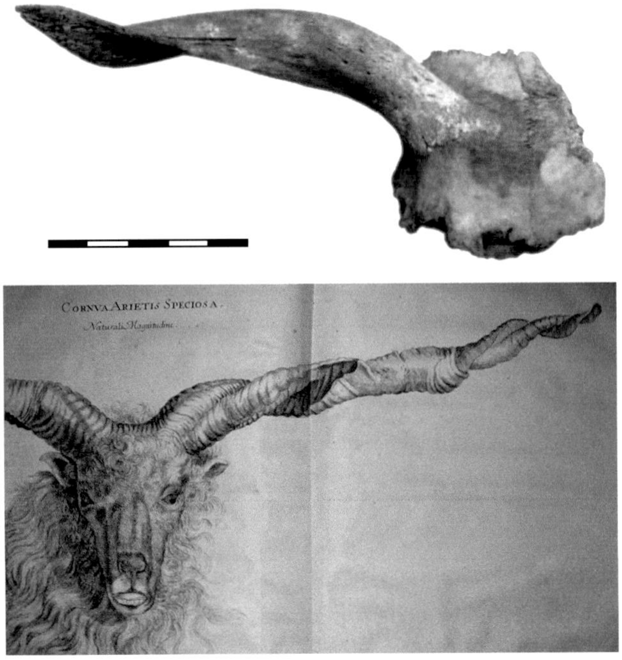

Fig. 8. Late medieval *"Racka"*-type sheep horn core from Vác (top) and a drawing of the same type observed in the Lower Danube region by Marsigli (1726). (Photo Tibor Kádas; Library source: Institute and Museum of Military History, Budapest).

Ottoman-Turkish expansion moving north and west prior to the occupation of Hungary itself. Others became allied with the conquerors to form a social class of their own within the Ottoman Millet system composed largely of Christian mobile pastoral communities of ethnic Vlachs and Serbs (Kursar 2013) engaged in long-distance transhumance. The economic effect of these political developments was an increasing circulation of livestock, intensifying the possibilities of cross-breeding and exchange thereby enhancing the phenotypic diversity of especially bovids in the region. A case in point is a form of sheep with V shaped, corkscrew horns which Buffon (1767) and Darwin (1868) described as 'Valachian'. The Hungarians call this breed '*Racka*' (the Romanian peasant's name it Serbian, the term *Rack*, meaning Serb: Drăgănescu 2007: 93). Gaál (1966: 79) perpetuated the popular 19th century view that 10th century conquering Hungarians brought this breed to the Carpathian Basin. Early osteological evidence of this form however is entirely missing. On the other hand, Racka-type Wallachian, Macedonian and Moldavian sheep may well have reached medieval Hungary as the result of the aforementioned medieval human population movements preceding the Ottoman occupation of the Carphathian Basin (Matolcsi 1977: 94; Vörös 2003: 358, Ill. 5/4–5). Corkscrew-shaped horn cores became increasingly frequent among the diverse Ottoman Period sheep remains in Hungary (Fig. 8).

Evidence of another enigmatic breed, the long-horned Hungarian Grey cattle is completely absent during the Ottoman Period. Both the withers height and morphological diversity of cattle increased during this time (Vörös 2003: 336, Table 5), but long horn cores pre-dating the 18th century are yet to be found. As detailed in our earlier review on Ottoman Period animal exploitation (Bartosiewicz and Gál 2003: 365), long-distance cattle trade was of key importance in Hungarian economy between the 16th–18th centuries. Similarly to other areas peripheral to urbanized Central and Western Europe, extensive graze in the hinterlands made transcontinental meat provisioning a lucrative business. Poland was in competition with Hungary for markets in Germany. Scania and Denmark supplied beef to urban centres along the Rhine, in Brabant and Flanders and long distance droves were likewise documented in Britain (Blanchard 1986: 445; Bartosiewicz 2012a: 28). Post-Ottoman Period, 18th century large horn cores were recovered in Buda (Csippán 2009). By this time longhorn cattle seems fashionable in Europe (Switzerland: De Bluë 1985: 65; The Netherlands: ten Broeke *et al.* 2000: 138). While an Ottoman connection seems unlikely in the case of the Hungarian Grey cattle, the indirect effect of several factors (availability of free range grazing, large scale livestock trade, increasing market competition in German speaking areas and Italy) stimulated the exchange of breeding stock which on the broad basis for selection likely contributed to the emergence of modern breeds.

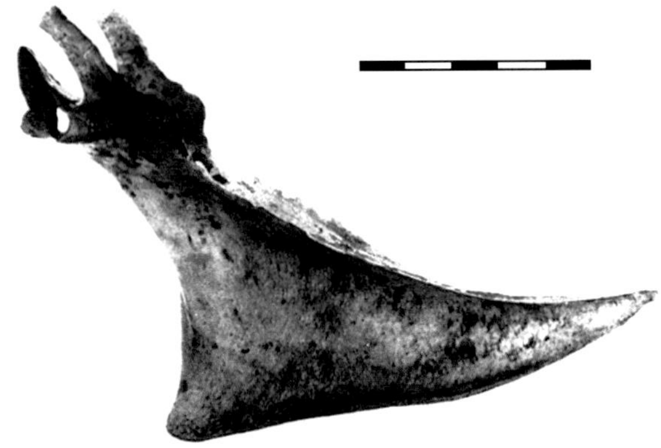

Fig. 9. Turkey sternum from the urban deposit of Székesfehérvár–Koch utca (Photo Krisztina Pálfay).

Luxury animals

Turkey, a bird originating in the New World, was introduced to Hungary during the Ottoman Period, although not necessarily through Ottoman mediation. A 16th century turkey sternum came to light during the excavations of the Székesfehérvár city wall (Fig. 9; Bartosiewicz 1997: 154). The bones from a mature as well as a young adult turkey were found at Vác–Zeneiskola, another urban site (Bartosiewicz 1995a: 59). The taxonomic position of a coracoid bone from the fort at Gyula still needs to be reconfirmed. A medieval deposit in the Buda Castle area

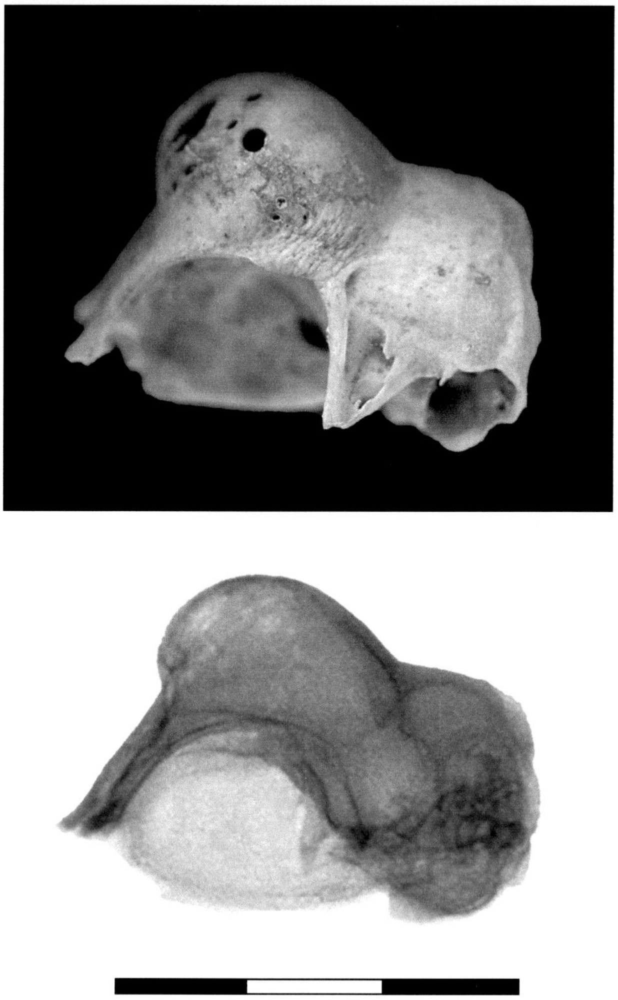

Fig. 10. Left lateral aspect of a crested hen neurocranium showing symptoms of cerebral hernia from Buda–Szent György tér, Királyi Istálló (top) and the X-ray image of the same bone (Photos Erika Gál, Attila Arany Tóth).

also yielded a turkey bone (Vörös 2003: 359) whose stratigraphic position is uncertain. Recently, a tarsometatarsus fragment of a male was identified in a post 1661–1670 deposit at the Christian stronghold of Szendrő–Felsővár (Upper castle) North of the Ottoman Empire (Daróczi-Szabó 2009: 129).

As of today there are two possible (not mutually exclusive) scenarios concerning the origins of these birds in Hungary. Occurrences in Christian-controlled areas suggest that they may have been imported from the West through Austria, Silesia and Bohemia (Mártha 1985) as indicated by contemporaneous finds. The first known mention of the Hungarian name is dated to 1589–1591 (Benkő 1976) and several references to this bird are known from 17th century inventories of Hungarian nobility (Vörös 2003: 360). The other possibility is Ottoman mediation from the East, illustrated by their occurrence in the occupied areas. Written sources mention 17th century turkey keeping in the Danube-Tisza Interfluve (Matolcsi 1975: 218), the core area of Ottoman occupation in Hungary.

Among poultry, crested hens were also identified at several sites in Ottoman Period Buda (Gál *et al.* 2010). This curious form is caused by a sub-pathological form of hereditary brain hernia marked by the feather crest. This feature has been promoted by conscious breeding. In domestic hen, cerebral hernia results from excess brain growth. The brain expands into an enlarged frontal area thinning the neurocranium. A protective crest and tuft compensating for the partially membranous skull has become a popular decorative feature. Two such deficient skulls are known from Roman provincial sites in Britain (Brothwell 1979) and Germany (Teegen 2008) respectively. In the absence of known medieval specimens the continuity of the Roman strand cannot be discussed. In Hungary, only three high status Ottoman Period sites associated with the Buda Castle area yielded archaeological evidence for this anomaly: Buda–Szent György tér, Királyi Istálló (Szent György Square, Royal Stables; Fig. 10), Buda–Teleki palota and Budapest–Rácz-fürdő (Rácz baths; Gál *et al.* 2010). Another post-medieval find was reported from Salzburg, Austria (Pucher 1989–1990).

Aside from the acquisition of exotic domestic birds, a possible import of a steppe species, tawny eagle (*Aquila rapax*) deserves attention. The tarsometatarsus of a female bird came to light from the late 16th century Bajcsavár on the Western frontier of the Ottoman Empire (Gál 2002: 104). Although this stronghold was manned by Christian mercenaries of the Habsburgs and was never lost to the Ottoman army, a camel maxilla in the archaeozoological material indicates access to animals typically kept by Turks, possibly as war-booty. Tawny eagle is not part of the avifauna of Hungary, but its occurrence was registered twice in the 20th century (Peterson *et al.* 1977: 84). It breeds in shrubby lowlands and open grassland habitats in most of Africa and across tropical southwestern Asia to India. This species is not easily trained for hawking as its diet consists largely of fresh carrion, i.e. passive prey. Various eagles, however, were often kept for the purposes of self-representation (Bartosiewicz 2012b). Naturally the chance occurrence of a 16th century straggler in Western Hungary cannot be ruled out. However, in addition to the osteological evidence of several raptor species (Gál 2002: 104), a small spherical rattle of 20 mm diameter, made from thin bronze sheeting was also recovered from this fort (Kovács 2002: 125). Such rattles have typically been used by falconers. It is thus possible that affluent mercenaries at the military post acquired the Bajcsa eagle from the neighbouring Ottoman Empire.

In addition to the circulation of live animals, exotic raw materials of animal origins can also be traced in the Ottoman archaeozoological

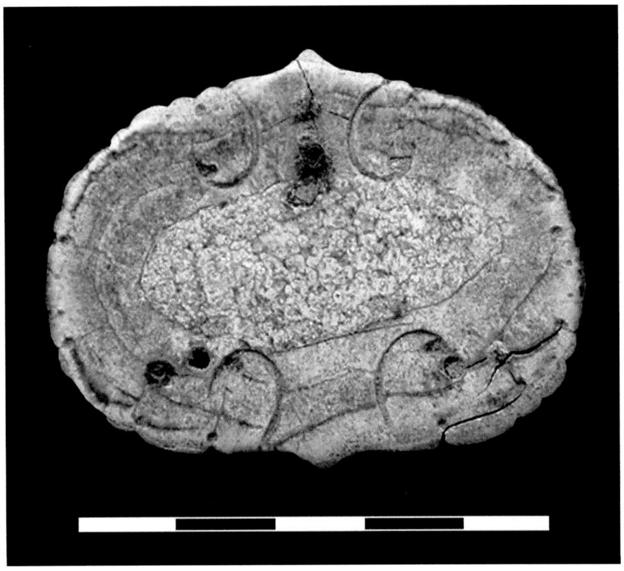

Fig. 11. The ornamental front surface of the walrus ivory belt plaque from Barcs–Palánkvár (Photo Erika Gál).

record. While, given the geographical extent of the Ottoman Empire, finds of elephant ivory would be unsurprising, a piece of manufactured walrus ivory from the Ottoman-Turkish fort of Barcs–Palánkvár (Fig. 11), identified on the basis of its amorphous, terrazzo-like internal structure, must be considered a special discovery (Gál and Kovács 2011). Beginning with the 17th century, tax records show payments by Russia to the Crimean Tartar Khanate. Among the valuable items thus contributed sometimes walrus ivory is mentioned. The precious raw material was then manufactured in the Crimea and passed on among others to the Sublime Porte in Istanbul. The walrus ivory belt plaque found at the Ottoman-Turkish fort of Barcs may have ended up in Hungary through Tartar and Ottoman mediation (Gál and Kovács 2011). This special archaeological walrus find in Hungary is paralleled in historical collections: several 18th century Balkan *yatagans* in the Museum of Military History in Budapest have hilts carved from walrus tusk (Sudár and Szőllősy 2014: 248).

African elephant ivory is preferred by carvers for its easy workability. Import sources used by medieval Europeans became readily available to the Turks following the mid-14th century expansion of the Ottoman Empire into North Africa. The dominance of Ottomans in the eastern Mediterranean ultimately rearranged established medieval ivory trade networks with Europe. Curiously enough, ivory objects in Hungary are directly inventoried in archaeological collections as 'bone' objects without further zoological analyses. Several references to 15th–16th century ivory combs are known from the Buda Castle area as well as sites in Visegrád (Choyke and Kováts 2010: 121–124). Since ivory raw material or workshop debitage have not yet been identified at sites in Hungary, one may assume that these combs were imported as manufactured goods or brought along by individuals as personal accessories.

Discussion and conclusions

The politically turbulent wake of the Middle Ages and the onset of 150 years of Ottoman-Turkish occupation of Hungary marked a time period of unusually intense cultural interaction in which animals played important and variegated roles. The agricultural landscape changed: free range animal keeping was a more viable form of production than sedentary crop-cultivation during times of recurrent warfare and political instability (Bartosiewicz 2000). Population displacement and commercial movements as well as alternating military occupation brought about changes in meat consumption by reducing a demand for pork in favour of mutton. It would be a particularly interesting topic for historical research to appraise how much the verticality of religious identities changed under the pressures of conversion to Islam and whether the first generation of those converted strictly observed pork prohibition as well.

Perpetual change also resulted in the introduction of new domestic animal species while new breeds of traditional domesticates also developed. Of the animals identified with Ottoman rule, camels disappeared with the Turks from central Europe (Bartosiewicz 1996b). Donkeys and mules have never become as highly esteemed animals among the locals as in Ottoman-Turkish culture.

The impact of the 150-year occupation was far more complex on autochthonous livestock. Precursors to modern breeds were equated with ethnic identities at the time. This was not only the case for horses, much in the forefront of thinking, traded, rustled and given away as generous presents. Other livestock were also identified by ethnic groups and/or associated geographical regions. Numerous population groups moved around as refugees, mercenaries and settlers, fleeing or serving the Ottoman Empire. This situation suited mobile pastoralists such as the Vlachs whose traditional way of life was key to temporary integration with Ottoman society. A spin-off of these social movements was an increasing circulation of livestock between regions within the territory under Ottoman control.

During the late 16th century cattle exports from Hungary under Ottoman occupation peaked (Bartosiewicz 1995a: 82, Table 32). According to Ottoman-Turkish tax rolls from Vác, herd sizes varied greatly although on average at least a dozen groups of 100–120 cattle crossed the Danube here every day. By this time, animals exerted a feedback effect on society through large-scale trade, stimulating the emergence of entrepreneurs and a vertical range of employees. Cattle were accompanied by half a dozen 'ox captains' per hundred beasts. These herders were bound by contracts rather than ethnic affiliation in a strongly androcentric sub-culture with shifting political connotations. In volatile times they are known to have formed paramilitary units and also provided the core of *heyducks*, mercenaries engaged in irregular warfare against both the Ottoman and Habsburg Empires (Bartosiewicz 2012b).

At the other end of the Ottoman Period social scale in Hungary, access to exotic animals evidently carried high status regardless of ethnic or religious affiliation. The control and ownership of curious animals, including domestic and tamed birds, was probably an important means of self-representation. This general phenomenon gained new dimensions as assigned markers of identities were also created in a world polarized by political interests and religious animosity. In the 1571 battle of Lepanto the joint fleet of European Catholic maritime states defeated the Ottoman fleet off the coast of Greece. The victory blocked Ottoman expansion into the Mediterranean and raised hopes in the west that Turks, previously invincible, could be beaten. It also forged a notable interest in the now less fearsome enemy: many things exotic were labelled 'Turkish' in Europe. Aldrovandi (1603: 316) referred to crested hen as *Gallina Turcica*. Based on the 1580 Venetian edition of the book by Nicolo de' Nicolay (1576), Jacopo Ligozzi (1547–1627) a miniaturist produced a spectacular series of Turkish genre pictures, showing people in the Ottoman Empire in ethnic costumes. However, in addition to de Nicolay's work Ligozzi, a seasoned painter of natural history, reinforced these stereotypes by adding animal attributes to the imagery. Thus, a clean-shaven Turkish scribe is shown in the company of a crested hen. A married woman from Greece is accompanied by a fat-tailed sheep, another curiosity in Europe often referred to by contemporaries as Turkish or Arabian sheep (Rumpolt 1581; Topsell 1607).

The flow of exotic animal raw materials also increased in the form of luxury goods from distant parts of the Ottoman Empire and even beyond. In addition to commercial circulation some such objects may tentatively be attributed to the movement of high-ranking persona such as civilian or military leaders and foreign envoys. Combs represent a particularly complex case. As objects, they are usually considered gender neutral or are associated with female identity in the Western tradition. Ottoman Period specimens made from elephant ivory found in the area of the Buda Castle and the nearby site of Budapest–Rácz-fürdő (where one of the crested hen skulls was also recovered) may have been owned by adult Moslem men who adhered to traditional beard grooming norms (Choyke and Kováts 2010: 125). These objects thus possibly represent pious and social behaviour in a private setting or in the public such as the baths: the site of Budapest–Rácz-fürdő was a Turkish *hamam* itself.

Archaeozoology is not simply a study of past relationships between animals and humans. Animal remains from archaeological sites can also elucidate human-human relationships in which animals have always served as media, conveying a broad range of messages. The Ottoman Period in Hungary, representing a prolonged, multi-generation exposure to influences from Southwest Asia offers excellent opportunities to explore the diverse meanings of animals at the time, laden with references to religion, ethnicity and social standing.

Acknowledgements

The authors gratefully thank Gyöngyi Kovács for her useful comments on archaeological aspects of the manuscript. Parts of the research presented in this paper were funded by project OTKA K 72231 of the Hungarian Research Fund, titled 'Studies on Settlement Archaeology and Environmental History in Southern Transdanubia, 1300–1700'.

References

Aldrovandi, U. 1603. *Ornithologiae: Hoc est de Avibus Historiae. Liber Decimusquartus*. Bononiae: apud Franciscum de Franciscis Senensem.

Almathen, F., Charruau, P., Mohandesan, E. Mwacharo, J. M., Orozco-ter Wengel, P., Pitt, D., Abdussamad, A. M., Uerpmann, M., Uerpmann, H.-P., De Cupere, B., Magee, P., Alnaqeeb, M. A., Salim, B., Raziq, A., Dessie, T., Abdelhadi, O. M., Banabazi, M. H., Al-Eknah, M., Walzer, C., Faye, B., Hofreiter, M., Peters, J., Hanotte, O. and Burger, P.A. 2016. Ancient and modern DNA reveal dynamics of domestication and cross-continental dispersal of the dromedary. *PNAS*, doi: 10.1073/pnas.1519508113

Bartosiewicz, L. 1995a. *Animals in the Urban Landscape in the wake of the Middle Ages*. Oxford: British Archaeological Reports, International Series 609.

Bartosiewicz, L. 1995b. Cattle trade across the Danube at Vác (Hungary). *Anthropozoologica* 21: 189–196.

Bartosiewicz, L. 1995c. Camel remains from Hungary. In,Buitenhuis, H. and Uerpmann, H-P. (eds) *Archaeozoology of the Near East II*: 119–125. Leiden: Backhuys Publishers.

Bartosiewicz, L. 1996a. Közép- és török kori állatmaradványok Segesdről (Medieval and Turkish Period animal bones from Segesd, Southwestern Hungary). *Somogyi Múzeumok Közleményei* 12: 183–222.

Bartosiewicz, L. 1996b. Camels in Antiquity: The Hungarian connection. *Antiquity* 70/268:

447–453.
Bartosiewicz, L. 1997. A Székesfehérvár Bestiary: Animal bones from the excavations of the medieval city wall. *Alba Regia* 26: 133–167.
Bartosiewicz, L. 1999a. Animal husbandry and medieval settlement in Hungary: A review. *Beiträge zur Mittelalterarchäologie in Österreich* 15: 139–155.
Bartosiewicz, L. 1999b. Turkish Period bone finds and cattle trade in south-western Hungary. In, Becker, C., Manhart, H., Peters, J. and Schibler, J. (eds) *Historia Animalum ex Ossibus*: 47–56. Beiträge zur Paläoanatomie, Archäologie, Ägyptologie, Ethnologie und Geschichte der Tiermedizin. Rahden/Westf.: Verlag Marie Leidorf GmbH.
Bartosiewicz, L. 2000. Állattartás és környezet a török-kori Kárpát-medencében (Animal keeping and environment during the Turkish Period in the Carpathian Basin). In, Füleky, Gy. (ed.) *A táj változásai a Kárpát-medencében a Történelmi Események Hatására*: 62–70. Budapest–Gödöllő: Kulturális Örökség Igazgatósága–Szent István Egyetem.
Bartosiewicz, L. 2009. Skin and Bones: Taphonomy of a Medieval Tannery in Hungary. *Journal of Taphonomy* 7/2–3: 95–111.
Bartosiewicz, L. 2012a. Régi kor árnya felé viszszamerengni mit ér? A skót hegyi marha és a romantika [Is it worth lamenting the past? The Scottish Highland cattle and romanticism]. In, Jávorka, L. (ed.) *Emberségről Példát". Válogatott Közlemények Bodó Imre 80. Születésnapjára*: 24–47. Budapest: a Magyar Szürke Marhát Tenyésztők Egyesületének és a Magyar Szürke Marhát Kedvelők Baráti Körének Kiadványa.
Bartosiewicz, L. 2012b. Show me your hawk, I'll tell you who you are. In, Raemaekers, D. C. M., Esser, E., Lauwerier, R. C. G. M. and Zeiler, J. T. (eds) *A Bouquet of Archaeozoological Studies, Essays in Honour of Wietske Prummel*: 181–190. Groningen: Barkhuis & University of Groningen Library.
Bartosiewicz, L. 2014. Camels in the Frontline. In, Mashkour, M. and Beech, M. J. (eds) *Old World Ancient Camelids between Arabia and Europe*. Proceedings of the 11th ICAZ conference in Paris. *Anthropozoologica* 49(2): 297–302.
Bartosiewicz, L., Csiky, G. and Gyarmati, J. 2008. Emberiességi szempontok és a hagyományos állatvágás két példája (Humane considerations and two examples of traditional slaughtering). *Animal welfare, etológia és tartástechnológia* 4/3: 130–149. [pdf] Available at: <http://www.animalwelfare.szie.hu/cikkek/200803/AWETH2008130149.pdf>.
Bartosiewicz, L. and Gál, E. 2003. Animal exploitation in Hungary during the Ottoman Era. In, Gerelyes, I. and Kovács, Gy. (eds) *Archaeology of the Ottoman Period in Hungary*: 365–376. Opuscula Hungarica III. Budapest: Magyar Nemzeti Múzeum.
Bartosiewicz, L. and Gál, E. 2007. Sample size and taxonomic richness in mammalian and avian bone assemblages from archaeological sites. *Archeometriai Műhely* 2007/1: 37–44. [pdf] Available at: <http://www.ace.hu/am/2007_1/AM-2007-01-BL.pdf>.
Bartosiewicz, L. and Gyöngyössy, M. 2006. The Khan's mule. In, Olsen, S. L., Grant, S., Choyke, A. M. and Bartosiewicz, L. (eds) *Horses and Humans: The Evolution of Human-Equine Relationships*: 289–299. Oxford: British Archaeological Reports, International Series 1560.
Benkő, L. 1976. (ed.) *A magyar nyelv történetietimológiai szótára*, Ö–Zs [Hungarian etymological dictionary, Ö–Zs]: 308–309. Budapest: Akadémiai Kiadó.
Blanchard, I. 1986. The Continental European cattle trades 1400–1600. *Economic History Review*, 2nd ser. XXXIX/3: 427–460.
Bökönyi, S. 1958. A Budai Várpalota ásatásának állatcsontanyaga. *Budapest Régiségei* XVIII: 455–484.
Bökönyi, S. 1974. *History of Domestic Mammals in Central and Eastern Europe*. Budapest: Akadémiai Kiadó.
Bökönyi, S. 1985. Tierknochenfunde aus dem Bereich der Werkstatt von Kamid el-Loz. In, Frisch, G., Mansfeld, G. and Thiele, W.-R. (eds) *Kamid el-Loz. 6. Die Werkstätten der spätbronzezeitlichen Palaste*: 199–205. Bonn: Saarbrücker Beiträge zur Altertumskunde 33.
Broeke, J. ten, Sabben, A. and Woudenberg, R. 2000. Antieke tegels–Antique tiles–Carreaux anciens. Oud-Beijerland: Van As.
Brothwell, D. 1979. Roman evidence of a crested form of domestic fowl, as indicated by a skull showing associated cerebral hernia. *Journal of Archaeological Science* 6: 291–294.
Buffon, G. L. L. 1767. *Histoire Naturelle, Générale et Particulière, avec la Description du Cabinet du Roy.* Paris: Imprimerie royale.
Choyke, A. M. and Kováts, I. 2010. Tracing the personal through generations: Late Medieval and Ottoman combs. In, Pluskowski, A., Kunst, G. K., Kucera, M., Bietak, M. and Hein, I. (eds) *Bestial Mirrors: Animals as Material Culture in the Middle Ages* 3: 115–127. Wien: ViaVias 3.
Csippán, P. 2009. 18. Századi szarvcsapleletek a budai Vízivárosból (Eighteenth century cattle horn core finds from the Víziváros District of Buda, Hungary). In, Bartosiewicz, L., Gál, E. and Kováts, I. (eds) *Csontvázak a Szekrényből. Válogatott Tanulmányok a Magyar*

Archaeozoológusok Visegrádi Találkozóinak Anyagából 2002–2009: 163–169. Budapest: Martin Opitz Kiadó.

Daróczi-Szabó, L. 2004. Állatcsontok a Teleki Palota törökkori gödréből [Animal Remains from the Turkish Era Well of the Teleki Palace]. *Budapest Régiségei* XXXVIII: 159–160.

Daróczi-Szabó, L., Daróczi-Szabó, M., Kovács, E. Zs., Kőrösi, A. and Tugya, B. 2015. Recent camel finds from Hungary. In, Mashkour, M. and Beech, M. J. (eds) *Old World Camelids. Anthropozoologica* 49(2): 265–280.

Daróczi-Szabó, M. 2009. Szendrő–Felsővár koraújkori állatcsontjainak vizsgálata (Early Modern Age animal remains from Szendrő–Felsővár, Hungary). In, Bartosiewicz, L., Gál, E. and Kováts, I. (eds) *Csontvázak a szekrényből. Válogatott tanulmányok a Magyar Archaeozoológusok Visegrádi Találkozóinak Anyagából 2002–2009*: 125–145. Budapest: Martin Opitz Kiadó.

Darwin, Ch. 1868. *The Variation of Plants and Animals under Domestication*. 2 vols. London: John Murray.

De Bluë, V. 1985. *Landauf-landab*. Bern–München: Edition Erpf.

Drăgănescu, C. 2007. A note on Balkan sheep breeds origin and their taxonomy. *Archiva Zootechnica* 10: 90–101.

Fekete, L. and Nagy, L. 1975. Budapest története a török korban [The history of Budapest during the Turkish Period]. In, Gerevich, L. and Kosáry, D. (eds) *Budapest Története* II. *Budapest Története a Késő͏bbi Középkorban és a Török Hódoltság Idején*. Budapest: Akadémiai Kiadó.

Gaál, L. 1966. *A Magyar Állattenyésztés Múltja* [The past of Hungarian animal breeding]. Budapest: Mezőgazdasági Kiadó.

Gál, E. 2002. Madárleletek a bajcsai várból (Bird remains from the Bajcsa fort). In, Kovács, Gy. (ed.) *Weitschawar/Bajcsa-Vár*: 101–105. Zalaegerszeg: Zala Megyei Múzeumok Igazgatósága.

Gál, E. and Bartosiewicz, L. 2016. Animal remains from the Ottoman-Turkish palisade fort at Barcs, Southwest Hungary. In, Kovács, Gy. and Zatykó, Cs. (eds) *Medieval and Ottoman Period in Southern Transdanubia, Hungary – The Natural Sciences*. Budapest: Institute of Archaeology, Research Centre for the Humanities, Hungarian Academy of Sciences.

Gál, E., Csippán, P., Daróczi-Szabó, L. and Daróczi-Szabó, M. 2010. Evidence of the crested form of domestic hen (*Gallus gallus f. domestica*) from three post-medieval sites in Hungary. *Journal of Archaeological Science* 37: 1065–1072.

Gál, E. and Kovács, Gy. 2011. A walrus-tusk belt plaque from an Ottoman-Turkish castle at Barcs, Hungary. *Antiquity Project Gallery*. [online] Available at: <http://antiquity.ac.uk/projgall/gal329/>.

Galik, A., Mohandesan, E., Forstenpointner, G., Scholz, U. M., Ruiz, E., Krenn, M. and Burger, P. 2015. A Sunken Ship of the Desert at the River Danube in Tulln, Austria. *PLOS ONE*, doi:10.1371/journal.pone.0121235

Gilbert, A. S. 1988. Zooarchaeological observations on the slaughterhouse of Meketre. *The Journal of Egyptian Archaeology* 74: 69–89.

Kovács, Gy. 1989. Juh astragalos-játékkockák a szolnoki vár területéről (Sheep astragalus gaming pieces from the area of the Szolnok fort). *Archeologiai Értesítő* 114 (1–2): 103–9.

Kovács, Gy. 2002. A bajcsai ásatás leletei (Finds from the Bajcsa excavations). In, Kovács, Gy. (ed.) *Weitschawar/Bajcsa-Vár*: 107–257. Zalaegerszeg: Zala Megyei Múzeumok Igazgatósága.

Kovács, Zs. E. 2012. Dispersal history of an invasive rodent in Hungary – subfossil finds of *Rattus rattus*. *Acta Zoologica Academiae Scientiarum Hungaricae* 58 (4): 379–394.

Kursar, V. 2013. *Being an Ottoman Vlach: On Vlach Identity (ies), Role and Status in Western Parts of the Ottoman Balkans (15th-18th Centuries)* (Bir Osmanlı Eflakı Olmak: Osmanlı Balkanlarının Batı Bölgelerinde Eflak Kimliği, Görevi ve Vaziyetine Dair (15.-18. Yüzyıllar)). *Ankara Üniversitesi Osmanlı Tarihi Araştırma* ve *Uygulama Merkezi Dergisi* 34: 115–161.

Marsigli, L. F. 1726. *Danubius Pannoni-Mysicus*. Hagae Comitum: apud P. Gosse, R. Chr. Alberts, P. de Hondt – Amstelodami: apud H.Uytwerf et F.Changuion.

Mártha, Zs. 1985. A pulyka rövid természetrajza és hagyományos tenyésztése Magyarországon [The short natural history of turkey and its traditional breeding in Hungary]. *A Magyar Mezőgazdasági Múzeumok Közleményei* 1984–1985: 261–291.

Matolcsi, J. 1975. *A háziállatok eredete* [The origins of domestic animals]. Budapest: Mezőgazdasági Kiadó.

Matolcsi, J. 1977. A budai várpalota északi előudvarában feltárt XIV–XV. századi állatcsontok (Os d'animaux des XIVe et XVe siècles mis au jour dans l'avant-cour nord du château royal de Buda). *Budapest Régiségei* XXIV: 179–197.

Nicolay, N. de 1576. *Le navigationi et viaggi fatti nella Turchia*. Scruta Mini, Anversa.

Peterson, R. T., Mountfort, G. and Hollom, P. A. D. 1977. *Európa madarai* [Birds of Europe]. Budapest: Gondolat.

Pucher, E. 1989–1990. Der frühneuzeitliche Knochenabfall eines Wirthauses neben der Salzburger Residenz. *Salzburger Museum Carolino Augusteum Jahresschrift* 35–36: 71–133.

Rumpolt, M. 1581. *Ein new Kochbuch*. Franckfort am Mayn, Verlag Sigmund Feyerabendt; Facsimile 1976, Hildesheim: Olms Presse.

Sudár, B. and Szőllősy, G. 2014. A Hadtörténeti Múzeum Hidegfegyver Gyűjteményében őrzött jatagánok egy sajátos csoportja (A Special Group of Yatagans in the Edged Weapons Collection of the Hungarian Military History Museum). *Acta Musei Militaris in Hungaria* 14: 219–237.

Takács, S. 1991. *Urban Architecture in Budapest*. Budapest: Officina Nova.

Teegen, W-R. 2008. A crested fowl from late antique *Augusta Treverorum*/Trier. In, Grupe, G., McGlynn, G. and Peters, J. (eds) *Limping Together Through the Ages. Joint Afflictions and Bone Infections*: 203–207. Rahden/Westf.: Verlag Marie Leidorf GmbH.

Topsell, E. 1607. *The Historie of Fovre-footed Beastes*. London: William Iaggard.

Tóth, A. J., Daróczi-Szabó, L., Kovács, Zs. E., Gál, E. and Bartosiewicz, L. 2010. In the Light of the Crescent Moon: Reconstructing Environment and Diet from an Ottoman-Period Deposit in Sixteenth to Seventeenth Century Hungary. In, VanDerwarker, A. M. and Peres, T. M. (eds) *Integrating Zooarchaeology and Paleoethnobotany. A Consideration of Issues, Methods, and Cases*: 245–286. New York–Dordrecht–Heidelberg–London: Springer.

Twigg, J. 2012. Animal remains from Visegrád-Lower Castle from the 16–17th century. *Acta Archaeologica Academiae Scientiarum Hungaricae* 63/1: 197–220. DOI: http://dx.doi.org/10.1556/AArch.63.2012.1.7.

Vörös, I. 1980. Előzetes jelentés a Békés–kastélyzugi törökkori palánkvár állatcsont leleteiről, 1975–78 (Vorbericht über die Tierknochenfunde der türkenzeitlichen Palisadenburg von Békés–Kastélyzug, 1975–78). *Archaeologiai Értesítő* 107: 112–114.

Vörös, I. 2003. Sixteenth and Seventeenth Century Animal Bone finds in Hungary. In, Gerelyes, I. and Kovács, Gy. (eds) *Archaeology of the Ottoman Period in Hungary*: 339–352. Opuscula Hungarica III. Budapest: Magyar Nemzeti Múzeum.

Wutke, S., Andersson, L., Benecke, N., Sandoval-Castellanos, E., Gonzalez, J., Hallsteinn Hallsson, J., Lõugas, L., Magnell, O., Morales-Muniz, A., Orlando, L., Hulda Pálsdóttir, A., Reissmann, M., Muñoz-Rodríguez, M. B., Ruttkay, M., Trinks, A., Hofreiter, M. and Ludwig, A. 2016. The origin of ambling horses. *Current Biology* 26/15: 697–699. DOI: http://dx.doi.org/10.1016/j.cub.2016.07.001.

Appendix I

Domestic mammals identified at the studied sites, NISP.

Site	**Cattle** *Bos taurus* L., 1758	**Sheep** *Ovis aries* L., 1758	**Goat** *Capra hircus* L., 1758	**Sheep or goat** Caprine	**Pig** *Sus domesticus* Erxl. 1777	**Horse** *Equus caballus* L., 1758	**Donkey** *Equus asinus* L., 1758	**Camel** *Camelus* sp.	**Water buffalo** *Bubalus bubalis* L., 1758	**Dog** *Canis familiaris* L., 1758	**Cat** *Felis catus* L., 1758	**Rabbit** *Oryctolagus cuniculus* L., 1758
Barcs–Palánkvár	7474	344	78	453	55	36				68	28	
Békés–Kastélyzug	1024			720	40	13	1			6	6	
Buda–Pasa palota	151			1245	5	2				7		
Buda–Teleki palota	190	1	1	892	3	2				12	60	2
Érd–Palánkvár	178			106	17	20				14	1	
Gyula–Vár	11	14	7		5	1						
Nagykanizsa–Vár	1267			305	234	73			1	42	9	
Segesd–Pékóföld	651	93	18	179	70	8				2		
Székesfehérvár–Koch utca	19	6		5	6	23				7		
Szekszárd–Palánk	1844	427	153	1753	452	227		14		117	20	
Szolnok–Vár	16	122				9						
Vác–Műszaki Főiskola	209	18	8	106	12	1				13	1	
Vác–Kossuth tér	86	4		5								
Vác–Zeneiskola	21			11						2		
Vác–Szentháromság tér	736			54	57	2	1		1	1		
Visegrád–Alsóvár	1037	21	6	936	132	14				26	9	
Visegrád–Salamon torony	3053			2005	534	56	2			79	84	100

Appendix II

Non-mammalian taxa and wild mammals identified at the studied sites, NISP.

Site	**Domestic hen** *Gallus domesticus* L., 1758	**Goose** *Anser* sp.	**Duck** *Anas* sp.	**Turkey** *Meleagris gallopavo* L., 1758	**Peacock** *Pavo cristatus* L., 1758	**Red deer** *Cervus elaphus* L., 1758	Roe deer *Capreolus capreolus* L., 1758	**Wild pig** *Sus scrofa* L., 1758	**Hare** *Lepus europaeus* Pall. 1778	**Fur-bearing game**	**Wild bird** Aves	**Fish** Pisces	**Pond turtle** *Emys orbicularis* L., 1758
Barcs–Palánkvár	929					39	31	8	25		17	4	1
Békés–Kastélyzug	32									4			
Buda–Pasa palota	27	8								15			
Buda–Teleki palota	854	9								36		6	
Érd–Palánkvár	7							1	1				
Gyula–Vár	1			1		2	4	1			12		
Nagykanizsa–Vár	65	7	1							40	3	2	
Segesd–Pékóföld	22		3							5			
Székesfehérvár–Koch utca	2									1			
Szekszárd–Palánk	387	60				8	21	94	33	35		440	7
Szolnok–Vár													
Vác–Műszaki Főiskola	4									2			
Vác–Kossuth tér		1										1	
Vác–Zeneiskola	4	1		2									
Vác–Szentháromság tér	8	9							1		1	2	3
Visegrád–Alsóvár										54			
Visegrád–Salamon torony	2040	59	2		1					224	11	89	

15 Striking new paths – Distinguishing ancient *Ovis orientalis* from its modern domestic descendant (Karakul breed) applying Geometric and traditional Morphometric approaches to the astragalus

Nadja Pöllath[1*], Paul Alibert[2], Renate Schafberg[3], Joris Peters[1,4]

1 ArchaeoBioCenter and Institute of Palaeoanatomy, Domestication research and the History of Veterinary Medicine, Ludwig Maximilian University Munich, Kaulbachstr. 37, D-80539 Munich, Germany; nadja.poellath@palaeo.vetmed.uni-muenchen.de
2 UMR CNRS 6282 Biogéosciences, Université de Bourgogne Franche-Comté, Bât. Gabriel 6, Boulevard Gabriel, 21000 Dijon, France
3 Martin-Luther University of Halle-Wittenberg, Museum of Domesticated Animals „Julius Kühn", Domplatz 4, 06108 Halle, Germany
4 Bavarian State Collection of Anthropology and Palaeoanatomy, 80333 Munich, Germany
* corresponding author

Abstract
In this contribution, Geometric Morphometrics and Log Shape Ratio are applied to the astragali of domestic Karakul sheep (*Ovis aries*) and wild sheep (*Ovis orientalis*) to test the suitability of these methods for distinguishing these two sheep populations. Both methods successfully separated the wild sheep excavated at Göbekli Tepe and Gusir Höyük (10th – 9th millennium BC) from modern domestic sheep of the Karakul breed. Although additional tests are necessary to exclude other factors, it can be hypothesised that the divergence between the astragali of the two populations most likely results from the process of domestication.

Keywords
Domestication, *Ovis orientalis*, astragalus, Geometric Morphometrics, Log Shape Ratio

Introduction
Wild sheep (*Ovis orientalis*) are among the earliest candidates for becoming one of the main farmyard animals, along with goat, pig and cattle (Peters *et al.* 1999; Zeder 2011; Peters *et al.* 2013, 2014a; Vigne 2015). The process of domestication began in the wider region of northern Mesopotamia during the 9th millennium cal. BC, which broadly corresponds to the early Pre-Pottery Neolithic B (PPN B) phase of the Near Eastern chronology (dating 8,800-8,200 cal. BC). During this period, prehistoric foragers started keeping caprines and pigs inside human settlements at more or less the same time in different parts of the region (Peters *et al.* 2013). However, in contrast to wild boar, whose management likely began as a mutualistic relationship with humans benefiting from the accumulation of food residues around human dwellings (Rosenberg *et al.* 1998), initial keeping of wild sheep and goat within human settlements required the conscious act of removing animals from their natural environment, controlling their mobility inside the anthropogenic environment, and deciding on (and caring for) their nutrition and reproduction. Whereas the earliest of these animal-keeping foragers targeted meat and fat, their successors soon thereafter also exploited caprines for their milk (Vigne and Helmer 2007).

The documentation of the early stages of domestication for *Bos*, *Sus*, *Ovis*, and *Capra* in the Near Eastern archaeofaunal record is the first step, before discussing the 'where' and 'how' and 'why'. In the past, osteometric data were regarded of particular importance since the reduction in size was considered a good marker for human management (= intention) and ongoing domestication (= biological response of the animal) (Meadow 1984; Peters *et al.* 1999). This assumption, however, has been challenged in recent years (e.g. Zeder 2011). While some researchers advise caution since the metrical profile of a population can be influenced by many factors such as climatic changes or domination of the sample by immature animals or by females, others see size reduction as a delayed effect of ungulate domestication, thus rendering approaches based on linear measurements inappropriate for recording the initial stages of the process (e.g., Legge 1996; Zeder 2011).

However, 'size' is only one morphometric property of a bone (or any other biological structure), the other being 'shape' (Kendall 1977; Bookstein 1991). Geometric Morphometrics (GM) is a collection of mathematical tools for the statistical analysis of shape, which has been successfully applied to teeth and crania from prehistoric contexts and is particularly appropriate when differences between populations are expected to be minimal (Cucchi *et al.* 2011; Evin *et al.* 2014; Seetah *et al.* 2014).

Most GM studies concerned with ungulate management, domestication processes and discrimination of populations are based on dental remains (Larson *et al.* 2007; Cucchi *et al.* 2011; Evin *et al.* 2013; 2014; Seetah *et al.* 2014). One reason for this is the fact that teeth are less affected by environmental factors while bones of the appendicular skeleton are remodelled throughout life influenced mainly by nutrition and physical activity (Zeder *et al.* 2006). Phenotypic plasticity, i.e., the response to environmental conditions, can lead to alternative expressions in terms of behaviour, physiology and, important in our case, morphology in a single genotype (West-Eberhard 1989; Price *et al.* 2002). Teeth show less phenotypic plasticity and are considered more appropriate in studies of genetic divergence in animal populations (Larson *et al.* 2007; Seetah *et al.* 2014). Investigating the phenotypic plasticity of bones, however, can be an advantage when dealing with mixed wild and managed populations during the early phase of domestication, for which little genetic divergence between early domestic, managed and true wild animals is to be expected due to ongoing gene flow. During this phase, the most radical and basic changes from living in the wild to being cared for by humans are restrictions regarding mobility and food choice. Our working hypothesis is that restricted mobility can cause changes in the shape of bones of the appendicular skeleton in the initial stage of the domestication process. Having said that, the reproductive and therefore genetic separation of early managed stock from their free-ranging wild relatives will eventually lead to genetic divergence between both populations, which will also affect the shape of bones. If so, GM should enable us to separate wild and domestic populations in general. Future work applying the same methodology could focus on goats as well, opening up the possibility of intra-species comparison by analysing diachronically domestic sheep and goats collected from the same early Neolithic to historic contexts. With these data at hand, it would be possible to verify whether over time the exploitation of the two species in mixed herds performing similar annual rounds resulted in convergence in morphology.

Cranial and dental remains are very rare at the two sites in our study, in particular those of *Ovis*; the second reason for considering postcranial skeletal elements for this GM analysis. Therefore, an element had to be selected that is present in reasonable numbers, is usually well-preserved, and offers enough structure for defining landmarks and outlines. With postcranial elements two issues arise: (1) only fully fused bones can be used; (2) most of the long bones are highly fragmented due to processing for marrow extraction. For long bones, only parts with early fusing epiphyses such as the distal humerus are likely to be present in numbers allowing statistical analyses. Another possibility are compact bones from largely meatless body parts that are usually hardly processed and therefore more often preserved and complete. In our assemblages, the astragalus (= talus) meets these conditions and so was chosen for this study. Apart from the fact that the astragalus offers a number of good features to distinguish between *Ovis* and *Capra* (Boessneck *et al.* 1964; Zeder and Lapham 2010), studies in ecomorphology and functional morphology also frequently use astragali since the astragalus reflects differences in the mode of locomotion. Thus, the astragalus is particularly suitable for this kind of analysis and allows for interpretation and discussion of the results of this study as compared to previous research (DeGusta and Vrba 2003; Plummer *et al.* 2008; Curran 2012; Haruda 2014; Barr 2014; Barak *et al.* 2011). While most of these studies are based on linear and surface measurements, Curran (2012) and Haruda (2014) applied 3D GM approaches to postcranial elements.

Since GM requires more resources and equipment and is more difficult to apply systematically in the field, it will probably remain a complementary technique (Evin *et al.* 2014; Vigne 2015). In contrast to GM, traditional osteometric methods rely on distance measurements that can be taken with callipers when recording data in the field. The second aim of our pilot study was therefore to test whether it was possible to apply multivariate methods on 'traditional' linear measurements, which separate size from shape more or less equally well. More widely used in biological research, the Log Shape Ratio (LSR) proposed by Jolicoeur (1959) and Mosiman (1970) is an approach to achieve this (e.g., Claude 2008, 2013; Barr 2014; Denys and Winkler 2015). Recently Vigne (2015) also recommended LSR for archaeozoological data for these reasons. In the following we address whether GM and LSR are able to detect shape differences in the astragalus of spatio-temporally distinct populations and discuss the biological foundation of these differences.

Material

This study is based on a total of 62 sheep and 9 gazelle astragali from two archaeological sites in

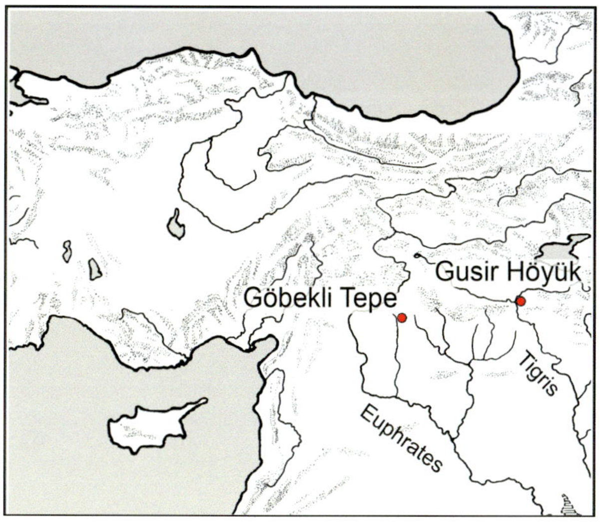

Fig. 1. Map of the region indicating the location of the two archaeological sites mentioned in the text.

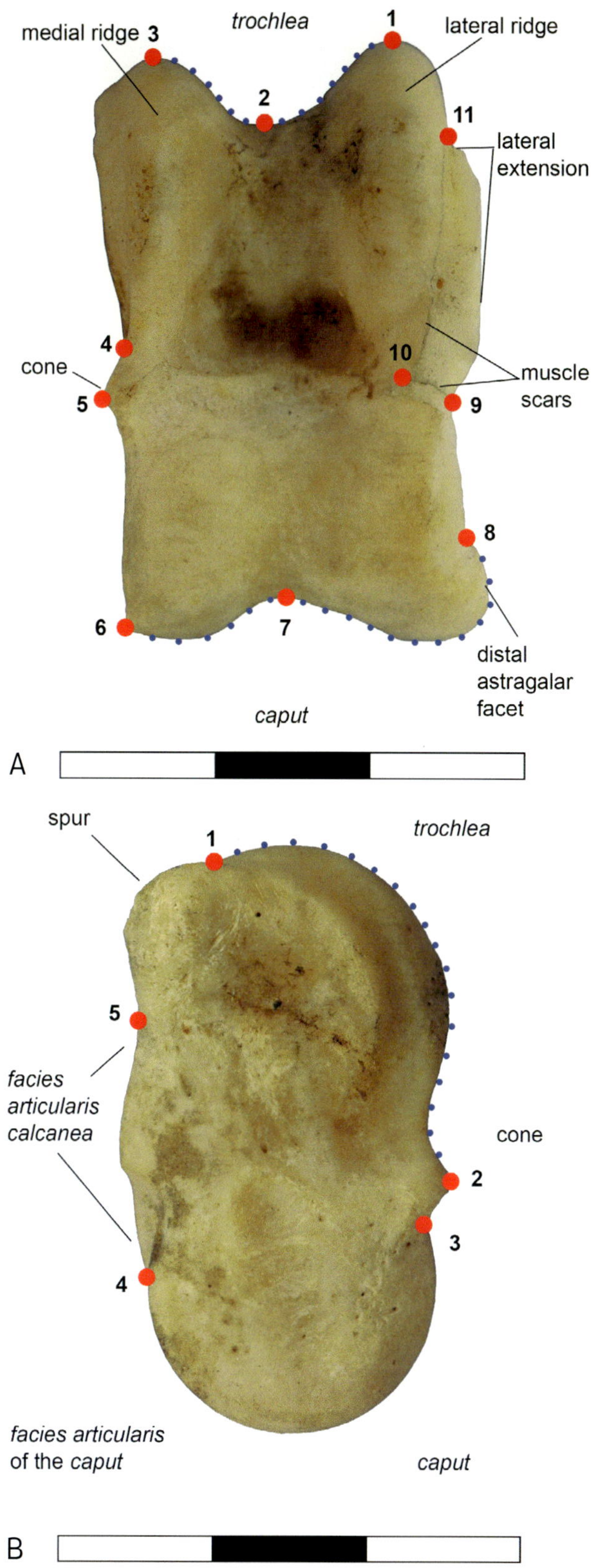

Fig. 2. Astragalus of sheep with the localisation of landmarks (red) and sliding semi-landmarks (blue) and anatomical terminology used in the text. A) Dorsal view; B) Medial view.

Southeast Anatolia (Fig. 1) and 41 astragali from a modern reference collection of sheep (Table 1). Although only well-preserved specimens were chosen for analysis, some are damaged on either the medial (for example, at the spur, see Fig. 2b) or dorsal side. On the other hand, linear measurements can be taken even in most cases where GM is not possible. For the LSR approach samples available in our lab had to be re-measured, including specimens which were not suitable for GM analyses. For this reason, datasets differ according to the methods applied.

For the two sites in Southeast Anatolia, caprine astragali were identified as pertaining to *Ovis* or *Capra* using the criteria developed by Boessneck *et al.* (1964) while considering the reliability of criteria as assessed by Zeder and Lapham (2010). Only specimens with an unambiguous identification as sheep were included in this study. As shown by Zeder and Lapham (2010: Table 5) correct assignation rates for astragali were high (97% in sheep, 100% in goat) with only sheep astragali misidentified as goats but no goat astragalus among sheep. It can therefore be safely assumed that no goat astragali are included in the archaeological samples. All parameters, such as taxonomic composition, demographic profiles, and stable isotope analysis, show that the faunal remains exclusively represent wild animals (Peters *et al.* 2013, 2014a). The sheep remains excavated at these sites are therefore considered as originating from populations of wild sheep (*Ovis orientalis*), which were contrasted with a modern population of domestic sheep (*Ovis aries*). Additionally, we included astragali from Goitered gazelles (*Gazella subgutturosa*) as an outgroup.

The size and shape of a given bone is influenced by several factors, some being internal or genetic such as sex and age, while others are external, such as nutrition, environment and living/husbandry conditions. The growth of animals is largely influenced by hormones. High levels of testosterone are responsible for higher growth rates in male mammals, particularly in polygamous species (Lawrence and Fowler 2002). Apart from other expressions such as the presence or size of horns,

Table I. Archaeological and modern samples studied.

Taxon	Origin/Site	Geometric Morphometrics			Traditional Morphometrics	Total (specimens)
		dorsal view	medial view	both views available (for PLS)		
Archaeological specimens						
Goitered gazelle, *Gazella subgutturosa*	Göbekli Tepe	9	-	-	9	**9**
Wild sheep, *Ovis orientalis*	Göbekli Tepe	36	33	27	23	**42**
Wild sheep, *Ovis orientalis*	Gusir Höyük	17	16	14	17	**20**
Modern specimens						
Karakul, *Ovis aries*	Museum of Domesticated Animals, Halle	41	41	41	40	**41**
Total		**103**	**90**	**82**	**89**	**112**

sexual dimorphism in these species is most obvious in the bimodal distributions of size variables such as height at the withers or bone widths and depths corresponding to body mass. However, the influence of sex on the form of bones is not the same in all skeletal elements. In early fusing (long bones) or early ossifying elements (short bones), sexual dimorphism is often less pronounced (Popkin *et al.* 2012: 1786). Nevertheless astragali of ewes can be expected to be smaller in general compared to those of rams. Whether sexual dimorphism also affects shape in the same degree needs to be tested.

After birth the astragalus continues growing in length, breadth, and width in sheep up until *c.* 9 months when the bone is completely ossified (n= 28 foetal to juvenile and 5 adult sheep of a single breed; personal unpublished data). While the growth in length stops at this age, appositional growth in breadth can occur throughout life, changing the shape of the bone in certain parts depending mainly on factors such as weight or strain (Popkin *et al.* 2012). The size and shape of bones are also affected by living conditions including nutrition, mobility, labour or landscape. (Bartosiewicz *et al.* 1997; Barak *et al.* 2011; Popkin *et al.* 2012). For the archaeological specimens most of the relevant information, such as sex and age is lacking. Relative to living conditions, the location of the site and the orography of its catchment area implies a sheep population well-adapted to varied mountainous habitat ranging from rolling hills to steep rugged terrain. For the modern specimens, sex, age and living conditions are known, but the sample includes only few immature animals. Currently more samples from suitable populations, including free-ranging feral and wild sheep with known sex and approximate age, as well as pastured domestic animals with known sex and exact age, are analysed. The factors mentioned above are omitted from this study but will be taken up in future papers.

Archaeological specimens

The remains of wild sheep were recovered from two sites in Upper Mesopotamia in what is now modern Turkey: Göbekli Tepe in the Şanlıurfa Province and Gusir Höyük in the Siirt Province (Fig. 1). Göbekli Tepe is a megalithic site situated on one of the southernmost elevations of the Germuş Mountains overlooking the fertile Harran plain to the south, some 18 km northeast of Şanlıurfa (Peters *et al.* 2014b). Radiocarbon dates suggest construction and use of the complex from the mid-10th until probably the early 9th millennium BC (radiocarbon dates between *c.* 9,500 to 8,000 cal. BC; Dietrich *et al.* 2013: Table 1). Besides the nine astragali of gazelles, a total of 42 sheep astragali was studied (Table 1). For the GM analyses, 36 specimens were available for the dorsal view and 33 for the medial view, with 27 specimens being well-preserved on both views. For the LSR the number of astragali totals 23.

Gusir Höyük is a Pre-Pottery Neolithic tell site on the shore of Lake Gusir, located some 40 km south of Siirt and close to the confluence of the Tigris and Botan rivers (Karul 2011). While the flint inventories and radiocarbon dates suggest occupation mainly during the PPNA there are also indications that the site was inhabited during the initial stage of PPN B (radiocarbon dates between *c.* 9,750 and 8,800 cal. BC; Karul 2011). The site is characterised by round to oval architectural structures mainly representing remains of houses and a large communal building with a stone stela positioned in its centre. This building probably served special purposes. The faunal remains hitherto analysed by us originate from this context as well as a large midden located at the edge of the inhabited area. From the total assemblage of 19 specimens, 17 astragali were available for GM analysis of the dorsal view, 16 for medial, and 14 for both (Table 1). For LSR 17 specimens could be employed.

Table 2. Definition of landmarks on the dorsal view.

No.	Definition
1	Most proximal point of the lateral ridge of the *trochlea*
2	Deepest point in the *sulcus* between the two ridges of the trochlea
3	Most proximal point of the medial ridge of the *trochlea*
4	Intersection of the lateral margin with the outward inflected medial ridge of the *trochlea*
5	Outermost point of the medial cone
6	Outermost medial point of the articular surface of the *caput*
7	Deepest, most proximal point in the depression of the *caput*
8	Proximal end of the distal astragalar facet
9	Lateral end point of the horizontal muscle scar
10	Junction of the horizontal and the diagonal muscle scars
11	End point of the lateral extension

Table 3. Definition of landmarks on the medial view.

No.	Definition
1	Point where the proximo-plantar protuberance of the *trochlea* protrudes the proximal curve of the medial ridge of the *trochlea*
2	Dorsal most point of the medial cone
3	Deepest point below the medial cone
4	Intersection of the plantar part of the *facies articularis* of the *caput* and the *facies articularis calcanea* of the plantar view
5	Deepest point below the posterior protuberance of the *trochlea*

Modern reference collection

Founded in 1865, breeds of many kinds of livestock species and their wild ancestors were kept and bred at the 'Haustiergarten' (garden of livestock species) of the Institute of Agricultural Sciences at the Martin-Luther University Halle-Wittenberg in Germany (Kühn 1888). The macerated skeletons of slaughtered (or deceased) animals were systematically collected and are housed in the 'Museum of Domesticated Animals "Julius Kühn"' (Schafberg and Wussow 2010). From the many sheep breeds and types present in this collection, Karakul sheep, a fat-tailed breed, was chosen since it represents numerically one of the largest pure-bred populations in the collection. A population of domestic sheep originating from the Near East seems probably more appropriate since it comes from the same environment once inhabited by the wild sheep studied here. However, such specimens were not available in sufficient numbers. Moreover, shape differences should become visible when comparing modern Karakul sheep to wild sheep if living in a anthropogenic environment is the driving factor. A total of 41 astragali were recorded. In contrast to the other collections, for all these specimens (21 female and 20 male animals; 5 sub-adult and 36 adult animals) living conditions, sex and age at death are known.

Methods

For each of the methods described above, two runs were carried out. The first combined gazelle and sheep data while the second focused on intra-species comparison of sheep. If Geometric Morphometric and Log Shape Ratio are not able to separate the astragali of gazelle from those of sheep, this would nullify the second run.

Geometric Morphometrics (GM)

All specimens were positioned in a sand-filled box with the surface horizontal and parallel to the lens of the camera, controlled by a bubble level. 2D digital photographs were taken using a light box and a reflex camera (Nikon D300) coupled with a micro lens (AF-S Micro Nikkor 60mm) and attached to the desk via an articulating arm. Three images with differing lighting were taken of the dorsal, medial, plantar, and lateral views of the astragalus.

The coordinates of landmarks and sliding semi-landmarks were measured using TpsDig2 v2.16 (Rohlf 2010). First we chose landmarks on each view which would capture the shape of the particular side best by keeping their number at a minimum. The astragalus acts as a double hinge joint between the tibia and the os centrotarsale (cubonavicular) with rather tight fittings of the joints resulting in restricted flexibility in artiodactyls. If changes in living conditions affect the morphology of the astragalus it is expected to be visible in the shape of the proximal and distal articulations in particular. Curves of sliding semi-landmarks were recorded to capture the shape of these articulations. We then tested repeatability of the landmarks and curves

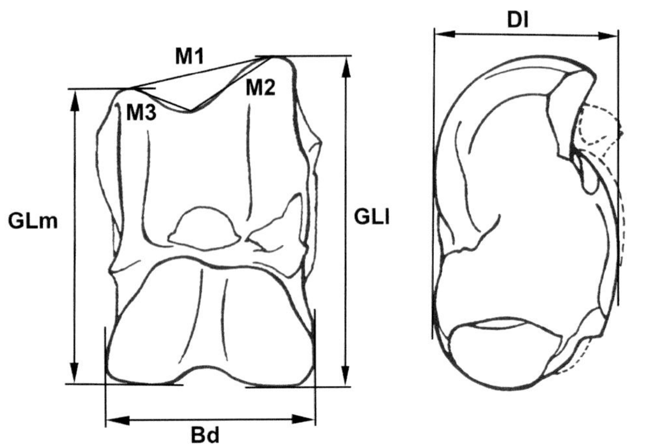

Fig. 3. Astragalus of sheep. Traditional measurements following von den Driesch (1976): GLl - Greatest Length lateral; GLm - Greatest Length medial; Dl _ Depth lateral; Bd - Breadth distal. Additional measurements M1, M2 and M3 defined in the text (illustration adapted from von den Driesch 1976).

with sliding semi-landmarks using one-way ANOVA for repeated measurements on a subsample of 41 specimens. This allowed excluding those landmarks or complete configurations with low repeatability rates (see the overview given in Fruciano 2016). This procedure resulted in a landmark configuration for the dorsal and medial view each, with eleven landmarks for the first and five for the latter (Tables 2 and 3). With respect to the dorsal view, one curve comprising 14 sliding semi-landmarks between landmarks (1) and (3) as well as a second curve of 18 sliding semi-landmarks between landmarks (6) and (8) were recorded to capture the shape of the proximal and distal articulation (Fig. 2a). For the medial view, the coordinates of 18 sliding semi-landmarks on a curve between landmarks (1) and (2) were measured to detect possible differences in the dorsoplantar direction of the bone (Fig. 2b).

The landmarks and sliding semi-landmarks were subjected to a Generalized Procrustes Analysis (GPA), which translates all specimens to the centroid of the landmark configurations, rotates them in order to minimize the distances between corresponding landmarks and scales all specimens to the unit centroid size (Rohlf and Slice 1990; Bookstein 1991). During this procedure, the semi-landmarks slide relative to each other so that the sum of Procrustes distances between each specimen and the mean shape is minimized (Bookstein 1997). These procedures were performed using the functions of the package 'geomorph' (version 3.0.1; Adams and Otarola-Castillo 2013) of R (version 3.2.4 revised; R Core Team 2016).

For the first run regarding the separation of gazelle from sheep, we considered only the complete configuration of the dorsal view of a smaller subsample. During the second run dealing with sheep alone all configurations were tested for each view, i.e. for the dorsal view landmarks and both curves, landmarks with curve 1 or curve 2, landmarks only, each curve separately, both curves combined; for the medial view we analysed landmarks and curve together as well as separately.

Log Shape Ratio (LSR)

The Log Shape Ratio (LSR) as proposed by Mosiman (1970) calculates size-free shape variables by scaling the original distance measurements. For the scaling factor there are different options such as the square root of the sum of squared distances, the arithmetic or the geometric mean (Claude 2008: 98-99). Whereas Vigne (2015: 135) proposed the sum of distances we followed Claude (2008) by considering the geometric mean of the measurements of each specimen as scaling factor:

$$g(x) = \prod_{i=1}^{p} x_i^{1/p} \quad ,(1)$$

with x being the vector of measurements and p the number of measurements. Dividing every measurement of a specimen by its geometric mean results in the so-called shape ratios:

$$z = \frac{1}{g(x)} x \quad ,(2)$$

with x being the vector of linear measurements, z the shape vector and $g(x)$ the geometric mean as size vector in this case. Finally, the logarithmic values of these shape ratios will be used as new variables (see also Claude 2008: 2013).

For the LSR a set of seven linear measurements was recorded: GLl, GLm, DL, Bd, M1, M2, M3 (Fig. 3). The first four are the traditional measurements as proposed by Von den Driesch (1976). The latter three were defined based on the GM analyses illustrating that the shape of the proximal end is well-suited for separating the different sheep populations (see below): M1 is the distance between the most proximal (highest) points of the lateral and the medial ridge of the *trochlea*; M2 refers to the distance between the most proximal point on the lateral ridge of the *trochlea* and the deepest point of the furrow (*sulcus*) between the ridge of the *trochlea*, and M3 represents the distance between the most proximal point of the medial ridge of the *trochlea* and the deepest point of the furrow (*sulcus*). These measurements were obtained using callipers.

As in GM, the first run considered only a subsample. Provided that this approach separated gazelle from sheep reasonably well, LSR would be applied to the entire sheep dataset. We decided to use a

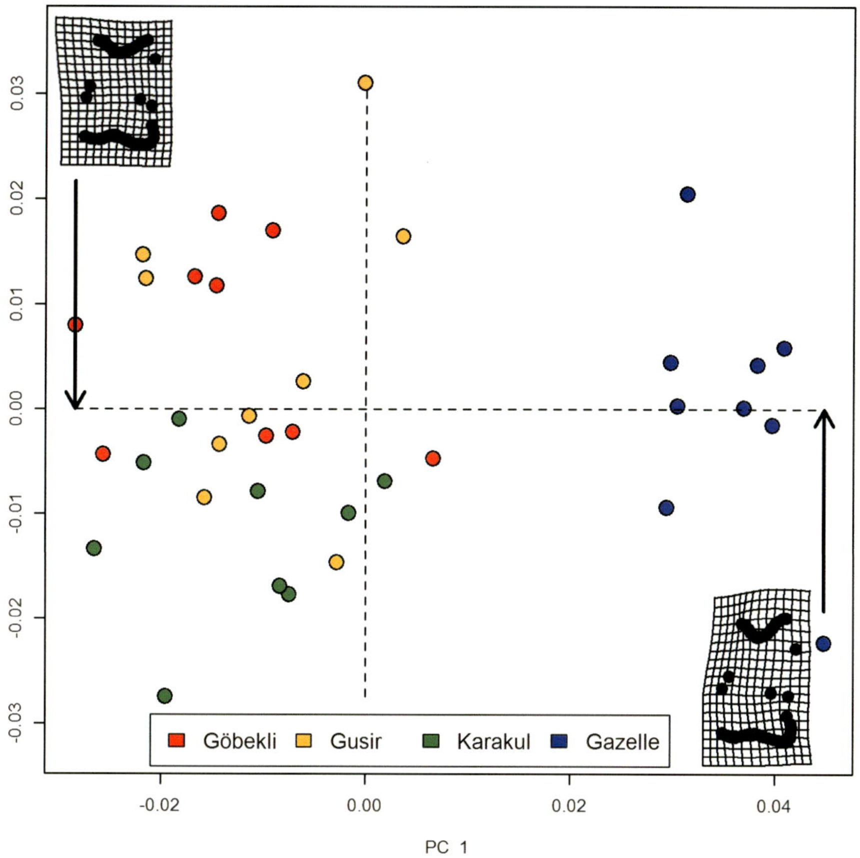

Fig. 4. Principal component analysis, shape differences between gazelle and sheep on the principal component axis I (PCI) and 2 (PC2). Geometric Morphometrics approach with thin plate spine deformation grids showing the maximum transformation on PCI.

two-step procedure, (1) to verify whether the set of the four 'Von den Driesch' measurements suffice to discern shape differences between the wild and domestic sheep populations, followed by step (2) during which the complete dataset including the 'new' measurements were tested.

Measurement error (ME)

Exploring measurement error (ME) is vital in morphometric studies in general and in intra-specific studies in particular, since high rates of measurement error might produce 'noise', which blurs existing subtle differences between groups (Claude 2008; see also summary and review by Fruciano 2016). In archaeozoology, measurement error procedures are almost never published and discussed and probably also very rarely performed at all, while the awareness of this issue is somewhat higher in researchers applying GM to archaeozoological materials. In this contribution, measurement error for both approaches has been examined. For LSR a model II ANOVA (one-way ANOVA on repeated measurements), which compares the variance between and within groups based on two sessions, was used to estimate the ME (Bailey and Byrnes 1990, Yezerinac *et al.* 1992; Claude 2008: 65-66). For GM, repeatability (or measurement error) was tested using Procrustes ANOVA as proposed by Klingenberg and McIntyre (1998) and based on two digitisation sessions as well.

Statistical analyses

Size variation of the populations was tested with pairwise ANOVA using post-hoc Scheffé's test due to the fact that sample sizes between the three sheep populations considered (Göbekli Tepe, Gusir Höyük, Karakul) are unequal. Size is visualised by boxplots (1) of the log of geometric mean of linear measurements and (2) of the log of centroid size of the medial and the dorsal views.

In applying both procedures, the calculation of log shape ratios as well as Procrustes superimposition, data become redundant. A principal component analysis (PCA) was therefore applied to the new variables to explore the variation present in the datasets and to extract meaningful principal components for further statistical analyses. For GM differences in shape are visualised with thin-plate spline deformation grids. Differences between populations are tested using pairwise ANOVA and permutation tests.

For GM a partial least squares regression (PLS) was performed on the Procrustes-aligned coordinates of the dorsal and the medial views. Since several of the astragali from the archaeological sites are damaged on one view or the other only a reduced dataset of 82 specimens with both views well-preserved was available for this procedure (see Table 1). Partial least square regression explores the covariation of two blocks of variables and quantifies the degree of association (for details see Rohlf and Corti 2000). Sample size (i.e. number of specimens) is a key issue in partial least square regression

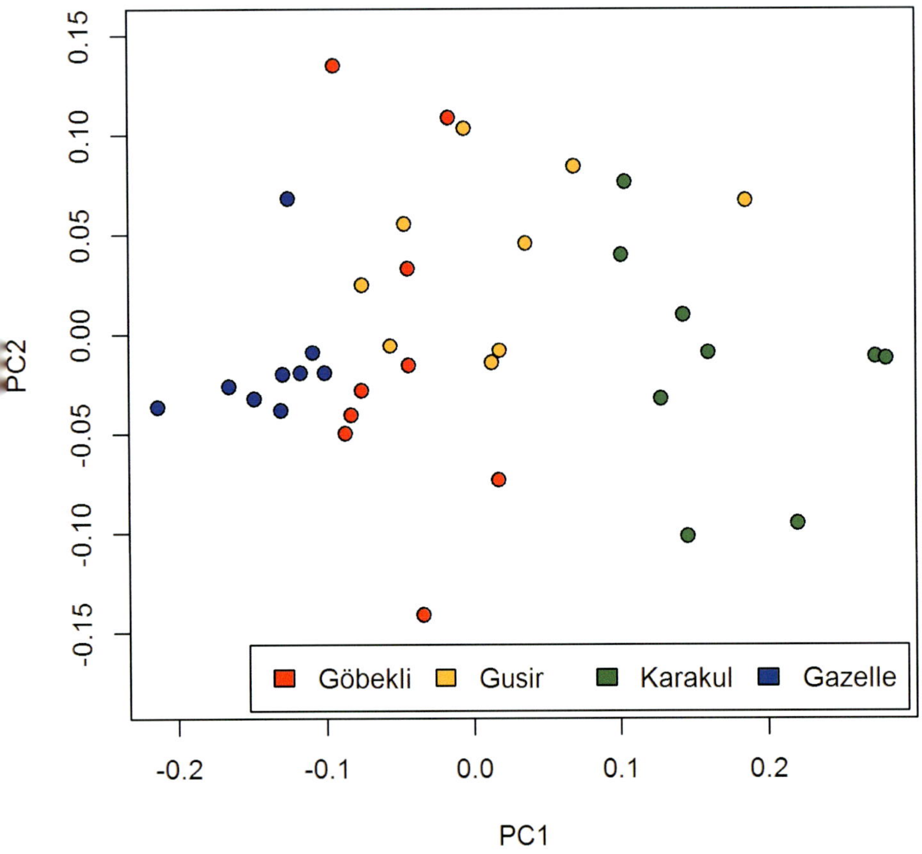

Fig. 5. Principal component analysis, shape differences between gazelle and sheep on the principal component axis I (PCI) and 2 (PC2). Log Shape Ratio approach.

(Mitteroecker and Bookstein 2007). To reduce the number of variables, the configurations with the best separation in the PCA for both views, in our case the landmark configuration and curve 1 for the dorsal view and the landmarks for the medial view were chosen for this type of analysis. Since each 2D landmark is defined by two coordinates the number of variables is twice the number of landmarks. Landmarks and curve 1 of the dorsal view contribute 50 variables (11 landmarks and 14 sliding semi-landmarks) in block 1 of the partial least square analysis, while the 5 landmarks of the medial view in block 2 provide 10 variables. The sample size is 82 (Table 1) and therefore only slightly larger than the number of variables. Following Klingenberg (2013) sample size can be considered sufficient for the partial least square analysis if the number of principal components accounting for most of the total shape variance is decidedly smaller than the sample size. In our case 95% of the total shape variation is contained in the first five principal components, which is clearly smaller than the sample size of 82.

Differences between sheep populations (Gusir Höyük, Göbekli Tepe, Karakul) and between groups (wild or Karakul) were tested using pairwise Procrustes ANOVAs with permutation tests for significance. If both wild populations show no significant differences in the Procrustes ANOVA, they are merged for further analyses. Shape changes are visualised in comparison to the mean shape for each view. Linear discriminant analysis is used to describe the differences in shape between pre-defined groups. In the case of GM, the number of principal components to be used in linear discriminant analysis was determined by stepwise including more principal components until the highest accuracy in the leave-one-out cross-validation was reached following Evin *et al.* (2013). For this approach, the same reduced dataset with matched pairs for both views as for partial least square analysis was chosen.

Turning to LSR, linear discriminant analysis was performed on the complete LSR dataset. Its reliability was tested with leave-one-out cross-validation (1000 permutations). Statistical analyses were carried out using basic functions of R and the packages 'geomorph', 'mass' and 'asbio' (Aho 2014) in R (version 3.2.4 revised; R Core Team 2016).

Results

Measurement error (ME)

The measurement error for GM analysis of the dorsal view of the astragalus was 9% and for the medial view 4%. This suggests that ME is small enough to produce meaningful results. Nonetheless, a ME just below 10% admonishes to not over-interpret results. For LSR, measurement error was 0.2% and therefore decidedly smaller than in GM. These results suggest there is a low risk of over-interpretation due to measurement error.

Shape differences between gazelle and sheep

As explained above, this run served to test whether the two methods (GM, LSR) allow the separation of

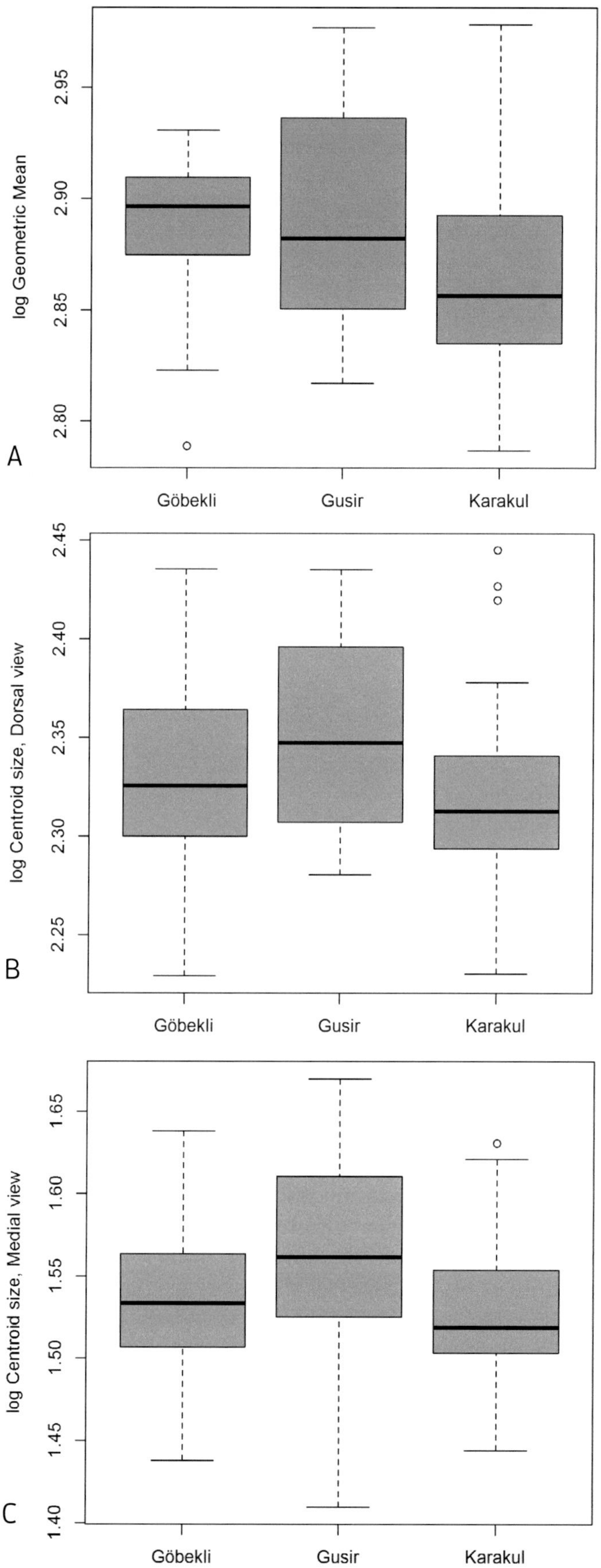

Fig. 6. Boxplots of (a) the log-transformed geometric mean (of LSR), (b) the log-transformed centroid size of the dorsal view (GM), (c) the log-transformed centroid size of the medial view (GM) for the three populations.

the astragali of the two species. The graphs for the first two principal component axes show that this is indeed the case (Figs. 4 and 5). For both GM and LSR the differences between sheep and gazelle are significant ($p < 0.001$ for LSR and GM). Therefore it seemed reasonable to explore the possible differences between Karakul and wild sheep astragali in a second run.

Variation in size and shape between the different populations of sheep

The ANOVA statistics for centroid size of both views show no significant differences between the means of the three populations (dorsal $p= 0.0545$; medial $p= 0.0789$). For the Geometric Mean, however, the ANOVA test detected a small but significant difference between the sheep populations at the 0.05 level ($p= 0.0375$). Scheffé's test, however, revealed that none of the differences between pairs of populations are significant (Göbekli-Gusir $p=0.94928$; Göbekli-Karakul $p=0.124612$; Gusir-Karakul $p=0.090542$). The boxplots show that all three populations largely overlap (Figs. 6a-c).

Turning to the Geometric Morphometric data, Procrustes ANOVA analyses found highly significant differences between the sheep populations for the complete configurations of both views (dorsal and medial $p<0.001$). Separate analyses for each configuration also revealed significant differences. Since the curve on the medial view did not differ significantly ($p=0.32$), we concluded that this part of the bone is not suitable for differentiating between the populations and it was excluded from further analyses. Pairwise tests consistently revealed that there were no significant differences between the wild sheep populations from Göbekli Tepe and Gusir Höyük, while there were always significant differences between the first two and the population of Karakul sheep (Table 4a). This suggests that for further statistical analyses of the GM data, the wild sheep populations of the archaeological sites can be merged. As such, we consider this assemblage representative of the wild sheep confined to early Neolithic Southeast Anatolia. In this respect, curve 2 of the dorsal view is an exception since the test was significant for the pairing of Göbekli and Karakul but not for Gusir and Karakul. We therefore excluded curve 2 from further considerations.

Based on the Procrustes ANOVA the following configurations were subjected to the principal component analysis (PCA): for the dorsal view the landmark configuration together with curve 1 and for the medial view the landmark configuration only. The variation in the dataset is displayed by the scatterplot for the first two principal component axes, which account for 54.7% of the total variation for the dorsal view and for 72.5% for the medial view

Table 4. Pairwise comparison of the groups for each ANOVA. Significant results are in bold.

a) Geometric Morphometrics

Dorsal

	Landmarks + Curves			**Landmarks**			**Curve 1**			**Curve 2**		
	Göbekli	Gusir	Karakul	Göbekli	Gusir	Karakul	Göbekli	Gusir	Karakul	Göbekli	Gusir	Karakul
Göbekli	1			1			1			1		
Gusir	0.431	1		0.243	1		0.705	1		0.378	1	
Karakul	**0.001**	**0.001**	1	**0.001**	**0.001**	1	**0.001**	**0.001**	1	**0.029**	0.144	1

Medial

	Landmarks + Curve			**Landmarks**			**Curve**		
	Göbekli	Gusir	Karakul	Göbekli	Gusir	Karakul	Göbekli	Gusir	Karakul
Göbekli	1			1			1	0.683	0.231
Gusir	0.353	1		0.219	1		0.683	1	0.233
Karakul	**0.001**	**0.001**	1	**0.001**	**0.001**	1	0.231	0.233	1

b) Log Shape Ratio

	Traditional measurements			**Complete dataset**		
	Göbekli	Gusir	Karakul	Göbekli	Gusir	Karakul
Göbekli	1			1		
Gusir	0.406	1		0.422	1	
Karakul	**0.001**	**0.001**	1	**0.001**	**0.001**	1

(Fig. 7 and 8). Both graphs illustrate that wild sheep from Gusir Höyük and Göbekli Tepe separate well from Karakul sheep. The differences between the mean shape of wild and the mean shape of Karakul sheep astragali are visualised in Fig. 9 (dorsal) and Fig. 10 (medial). The most striking differences are located at the proximal end in the dorsal view, where the curve of sliding semi-landmarks is rather shallow and asymmetrical with a marked bump on the lateral slope in wild sheep, while the typical astragalus of Karakul sheep tends to have a deeper and more symmetrical curve either with a less noticeable or no bump at all. Seen dorsally, shape differences are also visible on the muscle scars on the lateral ridge of the *trochlea*. The junction of both muscle scars tends to be more pronounced and situated more to the middle of the bone in wild sheep as compared to Karakul sheep. In the medial view, the change in shape is best visible in the position of landmarks 4 and 5 in relation to landmarks 3 and 1: astragali of wild sheep are much deeper and appear to be of more compact built as compared to the more elongated and slender astragali of the domestic Karakul sheep.

The partial least square analysis for block 1, i.e., landmark configuration and curve 1 of the dorsal view, and for block 2, i.e. the landmark configuration of the medial view, shows a significant covariation between both blocks (correlation coefficient r=0.568, $p<0.001$). The first two axes of the partial least square analysis account for 93.3% of the shared variance (Fig. 11). According to the results of the leave-one-out cross-validation procedure, linear discriminant analysis correctly identifies 97.6% of the specimens (Fig. 12).

The LSR analysis of the traditional linear measurements GLl, GLm, Dl, and Bd as defined by Von den Driesch (1976) shows some separation on the first two principal component axes (Fig.13a). This is reflected by the ANOVA statistic, which reports highly significant differences between the two SE Anatolian populations on the one hand and the Karakul sheep on the other, but no significant difference between the sheep from Gusir Höyük and Göbekli Tepe (Table 4b). The first two principal component axes account for 92.6% of the total variation. The leave-one-out cross-validation, however, reveals that the linear discriminant analysis only assigns 83.8% correctly implying several cases of misclassification mainly in the merged group of wild sheep (Fig. 14a).

Since in the GM analyses the proximal end exhibited the most considerable/clearest shape differences, three new linear measurements were defined to document the particular shape of the proximal part of the *trochlea tali*. They were included in the LSR dataset together with the traditional measurements. Applying principal component analysis on the enlarged LSR dataset shows a much better separation compared to the principal component analysis with the LSR based on the dataset of traditional measurements. Moreover, the separation turned out similarly well in comparison to those emerging from the principal component analyses of the GM dataset. While the Göbekli Tepe and Gusir Höyük populations cover approximately the same range,

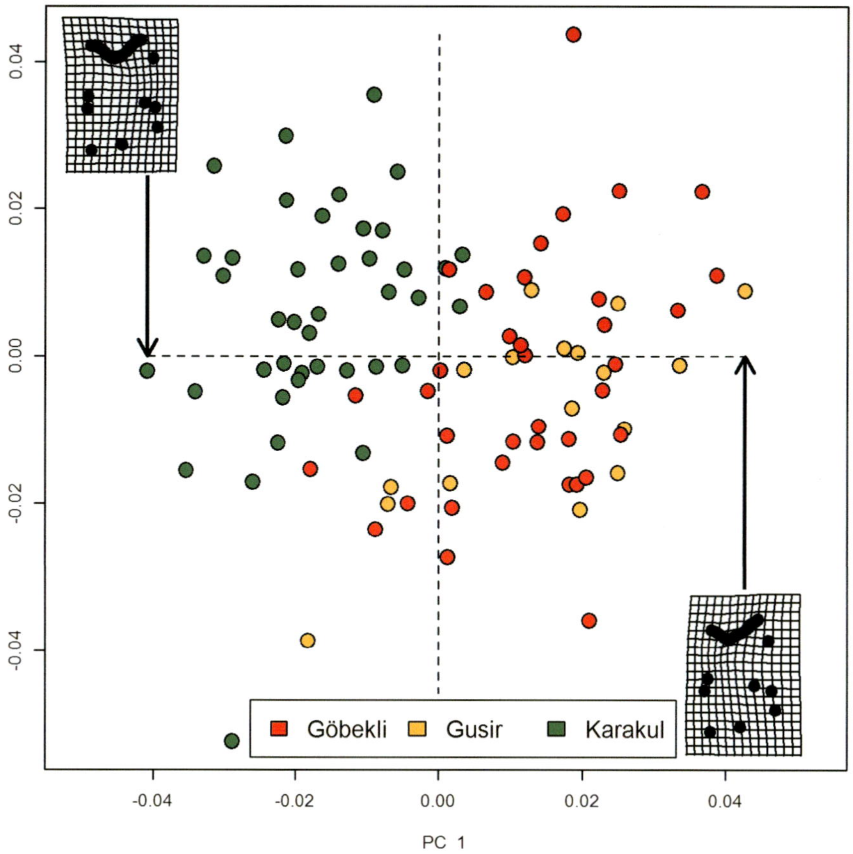

Fig. 7. Geometric Morphometrics. Principal component analysis of the dorsal view illustrating shape differences between the sheep populations from Göbekli Tepe, Gusir Höyük and the Karakul breed on principal component axis I (PCI) and principal component axis 2 (PC2) with thin-plate spine deformation grids showing the maximum transformations on principal component axis I.

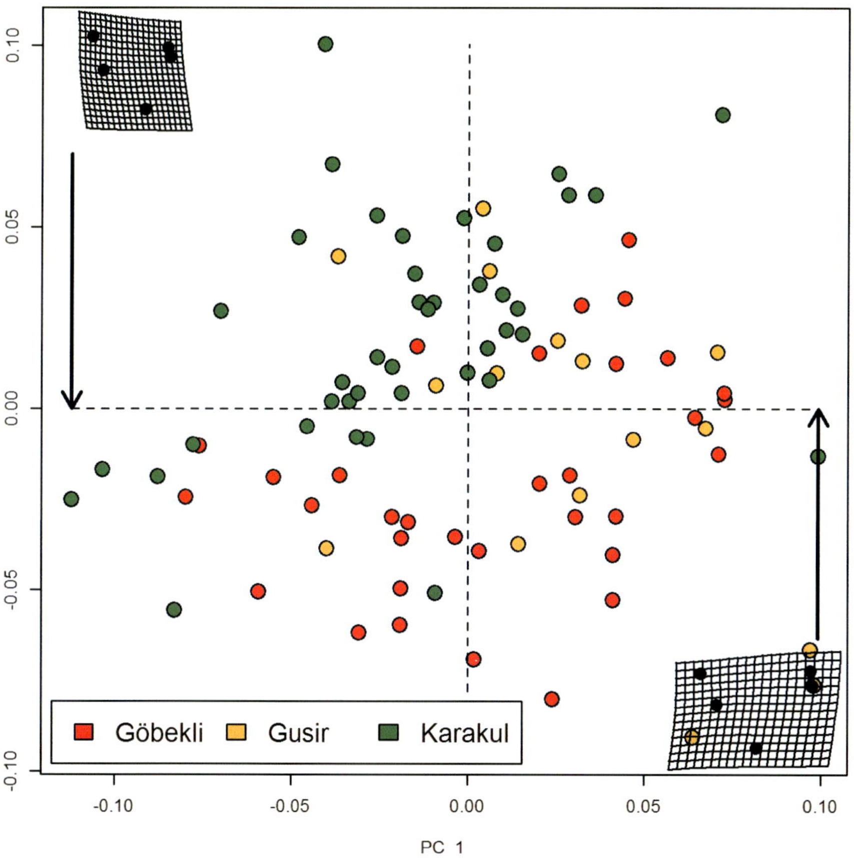

Fig. 8. Geometric Morphometrics. Principal component analysis of the medial view illustrating shape differences between the sheep populations from Göbekli Tepe, Gusir Höyük and the Karakul breed on principal component axis I (PCI) and principal component axis 2 (PC2) with thin-plate spine deformation grids showing the maximum transformations on principal component axis I (PCI).

the data points of the Karakul sheep position differently, separating them quite well from the archaeological populations (Fig. 13b). Overlap between the three populations is present, but small. The first two principal component axes account for 91.3% of the total variation. Differences between populations are significant based on the ANOVA statistics ($p<0.001$). Again, the pairwise comparisons revealed that the sheep from Göbekli Tepe and Gusir Höyük are obviously similar enough to be considered statistically a

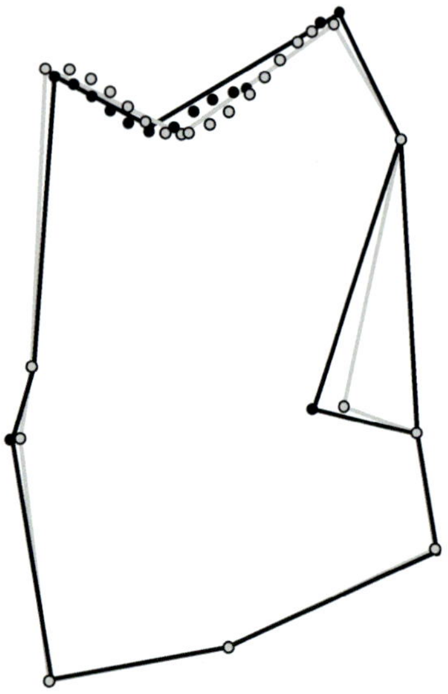

Fig. 9. Geometric Morphometrics. Dorsal view. Differences between the mean shape of wild sheep (black) and the mean shape of Karakul sheep (light grey), 2 x magnification.

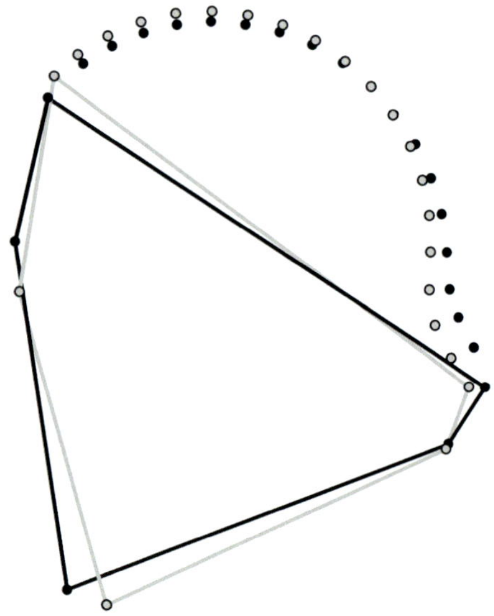

Fig. 10. Geometric Morphometrics. Medial view. Differences between the mean shape of wild sheep (black) and the mean shape of Karakul sheep (light grey), 2 x magnification.

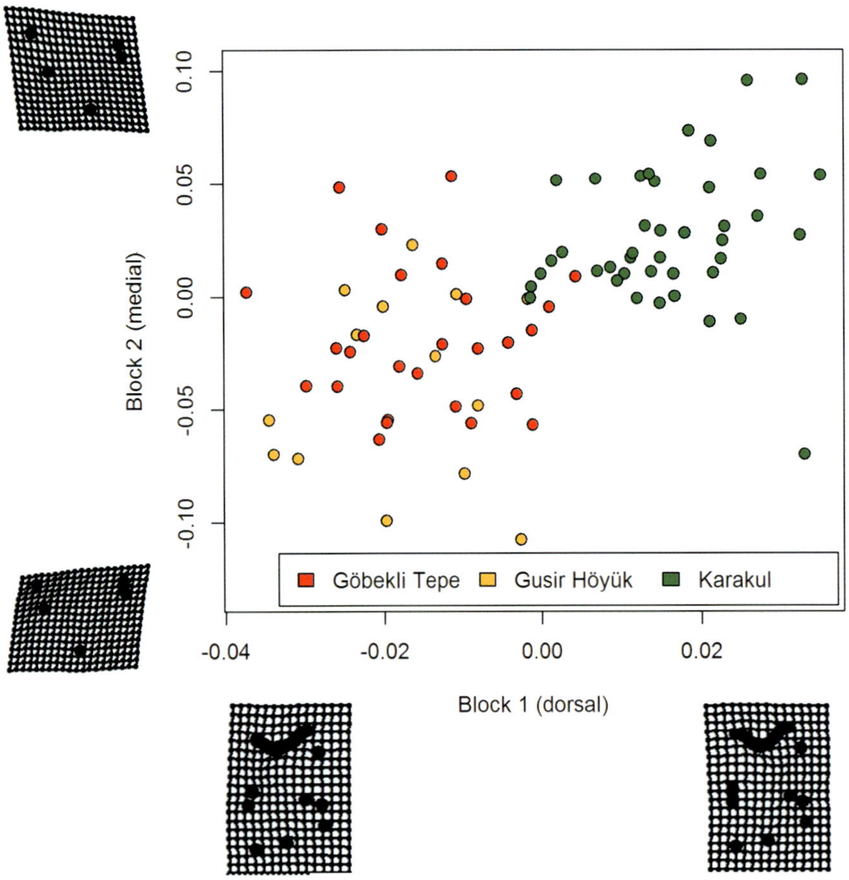

Fig. 11. Geometric Morphometrics. Partial least square analysis for the dorsal view as block 1 (landmarks and curve 1) and the medial view as block 2 (landmarks) with thin-plate spline transformation grids illustrating the maximum deformation for each end of each axis.

single population while the Karakul sheep differ significantly (Table 4b). Finally, for the following linear discriminant analysis the datasets of the sheep populations from Göbekli Tepe and Gusir Höyük were merged to form the Southeast Anatolian wild sheep population. The leave-one-out cross-validation showed that linear discriminant analysis is able to correctly assign 97.5% of cases to their respective group (Fig. 14b).

Discussion and concluding remarks

Detecting shape differences through GM and LSR

The first aim of this study was to test whether geometric morphometric and traditional morphometric techniques could separate groups of animals based on postcranial skeletal elements, in this case the astragalus. The results of the first run contrasting

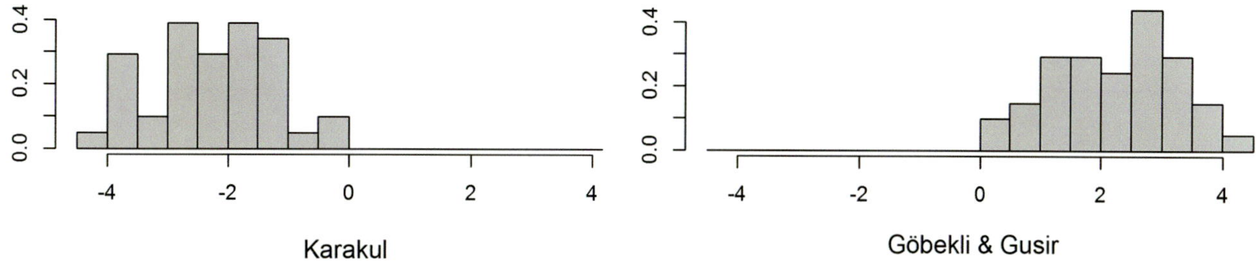

Fig. 12. Geometric Morphometrics. Histograms showing the values of the discriminant scores for the leave-one-out cross-validation.

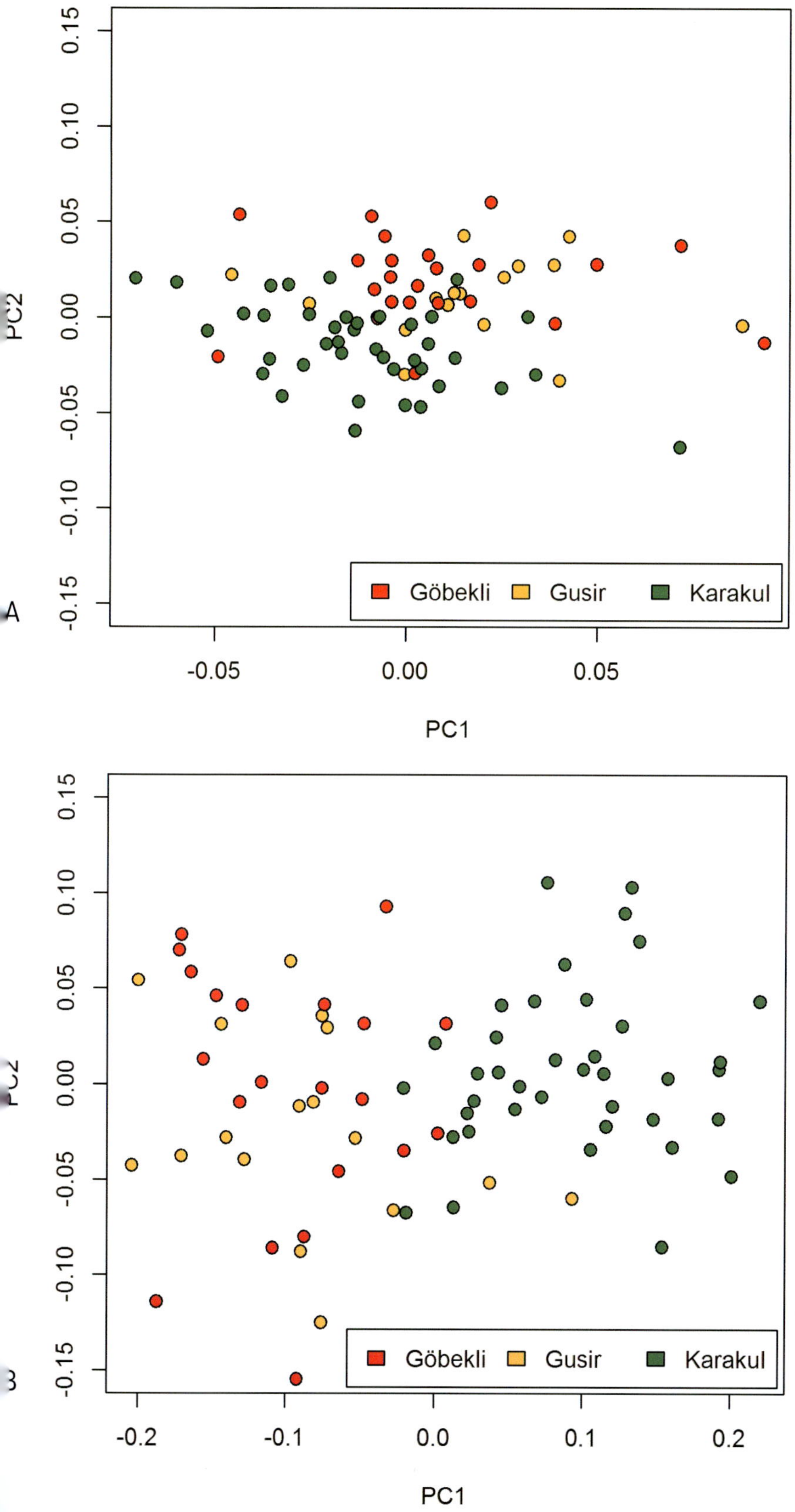

Fig. 13. Log Shape Ratio. Principal component analysis (a) for the datasets comprising the traditional measurements only and (b) for the complete dataset with traditional and additional linear measurements.

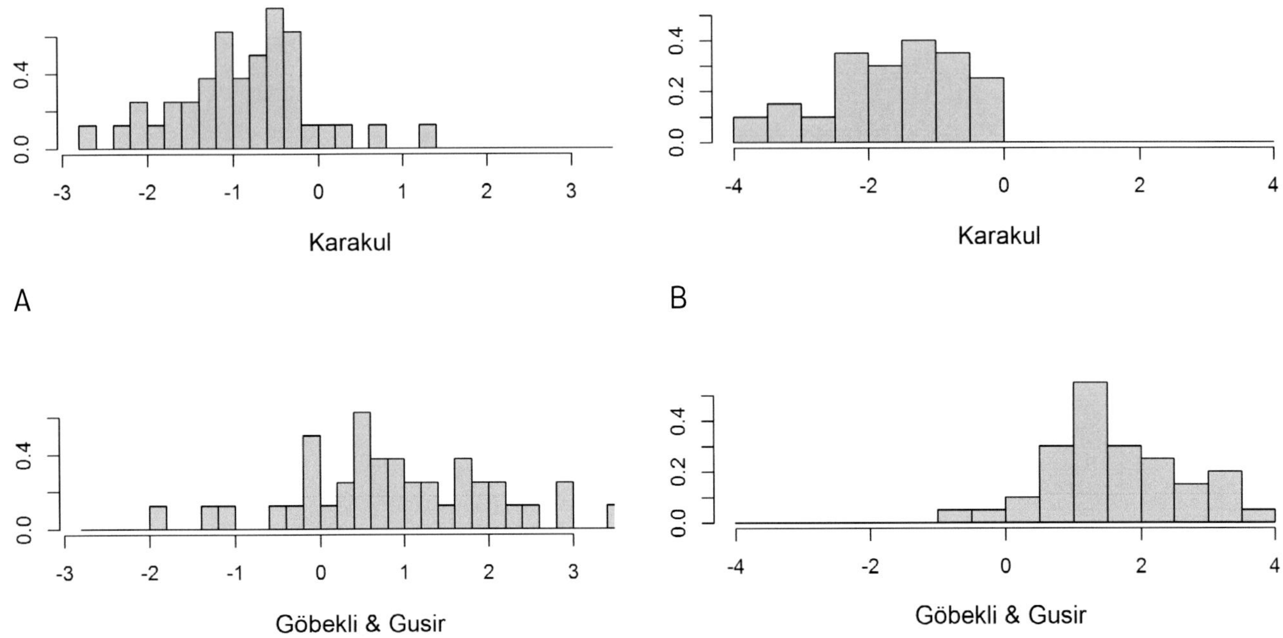

Fig. 14. Log Shape Ratio. Histograms showing the values of the discriminant scores for the leave-one-out cross-validation (a) for the datasets containing the traditional measurements only and (b) with the complete dataset with traditional and additional linear measurements.

gazelle and sheep show that both methods indeed differentiate reliably between the two species. The second run, restricted to the three populations of sheep, addressed a more challenging and delicate task: the separation of different populations within a single species. The results for the GM analyses show that this method is suitable for achieving this goal, considering the good separation between the Southeast Anatolian wild sheep populations and the Karakul breed. We also noted during the study that traditional measurements as given by Von den Driesch (1976) do *not* provide sufficient detail to allow for a reliable classification in the linear discriminant analysis. Since the GM identified the shape of the proximal part of the astragalus, more precisely the *sulcus* between the two ridges of the *trochlea tali,* as notably suitable for the differentiation between sheep populations, three new measurements were defined on this section of the astragalus and are included in the datasets of linear measurements. Using this enlarged dataset Linear Discriminant Analysis on the LRS data proved equally successful as GM. This suggests that in future research, whether GM can also be used independently to classify other postcranial elements, ideally with a focus on identifying regions where the separation of groups of specimens is particularly successful, should be tested so that additional linear measurements can be defined to document these. Taking linear measurements is much easier to accomplish in the field where time-consuming and labour-intensive techniques are difficult to apply. The approach taken here thus intends to encourage more researchers to address similar research questions with these methods. The nuances of shape changes, however, cannot be visualised in such detail using LSR analyses as with GM methods. GM can therefore be seen as a complementary technique which helps to improve traditional morphometric methods, thereby allowing a more detailed insight into biological processes which cannot be enlightened by the latter (Vigne 2015).

Shape differences between two populations of sheep

While there are no significant size differences between the means of the three sheep populations, the analyses of shape revealed that both methods allow for the differentiation of the wild sheep populations from SE Anatolia from modern Karakul sheep. The wild sheep hunted for food near Göbekli Tepe in the Euphrates basin and Gusir Höyük in the Tigris basin seem to originate from wild *Ovis* populations that do not exhibit differences in the shape of astragali. Whether this conclusion holds true for other skeletal elements must be resolved by future research. At the moment we assume that they are similar enough to be pooled into a single population – at least from an osteomorphometrical perspective that can be compared with other populations of sheep.

As such, both methods confirm statistically highly significant differences between the prehistoric SE Anatolian wild sheep and the modern Karakul sheep breed. The shape differences between the two sheep populations are best illustrated by comparing the mean shapes of both populations (Figs. 9 and 10). The wider *sulcus* on the proximal articulation in wild sheep results in a less tight interlocking of the tibio-talar articulation rendering this joint more flexible as compared to a deeper *sulcus*, which results in tighter interlocking. The latter feature is considered more characteristic for cursorial animals (Hildebrand and Goslow 2001). In contrast, a less tight interlocking allows the tibio-talar joint greater flexibility in medio-lateral movements (Knigge et al. 2015), which is favourable for animals that occupy mountainous habitats and need to climb and jump, a locomotor behaviour seen in wild sheep (and goat) (Groves and Leslie 2011). Muscle scars are pronounced in animals whose muscles are heavily strained, for example, by climbing, jumping, or fast movements in rugged terrain. More robust astragali, i.e. those which are deeper on the medial view, allow a greater range of movement in the distal articulation resulting in longer stride length and higher locomotor speed in the movement (Hildebrand and Goslow 2001). The morphological adaptations observed in the astragalus of wild sheep are consistent with their assumed way of life, i.e. constantly in search of pasture in a diverse landscape, necessitating climbing and even running or stampeding when chased by predators. The absence of pronounced muscle scars in Karakul sheep on the other hand agrees well with its late 19th and early 20th century living conditions at Halle, with animals being confined to small compounds on horizontal terrain. Arguably, the deep *sulcus* facilitated a tight interlocking of the astragalus in the *cochlea tibiae*, resulting in diminished joint rotation thereby reducing muscle effort of cursorial animals in motion (Hildebrand and Goslow 2001). This locomotor adaptation would be in agreement with the living conditions of many breeds of domestic sheep pastured by herdsmen in landscapes that lack the typical orographic features characterising the native distribution of *Ovis orientalis* in the Near and Middle East.

Inter- and intraspecific studies of the functional morphology of bones primarily of the hind extremities confirm that and illustrate how morphology is linked to an animal's locomotion and therefore to its habitat. Here we refer to the studies, e.g. of astragalus of domestic sheep (Haruda 2014), wild bovids (DeGusta and Vrba 2003, Plummer *et al.* 2008, Barr 2014), and three gorilla populations (Knigge *et al.* 2015) as well as the calcaneus, femur and tibia of cervids (Curran 2012). While closed-habitat bovids tend to have more elongated and slender astragali, species inhabiting open country possess shorter, more compact astragali, a feature related to their flight behaviour (DeGusta and Vrba 2003; Barr 2014). The influence of the environment on astragalar shape becomes also visible in different populations of a single species as shown in gorillas (Knigge *et al.* 2015) and domestic sheep from archaeological sites (Haruda 2014). In both studies the habitat of each population determines its locomotor behaviour and therefore the shape of the astragalus. Based on a sample from three sites totalling 38 specimens, the study on Late Bronze Age sheep astragali from Central Asia shows a clear separation of sheep populations kept in a mountainous environment from those living in plain and steppe habitats (Haruda 2014: 262). Haruda's interpretations regarding the link between morphology and environment are similar to the results presented here. Other factors affecting the shape of a bone such as sex and nutrition were considered ineffective for the astragalus mainly based on the results of the controlled feeding experiment conducted by Popkin *et al.* (2012), concluding that differences in astragalus morphology are (eco)phenotypic (Haruda 2014: 265). However, whether phenotypic variation is indeed the only effective variable once size is removed from the astragalus dataset analyses (Haruda 2014: 24, 260) needs to be tested further in our opinion.

On the other hand, experiments of joint loadings in (young) sheep indicate that different housing conditions or husbandry systems directly result in the adjustment of the trabecular bone architecture, which to some extent also changes the external shape of the bone (Barak *et al.* 2011). For the astragalus the changes are expected to appear in the parasagittal plane (Barak *et al.* 2011), which might correspond to the differences on the proximal articulation found in this study. Therefore a direct adaptive response of the bones to the drastic changes imposed by the living conditions of the anthropogenic environment has to be considered as well.

Controlled experiments with livestock species, however, are rare but would be necessary to evaluate the response of bones to different living conditions. The causes for the differences between prehistoric wild sheep and modern Karakul sheep are therefore difficult to determine: Are we dealing with a plastic response to the changes imposed by living in an anthropogenic environment or with phenotypes resulting from genetic selection towards less explorative and agile sheep that were easier to handle due to locomotor adaptation (here: slender shape of the astragalus)? Phenotypic plasticity is rarely discussed in archaeozoological contexts related to the domestication process, which is in clear contrast to similar research into plant domestication (see for example Larson *et al.* 2014, Zeder 2016). One reason for this is that in plants, plastic responses to the environment are easier to detect and test in experiments. Due to the comparably long generation time in livestock species, comparable experiments are inexpedient and impractical. Experiments with mice, however, proved that selective breeding for high runner mice eventually led to larger joint surfaces and thicker shafts, which are already present in very young mice suggesting that these characteristics are genetically fixed after some 20 generations of selective breeding, while there was also a plastic response in shaft thickness in the same direction (Garland and Kelly 2006; Wallace and Garland 2016). Finally, several experiments and studies showed that epigenetic modifications leading to phenotypic changes can be inherited by the next generations (Jensen 2015). The question as to whether these epigenetically fixed changes can influence the shape of bones has to our knowledge not been included in GM studies on domestication until now. In sum, whether the morphological divergence observed between the sheep populations studied can be explained by genetic selection towards animals with a reduced escape reflex, whether epigenetic effects are involved or whether this is due to phenotypic plasticity as a response to conditions of housing and/or husbandry cannot be answered with the dataset at hand. Although human management and subsequent domestication likely account for the morphological differences observed, it is clearly too early to present this as the definitive and only cause. Other factors that might influence shape, such as sex, age, allometry, geographic origin, habitat etc., need to be investigated with appropriate datasets. Furthermore it needs to be confirmed if the divergence between SE Anatolian wild sheep and the Karakul breed can be generalised, i.e. if other domestic sheep breeds or types show the same kind and degree of morphological adaptation. The genetic history of sheep is still poorly understood, in part due to the fact that ancient DNA in bones from arid regions like SE Anatolia is too degenerated to provide results (Bollognino and Vigne 2007). It is, for instance, still unclear whether or to what extent the SE Anatolian wild and early SE Anatolian domestic sheep were included in lineages composing populations that were later introduced to Europe or which lineages gave way to fat-tailed or the wool sheep. In absence of ancient DNA analyses the LSR and GM approaches presented here can help to illuminate this issue with additional datasets, which are currently being analysed.

In close cooperation with the Institute for Computer Sciences at the LMU Munich, a computer supported 3D-shape analytics approach is currently being developed that uses data-mining techniques in order to semi-automatically identify significant features in modern and ancient populations of wild and domestic sheep in order to enhance the possibilities of differentiating between the two. With this informatics tool we hope to identify additional features of the astragalus as well as in other skeletal elements, where intraspecific differences seem manifested but where the GM approach using 2D imagery either fails to define secure landmarks or because the structures are in positions not captured with the 2D images.

Finally, if ongoing work on larger and contemporary datasets shows that GM and/or LSR allow the separation of wild and domestic sheep, the resulting linear discriminant function will be applied to sheep astragali from archaeological sites containing mixed populations of hunted wild as well as domestic sheep and possibly offspring of wild and domestic parents that currently cannot be sorted with any certainty based on raw linear measurements. Provided that the separation achieved by the linear discriminant function is good enough (>90%), shape analyses might allow distinguishing wild from early domestic morphotypes amongst the sheep in the faunal assemblages from early farming and domestication sites in Anatolia dating to the 9th – 8th millennium BC, such as Nevalı Çori and Gürcütepe in SE Anatolia and Aşıklı Höyük in Central Anatolia.

Acknowledgements

This study was funded by the Deutsche Forschungsgemeinschaft under Grant PE 424/10-2. We are indebted to Necmi Karul, Istanbul University, for generously giving permission to study the faunal remains from Gusir Höyük. We would also like to thank Rémi Laffont for his highly valuable help during GM analyses with 'geomorph' R package. Finally, the comments, advice and corrections by two reviewers considerably helped to improve this article, for which we are particularly grateful.

References

Adams, D. C. and Otarola-Castillo, E. 2013. Geomorph: an R package for the collection and analysis of geometric morphometric shape data. *Methods in Ecology and Evolution* 4: 393-399.

Aho, K. 2014. Foundational and applied statistics for biologists using R. Boca Raton, FL: CRC/Taylor and Francis.

Bailey, R. C. and Byrnes, J. 1990. A new, old method for assessing measurement error in both univariate and multivariate morphometric studies. *Systematic Zoology* 39(2): 124-130.

Barak, M. M., Lieberman, D. E. and Hublin, J.-J. 2011. A Wolff in sheep's clothing: trabecular bone adaptation in response to changes in joint loading orientation. *Bone* 49: 1141-1151.

Barr, W. A. 2014. Functional morphology of the bovid astragalus in relation to habitat: controlling phylogenetic signal in ecomorphology. *Journal of Morphology* 275: 1201-1216.

Bartosiewicz, L., Van Neer, W. and Lentacker. A. 1997. *Draught Cattle: Their Osteological Identification and History*. Annals of Scientific Zoology. Tervuren: Royal Museum of Central Africa.

Boessneck, J., Müller, H.-H., and Teichert, M. 1964. Osteologische Unterscheidungmerkmale zwischen Schaf (*Ovis aries* Linné) und Ziege (*Capra hircus* Linné). *Kühn-Archiv* 78: 5-129.

Bollognino, R. and Vigne, J.-D. 2007. Temperature monitoring in archaeological animal bone samples in the Near East arid area, before, during and after excavation. *Journal of Archaeological Science* 35(4): 873-881.

Bookstein, F. L. 1991. *Morphometric Tools for Landmark Data: Geometry and Biology*. New York: Cambridge University Press.

Bookstein, F. L. 1997. Landmark methods for forms without landmarks: morphometrics of group differences in outline shape. *Medical Image Analysis* 1(3): 225-243.

Claude, J. 2008. *Morphometrics with R*. New York: Springer.

Claude, J. 2013. Log-shape ratios, Procrustes superimposition, elliptic Fourier analysis: three worked examples in R. *Hystrix* 24(1): 94-102.

Cucchi, T., Hulme-Beaman, A., Yuan, J. and Dobney, K. 2011. Early Neolithic pig domestication at Jiahu, Henan Province, China: clues from molar shape analyses using geometric morphometric approaches. *Journal of Archaeological Science* 38: 11-22.

Curran, S., 2012. Expanding ecomorphological methods: geometric morphometric analysis of Cervidae post-crania. *Journal of Archaeological Science* 39: 1172-1182.

DeGusta, D. and Vrba, E. 2003. A method for inferring paleohabitats from the functional morphology of bovid astragali. *Journal of Archaeological Science* 30: 1009-1022.

Denys, E. and Winkler, A. 2015. Advances in integrative taxonomy and evolution of African murid rodents: How morphological trees hide the molecular forest. In, Cox, P. and Hautier L. (eds) *Evolution of the rodents. Advances in Phylogeny, Functional Morphology and Development*: 186-220. Cambridge: Cambridge University Press.

Dietrich, O., Köskal-Schmidt, Ç., Notroff, J. and Schmidt, K. 2013. Establishing a radiocarbon sequence for Göbekli Tepe. State of Research and New Data. *Neo-Lithics* 13(1): 36-41.

Evin, A., Cucchi, T., Cardini, A., Vidarsdottir, U. S., Larson, G. and Dobney, K. 2013. The long and winding road: identifying pig domestication through molar size and shape. *Journal of Archaeological Science* 40: 735-743.

Evin, A., Cucchi, T., Escarguel, G., Owen, J., Larson, G., Vidarsdottir, U. S. and Dobney, K. 2014. Using traditional biometrical data to distinguish west Palearctic wild boar and domestic pigs in the archaeological record: new methods and standards. *Journal of Archaeological Science* 43: 1-8.

Fruciano, C. 2016. Measurement error in geometric Morphometrics. In, Posnien, N. and Prpic, N.-M. (eds) *Size and Shape: Integration of Morphometrics, Mathematical Modelling, Developmental and Evolutionary Biology*. Development Genes and Evolution 226 (3): 139-158. http://dx.doi.org/10.1007/s00427-016-0537-4.

Garland, T. Jr. and Kelly, S. A. 2006. Phenotypic plasticity and experimental evolution. *Journal of Experimental Biology* 209: 2344-2361.

Groves, C. P. and Leslie, D. M. Jr. 2011. Family Bovidae (hollow-horned ruminants). In, Wilson, D. E. and Mittermeier, R. A. (eds) *Handbook of the Mammals of the World* 2. Hoofed Mammals: 585-588. Barcelona: Lynx Edicions.

Haruda, A. F. 2014. *Central Asian Economies and Ecologies in the Late Bronze Age: Geometric Morphometrics of the Caprid Astragalus and Zooarchaeological Investigations of Pastoralism*. PhD. University of Exeter.

Hildebrand, M. and Goslow, T. 2001. *Analysis of Vertebrate Structure*. 5th ed. New York: JohnWiley and Sons.

Jensen, P. 2015. Adding 'epi-' to behaviour genetics: implications for animal domestication. *The Journal of Experimental Biology* 218: 32-40.

Jolicoeur, P. 1959. Multivariate geographical variation in the wolf *Canis lupus* L. *Evolution* 13: 283-299.

Karul, N. 2011. Gusir Höyük. In, Özdoğan, M., Başgelen, N. and Kuniholm, P. (eds) *The Neolithic in Turkey* 1: 1-19. Istanbul: Archaeology and Art Publications.

Kendall, D. 1977. The diffusion of shape. *Advances in Applied Probability* 9: 428-430.

Klingenberg, C. P. 2013. Cranial integration and modularity: insights into evolution and development from morphometric data. *Hystrix* 24(1): 43-58.

Klingenberg, C. P. and McIntyre, G. S. 1998. Geometric morphometrics of developmental instability: analyzing patterns of fluctuating asymmetry with Procrustes methods. *Evolution* 52: 1363-1375.

Knigge, R. P., Tocheri, M. W., Orr, C. M. and McNulty, K. P. 2015. Three-Dimensional Geometric Morphometric Analysis of Talar Morphology in Extant Gorilla Taxa from Highland and Lowland Habitats. *The Anatomical Record* 298: 277-290.

Kühn, J. 1888. *Das Studium der Landwirthschaft an der Universität Halle. Geschichtliche Entwicklung und Organisation Desselben*. Eine Festschrift zur Feier des 25 jährigen Bestehens des landwirthschaftlichen Insitutes der Universität. Halle a. S.: Plötz'sche Buchdruckerei.

Larson, G., Cucchi, T., Fujita M., Matisoo-Smith,

E., Robins, J., Anderson, A., Rolett, B., Spriggs, M., Dolman, G., Kim, T.-H., Thi Dieu Thuy, N., Randi, E., Doherty, M., Awe Due, R., Bollt, R., Djubiantono, T., Griffin, B., Intoh, M., Keane, E., Kirch, P., Li, K.-T., Morwood, M., Pedriña, L. M., Piper, P. J., Rabett, R. J., Shooter, P., Bergh, G. van den, West, E., Wickler, S., Yuan, J., Cooper, A. and Dobney, K. 2007. Phylogeny and ancient DNA of *Sus* provides new insights into Neolithic expansion in Island Southeast Asia and Oceania. *Proceedings of the National Academy of Science* 104: 15276-15281.

Larson, G., Piperno, D. R., Allaby, R. G., Purugganan, M. D., Andersson, L., Arroya-Kalin, M., Barton, L., Climer Vigueira, C., Denham, T., Dobney, K., Doust, A. N., Gepts, P., Gilbert, M. T. P., Gremillion, K. J., Lucas, L., Lukens, L., Marshall, T. P., Olsen, K. M., Pires, J. C., Richerson, P. J., de Casas, R. R., Sanjur, O. I., Thomas, M. G. and Fuller, D. Q. 2014. Current perspectives and the future of domestication studies. *Proceedings of the National Academy of Science* 111: 6139-6146.

Lawrence, T. L. J. and Fowler, V. R. 2002. *Growth of Farm Animals*. 2nd ed. Wallingford, Oxon, UK: CABI Publishing.

Legge, A. J. 1996. The beginning of caprine domestication in Southwest Asia. In, Harris, D. R. (ed.) *The Origins and Spread of Agriculture and Pastoralism in Eurasia*: 238-262. London: UCL Press.

Meadow, R. H. 1984. Animal domestication in the Middle East, a view from the eastern margin. In, Clutton-Brock, J. and Grigson, C. (eds) *Animals and Archaeology*, Vol. 3: Early herders and their Flocks. BAR International Series 202: 309-337. Oxford: British Archaeological Reports.

Mitteroecker, P. and Bookstein, F. L. 2007. The conceptual and statistical relationship between modularity and morphological integration. *Systematic Biology* 56: 818-836.

Mosiman, J. E. 1970. Size allometry: size and shape variables with characterizations of the lognormal and generalized gamma distributions. *Journal of the American Statistical Association* 65: 930-948.

Peters, J., Helmer, D., Von den Driesch, A. and Saña Segui, M. 1999. Early Animal Husbandry in the Northern Levant. *Paléorient* 25: 27-48.

Peters, J., Buitenhuis, H., Grupe, G., Schmidt, K. and Pöllath, N. 2013. The long and winding road: Ungulate exploitation and domestication in Early Neolithic Anatolia (10000-7000 cal BC). In, Colledge, S., Conolly, J., Dobney, K., Manning K. and Shennan S. (eds) *The Origins and Spread of Domestic Animals in Southwest Asia and Europe*: 83-114. Walnut Creek, Cal.: Left Coast Press.

Peters, J., Arbuckle, B. S. and Pöllath, N. 2014a. Subsistence and beyond: animals in Neolithic Anatolia. In, Özdoğan, M., Başgelen, N. and Kuniholm, P. (eds) *The Neolithic in Turkey* 6: 135-203. Istanbul: Archaeology and Art Publications.

Peters, J., Schmidt, K., Dietrich, O. and Pöllath, N. 2014b. Göbekli Tepe: agriculture and domestication. In, Smith, C. (ed.) *Encyclopedia of Global Archaeology*: 3065-3068. New York: Springer.

Plummer, T. W., Bishop, L. C. and Hertel, F. 2008. Habitat preference of extant African bovids based on astragalus morphology: operationalizing ecomorphology for palaeoenvironmental reconstruction. *Journal of Archaeological Science* 35: 3016-3027.

Popkin, P. R. W., Baker, P., Worley, F., Payne, S. and Hammon, A. 2012. The Sheep Project (1): determining skeletal growth, timing of epiphyseal fusion and morphometric variation in unimproved Shetland sheep of known age, sex, castration status and nutrition. *Journal of Archaeological Science* 39: 1775-1792.

Price, T. D., Qvarnström, A. and Irwin, D. E. 2002. The role of phenotypic plasticity in driving genetic evolution. *Proceedings of the Royal Society B: Biological Sciences 270 (1523): 1433-1440.*

R Core Team 2016. *R: A Language and Environment for Statistical Computing.* [online] Available at: <https://www.R-project.org/> [Accessed 20 April 2016]. Vienna, Austria: R Foundation for Statistical Computing.

Rohlf, F. J. 2010. *tpsDig. Version 2.16*. [online] Available at: <http://life.bio.sunysb.edu/morph> [Accessed 20 April 2016]. Stony Brook: Department of Ecology and Evolution, State University of New York.

Rohlf, F. J. and Corti, M. 2000. The use of two-block partial least-squares to study covariation in shape. *Systematic Biology* 49: 740-753.

Rohlf, F. J. and Slice, D. E. 1990. Extensions of the Procrustes method for the optimal superimposition of landmarks. *Systematic Biology* 39: 40-59.

Rosenberg, M., Nesbitt, R., Redding, R. and Peasnall, B. L. 1998. Hallan Çemi, pig husbandry, and post-Pleistocene adaptations along the Taurus-Zagros arc (Turkey). *Paléorient* 24: 25-41.

Schafberg, R. and Wussow, J. 2010. Julius Kühn. Das Lebenswerk eines argrarwissenschaftlichen Visionärs. *Züchtungskunde* 82(6): 468-484.

Seetah, K., Cucchi, T., Dobney, K. and Barker, G. 2014. A geometric morphometric re-evaluation of the use of dental form to explore differences in horse (*Equus caballus*) populations and its potential zooarchaeological application. *Journal of Archaeological Science* 41: 904-910.

Vigne, J.-D. 2015. Early domestication and farming:

What should we know or do for a better understanding. *Anthropozoologica* 50(2): 123-150.
Vigne, J.-D. and Helmer, D. 2007. Was milk a secondary product in the old world Neolithisation process? Its role in the domestication of cattle sheep and goats. *Anthropozoologica* 42(2): 9-40.
Von den Driesch, A. 1976. *A Guide to the Measurement of Animal Bones from Archaeological Sites*. Peabody Museum Bulletins 1. Cambridge, MA: Harvard University.
Wallace, I. J. and Garland, T. Jr. 2016. Mobility as an emergent property of biological organization: Insights from experimental evolution. *Evolutionary Anthropology* 25: 98-104.
West-Eberhard, M. J. 1989. Phenotypic plasticity and the origins of diversity. *Annual Review of Ecology and Systematics* 20: 249-278.
Yezerinac, S. M., Lougheed, S. C. and Handford, P. 1992. Measurement error and morphometric studies: statistical power and observer experience. *Systematic Biology* 41: 471-482.
Zeder, M. 2011. The origin of agriculture in the Near East. *Current Anthropology* 52(4): 221-235.
Zeder, M. 2016. Domestication as a model system for niche construction theory. *Evolutionary Ecology* 30: 325-348.
Zeder, M. A., Emshwiller, E., Smith, B. D. and Bradley, D. G. 2006. Documenting domestication: the intersection of genetics and archaeology. *Trends in Genetics* 22(3): 139-155.
Zeder, M. A. and Lapham, H. A. 2010. Assessing the reliability of criteria used to identify postcranial bones in sheep, *Ovis* , and goats, *Capra*. *Journal of Archaeological Science* 37: 2887-2905.

16 Caprine age estimations: methodological considerations through the study of Cretan Agrimi goat mandibles (*Capra aegagrus cretica*)

Moussab Albesso[1]

1 PhD student at Archéorient (UMR 5133, Maison de l'Orient et de la Méditerranée), Lyon 2 University

Abstract

This article presents the study of present Agrimi goat remains (*Capra aegagrus cretica*) collected in the Samariá Gorge in Crete during the 2011 survey organized by A. Gardeisen (CNRS, UMR5140).
The study compares morphological data taken from the craniums, horn cores and mandibles of this population with data from other better-known sub-species. Mortality ages are established from the mandibles of nearly forty individuals. Age estimation methods using modern domestic species are discussed. The question is whether these can be directly applicable to all caprine populations, especially small sized animals.

Keywords

Agrimi goats, ageing, tooth wear and eruption, zooarchaeology, goat morphology

The Agrimi goat (*Capra aegagrus cretica*) is a wild goat that lives only in Crete on the Samariá Gorge and on the surrounding Dia, Agioi Theodoroi and Agioi Pantes islands. A nature reserve was created in 1962 in the Samariá Gorge to protect this population which is in constant demographic decline. During a 1996 census around 500 individuals were counted in the gorges (Bar-Gal *et al.* 2002: 369).

This species (Fig. 1) is a medium sized ungulate that has a brown fur with a central black stripe along its back. Archaeological and historical evidence shows that they were hunted on Crete since the Neolithic period (ca. 6000 cal. B.C.; Bar-Gal *et al.* 2002: 369; Wilkens 1996). Their absence from Pleistocene sites indicates this species was introduced to the island by man during neolithisation (Bar-Gal *et al.* 2002: 369). The phenotype of this Cretan population resembles those of the wild goat or bezoar ibex (*Capra aegagrus aegagrus*). However, recent DNA studies suggest that they are closer to Middle Eastern domestic goats, and would have been introduced to the island and then became feral (Bar-Gal *et al.* 2002). Nonetheless, this hypothesis is still in discussion. Another hypothesis suggests they may have been introduced as a wild animals, and their domestic traits would come from later genetic mixing with domestic goats brought to Crete afterwards (Horwitz and Bar-Gal 2006).

During autumn 2011, a scientific mission directed by A. Gardeisen (CNRS, UMR 5140, Lattes) collected Agrimi goat skeletons in the Samariá Gorge in collaboration with the forestry department of Chania (Fig. 2).

The team brought back 21 complete skeletons belonging to three males, nine females and nine of unknown sex (five skeletons measured, see appendix 1), 44 skulls (only three with mandibulae) from 11 males, 13 females and 16 unknown sex due to their young age (16 measured skulls, see appendix 2), as well as 63 isolated mandibulae pertaining to animals of indeterminate sex (58 measured mandibulae, see appendix 3). The majority of the complete or near-complete skeletons was discovered in caves and are female. No whole adult male skeleton was found, the only three complete male skeletons belong to juvenile individuals. Within the sample of male skulls collected, two were given to the team by one of the Reserve's foresters. The ethological observations made during the mission seem to indicate that male goats tend to live in less accessible zones at higher altitudes. It is difficult to determine the cause of death of these individuals. The ones discovered in the cave (whole or partial) seem to have died of old age or illness. For the others, discovered in the gorges undergrowth (skulls or sections of limbs) the death of the animals may be linked to mortal falls. The site, a natural reserve, strictly prohibits hunting.

Fig. 1. Female Agrimi goat (*Capra aegagrus cretica*) – Samariá Gorge 2011.

Fig. 2. Agrimi goat skeleton - Samariá Gorge 2011.

Osteological morphology of Agrimi goat (*Capra aegagrus cretica*)

The morphological and metrical study of this bone sample revealed differences with other goat species.

Some morphological characteristics, such as the shape of the male horns (Fig. 3), which are of large size, curved towards the back as a « scimitar shape », and a rounded shape for the anterior part, are similar to those of wild goats (*Capra aegagrus*; Helmer 1992: 35-36). The cross section of both male and female Agrimi goat cornual processes are also very similar to those of the Middle Eastern wild goats (Fig. 4; Helmer 1992: 31-37; Helmer and Gourichon 2008: 129-130; Vigne 1988: 3-5; Vila 1998: 97).

However, the size of this animal (average wither height: 50.8 cm; Schramm 1967: 341) is very small compared to any other contemporary wild goat, such as the Zagros wild goat (north-western Iran and north-eastern Iraq; Zeder 2008: 271-277). The male specimens collected in the Samariá Gorge being juveniles or sub-adults, it has not been possible to compare them to male Zagros wild goats. However the comparisons made with the female bones clearly underline the smaller size of the Samariá wild goats (Fig. 5). Also, some of the bones seem to possess morphological characteristics closer to *Ovis* than *Capra* genera, notably the medial epicondyle of the humerus (Boessneck 1969, Fernandez 2001, Zeder and Lapham 2010; Fig. 6).

We have also noted morphological differences within the species but they can simply be due to sexual dimorphism. The proximal extremities of the ulnas show differences around the olecranon tuberosity. Female ulna displays a much more prominent tuberosity (Fig. 7).

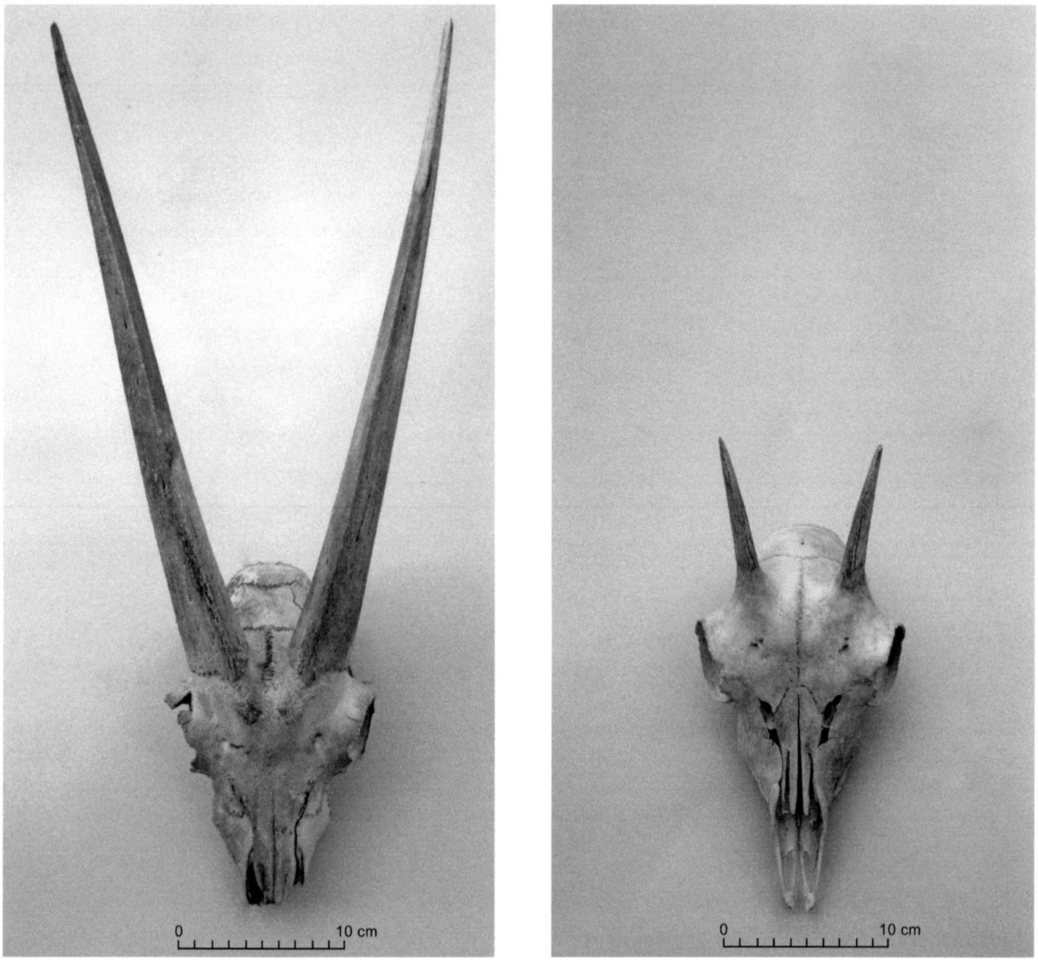

Fig. 3. Agrimi goat (*Capra aegagrus creticus*) skulls: male (left) and female (right).

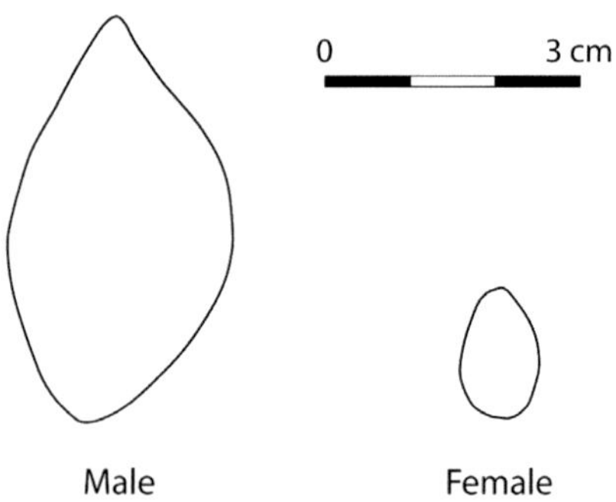

Fig. 4. Cross-section of the base of male and female Agrimi goat cornual processes.

Age estimations of Agrimi goats (*Capra aegagrus cretica*)

Numerous archaeozoological studies about caprine remains use occlusal mandibular tooth wear stages to determine the individual age. Deniz and Payne (1982) proposed protocols to describe the different stages of tooth wear on caprine mandibulae, and to attribute each mandibula to an age group (Payne 1973; Deniz and Payne 1982). Payne's 1973 work put the data into nine age groups (A, B, C,...) based on the tooth wear of an archaeological population established from comparisons with modern Turkish sheep of approximate ages, all under two years of age. This protocol was then improved by Deniz and Payne in 1982, thanks to the use of absolute age correlations, based on Turkish angora goats for which the ages and sexes were known precisely. This method, with the study of complete dental rows, results in dental wear stage estimations, that are relatively coherent age groups for young individuals, but not for more advanced individuals. Also, it's delicate to apply to isolated teeth, frequently found in archaeology, because it isn't precise enough. Tooth wear stages are difficult to determine once all the cusps are worn down, as they can remain unaltered for several years. To determine a method applicable to isolated teeth Ducos (1968) suggested an index (i) calculated from the height and transversal diameter of the tooth (H/TD) to precisely estimate all dental remains and give each tooth an age group. This index theoretically reduces errors linked to the size of teeth, which can vary depending on the population studied. The age/index correspondences were then improved by Helmer with a study of a population of approximately thirty modern sheep from the south of France (Helmer 2000b, see also Helmer and Vigne 2004). Other archaeozoologists also worked on age profiles using this index and adapted the correspondences and age groups to specific populations (Vila 1998, Gourichon 2004).

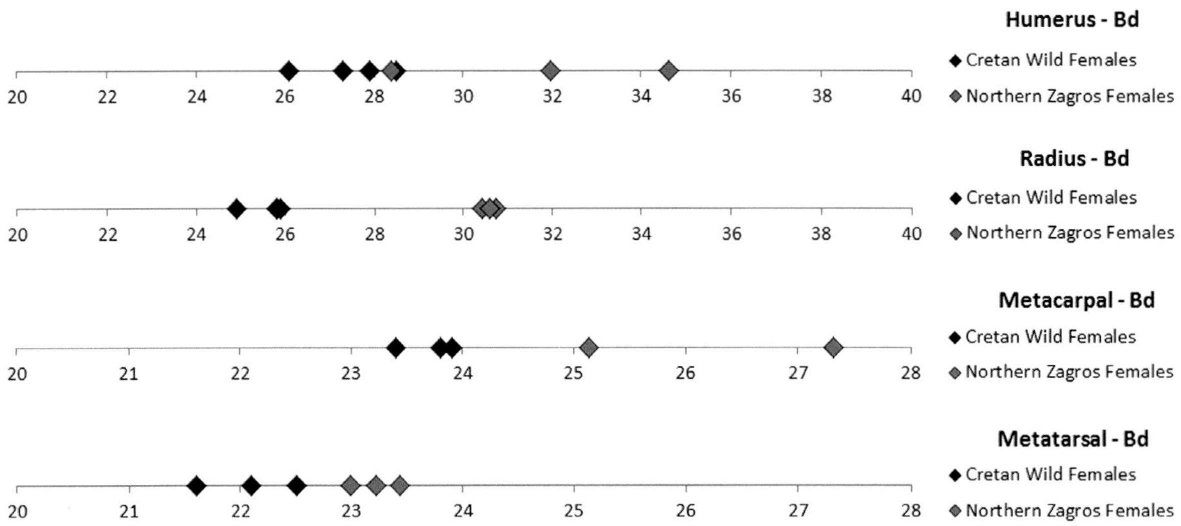

Fig. 5. Comparison of the maximal distal breadth (Bd) on Agrimi goats and Zagros wild goats long bones (Zeder 2008: 271-277).

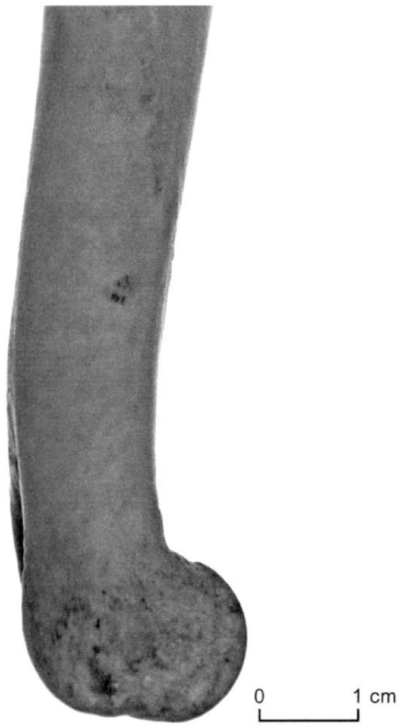

Fig. 6. Medial view of the distal extremity of a female humerus *Capra aegagrus cretica.*

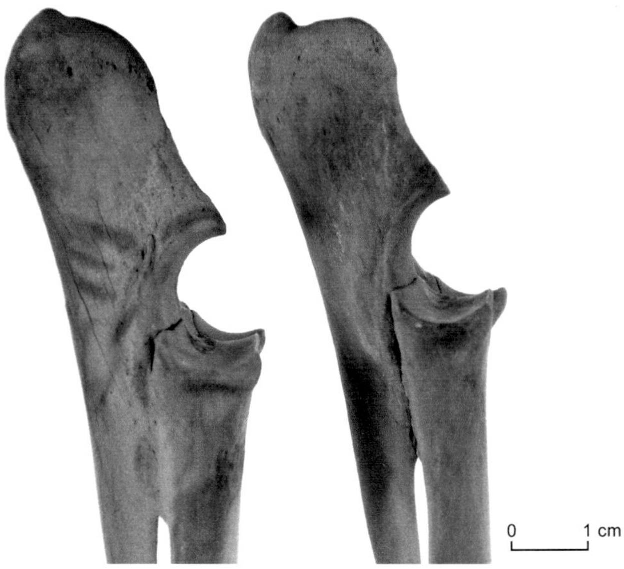

Fig. 7. Medial view of the radius and ulna of a male individual (on the left) and a female individual (on the right) *Capra aegagrus cretica.*

All the caprine age estimation models based on visual observation of dental wear or index (i) calculations are carried out on domesticated populations of different regions of Europe and the Middle East. The Agrimi goat remains collected during the study constitute an interesting assemblage allowing us to see if these methods can be applied to a small body-sized wild population from a Mediterranean island. The dental referential of this study is based on 54 mandibulae, six linked to skulls or skeletons, 48 isolated.

A first estimation of dental ages of this assemblage was conducted following Deniz and Payne's method, based on Fig. 20 to 24 of their paper (Deniz and Payne 1982: 173-180) Simplified to numbers by Gourichon (2004: Annexes, table 2.10, 222). The age groups used are those of Payne's (1973) works, they are referenced with letters from A to I. (A: 0 to 2 months, B: 2 to 6 months, C: 6 to 12 months, D: 1 to 2 years, E: 2 to 3 years, F: 3 to 4 years, G: 4 to 6 years, H: 6 to 8 years, I: over 8 years). For each mandibula, the age is estimated from dental wear stages of the occlusal surface of the teeth and dental eruptions (Fig. 8). As it is difficult to attribute precise wear stages for groups E and F, and H and I because they evolve very little in between these two intervals they have been merged into EF and HI groups. We note the absence of the first age group A. Two mandibulae (C8 and C9) are linked to age group B. Although tooth D4 of the C9 mandibula can be attributed to group C, M1 is only just starting its eruption, therefore classing it as a B age group item. The post cranial skeletal analysis of both these individuals confirm their attribution to this category. Fifteen mandibulae are put into age group C, within this group the C10 mandibula can be discussed: teeth D4 and M1 of this item fit age group D but M2 is still

Identification		Age classes		Wear stages			
Mandible	Laterality	Years	Deniz/Payne	D4/P4	M1	M2	M3
C8	L	2-6 months	B	BC	AB		
C9	L			C	AB		
CA25	L			CD	C		
CA23	L			CD	C		
CA24	L			CD	C		
CA26	R			CD	C		
CA29	R			CD	C		
CA28	L			CD	C		
CA27	R			CD	C		
CA30	L	6-12 months	C	CD	C		
CA31	L			CD	C		
CA33	L			D	C		
CA34	R			D	C		
CA35	R			D	C		
C1	L			D	C		
C2	L			D	C		
C10	L			D	DEF	BC	
C11	L			D	DEF	D	
C6	L			D	DEF	D	
CA36	R			D	DEF	D	
CA37	L	1-2 years	D	D	DEF	D	
CA38	R			D	DEF	D	
C12	L			D	DEF	D	
CA39	R			D	DEF	D	
CA40	L			D	DEF	D	CD
CA18	L			EF	DEF	D	CD
C13	L			EF	DEF	D	CD
CA22	R			EF	DEF	D	E
CA15	L			EF	DEF	D	E
CA20	L	2-4 years	EF	EF	DEF	D	E
CA21	R			EFG	DEF	D	E
C3	L			EFG	DEF	D	E
CA2	R			GHI	GHI	EFG	F
CA6	R			GHI	GHI	EFG	F
C4	L			GHI	GHI	EFG	F
CA16	L			GHI	GHI	EFG	FGH
C14	L			GHI	GHI	EFG	FGH
C7	L			GHI	GHI	EFG	FGH
CA1	R	4-6 years	G	GHI	GHI	EFG	FGH
CA17	L			GHI	GHI	EFG	FGH
CA11	R			GHI	GHI	EFG	FGH
CA4	R			GHI	GHI	EFG	FGH
CA7	R			GHI	GHI	H	FGH
CA13	L			GHI	GHI	H	FGH
C18	L			GHI	GHI	H	FGH
CA10	R			GHI	GHI	HI	FGH
CA8	R			GHI	GHI	HI	FGH
CA19	L			GHI	GHI	HI	FGH
CA12	R	> to 6 years	HI	GHI	GHI	HI	I
CA5	R			GHI	GHI	HI	I
CA9	R			GHI	GHI	HI	I
C16	L			GHI	GHI	HI	I
CA14	L			GHI	GHI	HI	I
C5	L			GHI	GHI	HI	I

Fig. 8. Dental age profiles of Agrimi goats mandibulae based on the stages of tooth wear according to Deniz and Payne (1982) and Payne (1973).

Table I. Comparison between Agrimi goat age estimations following Deniz and Payne's (1982) method (occlusal dental wear observation) and Helmer's method (index H/TD of teeth).

Identification	*Age class*		*M1*				*M2*				*M3*			
Mandible	**Age**	**according to E. Deniz & S. Payne**	**according to Helmer's index**	**H**	**DTd**	**Index H/Dtd**	**according to Helmer's index**	**H**	**DTd**	**Index H/Dtd**	**according to Helmer's index**	**H**	**DTd**	**Index H/Dtd**
C8	2-6 months	B												
C9														
CA26	6-12 months	C												
CA29			D	28.2	7.3	3.86								
CA27			D	26.0	7.1	3.66								
CA35			D	31.1	7.7	4.04								
C1			D	27.9	7.3	3.82								
C2			D	27.5	7.8	3.53								
C10			D	26.4	7.4	3.57								
C11	1-2 years	D	D	25.4	7.1	3.58								
C6			D	27.5	7.7	3.57								
CA37			D	25.9	7.2	3.60	EF	31.8	7.6	4.18				
CA38			D	24.8	6.8	3.65	EF	34.3	7.9	4.34				
C12			EF	24.7	7.3	3.38	EF	34.3	8.1	4.23				
C13	2-4 years	EF	EF	22.1	7.3	3.03	EF	31.6	7.8	4.05				
CA22			EF	22.8	8.1	2.81	EF	32.2	8.2	3.93				
CA15							EF	31.2	8.1	3.85				
CA20			EF	20.6	7.4	2.78	EF	30.1	8.1	3.72				
CA21			EF	20.2	7.4	2.73	EF	32.3	9.1	3.55				
C3			EF	19.9	7.6	2.62	EF	29.8	8.3	3.59				
CA2											G	29.2	8.1	3.60
CA6							HI	17.1	7.2	2.4	G	25.8	7.6	3.39
C4			HI	8.7	7.1	1.23	HI	16.8	7.4	2.27	HI	24.4	7.4	3.30
CA16		G					G	23.6	7.8	3.03	G	28.5	8.1	3.52
C14			HI	8.8	6.9	1.28	G	19.4	6.8	2.85	G	26.2	7.3	3.59
C7			HI	9.7	6.7	1.45	HI	17.8	7.5	2.37	G	28.7	7.7	3.73
CA1							HI	17.4	7.4	2.35	HI	24.7	7.4	3.34
CA17							HI	15.6	7.2	2.17	HI	23.5	7.5	3.13
CA11							HI	15.9	7.3	2.18	HI	19.9	7.1	2.80
CA4							HI	14.7	7.6	1.93	HI	23.1	7.5	3.08
CA7	> to 6 years	HI	HI	3.9	7.6	0.51					HI	21.7	7.6	2.86
CA13			HI	5.9	7.1	0.83					HI	19.9	7.3	2.73
C18			HI	5.7	6.9	0.83	HI	13.7	7.2	1.90	HI	19.4	7.3	2.66
CA10			HI	5.7	7.3	0.78					HI	20.5	7.8	2.63
CA8											HI	16.7	7.6	2.20
CA9			HI	5.8	5.9	0.98					HI	15.3	7.8	1.96
C16			HI	5.5	6.3	0.87	HI	8.1	7.3	1.11	HI	14.2	7.9	1.80
CA14			HI	3.9	5.6	0.70					HI	13.3	7.7	1.73
C5			HI	5.5	6.9	0.80	HI	8.2	7.4	1.11	HI	12.6	7.9	1.59

Fig. 9. Agrimi goat left mandibula sawed in order to expose the teeth (individual CI2).

erupting thus putting it more into age group C. The D category regroups 8 mandibulae. One can note, for individual CA40, that M3 is erupting whereas D4 is at its last wear stage. According to Deniz and Payne this combination is particularly unusual (Deniz and Payne 1982: 181). According to them M3 erupts after P4. The age estimation of individuals C11 and C6 from post-cranial skeletal analysis confirms that both are under two years of age. Ten mandibulae pertain to group EF, although the P4 and M1 teeth of mandibulae CA2, CA6 and C4 are in the last tooth wear category, and could be classed as group G, their M3 are not occluded, which explains why they are attributed to age group EF. Finally age group G and HI contain 7 and 12 mandibulae respectively.

The correlation of dental wear stages and dental eruptions seems to show a particularly fast tooth wear for this population. Some M1 molars for example show advanced wear while M3 is hardly erupting.

Deniz and Payne's age estimation method was used as a reference and comparative model to check whether other methods, in particular the one developed by Helmer following Ducos (1968), could be efficient on this population. For that purpose, 39 mandibulae were sawed in order to extract and measure the teeth (Fig. 9). Some teeth could not be measured due to bad states of preservation, or because of irregular tooth wear. This study took place in the Archéorient laboratory (UMR 5133) of the Maison de l'Orient and at the Archaeology of Mediterranean Societies laboratory at Lattes (UMR 5140). It also permitted to check for the presence of discriminating morphological criteria of goat versus sheep on the teeth of this population (Payne 1985, Halstead *et al.* 2002, Helmer 2000a). The D4, M2 and M3 teeth clearly show the presence of these criteria, contrary to the M1 for which these criteria remain very little visible or identifiable. The criteria are attested for: D4 (criterion 2 and 4 in Zeder and Pilaar 2010: Fig. 2, 227), M2 (criterion 1, 3, 4, 5 and 7 from Zeder and Pilaar 2010: Fig. 5, 229) and M3 (criteria

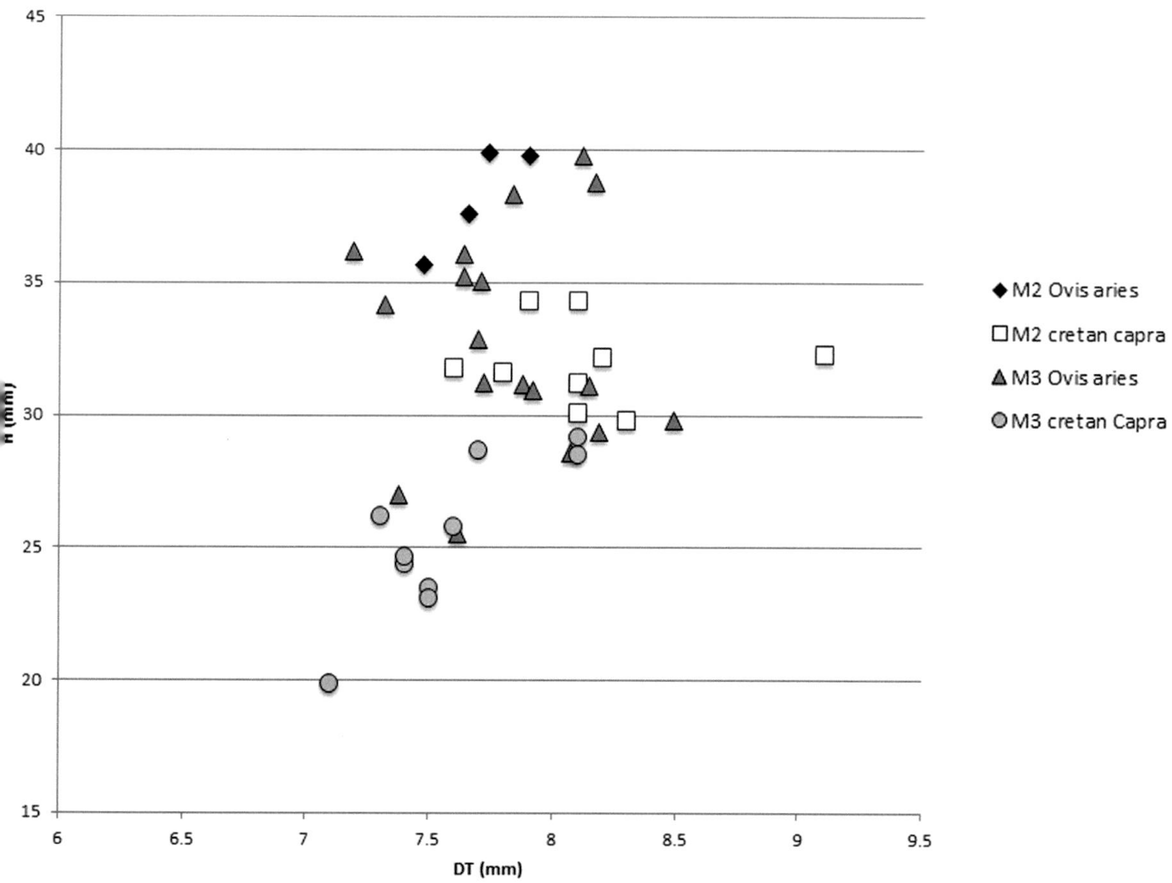

Fig. 10. Dental heights/Transversal diameters of M2 and M3 of non-worn Agrimi goat teeth and current sheep's teeth from the south of France (Blaise 2006: 212).

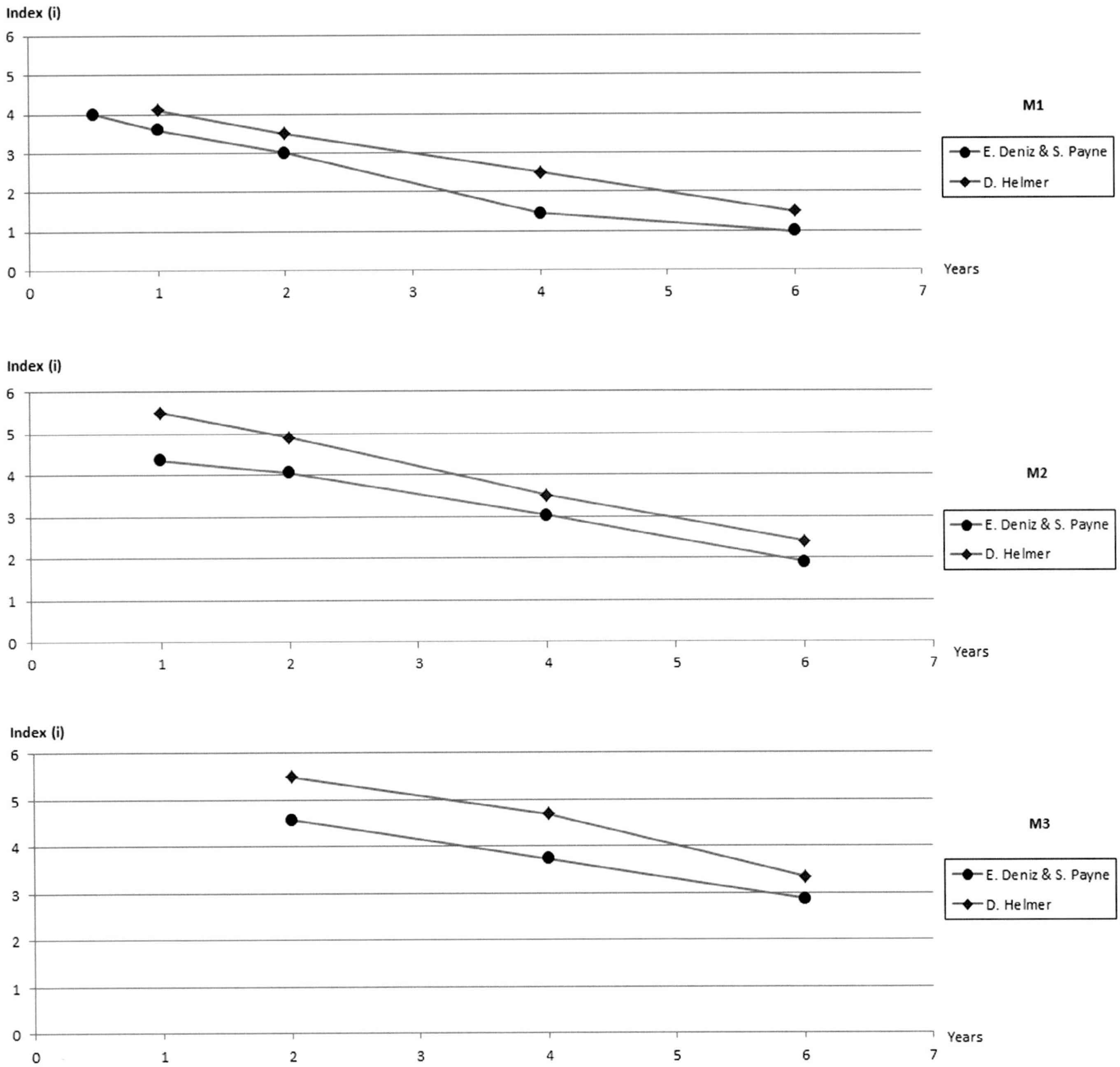

Fig. II. Comparison between the indexes (i) of Agrimi goat teeth aged using the dental wear method (Deniz and Payne 1982) (black circles) and the indexes (i) proposed by Helmer (2000b) for domesticated goats.

1 to 5 Zeder and Pilaar 2010: Fig. 5, 229), however they are hardly visible or identifiable for M1 (criteria 1, 3, 4, 5 and 7 in Zeder and Pilaar 2010: Fig. 6, 230).

The confrontation of the results of the age profiles based on dental wear and tooth measurements show different age groups depending on the method applied (Table 1).

If we consider Helmer's (2000b) index for M1, all individuals of age group C in Deniz and Payne's (1982) method, aged from 6 to 12 months, are moved to group D. The D stage only varies slightly with methods using M1 teeth to age individuals, a single individual changes category from D to EF. However the use of M2 teeth with the index method leads to moving them to group EF (2 to 4 years).

Likewise, the use of different methods using M1 does not lead to any variance in results for EF. For M2 however some differences can be noted although they are minor: two individuals move to HI (over 6 years) with Helmer's method. M3s are more variable, using Helmer's protocol, no animal is part of EF and all are moved to group G (4 to 6 years).

Very few individuals attributed to group G with Deniz and Payne's method stay in that category once submitted to the dental height index method, all M1 move over to stage HI, only a few remain in group G (two M2 and three M3).

We therefore note the absence of younger age groups for each molar (C for M1, D for M2 and EF for M3) when tooth height index methods are applied. Also the comparison of results for each age group

Table2. Dental wear index limits of Agrimi goats (*Capra aegagrus cretica*) according to their age groups.

Age classes		*Lower teeth*			
Payne	**Age**	**M1**	**M2**	**M3**	**Corr.**
	0 birth				
A					x 6
	2 months				
B		no wear			x 3
	6 months	4.0			
C			no wear		x 2
	12 months	3.6	4.4		
D				no wear	x 1
	2 years	3.0	4.1	4.6	
EF					x 0.5
	4 years	1.5	3.0	3.7	
G					x 0.5
	6 years	1.0	1.9	2.9	
HI					x 0.25

shows significant difference between Deniz and Payne's method versus Helmer's one. An offset of one to two age groups is often observed over the whole sample of teeth.

In conclusion, the application of Payne's works to our sample of Agrimi goats shows the evidence of a population ranging from two month to over six years of age. The study of the growth stages of the post-cranial skeletal elements (Barone 1976: 53, Silver 1969) also shows the presence of goats younger than 6 months old and advanced individuals, according to the estimations based on dental wear stages.

Finally, as we have seen throughout this study, the age estimation methods based on dental wear indexes applied to definitive molars sometimes show considerable discrepancies, especially with M2 and M3, compared to the study of wear of full occlusal dental rows, that also reveal they aren't appropriate for this specific population.

These results don't compromise the validity of the different methods in their globality but simply show they aren't applicable to every population. Indeed, studies have already shown their validity on recent populations of known ages (Blaise 2006, Greenfield and Arnold 2008).

Several factors can take part in compromising the reliability of established methods. The first is very quick dental wear. As a matter of fact several individuals have particularly advanced dental wear. There therefore are signs of strong abrasion that may be linked to their habitat's environment, unless it is a specific biological characteristic of this particular population.

The second factor concerns the morphological attributes of this population. The index (i) was developed to get rid of the size differences of different caprine populations. However, this index considers that the height and diameter of a tooth varies proportionally depending on the size of the goats, but the comparison of H and DT measures of non-worn teeth of the Cretan goats and of sheep from contemporary sheep from the south of France (Blaise 2006) seem to show that for an equal transversal diameter the height of the teeth of the Cretan goats remain inferior (Fig. 10). However, it will be necessary to confirm these observations by comparing measures taken from Cretan Agrimi goats to other caprine populations of known ages.

The teeth of lesser height compared to their diameter presenting fast dental wear therefore have a direct influence on the results obtained when calculating the index (i). For each molar the comparison of reference indexes with those calculated for each age group using the maximum indexes calculated for each age group following Deniz and Payne's method (Fig. 11) show that the Agrimi goat indexes are always smaller than all the others.

The reference tables currently used and based on the index (i) are not appropriate for this population. To determine the age of Agrimi goats from isolated teeth it is necessary to establish a new reference adapted to this type of goat. We therefore suggest the reference presented as table 2.

Acknowledgements

I wish to thank A. Gardeisen, the archaeological laboratory Archéologie des Sociétés Méditerranéennes (UMR 5140), the Ecole Française d'Athènes, the Labex Archimède in Montpellier as well as the head forestry department of Chania in Crete and specifically M. I. Fotakis. I also wish to thank E. Vila and L. Gourichon for the many discussion on the subject, as well as the Archéorient laboratory (UMR 5133) and the doctoral school ED 483 of the University of Lyon 2 for their funding to take part in ASWA 2015.

References

Bar-Gal, G. K., Smith, P., Tchernov, E., Greenblatt, C., Ducos, P., Gardeisen, A. and Horwitz, L. K. 2002. Genetic evidence for the origin of the Agrimi goat (*Capra aegagruscretica*). *Journal of Zoology* 256: 369-377.

Barone, R. 1976. *Anatomie Comparée des Mammifères Domestiques. II. Atlas*. Paris: Vigot Frères.

Blaise, E. 2006. Référentiel actuel de brebis « Préalpes-du-Sud » (Digne, Alpes-de-Haute-Provence, France): pratiques d'élevage et âges dentaires. *Anthropozoologica* 41 (2): 191-214.

Boessneck, J. 1969. Osteological differences between sheep (*Ovis aries L.*) and goat (*Capra hircus L.*). In, Brothwell, D. R. and Higgs, E. S. (eds) *Science in Archaeology, a Survey of Progress and Research*: 331-358. London: Thames and Hudson.

Deniz, E. and Payne, S. 1982. Eruption and wear in the mandibular dentition as a guide to ageing Turkish Angora goats, In, Wilson, B., Grigson, C. and Payne, S. (eds) *Ageing and Sexing Animal bones from Archaeological Sites.* British Archaeological Reports, International Series, 109*:* 155-205. Oxford, Archaeopress.

Ducos, P. 1968. *L'origine des Animaux Domestiques de Palestine*. Université de Bordeaux (Mémoire de Institut de Préhistoire de l'Université de Bordeaux 6).

Fernandez, H. 2001. *Ostéologie Comparée des Petits Ruminants Eurasiatiques Sauvages et Domestiques (genres* Rupicapra, Ovis, Capra *et* Capreolus*), Diagnose Différentielle du Squelette Appendiculaire, volume II, planches et figures*. PhD. Université de Genève: Département de zoologie et biologie animale.

Gourichon, L. 2004. *Faune et Saisonnalité: l'Organisation Temporelle des Activités de Subsistance dans l'Epipaléolithique et le Néolithique Précéramique du Levant Nord (Syrie)*. PhD. Lyon: Université Lumière – Lyon 2.

Greenfield, H. J. and Arnold, E. R. 2008. Absolute age and tooth eruption and wear sequences in sheep and goat: determining age-at-death in zooarchaeology using a modern control sample. *Journal of Archaeological Science* 35: 836-849.

Halstead, P., Collins, P. and Isaakidou, V. 2002. Sorting the sheep from the goats: morphological distinction between the mandibles and mandibular teeth of adult *Ovis* and *Capra*. *Journal of Archaeological Science* 29: 545-553.

Helmer, D. 1992. *La Domestication des Animaux par les Hommes Préhistoriques*. Paris: Masson.

Helmer, D. 2000a. Discrimination des genres *Ovis* et *Capra* à l'aide des prémolaires inférieures 3 et 4 et interprétation des âges d'abattages: l'exemple de Dikili Tash (Grèce). *Anthropozoologica* 31: 29-38.

Helmer, D. 2000b. Étude de la faune mammalienne d'El Kowm 2 (Syrie). In, Stordeur, D. (ed.) *Une Île dans le Désert: El Kowm 2 (Néolithique Précéramique, 8000-7500 BP Syrie)*: 233-264. Paris: CNRS Éditions.

Helmer, D. and Gourichon, L. 2008. Premières données sur les modalités de subsistance à Tell Aswad (Syrie, PPNB Moyen et Récent, néolithique céramique ancien). Fouilles 2001-2005. *Archaeozoology of the Near East VIII, Travaux de la Maison de L'Orient et de la Mediterranée* 49 (1): 119-151.

Helmer, D. and Vigne, J. D. 2004. La gestion des caprinés domestiques dans le midi de la France. In, Bodu, P. and Constantin, C. (eds) *Approches Fonctionnelles en Préhistoire. Actes de XXVème Colloque du Congrès Préhistorique de France, Nanterre, novembre 2000*: 397-407. Paris: Société Préhistorique Française.

Horwitz, L. K. and Bar-Gal, G. K. 2006. The Origin and Genetic Status of Insular Caprines in the Eastern Mediterranean: A Case Study of Free-Ranging Goats (*Capra aegagruscretica*) on Crete, *Human Evolution* 21: 123-138.

Payne, S. 1973. Kill off Pattern in Sheep and Goats: the mandibles from Asvan kale, *Anatolian Studies* 23: 281-303.

Payne, S. 1985. Morphological Distinctions between the Mandibular Teeth of Young Sheep, *Ovis*, and Goats, *Capra*. *Journal of archaeological Science* 12: 139-147.

Schramm, Z. 1967. Long bones and heights in withers of goat. *Roczniki Wyzszej Szkoly Rolniczej w Poznaniu* 36: 89-105.

Silver, I. A. 1969. Ageing of domestic animals, In, Brothwell, D. R. and Higgs, E. S. (eds) *Sciences in Archaeology:* 273-277. London: Thames and Hudson.

Vila, E. 1998. *L'exploitation des Animaux en Mésopotamie aux IVe et IIIe Millénaires avant J.-C.* Monographie du CRA 21. Paris: CNRS Editions.

Vigne, J. D. 1988. Les grandes étapes de la domestication de la chèvre: une proposition d'explication de son statut en Europe occidentale. *La Chèvre, Ethnozootechnie* 41: 1-13.

Von den Driesch, A. 1976. A Guide to the measurement of animal bones from archaeological sites. *Peabody Museum Bulletin* 1: 1-136.

Wilkens, B. 1996. Faunal Remains from Italian Excavations on Crete. In, Reese, D. (ed.) *Pleistocene and Holocene Fauna of Crete and its First Settlers* : 241-261. Madison Wisconsin: Prehistory Press.

Zeder, M. 2008. Animal domestication in the Zagros: an update and directions for future research. *Archaeozoology of the Near East VIII, Travaux de la Maison de L'Orient et de la Mediterranée* 49 (1): 243-277.

Zeder, M. and Lapham, H. 2010. Assessing the reliability of criteria used to identify postcranial bones in sheep, *Ovis*, and goats, *Capra*. *Journal of Archaeological Science* 37: 2887-2905.

Zeder, M. and Pilaar, S.E. 2010. Assessing the reliability of criteria used to identify mandibles and mandibular teeth in sheep, *Ovis*, and goats, *Capra*. *Journal of Archaeological Science* 37(2): 225-242.

Appendix I

Measurements of Agrimi skeletons, according to von den Driesch 1976.

	Specimen	Sex	GB	GL	BFcc	GLF	H						
V.Atlas	C6	Male	40.7	31.3	35.9		x						
	C5	Female	55.6	43.4	40.1	38.2	28.3						
	C17	Female	55.2	46.7	37.7	39.2	29.8						
	CA65	Female	X	45.8	40.6	42.5	28.3						
	CA66	Female	60.6	55.3	42.1	44.3	32.1						
V.Axis	Specimen	Sex	LCDe	LAPa	BFcr	BPacd	BPtr	SBV	BFcd	H			
	C6	Male	x	x	x	x	x	x	x	x			
	C5	Female	51.2	45.7	39.2	19.7	37.3	21.9	28.5	42.4			
	C17	Female	45.5	40.1	37.3	26.7	34.9	19.7	19.9	37.8			
	CA65	Female	49.6	35.8	39.9	26.5	X	21.8	19.6	X			
	CA66	Female	54.5	42.3	41.2	19.5	37.5	22.2	30.8	40.6			
V.Cervical	Specimen	Sex	Bpacr	GLPa	Bpacd								
	C6	Male	24.4	24.3	19.8								
	C5	Female	28.6	41.4	28.5								
	C17	Female	27.8	35.7	28.3								
	CA65	Female	26.9	36.1	26.1								
	CA66	Female	31.6	38.8	30.3								
V.Thoracic	Specimen	Sex	H										
	C6	Male	36.5										
	C5	Female	88.6										
	C17	Female	69.5										
	CA65	Female	x										
	CA66	Female	77.7										
V.Lumbar	Specimen	Sex	H	BPtr	BFcr	HFcr							
	C6	Male	31.7	57.3	16.3	12.5							
	C5	Female	41.4	84.2	17.4	13.9							
	C17	Female	38.7	37.2	17.5	13.8							
	CA65	Female	39.2	X	17.6	13.8							
	CA66	Female	41.3	82.9	19.8	13.7							
V.Sacrum	Specimen	Sex	GL	PL	GB	BFcr	HFcr						
	C6	Male	x	x	45.7	x	x						
	C5	Female	75.3	72.2	68.8	26.4	10.3						
	C17	Female	85.5	81.8	66.1	25.2	9.7						
	CA65	Female	62.7	56.7	63.1	26.3	10.3						
	CA66	Female	x	x	x	x	x						
Scapula	Specimen	Sex	DHA	HS	Ld	SLC	GLP	LG	BG				
	C6	Male	106.6	96.4	55.8	10.7	19.6	16.4	17.1				
	C5	Female	139.7	130.5	92.9	15.9	28.3	23.5	20.3				
	C17	Female	131.2	120.2	34.3	14.5	25.8	22.5	17.3				
	CA65	Female	x	x	x	x	x	x	x				
	CA66	Female	x	x	x	x	x	x	x				
Humerus	Specimen	Sex	GL	BP	Dp	SD	Bd	BT					
	C6	Male	99.7	20.5	28.4	9.8	24.2	23.6					
	C5	Female	147.9	35.4	41.2	13.4	28.5	27.1					
	C17	Female	131.8	32.9	37.4	12.7	26.1	25.4					
	CA65	Female	137.9	33.1	38.4	13.2	27.9	26.9					
	CA66	Female	146.8	33.1	39.2	13.8	27.3	26.7					
Radius	Specimen	Sex	GLl	GL	PL	BP	BFp	SD	Bd	BFd			
	C6	Male	130.6	102.3	101.3	23.5		12.3	21.3	x			
	C5	Female	174.9	139.4	134.6	27.9	26.8	15.4	25.9	24.4			
	C17	Female	162.2	131.6	128.6	26.6	25.2	15.4	25.8	23.4			
	CA65	Female	131.5	134.1	132.9	27.6	26.6	14.7	24.9	23.9			
	CA66	Female	179.3	139.8	137.7	27.5	26.8	15.5	25.8	24.8			
Ulna	Specimen	Sex	GL	LO	DAP	SDO	BPC						
	C6	Male	120.3	17.6	19.5	17.1	18						
	C5	Female	147.6	36.5	23.4	20.1	21						
	C17	Female	160.5	33.4	22.4	19.8	20.3						
	CA65	Female	164.1	36.4	23.2	19.7	21.2						
	CA66	Female	177.2	37.3	22.8	20.4	21.9						
Metacarpal	Specimen	Sex	GL	BP	Dp	SD	Bd						
	C6	Male	70.9	19.5	14	10.7	20.7						
	C5	Female	101.8	21.9	15.2	13.5	23.8						
	C17	Female	97.3	20.3	14.3	13.6	23.4						
	CA65	Female	100.1	21.7	15.9	13.8	23.8						
	CA66	Female	99.8	21.5	16.4	15.3	23.9						
Pelvis	Specimen	Sex	GL	LA	LAR	LS	SH	SB	Lfo	GBTc	GBA	GBTi	SBI
	C6	Male	137.3	19.9	18.8	40.5	30.4	5.9	31	x	x	x	x
	C5	Female	188.2	26.2	21.8	56.1	14.1	6.6	39.1	117.7	88.2	69.8	23.1
	C17	Female	171.4	22.9	20.4	49.9	12.5	5.6	34.5	111.9	x	65.8	51.5
	CA65	Female	168.5	23.3	21.9	45.8	13.2	7.6	35.4	x	84.1	65.3	51.9
	CA66	Female	176.4	25.1	23.1	52.8	15.1	7.8	33.2	111.9	90.4	75.5	57.1
Femur	Specimen	Sex	GL	GLC	BP	DC	SD	Bd					
	C6	Male	x	119.4	27.5	14.7	10.7	21.2					
	C5	Female	165.1	168.4	38.2	15.1	14.9	32.9					
	C17	Female	155.3	150.9	34.4	18.4	14.8	31.2					
	CA65	Female	160.8	159.4	37.9	18.5	14.5	33.7					
	CA66	Female	171.3	166.7	83.8	18.7	16.2	33.1					
Tibia	Specimen	Sex	GL	Bp	SD	Bd	Dd						
	C6	Male	145.9	29.9	10.2	18.7	14.2						
	C5	Female	197.5	37.1	12.1	23.2	16.8						
	C17	Female	186.4	35.2	13.1	20.7	16.9						
	CA65	Female	188.7	36.8	12.2	22.3	18.2						
	CA66	Female	196.5	35.8	13.7	23.9	17.4						

Appendix I (cont.)

	Specimen	Sex	GL	GB			
Patella	C5	Female	24.1	19.5			
	C17	Female	19.8	15.3			
	CA65	Female	x	x			
	CA66	Female	x	x			
	Specimen	**Sex**	**GLl**	**GLm**	**Dl**	**Dm**	**Bd**
Astragalus	C6	Male	25.3	23.3	14.2	13.7	15.9
	C5	Female	26.1	24.5	14.2	14.5	16.9
	C17	Female	25.2	23.2	14.1	14.1	16.1
	CA65	Female	26.8	25.4	14.8	15.1	17.6
	CA66	Female	26.9	25.4	14.3	14.9	17.2
	Specimen	**Sex**	**GL**	**GB**			
Calcaneus	C6	Male	43.3	16			
	C5	Female	53.1	17.3			
	C17	Female	51.1	17.1			
	CA65	Female	55.3	17.9			
	CA66	Female	54.8	18.1			
	Specimen	**Sex**	**GB**				
Navicular	C6	Male	19.5				
	C5	Female	20.8				
	C17	Female	19.9				
	CA65	Female	21.3				
	CA66	Female	21.5				
	Specimen	**Sex**	**GL**	**Bp**	**SD**	**Bd**	**Dd**
Metatarsal	C6	Male	82.1	17.3	8.4	19	11
	C5	Female	112.5	18.7	11.1	21.3	13.8
	C17	Female	107.3	17.3	17.9	11.2	22.1
	CA65	Female	108.3	17.5	17.4	10.8	21.6
	CA66	Female	110.1	18.7	17.9	12.2	22.5
	Specimen	**Sex**	**GL**	**Bp**	**SD**	**Bd**	
Phalanx I	C6	Male	26.3	8.1	6.2	8.7	
	C5	Female	23.5	11.6	8.6	9.2	
	C17	Female	32.8	10.9	7.8	10.3	
	CA65	Female	34.6	10.6	8.2	10.3	
	CA66	Female	x	x	x	x	
	Specimen	**Sex**	**GL**	**Bp**	**SD**	**Bd**	
Phalanx II	C6	Male	16.6	7.9	5.5	7.1	
	C5	Female	13.3	11.6	8.9	12.5	
	C17	Female	22.2	9.9	7.2	8.3	
	CA65	Female	24.2	9.9	6.9	8.1	
	CA66	Female	x	x	x	x	
	Specimen	**Sex**	**DLS**	**Ld**	**MBS**		
Phalanx III	C6	Male	24.5	19.1	3.2		
	C5	Female	30.1	24.3	4.4		
	C17	Female	27.1	21.8	3.8		
	CA65	Female	25.9	20.6	4.5		
	CA66	Female	x	x	x		

Appendix II

Measurements of Agrimi skulls, according to von den Driesch 1976.

Skull	Sex	1	2	3	4	5	6	7	8	9	10	11	12	13	14	15	16	17	18	19	20	21	22
C8	x	x	x	x	x	x	x	x	85.4	51.2	52.9	45.3	x	16.9	24.1	x	x	x	x	x	x	46.1	16.1
CA71	Male	X	X	x	90.1	x	87.7	x	86.8	52.2	52.3	38.9	x	105.9	22.5	x	x	97.5	x	x	x	43.2	15.2
CA68	Male	x	x	x	104.3	x	107.3	x	100.3	53.7	58.9	97.6	x	122.1	28.3	x	x	112.5	x	x	x	57.3	30.5
C62	Male	x	x	x	107.2	x	115.8	x	100.2	52.4	60.3	96.1	x	123.1	30.1	53.2	x	115.2	x	x	x	56.9	29.3
C6	Male	x	x	x	x	x	x	x	x	x	x	x	x	x	x	x	x	x	x	x	x	x	x
C11	Male	162.2	153.4	137.9	96.2	42.4	93.2	89.6	92.9	51.4	57.1	42.8	105.6	111.6	27.4	47.2	93.1	105.2	84.2	67.4	58.9	53.4	29.5
C61	Male	x	x	x	x	x	x	x	107.3	52.2	67.1	106.1	x	127.8	34.5	x	x	x	x	x	x	55.3	30.1
CA72	Male	x	x	x	137.6	x	132.9	x	123.9	63.4	70.9	124.2	x	147.3	34.1	70.1	x	143.6	x	x	x	62.8	41.5
CA67	Male	218.6	207.9	191.3	134.7	58.8	135.9	126.3	127.6	66.1	75.4	128.1	139.1	152.1	40.1	73.5	128.9	143.3	118.1	93.79	80.1	60.1	40.9
C60	Male	x	x	x	150.8	x	144.1	x	130.1	61.9	78.1	71.5	x	162.4	42.3	76.3	x	157.2	x	x	x	67.3	44.2
C5	Female	179.5	169.1	151.8	117.2	35.9	109.9	95.5	108.8	69.9	64.4	44.2	116.1	135.6	34.2	x	100.3	118.1	84.9	65.3	44.3	35.1	24.1
CA66	Female	190.9	191.8	172.8	119.2	54.8	113.4	118.5	104.6	55.5	63.7	95.7	140.1	137.3	35.3	65.8	123.6	131.7	107.5	87.9	70.9	57.6	37.9
CA70	Female	x	x	x	122.7	x	111.5	x	107.8	61.3	91.8	100.8	x	140.5	33.5	x	x	130.5	x	x	x	54.4	36.5
CA69	Female	x	x	x	115.7	x	99.7	x	99.8	57.7	58.8	93.2	x	132.8	33.5	65.8	x	121.1	x	x	x	53.8	35.6
CA65	Female	x	x	x	52.7	x	102.3	x	101.1	55.8	61.8	93.4	x	132.1	25.4	x	x	123.7	x	x	x	53.2	36.1
C17	Female	189.1	180.8	157.2	110.1	52.3	103.2	106.8	102.4	57.5	60.8	93.1	127.3	132.9	32.1	60.8	112.9	120.7	101.6	82.9	62.2	47.3	29.9

Skull	Sex	23	24	25	26	27	28	29	30	31	32	33	34	35	36	37	38	39	40	41	42	Length of the hornecore
C8	x	31.7	31.8	29.6	x	x	x	x	x	36.1	53.9	62.8	80.8	52.4	44.7	19.7	27.1	45.8	19.5	11.9	7.4	14.8
CA71	Male	29.4	31.1	31.2	58.1	38.1	48.4	38.8	16.8	32.8	55.9	62.9	84.7	57.1	27.5	19.5	x	45.2	30.1	11.4	14.2	25.7
CA68	Male	29.4	32.5	34.8	66.9	42.4	57.2	22.1	20.7	37.3	72.5	66.9	97.3	69.1	53.7	24.2	34.1	50.6	76.2	27.2	19.3	109.1
C62	Male	29.5	34.1	35.6	64.1	43.4	55.2	21.4	17.9	34.9	68.7	67.2	93.6	68.5	53.6	23.7	31.2	50.2	72.1	17.8	24.9	99.6
C6	Male	x	31.4	30.7	x	37.5	49.1	17.3	16.8	x	56.2	x	87.1	75.4	x	x	x	46.9	37.8	x	6.5	36.8
C11	Male	26.6	31.9	32.9	58.2	39.1	51.2	20.1	18.1	42.8	61.8	61.4	90.1	63.2	54.4	21.6	32.1	50.4	13.2	12.1	53.6	64.5
C61	Male	25.2	33.6	38.4	66.9	x	x	x	x	35.4	68.7	68.5	101.3	72.9	58.6	25.9	33.2	54.1	85.2	30.2	19.7	140.2
CA72	Male	21.7	34.2	38.6	77.8	50.1	66.8	20.3	20.3	33.1	88.1	70.7	117.1	84.1	65.2	28.2	41.3	64.9	111.7	31.1	43.9	240.2
CA67	Male	19.3	34.2	36.9	79.7	49.3	69.8	22.4	19.3	33.9	88.7	74.9	116.8	86.8	66.1	26.8	39.8	63.6	122.8	46.8	29.8	195.3
C60	Male	24.5	35.8	40.1	84.2	50.8	79.1	29.8	29.8	28.9	89.4	73.4	117.3	89.5	67.4	33.9	42.3	64.5	134.1	57.2	33.6	392.6
C5	Female	33.2	33.8	37.4	41.8	60.8	22.4	20.3	42.9	56.6	70.4	107.3	70.4	60.3	34.6	35.4	57.6	65.4	22.6	16.1	91.4	x
CA66	Female	23.6	32.8	34.5	74.4	45.8	63.2	19.2	23.4	30.2	68.6	67.4	109.5	81.2	63.9	22.3	34.6	59.2	63.3	20.7	15.9	75.3
CA70	Female	19.2	36.3	38.7	73.6	42.9	61.7	20.7	20.4	27.9	63.1	68.9	109.7	80.5	63.2	24.2	10.9	60.3	63.3	20.4	16.1	85.3
CA69	Female	19.6	34.3	35.8	62.3	36.8	56.7	17.7	18.1	31.3	61.9	67.3	100.1	71.8	58.5	23.9	38.4	52.9	49.6	12.4	18.1	59.3
CA65	Female	18.1	35.8	36.9	67.8	41.2	56.9	19.9	19.7	31.7	67.1	65.9	105.1	73.4	56.9	22.4	35.2	56.2	60.8	16.2	20.5	83.1
C17	Female	18.1	34.2	33.1	66.6	56.8	39.9	21.4	20.6	30.8	61.2	65.9	104.4	76.2	61.1	23.4	35.3	53.4	55.7	17.1	11.1	66.9

Appendix III

Measurements of Agrimi mandibulae, according to von den Driesch 1976.

	Age class		Identification																	
N	Years	Deniz & Payne	Mandible	Laterality	1	2	3	4	5	6	7	8	9	11	12	13	14	15a	15b	15c
1	0-6 months	B	C8	L	94.8	100.3	25.4	71.2	64.1	47.3	41.1	x	31.1	22.7	40.1	40.1	49.9	21.7	16.3	8.1
2			C9	L	102.5	109.7	27.4	74.1	69.5	81.1	43.6	x	28.6	24.6	45.8	47.8	67.4	25.4	16.6	9.4
3			CA25	L	98.1	102.4	26.6	70.7	66.8	78.1	40.5	25.4	13.8	22.8	32.4	31.4	46.6	23.2	12.8	9.6
4			CA23	L	96.8	102.9	24.9	73.6	65.8	77.1	42.9	27.6	13.4	25.9	34.2	35.1	51.4	23.7	13.2	9.1
5			CA24	L	103.7	111.2	29.9	76.7	71.3	82.7	43.6	28.1	13.6	25.1	34.7	33.7	X	25.5	13.2	9.2
6			CA26	R	103.8	110	29.3	75.2	71.3	82.9	44.1	28.5	13.7	25.9	35.7	32.3	52.5	25.4	13.2	10.3
7			CA29	R	105.6	112.6	30.1	76.6	71.1	83.1	43.2	28.2	13.5	26.9	36.4	35.2	51.4	25.9	18.3	10.4
8	6-12 months	C	CA28	L	103.8	110.1	28.3	77.1	69.8	81.4	43.3	30	13.2	28.3	36.7	36.5	x	25.6	14.1	10.2
9			CA27	R	105.6	112.5	30.2	75.7	69.8	83.4	40.4	26.9	12.4	29.2	38.6	35.7	53.2	22.4	17.2	10.1
10			CA30	L	105.2	112.6	30.4	76.1	69.3	82.7	40.3	26.4	12.9	30	36.1	35.8	54.3	23.1	14.1	11.1
11			CA31	L	118.3	123.2	36.6	82.3	79.4	93.2	44.1	28.6	13.8	30.2	40.9	40.1	X	24.5	15	10.9
12			CA33	L	108.9	115.8	33.8	75.7	74.8	84.7	42.9	27.6	13.5	25.4	41.2	40.5	56.2	24.2	14.4	10.1
13			CA34	R	108.7	115.8	32.8	75.1	74.8	84.8	43.1	28.6	13.9	25.8	36.9	35.1	54.9	25.6	13.7	9.9
14			CA35	R	x	x	33.3	x	83.5	94.9	53.2	39.1	15.6	x	48.1	x	x	31.6	17.4	11.7
15			C1	L	x	x	30.3	x	71.9	83.8	42.1	x	27.5	26.5	47.7	48.9	x	24.5	17.6	9.5
16			C2	L	107.9	110.8	30.3	68.2	71.3	84.1	41.6	x	28.4	26.5	46.5	48.6	68.2	24.5	17.4	8.3
17			C10	L	107.4	114.7	27.3	84.6	75.7	83.6	53.9	x	28.5	26.2	47.9	48.6	x	27.6	16.3	9.8
18			C11	L	115.9	124.7	31.7	86.1	80.5	91.3	53.8	x	39.8	28.6	56.8	52.5	71.7	32.5	21.9	10.8
19			C6	L	117.4	124.1	30.2	86.8	84.2	95.4	55.9	x	28.9	x	52.3	50.3	72.4	29.2	15.8	10.4
20	1-2 years	D	CA36	R	X	X	27.9	X	76.6	85.5	53.1	37.9	14.5	25.8	47.2	34.1	64.2	28.1	13.9	10.6
21			CA37	L	126.9	136.3	34.9	93.2	87.8	97.9	53.4	34.3	15.1	32.6	48.2	45.3	71.6	29.7	15.8	11.9
22			CA38	R	128.5	137.2	34.2	96.3	87.8	102.5	56.8	41.3	14.6	33.2	47.8	43.5	66.8	31.3	15.9	12.5
23			C12	L	127.9	136.3	34.7	95.5	88.1	101.8	55.5	x	25.6	33.8	55.1	54.1	x	29.8	18.7	10.8
24			CA39	R	128.8	137.7	34.7	94.2	87.4	99.9	53.2	38.7	13.8	33.2	49.9	45.7	68.7	28.9	15.8	12.5
25			CA40	L	x	x	38.1	x	x	x	x	43.3	x	x	49.5	48.7	73.3	31.2	18.5	x
26			CA18	L	136.4	144.1	41.7	94.3	95.8	110.1	55.4	29.5	26.1	33.6	52.1	54.8	80.1	29.5	17.4	13.8
27			C13	L	x	x	35.9	x	95.3	108.8	65.4	40.1	22.1	x	57.4	54.8	x	32.3	17.9	11.7
28			CA22	R	x	x	41.1	x	94.1	111.1	54.1	29.8	24.1	x	64.5	54.9	x	31.2	18.4	14.1
29	2-4 years	EF	CA15	L	140.3	149.1	34.1	102.2	99.1	115.7	68.5	44.6	23.6	33.1	55.7	50.1	73.6	34.9	17.6	13.6
30			CA20	L	135.4	143.9	34.2	103.4	95.8	108.1	64.8	41.4	24.1	31.5	52.7	48.1	x	32.1	17.2	13.3
31			CA21	R	x	x	33.6	x	95.8	109.1	65.9	42.8	25.1	34.4	54.7	49.1	x	32.6	16.9	14.5
32			C3	R	130.7	139.6	31.1	99.5	92.7	109.4	65.2	41.3	23.2	33.9	60.5	57.5	x	33.3	17.2	12.2
33			CA2	R	155.6	167.1	42.2	114.6	108.2	122.2	65.4	44.3	23.2	38.4	60.3	54.1	83.5	32.1	18.8	14.9
34			CA6	R	144.4	151.9	36.1	105.9	100.1	119.4	61.9	42.4	20.2	37.7	55.1	51.6	71.3	25.7	16.2	14.
35			C4	L	144.1	152.9	35.7	10.9	98.9	108.4	63.8	41.1	30.1	37.8	54.5	52.1	71.6	26.4	16.8	13.1
36			CA16	L	155.8	167.3	42.3	113.4	108.1	125.4	65.4	34.3	21.6	37.4	61.7	55.9	81.6	33.1	18.9	14.4
37			C14	L	164.7	170.4	50.8	116.7	112.2	132.7	65.4	44.8	20.2	43.1	58.9	56.1	81.1	31.9	19.4	14.1
38			CA72	L	160.1	172.8	41.3	119.6	109.5	132.1	68.5	46.4	21.9	40.1	59.6	56.6	x	30.8	19.1	15.1
39			C7	L	x	x	34.2	x	98.3	116.1	64.8	43.5	19.8	35.5	51.8	48.5	69.6	28.6	16.5	13.8
40	4-6 years	G	CA1	R	163.7	170.1	51.1	111.4	115.5	131.6	64.1	43.6	20.2	39.4	59.9	53.2	x	30.2	19.2	14.2
41			CA17	L	152.4	159.3	41.6	109.9	106.2	120.8	62.7	42.8	21.1	39..1	58.5	53.9	80.5	28.5	17.5	12.
42			CA67	L	162.7	168.9	48.8	115.1	112.7	134.4	64.5	44.8	20.3	41.4	62.7	59.2	85.2	31.9	19.6	16.
43			CA11	R	136.9	146.3	36.7	101.6	95.3	111.9	61.7	41.4	20.7	34.5	50.9	46.7	68.3	29.9	16.7	12.4
44			CA4	R	160.1	168.8	46.1	113.4	112.4	129.2	65.7	45.3	20.7	39.4	61.7	55.9	83.6	28.9	18.7	16.
45			CA7	R	145.5	154.1	38.2	108.1	99.6	118.2	62.9	43.1	19.5	19.1	54.5	48.3	71.8	26.3	16.3	13.
46			CA13	L	137.9	144.5	36.1	102.8	95.5	112.4	59.7	40.9	20.1	32.7	50.1	47.6	x	27..7	16.5	11.
47			C18	L	x	x	38.3	x	97.3	115.9	59.9	39.5	19.2	x	51.7	49.9	71.1	22.9	16.8	18.
48			CA10	R	164.8	174.7	48.1	115.1	116.1	134.3	66.5	44.6	21.2	41.8	61.9	55.8	86.1	29.1	20.6	15.
49			CA8	R	145.5	152.1	38.8	104.8	102.3	120.1	62.7	41.4	20.3	33.1	55.5	50.5	x	27.2	17.5	12.
50			CA19	L	144.8	149.9	36.5	106.3	99.1	117.9	62.5	43.5	19.4	38.1	54.8	49.8	75.3	26.1	17.5	13.
51			CA5	R	x	x	33.7	109.4	94.1	115.2	68.4	45.8	24.5	34.9	56.9	52.1	x	34.5	18.3	13.
52	> to 6 years	HI	CA9	L	141.7	148.6	43.4	96.9	101.5	115.1	55.9	34.9	19.8	36.4	57.6	53.8	x	25.7	17.6	13.
53			C16	L	141.8	151.9	37.8	102.2	101.9	117.2	65.1	43.5	20.8	35.4	54.9	50.5	75.8	26.1	14.9	14.
54			CA14	L	142.6	151.7	41.9	99.7	97.9	116.5	57.2	36.2	19.1	35.8	58.8	54.2	79.8	26.5	15.7	14.
55			C5	L	151.2	156.9	43.4	107.3	104.9	121.5	60.4	40.5	20.9	37.8	58.5	54.9	80.3	26.6	19.5	12.
56			C15	L	143.8	153.3	37.9	104.7	101.5	118.3	63.4	42.9	20.7	34.2	55.8	51.9	x	24.8	17.5	12.
57			CA65	L	138.4	147.7	35.4	98.8	95.1	110.8	56.5	36.2	19.2	43.5	50.5	50.7	71.5	20.7	15.5	11.
58			C17	L	140.9	147.7	41.5	98.2	97.5	114.7	55.1	35.4	19.1	37.8	48.2	43.1	64.1	23.9	14.2	12.

Bibliography of Dr. Hijlke Buitenhuis

Esther E. Scheele

1978

Clason, A. T. & H. Buitenhuis, 1978. Archeozoölogisch onderzoek in het Midden-Oosten. *Spiegel Historiael* 13, 677-687.

Clason, A. T. & H. Buitenhuis, 1978. A preliminary report on the faunal remains of Nahr el Homr, Hadidi, and Ta'as in the Tabqa Dam Region in Syria. *Journal of Archaeological Science* 5, 75-83

1979

Buitenhuis, H., 1979. The faunal remains from Tell Hadidi. In: M. Kubasiewicz (ed.). *Archaeozoology. Volume* I, 164-175

1983

Zeist, W. van & H. Buitenhuis, 1983. A Palaeobotanical study of Neolithic Erbaba, Turkey. *Anatolica* 10, 47-89. Buitenhuis, H., 1983. The animal remains of Tell Sweyhat, Syria. *Palaeohistoria* 25, 131-144

1984

Buitenhuis, H., 1984. Some remarks on the possible developments of Bos taurus breeds in the Near East. In: J. Clutton-Brocj & C. Grigson (eds.), *Animals and Archaeology* 3. *Early Herders and their flocks*. Oxford, 213-221 (B.A.R. Intern. Series 202).

Buitenhuis, H., 1984. Archeozoölogisch onderzoek in het gebied van de boven- en middenloop van de Euphraat in Turkije en Syrië, van het Epipaleolithicum (ca. 10.000 v. Chr.) tot aan de Byzantijns – Arabische periode (ca. 1200 n. Chr.). *Jaarverslag ARCHON* 1984, 110-115.

1985

Buitenhuis, H., 1985. Preliminary report on the faunal remains of Hayaz Höyük from the 1979-1983 seasons. *Anatolica* 12, 61-69.

Buitenhuis, H., 1985. The animal remains of Tell es-Sweyhat, Syria. *Palaeohistoria* 25, 131-144.

1988

Buitenhuis, H., 1988. *Archeozoölogisch onderzoek langs de Midden-Eufraat, onderzoek van het faunamateriaal uit zes nederzettingen in Zuidoost-Turkije en Noord-Syrië daterend van ca. 10.000 BP tot 1400 AD. PhD-Thesis.* University of Groningen.

1989

Clason, A.T. & H. Buitenhuis, 1989. III.b Animals. In: G. van der Kooi & M.M. Ibrahim (eds.). *Picking up the threads... A continuing review of excavations at Deir Alla, Jordan.* Leiden, 38-42.

Clason, A.T. & H. Buitenhuis, 1989. III.b Dieren. In: G. van der Kooi & M.M. Ibrahim (eds.). *Een verhaal voor het oprapen. Opgravingen te Deir Alla in de Jordaanvallei.* Leiden 38-42.

1990

Roodenberg, J., L. Thissen & H. Buitenhuis, 1990. Preliminary report on the archaeological investigations at Ilıpınar in NW Anatolia. *Anatolica* 34, 61-144.

Buitenhuis, H., 1990. Archaeozoological aspects of Late Holocene economy and environment in the Near East. In: S. Bottema, G. Entjes-Nieborg & W. van Zeist. *Man's role in the shaping of the Eastern Mediterranean landscape,* 195-206.

1991

Buitenhuis, H., 1991. Some equid remains from South Turkey, North Syria and Jordan. In: R.H. Meadow & H.-P. Uerpmann (eds.). *Equids in the ancient world. Volume* II, 34-74.

1992

Buitenhuis, H., 1992. Archeozoölogisch onderzoek bij de zee van Marmara, Turkije. *Paleo-aktueel* 3, Groningen, 50-54.

1993

Buitenhuis, H. & A.T. Clason (eds.), 1993. *Archaeozoology of the Near East. Proceedings of the First International Symposium on the Archaeozoology of Southwestern Asia and adjacent areas.* Leiden

1994

Buitenhuis, H., 1994. Note on archaeozoological research around the sea of Marmara. *Anatolica* 20, 141-144.

1995

Buitenhuis, H. & H.-P. Uerpmann (eds.), 1995. *Archaeozoology of the Near East II. Proceedings of the Second International Symposium on the Archaeozoology of Southwestern Asia and adjacent areas.* Leiden.

Buitenhuis, H., 1995. A quantative approach to species determination of Ovicapridae. In: H. Buitenhuis & H.-P. Uerpmann (eds.). *Archaeozoology of the Near East* II. *Proceedings of the Second International Symposium on the Archaeozoology of Southwestern Asia and adjacent areas.* Leiden, 140-155.

Buitenhuis, H. 1995. The faunal remains. In: Roodenberg, J. (ed.), *The Ilıpınar Excavations I: Five seasons of fieldwork in NW Anatolia* 1987-91. Leiden and Istanbul, 151-157.

1996

Buitenhuis, H., 1996. Archaeozoology of the Holocene in Anatolia: a review. In: Ş. Demirci, A.M. Özer & G.D. Summers (eds.), 1996. *Archaeometry* 94. *The proceedings of the 29th Int. Symposium on Archaeometry (Ankara 9-14 May, 1994),* 411-421.

Buitenhuis, H., 1996. A quantitative approach to species determination of Ovicapridae. In: H. Buitenhuis & H.-P. Uerpmann (eds.), *Archaeozoology of the Near East* II. Leiden, 140-155.

Buitenhuis, H. & A.F.L. van Holk, 1996. Botten aan boord. *Paleo-aktueel* 7, 113-116.

1997

Clason, A. & H. Buitenhuis, 1997. Change and continuity in the animal food resources in Bronze Age towns of the Orient. In: G. Wilhelm (eds.), 1997. *Die Orientalische Stadt: Kontinuität, Wandel, Bruch. 1. Internationales Colloquim der deutschen Orient-Gesellschaft 9.-10. Mai 1996 in Halle/Saale.* Saarbrücken, 199-219.

Buitenhuis, H., 1997. Aşıklı Höyük: a "protodomestication" site. In: M. Kobaki & J. Wahl (eds.). *Anthropozoologica* 25-26, 655-670.

Buitenhuis, H., 1997. Aşıklı Höyük: vroege domesticatie en veeteelt. *Paleo-aktueel* 8. Groningen, 38- 42.

1998

Buitenhuis, H., L. Bartosiewicz & A.M. Choyke (eds.), 1998. *Archaeozoology of the Near East III. Proceedings of the Third International Symposium on the Archaeozoology of Southwestern Asia and adjacent areas.* Groningen (ARC publication 18).

Clason, A.T. & H. Buitenhuis, 1998. Patterns in animal food resources in the Bronze Age in the Orient. In: H. Buitenhuis, L. Bartosiewicz & A.M. Choyke (eds.). *Archaeozoology of the Near East* III. *Proceedings of the Third International Symposium On the Archaeozoology of Southwestern Asia and adjacent areas.* Groningen, 233-242 (ARC publication 18).

Buitenhuis, H. & I. Caneva, 1998. Early animal breeding in South-Eastern Anatolia: Mersin Yumuktepe. In: P. Anreiter, L. bartosiewicz, E. Jerem & W. Meid. *Man and the animal world. Studies in Archaeozoology, Archaeology, Anthropology and Palaeolinguistics in memoriam Sándor Bökönyi,* 121-130.

1999

Vigne, J.-D., H. Buitenhuis & S. Davis, 1999. Les premiers pas de la domestication animale à l'ouest de l'Euphrate : Chypre et l'Anatolie central. *Paléorient,* Vol. 25, No. 2 (1999), 49-62.

Buitenhuis, H., 1999. A first note on the faunal remains of Güvercinkayası. *Anatolica* 25, 64-69.

Buitenhuis, H., 1999. Archeozoölogie. In: Y. Dijkstra, O. Brinkkemper & C.W. Koot (red.). *Archeologie in de Betuweroute. Papendrechtse Stroomrug.* Amersfoort, 79-92 (Rapportage Archeologische Monumentenzorg 80).

2000

Mashkour, M., A.M. Choyke, H. Buitenhuis & F. Poplin (eds.), 2000. *Archaeozoology of the Near East IVB. Proceedings of the Fourth International Symposium on the Archaeozoology of Southwestern Asia and adjacent areas.* Groningen (ARC publication 32).

2001

Balkan-Atlı N., D. Binder, V. Beugnier, H. Buitenhuis, C. Kuzucuoğlu, D. Mouralis, L. Slimak, I. Thery, 2001. Les ateliers de taille d'obsidienne. Fouilles de Kömürcü - Kaletepe 2000. *AnatoliaAntiqua*, Tome 9, 193-205

Buitenhuis, H., 2001. De eerste boeren in Nederland / The first farmers in the northern Netherlands. In: L. Domboróczki &A. Ufkes (eds.), *Early Farmers in Europe*. Valencia, 107-111.

Buitenhuis, H., 2001. A first survey of the faunal remains from Tepecik Ciftlik, Turkey. Internal Report, Istanbul University

Buitenhuis, H., 2001. Archeozoölogie. In: J.S. Krist, J.B. de Voogd & J. Schoneveld, *Een vindplaats uit de Late IJzertijd en Vroeg-Romeinse Tijd aan de Schalkwijkse weg te Houten, terrein 14, Provincie Utrecht*. Groningen, 135-168 (ARC-Publicaties 48).

Buitenhuis, H., 2001. Bijlage 5. Archeozoölogisch onderzoek. In: J. Krist, *Aanvullend Archeologisch Onderzoek in het tracé van de Betuweroute, vindplaats 46, Nijmegen/Verloren Zeeg.* Amersfoort, 35-39 (Rapportage Archeologische Monumentenzorg 40).

Buitenhuis, H., 2001. Archeozoölogie. In: J. Schoneveld& E.F. Gehasse (red.), *Archeologie in de Betuweroute. Boog C-Noord, een vindplaats bij Meteren op de overgang van Neolithicum naar Bronstijd.* Amersfoort, 141-168 (Rapportage Archeologische Monumentenzorg 84).

2002

Buitenhuis, H., A.M. Choyke, M. Mashkour & A.H. Al-Shiyab (eds.), 2002. *Archaeozoology of the Near East V. Proceedings of the Fifth International Symposium on the Archaeozoology of Southwestern Asia and adjacent areas.* Groningen (ARC publication 62).

Özbaşaran, M. & H. Buitenhuis, 2002. Proposal for a Regional Terminology for Central Anatolia.In: F. Gerard & L. Thissen (eds.), *The Neolithic of Central Anatolia, Internal Developments and External Relations During the 9th-6th millenia cal. BC.* Istanbul, 67-77.

Buitenhuis, H., 2002. Archeozoölogie. In: J. Schoneveld & P.S.M. Kranendonk (red.*), Archeologie in de Betuweroute. Drie erven uit de Midden-Bronstijd bij Lienden.* Amersfoort, 191-226 (Rapportage Archeologische Monumentenzorg 89).

Buitenhuis, H. & H. Halici, 2002. Archeozoölogie. In: J. Milojkovic en E. Smits (red.), *Lage Blok. Een nederzettingsterrein uit de Midden-IJzertijd bij Meteren (gemeente Geldermalsen).* Amersfoort, 149-178 (Rapportage Archeologische Monumentenzorg 90).

Buitenhuis, H., 2002. Dierlijk botmateriaal. In: M.J.L.Th. Niekus, *Archeologisch onderzoek van een vroegmiddeleeuws terpje bij Sneek, gem. Sneek, Friesland.* Groningen, 27-30 (ARC-Publicaties 54).

Buitenhuis, H., 2002. Dierlijk botmateriaal. In: J.B. Hielkema, *Archeologisch onderzoek bij de Boltastate in Harlingen, provincie Friesland.* Groningen, 15-18 (ARC-Publicaties 55).

Buitenhuis, H., 2002. Archeozoölogisch onderzoek. In: J.S. Krist & C.G. Koopstra, *Aanvullend Archeologisch Onderzoek plangebied Hema terrein te Huissen, gemeente Bemmel (Gld.).* Groningen, 15-18 (ARC-Publicaties 56).

Buitenhuis, H., 2002. Faunaresten AAO en DO. In: J.S. Krist, *Archeologisch onderzoek te Huissen-Bloemstraat fase 3 (provincie Gelderland).* Groningen, 43-72 (ARC-Publicaties 57).

Buitenhuis, H., 2002. Faunaresten. In: J.B. Hielkema, *Aanvullend Archeologisch Onderzoek op bedrijvenpark Medel, vindplaats 1, gemeente Echteld-Tiel (Gld.).* Groningen, 25-30 (ARC-Publicaties 65).

2003

Özbaşaran, M., G. Duru, H. Buitenhuis, N. Kayacan & R. Christidou, 2003. Musular 2001 Yılı Kazısı. *24. Kazı Sonuçları Toplantısı* 24, 259- 268 Buitenhuis, H., 2003. Faunaresten. In: M.J.M. de Wit, *Een Definitief Archeologisch Onderzoek langs de Frieslandweg te Emmen, gemeente Emmen (Dr.).* Groningen, 119-122 (ARC-Publicaties 64).

Buitenhuis, H., 2003. Faunaresten. In: J.B. Hielkema, *Aanvullend Archeologisch Onderzoek op terrein 9 te Houten-Loerik, gemeente Houten (U.).* Groningen, 35-46 (ARC-Publicaties 69).

Buitenhuis, H., 2003. Faunaresten. In: P.B. Kooi & M.J.M. de Wit, *Een Definitief Archeologisch Onderzoek langs de Rijksweg N34 te Borger, gemeente Borger-Odoorn (Dr.).* Groningen, 49-50 (ARCPublicaties 71).

Buitenhuis, H., 2003. Faunaresten. In: J.S. Krist, *Een Aanvullend Archeologisch Onderzoek op vindplaats 10 binnen het plangebied 'Deest aan het Water' te Deest, gemeente Druten (Gld.).* Groningen, 31-38 (ARC-Publicaties 76).

Buitenhuis, H., 2003. Faunaresten. In: J.S. Krist & G.J. de Roller, *Aanvullend Archeologisch Onderzoek op de nieuwbouwlocatie 'Ledelplein 16' te Oostburg, gemeente Oostburg (Z.).* Groningen, 17-21 (ARCPublicaties 77).

Buitenhuis, H., 2003. Faunaresten. In: J.B. Hielkema, *Een boerenerf uit de Bronstijd. Een archeologische opgraving te Tiel-Medel-Lingewei, vindplaats 1, gemeente Tiel (Gld.).* Groningen, 75- 82 (ARC-Publicaties 79).

Buitenhuis, H., 2003. Faunaresten. In: C.G. Koopstra, *Een Definitief Archeologisch Onderzoek aan de Muggenlaan te Nieuwveen, gemeente Liemeer (Z.-H.).* Groningen, 31-33 (ARC-Publicaties 81).

Buitenhuis, H., 2003. Faunaresten. In: A. Ufkes, *Een archeologisch onderzoek op het terrein 'Alhambra' te Vlissingen (Z.).* Groningen, 135-144 (ARC-Publicaties 85).

Buitenhuis, H., 2003. Faunaresten. In: A. Ufkes & J.R. Veldhuis, *Nederzettingssporen uit de Bronstijd bij Zwaagdijk-Oost. Een definitief archeologisch onderzoek in het Plangebied Zwaagdijk-Oost, gemeente Wervershoof (N.-H.).* Groningen, 155-187 (ARC-Publicaties 113).

2004

Slimak, L., Roche, H., Mouralis, D., Buitenhuis, H., Balkan-Atli, N., Binder, C., Kuzucuoglu, C. and Grenet, M., 2004. Kaletepe Deresi 3 (Turquie). *Aspects archéologiques, chronologiques et paléontologiques d'une sequence Plestocène en Anatolie central,* Comptes-Rendus Palevol de l'Académie des Sciences de Paris 3, 411-420.

Buitenhuis, H., 2004. Faunaresten. In: A. Nieuwhof, *Een archeologisch inventariserend veldonderzoek (IVO) door middel van proefsleuven aan de Zuidweg te Naaldwijk (Z.-H.).* Groningen, 43-49 (ARCPublicaties 94).

Buitenhuis, H., 2004. Dierlijk bot. In: M.J.M. de Wit & A. Nieuwhof, *Een bijzonder ijzeren schild uit de gracht van de tweede Ripperdaborg. Een archeologisch inventariserend veldonderzoek (IVO) door middel van proefsleuven aan de Geert Reinderstraat te Winsum, gemeente Winsum (Gr.).* Groningen, 21-26 (ARC-Publicaties 106).

Buitenhuis, H., 2004. Faunaresten. In: S.J. Tuinstra & J. Vanden Borre, *Een archeologisch inventariserend veldonderzoek (IVO) door middel van boringen en proefsleuven bij het plangebied 't Slot Noord te Bergambacht, gemeente Bergambacht (Z.H.).* Groningen, 47-49 (ARC-Publicaties 107).

2005

Russel, N, L. Martin & H. Buitenhuis, 2005. Cattle Domestication at Çatalhöyük revisited. *Current Anthropology, Volume* 46, *Supplement, December* 2005, 101-133.

Buitenhuis, H., A.M. Choyke, L.A. Martin, L. Bartosiewicz & M. Mashkour (eds.), 2005. *Archaeozoology of the Near East VI. Proceedings of the Sixth International Symposium on the Archaeozoology of Southwestern Asia and adjacent areas.* Groningen (ARC publication 123).

Buitenhuis, H., 2005. Faunaresten. In: A. Ufkes, *Een nederzetting uit de Vroege Bronstijd te Tiel- Medel. Een archeologische opgraving te Tiel-Medel 'Oude Weiden', vindplaats 5, gemeente Tiel (Gld.).* Groningen, 101-111 (ARC-Publicaties 124).

Buitenhuis, H., 2005. Faunaresten. In: J.B. Hielkema, *Een bureau-onderzoek en een archeologische opgraving met beperkingen bij de aanleg van een persleiding tussen Ruinen en Echten, gemeente De Wolden (Dr.).* Groningen, 17-18 (ARC-Publicaties 126).

Buitenhuis, H., 2005. Faunaresten. In: S.J. Tuinstra, M.A. Huisman & A. Nieuwhof, *Een archeologische opgraving in het kader van de aanleg van de infrastuctuur in de VINEX-wijk Schuytgraaf te Arnhem-Zuid, gemeente Arnhem (Gld.).* Groningen, 55-59 (ARC-Publicaties 129).

2006

Buitenhuis, H., D.C. Brinkhuizen, C. van Loon en T. de Ridder (red.), 2006. *Gat in de Markt 1.101. Het dierlijk botmateriaal.* Groningen (VLAK-verslag 15.5).

Buitenhuis, H., 2006. Faunaresten. In: B. Silkens, *Een archeologisch inventariserend veldonderzoek (IVO) door middel van proefsleuven aan de Nedereindsestraat te Kesteren, gemeente Neder-Betuwe (Gld.).* Groningen, 39-41 (ARC-Publicaties 143).

Buitenhuis, H., 2006. Faunaresten. In: S.J. Tuinstra, *Een archeologisch inventariserend veldonderzoek (IVO) door middel van proefsleuven bij rijksmonument De Trije Terpen te Dokkum, gemeente Dongeradeel (Fr.).* Groningen, 73-75 (ARC-Publicaties 146).

Buitenhuis, H., 2006. Faunaresten. In: J.Y. Huis in 't Veld, *Midden-Bronstijd huizen onder een Romeinse akker. Een archeologische opgraving op het plangebied Hogewald II te Beuningen, gemeente Beuningen (Gld.).* Groningen, 97-102 (ARC-Publicaties 157).

2007

Pearson, J.A., H. Buitenhuis, R.E.M. Hedges, L. Martin, N. Russel & K.C. Twiss, 2007. New light on early caprine herding strategies from isotope analysis: a case study from Neolithic Anatolia. *Journal of Archaeological Science* 34, 2170-2179.

Buitenhuis, H., 2007. Faunaresten. In: A. Ufkes & B. Silkens, *Prehistorische boeren en laatmiddeleeuwse tollenaars langs de Oude Doetinchemseweg. Een archeologische opgraving bij Wijnbergen 'De Kap', gemeente Doetinchem (Gld.).* Groningen, 133-136 (ARC-Publicaties 161).

Buitenhuis, H., 2007. Faunaresten. In: B. Silkens, P.J.A. Stokkel & M.J.M. de Wit, *Nederzettingssporen uit de IJzertijd en de Middeleeuwen. Een archeologische opgraving aan de Kapelstraat te Loil, gemeente Montferland (Gld.).* Groningen, 85-88 (ARC-Publicaties 183).

Buitenhuis, H., 2007. Faunaresten. In: M.J.M. de Wit & G.M.A. Bergsma. *Een archeologische begeleiding (AB) in het trace van de persleiding van Midlaren tot de Plankensloot, gemeente Tynaarlo (Dr.),* 22-23 (ARC-Publicaties 176).

2008

Buitenhuis, H., 2008. Ilipinar: The faunal remains from the late Neolithic and early Chalcolithic levels. In: E. Vilà, L. Goucherin, A. Choyke & H. Buitenhuis (eds.). *Archaeozoology of the Near East* VIII. *Proceedings of the Eight International Symposium on the Archaeozoology of Southwestern Asia and adjacent areas.* Lyon: Travaux de la Maison de l'Orient et de la Méditerranée (TMO), 299-322.

Russel, A. & H. Buitenhuis, 2008. Tell Damishliyya faunal bone report. *Anatolica* 34, 315-337.

Slimak, L., S.L. Kuhn, H. Roche, D. Mouralis, H. Buitenhuis, N. Balkan-Atli, D. Binder, C. Kuzucuoğlu & H. Guillou, 2008. Kaletepe Deresi 3 (Turkeu): Archaeological evidence for early human settlement in Central Anatolia. *Journal of Human Evolution* 54, 99-111.

Buitenhuis, H., D.C. Brinkhuizen & J. Schoneveld, 2008. Faunaresten. In: A. Ufkes, *Een middeleeuws leerbewerkerskwartier in Sneek. Een archeologische opgraving aan het Martiniplein te Sneek, gemeente Sneek (Fr.).* Groningen, 155-220 (ARC-Publicaties 17).

Buitenhuis, H., 2008. Walvisbotten. In: P.J.A. Stokkel, *Een archeologische begeleiding langs de Oudelandertocht, gemeente Wieringermeer (N.-H.). Geldermalsen,* 4-6 (ARC-Rapporten 2008-128).

Buitenhuis, H., 2008. Faunaresten. In: G.J. de Roller, *De kazemat van de muurtoren aan de Weeshuisstraat te Buren. Een archeologisch inventariserend veldonderzoek (IVO) door middel van bureau-onderzoek, een proefsleuf en een archeologische opgraving in Buren, gemeente Buren (Gld.).* Groningen, 73-75 (ARC-Publicaties 150).

Buitenhuis, H., 2008. Faunaresten. In: B. Silkens & P.J.A. Stokkel, *Een archeologisch inventariserend veldonderzoek door middel van proefsleuven aan de Marathonweg te Vlaardingen (Z.-H.).* Groningen, 23-24 (ARC-Publicaties 186).

Buitenhuis, H., 2008. Overig vondstmateriaal. In: P.J.A. Stokkel, *Resten uit het Mesolithicum, IJzertijd/ Romeinse Tijd en de Nieuwe Tijd: een archeologische noodopgraving aan de Middegaal te Veghel, gemeente Veghel (N.-Br.).* Groningen, 47-50 (ARC-Publicaties 188).

Buitenhuis, H., 2008. Faunaresten. In: A. Ufkes & S.J. Tuinstra, *Soldatenbarakken op het Bolwerk Zeeland van de vesting Coevorden. Een archeologisch inventariserend veldonderzoek (IVO) door middel van proefsleuven op het Bogasterrein te Coevorden, gemeente Coevorden (Dr.).* Groningen, 53-56 (ARC-Publicaties 190).

Vilà, E., L. Goucherin, A. Choyke & H. Buitenhuis (eds.), 2008. *Archaeozoology of the Near East VIII. Proceedings of the Eight International Symposium on the Archaeozoology of Southwestern Asia and adjacent areas.* Lyon: Travaux de la Maison de l'Orient et de la Méditerranée (TMO).

2009

Buitenhuis, H., 2009. Faunaresten. In: M.J.M. de Wit, G.M.A. Bergsma & M. Essink, *Een archeologische opgraving op plangebied 'Daalkampen II', fase 1, te Borger, gemeente Borger-Odoorn (Dr.).* Groningen, 135-136 (ARC-Publicaties 189).

Buitenhuis, H., 2009. Faunaresten. In: A. Ufkes, *Een archeologische opgraving aan de Agnietenstraat te Tiel, gemeente Tiel (Gld.).* Groningen, 33-41 (ARC-Publicaties 199).

Buitenhuis, H., 2009. Faunaresten. In: P.J.A. Stokkel, *Middeleeuwse erven op Goeree Overflakkee. Een archeologische opgraving aan de Smalle Einde te Ouddorp, gemeente Goedereede (Z.-H).* Groningen, 83-96 (ARC-Publicaties 200).

Buitenhuis, H., 2009. Vondstmateriaal. In: M.J.M. de Wit, *Een archeologisch inventariserend veldonderzoek (IVO) door middel van proefsleuven op de Noordbargeres, in het kader van bestemmingsplan Centrumplan West te Emmen, gemeente Emmen (Dr.).* Groningen, 23-24 (ARCPublicaties 203).

2010

Buitenhuis, H., 2010. Faunaresten. In: A. Ufkes, *'De Hof'. Een bezit van klooster Echternach. Een archeologische opgraving van sporen uit de IJzertijd en de Volle Middeleeuwen op het plangebied 'Neerakker' te Bakel, gemeente Gemert-Bakel (N.-Br.).* Groningen, 265-268 (ARC-Publicaties 204).

Buitenhuis, H., 2010. Faunaresten. In: M.J.M. de Wit & P.J.A. Stokkel, *Middeleeuwse erven en Nieuwe Tijdse bebouwing op plangebied 'De Borch'. Een archeologische opgraving te Varsseveld, gemeente Oude IJsselstreek (Gld.).* Groningen, 67-70 (ARC-Publicaties 207).

Buitenhuis, H., 2010. Faunaresten. In: M.J.M. de Wit & P.J.A. Stokkel, *Een Romeinse nederzetting in Huissen. Een archeologische opgraving in het verlengde van de Hortensialaan te Huissen, gemeente Lingewaard (Gld).* Groningen, 143-150 (ARC-Publicaties 208).

Buitenhuis, H., 2010. Vondstmateriaal. In: M.J.M. de Wit & M.C. Botermans, *Een archeologische opgraving aan de Halstraat 6 - 10 in het centrum van Breda, gemeente Breda (N.-Br.).* Groningen, 19-24 (ARC-Publicaties 209).

Buitenhuis, H., 2010. Faunaresten. In: M.C. Botermans & N. van Malssen, *De archeologische begeleiding van de aanleg van rioolsleuven en een sloopbegeleiding op locatie Koningshof te Gouda, gemeente Gouda (Z-H).* Groningen, 83-86 (ARC-Publicaties 212).

Buitenhuis, H., 2010. Faunaresten. In: J.J. Lenting, *Archeologisch onderzoek naar het Magazijn van Oorlog aan de Oudeschans 2 te Delfzijl (Gr).* Groningen, 107-112 (ARC-Publicaties 214).

Buitenhuis, H., 2010. Fauna. In: A.R. Wieringa. *Wonen en werken op het veen. Een nederzetting uit de Romeinse IJzertijd aan de Matsloot, ten westen van Hoogkerk.* Groningen, 31-33 (Stadse Fratsen 20).

2011

Buitenhuis, H., 2011. Faunaresten. In: S.J. Tuinstra, J.R. Veldhuis & J.A.W. Nicolay, *Hallum, een welvarend dorp aan de monding van de Middelzee. Een archeologische opgraving te Hallum, gemeente Ferwerderadeel (Fr).* Groningen, 205-214 (ARC-Publicaties 205).

Buitenhuis, H., 2011. Faunaresten. In: P.J.A. Stokkel & G.M.A. Bergsma, *Een archeologische opgraving van een grafveld op het plangebied Agropark II te Huissen, gemeente Lingewaard (Gld.).* Groningen, 67-72 (ARC-Publicaties 215).

Buitenhuis, H., 2011. Faunaresten. In: M.J.M. de Wit & P.J.A. Stokkel, *Bronstijdboeren te Zwaagdijk-Oost. Een definitief archeologisch onderzoek in plangebied Zwaagdijk-Oost II, gemeente Wervershoof (N.-H.).* Groningen, 55-60 (ARC-Publicaties 221).

Buitenhuis, H. & G.M.A. Bergsma, 2011. Faunaresten en menselijk botmateriaal. In: J.J. Lenting & N. van Malssen, *Een archeologisch onderzoek naar de Bruggeschans aan de Winschoterhoogebrug te Winschoten (Gr).* Groningen, 69-74 (ARC-Publicaties 222).

Buitenhuis, H., 2011. Faunaresten. In: A. Ufkes, *Een archeologische opgraving in de vroegmiddeleeuwse ringwalburg van Domburg, gem. Veere (Z.).* Groningen, 149-160 (ARC-Publicaties 223).

Buitenhuis, H. & G.M.A. Bergsma, 2011. Dierlijk en menselijk botmateriaal. In: J.J. Lenting & N. van Malssen, *Een archeologische begeleiding (protocol opgraven) op deelgebied Bogas (Koesteeg/Oostersingel) van het Centrumplan Coevorden, gemeente Coevorden (D).* Groningen, 65-66 (ARC-Publicaties 224).

Buitenhuis, H., 2011. Faunaresten. In: M.J.M. de Wit, *Romeinse en middeleeuwse bewoning op een woonheuvel in het rivierengebied. Een archeologisch inventariserend veldonderzoek (IVO) door middel van proefsleuven op terrein 'De Wilmert', plangebied 'Medel II' te Echteld, gemeente Neder-Betuwe (Gld.).* Groningen, 103-116 (ARC-Publicaties 228).

2012

Özbaşaran, M., G. Duru, N. Kayacan, B. Erdoğu & H. Buitenhuis,2012 Musular: The 8th Mill cal. BC Sattelite Site of Aşıklı. In: M. Özdoğan, N. Başgelen & P. Kuniholm (eds.) *The Neolithic in Turkey, New Excavations & New Research. Central Turkey.* Istanbul, 159-180.

Buitenhuis, H., 2012. Faunaresten. In: A. Ufkes & S.J. Tuinstra, *Erf `Tanneke Blox' in de 16e-19e eeuw. Een definitieve archeologische opgraving aan de Haatlanderdijk in het plangebied Zuiderzeehaven te Kampen, gemeente Kampen (Ov.).* Groningen, 97-102 (ARC-Publicaties 167).

Buitenhuis, H., 2012. Faunaresten. In: A. Ufkes & S.J. Tuinstra, *Een archeologisch inventariserend veldonderzoek (IVO) en een archeologische opgraving van een vroegmiddeleeuwse terp bij de afrit 'Akkerwinde' te Sneek, gemeente Sneek (Fr.).* Groningen, 93-98 (ARC-Publicaties 191).

Buitenhuis, H., 2012. Faunaresten. In: M.J.M. de Wit, *Wonen en werken in het IJsseldal in de Midden-Bronstijd -- Vroege Middeleeuwen. Archeologisch onderzoek naar de nederzettingen op plangebied `Achter 't Holthuis' te Twello, gemeente Voorst (Gld).* Groningen, 345-349 (ARC-Publicaties 234).

Buitenhuis, H., 2012. Faunaresten. In: M.J.M. de Wit & M.C. Botermans, *Van boerenerf tot bankgebouw. Een archeologische begeleiding (protocol opgraven) op plangebied Kerckenzight aan de Steenstraat 48-52 te Boxmeer, gemeente Boxmeer (NB).* Groningen, 57-60 (ARC-Publicaties 238).

2013

H. Buitenhuis, J. Peters *et al.*, 2013. The long and winding road: ungulate exploitation and domestication in Early Neolithic Analtola (10,000-7,000 cal BC). In: S. Colledge, J. Connolly *et al.* (eds.), *The Origins and Spread of Domestic Animals in Southwest Asia and Europe*. Walnut Creek, CA, USA, 83-114.

Buitenhuis, H., 2013. Faunaresten. In: A.R. Wieringa. *Oude sporen, nieuw station. Een opgraving aan de Helperzoom te Groningen.* Groningen, 35 (Stadse fratsen 34).

Buitenhuis, H., 2013. Leder, keramische artefacten en faunaresten. In: J.J. Lenting & A. Ufkes. *Een verdronken 14e-eeuws handelskwartier van Biervliet. Een archeologische opgraving in het tracé van de N61 bij Biervliet, gemeente Terneuzen (Z.).* Groningen, 99-106 (ARC-Publicaties 211).

2014

Arbuckle, B.S., Kansa, S.W., Kansa, E., Orton, D., Çakırlar, C., Gourichon, L., Atici, L., Galik, A., Marciniak, A., Mulville, J. and Buitenhuis, H., 2014. Data sharing reveals complexity in the westward spread of domestic animals across Neolithic Turkey. *PloS one*, 9(6), p.e99845.

Buitenhuis, H., 2014. Bot en schelpmateriaal. In: T.N. Krol, *Archeologische opgraving Helperzoom, spoorverbreding, te Groningen, gemeente Groningen (GR).* 27-29 (MUG-publicatie 2013-81).

Stiner, M.C., Buitenhuis, H., Duru, G., Kuhn, S.L., Mentzer, S.M., Munro, N.D., Pöllath, N., Quade, J., Tsartsidou, G. and Özbaşaran, M., 2014. A forager–herder trade-off, from broad-spectrum hunting to sheep management at Aşıklı Höyük, Turkey. *Proceedings of the National Academy of Sciences*, p.201322723.